Kitāb al-Masālik wa l-mamālik by Abū Isḥāq al-Iṣṭakhrī

Bibliotheca Geographorum Arabicorum

VOLUME I-1

The titles published in this series are listed at *brill.com/bga*

Kitāb al-Masālik wa l-mamālik by Abū Isḥāq al-Iṣṭakhrī

Viae regnorum: descriptio ditionis Moslemicae / *auctore Abu Ishák al-Fárisí al-Istakhrí*

M.J. de Goeje's Classic Edition (1870)

with a biographical note on M.J. de Goeje
by Jan Just Witkam

BRILL

LEIDEN | BOSTON

This volume was originally published in 1870. For historical purposes, the original prelims have been retained.

Cover illustration: 'Mappemonde arabe', Manuscrits orientaux, Arabe 2214, Bibliothèque nationale de France.

ISSN 2214-5737
ISBN 978-90-04-25868-6 (paperback)

This book is printed on acid-free paper.

Portrait in charcoal of M.J. de Goeje (1836–1909), made in or around 1905 by Thérèse van Duyl-Schwarze (1851–1918). Collection Leiden University Library (Icones Leidenses No. 279)

Michael Jan de Goeje (1836–1909) and the Editing of Arabic Geographical Texts

*Jan Just Witkam**

On Sunday 25 August 1889, four Egyptian gentlemen, ʿAbdallāh Fikrī Pāshā and Muḥammad Amīn Fikrī Bey, father and son, and their two compatriots, Ḥamza Fatḥ Allāh and Maḥmūd ʿUmar al-Bāǧūrī, visited Leiden in the Netherlands. They were en route from Egypt, through Italy, France and England, to Stockholm and Christiania (Oslo), in order to attend the 8th International Congress of Orientalists as the official Egyptian delegation.[1] The old Pasha, who previously had held the post of *Nāẓir al-Maʿārif al-Miṣriyya* ('Supervisor of Education in Egypt')[2] was a dedicated educationalist and an accomplished poet and *littérateur*. At the congress he presented a sample of his own commentary on a *Qaṣīda* by Ḥassān b. Thābit.[3] His son, Muḥammad Amīn Fikrī, presented a paper on problems connected with the use of spoken Arabic vs. standard Arabic.[4] He also completed the travelogue, since his father, the intended author, died soon after their safe return to Egypt.[5]

---

* Leiden University Centre for the study of Islam and Society (LUCIS), <j.j.witkam@hum.leidenuniv.nl>. I wish to thank Dr. Arnoud Vrolijk (Leiden University Library) for his help and support during the final stages of the composition of the first version of this paper. It was read at the Fifth Conference of al-Furqān Islamic Heritage Foundation, London, 24–25 November 1999. An early version of the text was published in Arabic: 'M.J. de Goeje wa-Taḥqīq al-Nuṣūṣ al-Ǧughrāfiyya al-ʿArabiyya', in Brahim Chabbouh (ed.), *ʿUlūm al-Arḍ fil-Makhṭūṭāt al-Islāmiyya / The Earth and its Sciences in Islamic Manuscripts*. London (Al-Furqān Islamic Heritage Foundation) 1426/2005, pp. 161–201. The present version dates from October 2014.

1 Held 2–13 September 1889.

2 *Nāẓir al-Maʿārif al-Miṣriyya* or *al-ʿUmūmiyya*, or something similar, is what is usually added to his name on the title pages of his works. On the title page of his *Risāla fī Muqāranat Baʿḍ Mabāḥith al-Hayʾa bil-Wārid fī al-Nuṣūṣ al-Sharʿiyya* (Cairo, Maṭbaʿat al-ʿĀṣima, 1315/1897–1898), a comparative work on modern astronomy and Islamic law, he is referred to as *Wazīr al-Maʿārif al-Miṣriyya sābiqan*, formerly Minister of Education of Egypt.

3 The event is described in his *Irshād al-Alibbāʾ ilā Maḥāsin Urubbā*, Cairo (al-Muqtaṭaf) 1892, pp. 655–658. The full text of the commentary by Abdallāh Fikrī, entitled *ʿUǧālat al-Bayān ʿalā Dīwān Ḥassān* was published in a posthumous edition of some of his works in poetry and prose, *al-Āthār al-Fikriyya* (Cairo 1315/1897–1898), pp. 334–396.

4 *Nubdha fī Ibṭāl Raʾy al-Qāʾilīn bi-Taʿwīḍ al-Lugha al-ʿArabiyya al-Ṣaḥīḥa bil-Lugha al-ʿĀmiyya fil-Kutub wal-Kitāba*. The text is published in *Irshād al-Alibbāʾ*, pp. 674–701.

5 *Irshād al-Alibbāʾ*.

 | DOI 10.1163/9789004258686_CH1

They had come to Leiden, long a centre of Oriental studies, for several reasons. Its library housed, and still houses, an important and ever increasing collection of Islamic manuscripts. Its faculty had counted in the course of time many famous names in Oriental studies among its members. Moreover, Leiden was home to the publishing house of E.J. Brill, which published numerous important text editions, some of which are still in use.[6] These facts qualified, and still qualify, Leiden as a favourite venue for scholarly tourism, and that was exactly what the four Egyptians had come for. When they had settled in their hotel, the Lion d'Or in the Breestraat, they were told that as it was Sunday, the library was closed, as was the Faculty and 'La Maison Brill.' Seeing the disappointment on their faces, the manager of the hotel told them he would try to find a solution and sent an employee on an errand. The man came back with the professor of Arabic, Michael Jan de Goeje, who told the relieved Egyptians, that in an hour's time he would organize a visit to the library and later on to Brill's as well.[7] Thanks to their hotel manager they had met the very person who embodied the three points of interest that attracted so many learned tourists to Leiden.[8]

De Goeje had been a professor of Arabic in Leiden since 1869; he was the honorary curator of the Oriental collections in the University Library,[9] he was the co-author of the Catalogue of Oriental and Arabic manuscripts in Dutch collections[10] and his scholarly editions appeared mostly with Messrs. Brill, whose house experienced an unprecedented prosperity, not least because of De Goeje's editorial projects. De Goeje was a typical exponent of the philological and editorial period in Middle Eastern studies, when the numerous sources of our knowledge of Arabic and Islamic culture in the European libraries were,

6 Brill's just launched 'Classic Arabic Texts Online' Part 1 (CATO-1), which offers c. 19,000 pages of classic Brill editions of Arabic texts in a full-text searchable format. De Goeje's *Bibliotheca Geographorum Arabicorum* (BGA) is included in this first part.

7 There is an interesting and rarely seen double perspective of this encounter, first the account by father and son Fikrī in their travelogue, *Irshād al-Alibbā'*, pp. 523–529, and a year later by M.J. de Goeje in his review of that book, 'Een Egyptenaar in Europa', in *De Gids* 1893/I, pp. 566–577.

8 See also on this visit, albeit from a somewhat different angle, Arnoud Vrolijk, 'Proverbial misunderstandings: The sources of Christiaan Snouck Hurgronje's collection of Egyptian proverbs', in *Der Islam* 79 (2002), pp. 103–117, especially pp. 113–116.

9 With the arcane title *Interpres Legati Warneriani* ('interpreter of Warner's legacy'), an office that has been held by the present author from 1992–2010. Then this honorific title was de facto abolished by the director of Leiden University Library.

10 *Catalogus Codicum Orientalium Bibliothecae Academiae Lugduno-Batavae*. 6 vols. Leiden 1851–1877. *Catalogus Codicum Arabicarum Bibliothecae Academia Lugduno-Batavae*. Editio secunda. Vols. 1–2. Leiden 1888–1907.

for the first time, disclosed to the public in detailed catalogues and made known by a large number of text editions that could stand the touch of criticism. It was also the time in which the Lachmannian method gradually became introduced in the editing of texts.

In the academic world of the first half of the nineteenth century there was an overall revival of textual criticism, following new methodologies that are mostly associated with the work of the German scholar of classical and old-Germanic studies and philologist Karl Lachmann (1793–1851). He devised what later came to be known as the Lachmannian method of textual criticism. This can be summarized as to attempt to bring about the reconstruction of a text in a version that is closest to the supposed original of the author, by removing the mistakes and interpretations of later periods that had become attached to texts. The main instrument for this method is the establishment of the stemma, by which is meant a genealogy of manuscript witnesses of a text. With the help of such a stemma those manuscripts that are without value for the establishment of the text can be distinguished from the valuable ones and then be eliminated. Ideally, the result of the application of the stemmatical method would be that the archetypal version of the author's version of a text could be re-discovered. One of the better-known instruments of textual criticism, which were perfected by Lachmann, is the rule of the more difficult reading prevailing over the simple one. This served to reconstruct original readings, which had been trivialized by generations of copyists. Lachmann never wrote a textbook on his method, but he developed and demonstrated it in his impressive production of critical editions, studies on aspects of textual criticism and commentaries on editions. He wrote on, and also edited, important texts not only in Greek and Latin literature, but also in the old-German and old-English literatures, and the New Testament. This choice of texts and fields of scholarship made his influence widely felt. Till today, students of textual criticism are (mostly unwittingly) tributary to his findings and users of his method.[11]

Some further biographical details on Michael Jan de Goeje are now in place. His father was a protestant minister in Dronrijp, a village in Frisia, a northern province of the Netherlands, where De Goeje was born on 13 August 1836. Although he was an apt and eager pupil in primary school, the follow-up of his vocational choice proved to be complicated. If everything had gone according to plan, he might have become a pharmacist, either in the Netherlands or in

11 See on the Lachmannian method and Oriental editing my 'Establishing the stemma: Fact or fiction?' in: *Manuscripts of the Middle East* 3 (1988), pp. 88–101, and its follow-up 'The Philologist's Stone. The Continuing Search for the Stemma', in: *Comparative Oriental Manuscript Studies Newsletter* 6 (July 2013), pp. 34–38.

the Dutch East Indies. But these and similar plans failed, and in the end it was decided that he would study theology in Leiden, with the most probable prospect of becoming, like his father, a minister. He began his study in 1854; however, he quickly became intrigued by the study of Semitic languages, the knowledge of which was, at the time, a requirement for students of theology and, therefore, part of their academic curriculum. It was assumed, of course, that they were capable of reading the Scriptures in their original languages, Hebrew, Aramaic and Greek. In Leiden he also received a thorough training in classical philology from his teachers J. Bake (1787–1864) and C.G. Cobet (1813–1889). With them he may have been confronted with the critical methodology towards texts, following the approach of Lachmann, whose influence was increasingly being felt in Dutch scholarship as well.[12] However, the professor whom De Goeje came to value most as his teacher was the arabist R.P.A. Dozy (Reinhart Dozy, 1820–1883), who occupied the Leiden chair of History. Under Dozy's influence, De Goeje decided that a clergyman's career was not what he wanted out of life and he choose for a study in Semitic languages, which meant Hebrew, Syriac and Arabic. Soon he was completely under the spell of the latter language. He formulated his scholarly ideal as collecting Arabic texts that were sources for the development of human civilization, then critically assessing these sources and, subsequently, composing an attractive scholarly synthesis of these. These sources were in the fields of literature, history and geography.[13] In 1859 he was put in the daily charge of the Leiden Oriental manuscripts collection.[14] In 1860, he defended his PhD thesis, which was a partial edition of al-Yaʿqūbī's *Kitāb al-Buldān*. In 1866, he became extra-ordinarius professor, and in 1869 this appointment was changed into a full professorship for Arabic in Leiden. In 1906 he retired, internationally acclaimed as one of the greatest philologists of his time. He died on 17 May 1909.

12 The knowledge of Latin (which university students had already acquired in their grammar school) had another, practical purpose as well. Latin was the language in which most scholarly works in large parts of Europe were published, more or less the same as English is nowadays. Several of De Goeje's works directed to an international audience are written in Latin, which makes them almost entirely inaccessible, now that knowledge of Latin has become such a rare commodity.

13 See for an extensive survey of de Goeje's life and work, C. Snouck Hurgronje, *Levensbericht van Michael Jan de Goeje* [with a list of his writings by Th.W. Juynboll]. Amsterdam 1909 (in *Jaarboek der Koninklijke Akademie van Wetenschappen* 1909).

14 Then counting not less than circa two thousand volumes of mostly Arabic, Persian, Turkish and Hebrew texts, including the important collections of J.J. Scaliger (1540–1609), Jacobus Golius (1596–1667) and Levinus Warner (1619–1665).

At an early stage of his scholarly life, De Goeje realized that good editions were a first necessity and during his many-sided career, from 1856 till his death in 1909, editions of an impressive number of texts were his major and regular product. It is not inappropriate to call his office a text editing factory, for not only did he produce a considerable number of editions himself, he also had his students work on parts of his projects and carry out the proofreading, indexing, etc. De Goeje's own editions include several historical and geographical works by al- Yaʿqūbī (his thesis of 1860, to be repeated and augmented in 1892 as vol. 7 of the *Bibliotheca Geographorum Arabicorum* (*BGA*)), al-Idrīsī (1866, together with Dozy), al-Balādhurī (1863–1866), several volumes of the history of al-Ṭabarī (1879–1901), a project that was organized by him.[15] Furthermore, several literary texts, notably the *Dīwān* of Abū al-Walīd Muslim b. al-Walīd al-Anṣārī (1875) and Ibn Qutayba's *Kitāb al-Shiʿr wal-Shuʿarāʾ* (1904). A stream of articles on many subjects of Oriental studies appeared both in international scholarly journals, often with spin-offs in Dutch periodicals written for a wider audience. Never again would the Dutch general public be so informed about aspects of Middle-Eastern culture in such an erudite and simultaneously entertaining way.

One of the pillars of De Goeje's lasting fame was his single-handed editing of what he has called the 'Library of Arabic Geographers' (*Bibliotheca Geographorum Arabicorum*, *BGA*), a series of eight volumes containing geographical texts, which were published between 1870–1894. The texts in this major project were the *Kitāb Masālik al-Mamālik* by al-Iṣṭakhrī (1870), the *Kitāb al-Masālik wal-Mamālik* by Ibn Ḥawqal (1873), the *Kitāb Aḥsan al-Taqāsīm fī Maʿrifat al-Aqālīm* by al-Muqaddasī (1877, 2nd edition 1906), the *Kitāb al-Buldān* by Ibn al-Faqīh al-Hamadhānī (1885), the *Kitāb al-Masālik wal-Mamālik* by Ibn Khordādhbeh with excerpts of the *Kitāb al-Kharāǧ* by Qudāma b. Ǧaʿfar (1889), the seventh part of the *Kitāb al-Aʿlāq al-Nafīsa* by Ibn Rusta and the *Kitāb al-Buldān* by al-Yaʿqūbī (1892) and, finally, the *Kitāb al-Tanbīh wal-Ishrāf* by al-Masʿūdī (1894). All these publications were first-time 'critical' editions of the texts concerned, and all were provided with some information on the manuscripts used,[16] an extensive text-critical apparatus containing the

15 See J.T.P. de Bruijn, 'Collective studies of the Muslim world: Institutions, projects and collections' in: W. Otterspeer (ed.), *Leiden Oriental connections 1850–1940*, (Leiden, E.J. Brill, 1989), pp. 94–114, esp. pp. 99–100. See also Arnoud Vrolijk, 'The Leiden edition of Ṭabarī's Annals. The search for the Istanbul manuscripts as reflected in Michael Jan de Goeje's correspondence', in: *Quaderni di Studi Arabi* 19 (2001), pp. 71–86. De Goeje's vast correspondence with his collaborators in the Ṭabarī edition is kept in Leiden University Library, Or. 5585 e-f.

16 These are a bit short, however, according to present-day requirements.

variant readings of the major manuscripts known[17] and, finally, indexes and glossaries of technical terms and special vocabulary, which was not found in the usual dictionaries of classical Arabic. In one case (volume 5) a French translation was added. The German translation of all texts, which De Goeje had envisaged from the very beginning of the project, was never realized, however.[18]

Another feature in De Goeje's work must be mentioned here, namely his sense of the importance of teamwork. Certainly for the two larger projects in which he was involved, the *BGA* and the Ṭabarī edition, but equally for the other editions which he produced, he was all too aware of the fact that teamwork and international co-operation between scholars were of prime importance for the success of such vast enterprises. On the one hand, the mere size of a project made joining forces necessary and, on the other hand, it was important to have good connections, in the pre-photography era, in other European manuscript collections. Colleagues and friends frequently provided De Goeje with notes and quotations from manuscript sources, which they had to copy themselves. In the publication of the *BGA* this is illustrated by De Goeje's dedications in each volume to both his learned friends and academic colleagues.[19]

The intricate internal relationship between the different texts of the Balkhī school of geographers is an illustration of how these texts have fared in history. Even with the overview of the manuscripts that we now enjoy, it is virtually impossible to extricate the separate texts from one another. This is already shown by the way De Goeje had to handle his manuscripts. His point of departure was the classical, predominantly static approach of texts. He looked at a text as a stable and well-defined entity, which had been composed and completed by an author and then transmitted to later periods by scribes. Such a reductionist approach may be feasible and fruitful with texts from Classical Antiquity,[20] for a more recent and continuous textual tradition like Arabic literature, it is not always possible to maintain such a static appreciation of texts. With the early geographical texts of the Balkhī school such an approach clearly proved to be too simple. Even if De Goeje had wanted to distinguish between

---

17 See the appendix, hereafter.

18 Vol. 1 of the series (al-Iṣṭakhrī, 1870), introduction p. viii.

19 Most notably to Carlo de Landberg for his help in procuring the manuscript of Ibn Khordādhbeh's *Kitāb al-Masālik wal-Mamālik* (now MS Vienna, ÖNB, Mixt. 783).

20 This is not because of an intrinsic difference of classical Greek and Latin texts, but because of the long distance in time between the period in which the classical author lived and our own period. It is, in fact, our lack of knowledge about the early stages of that period, in which classical texts crystallized into a fixed shape, that creates the false impression of being static.

the works of al-Iṣṭakhrī and Ibn Ḥawqal, he could not escape using his findings with manuscripts of one text as a textual witness for establishing the other. So his two editions of the oldest texts of the Balkhī geographers, *BGA* 1 (al-Iṣṭakhrī) and *BGA* 2 (Ibn Ḥawqal), are so closely interrelated that each edition is used as a textual witness to the other.[21]

For the editing of purely literary texts the editor needs a sufficient amount of textual evidence, together with a fair knowledge of the literary usage of the language of the text. The term 'fair knowledge' is a recurrent theme in discussions about textual criticism. It is sometimes referred to by Middle-Eastern scholars as *al-Dhawq al-Salīm*, 'good taste.' This implies that the editor who has this good taste, knows the right level and register of language and, thus, he should be able to change the text for the better, *as the author might have done,* or even try to do it better than the author, if the editor falls into the trap of identifying himself too much with his object of research. The extent to which this 'good taste' is exercised distinguishes a good edition from a bad one. The use of 'good taste' is the personal element in textual criticism. However, caution is necessary when the editor feels tempted to use his 'taste', as its reckless use results in erroneously applying textual criticism and devising conjectures. Yet, the editing of texts is not a mechanical procedure, as some followers of Lachmann would have it.[22]

In the editing of scientific texts the editor's knowledge of the field of science becomes relatively more important. It is obvious that whoever edits an old mathematical text, should not only know about philology, codicology and paleography, but also about the ins and outs of the history of mathematics. Conjectures to improve corrupt texts should, then, not be made solely on the basis of philological arguments, but equally with mathematical considerations. And the same goes for other technical and scientific texts. To a certain extent,

---

21 See the appendix to this article, which gives a survey of the manuscripts used by De Goeje for his editions. Gerald R. Tibbetts has drawn a stemma of the texts of the Balkhī school of geography, demonstrating the complications of transmission of the material ('The Balkhi school of geographers', in J.B. Harley & David Woodward, Cartography in the traditional Islamic and South Asian societies II/1 (Chicago 1992), 108–136, on p. 111). See now also Fuat Sezgin, *Geschichte des arabischen Schrifttums*, vols. 10 and 12, both published in Frankfurt in 2000.

22 The work by Paul Maas, *Textual criticism*, which has been used widely by generations of classical scholars, is confusing in this respect. If it is at all applicable in practical textual criticism of Greek and Latin, in all its extreme consequences it is hardly useful for textual criticism of the Islamic literatures. I have expressed my doubts about the applicability of Maas' book in 'Establishing the stemma. Fact or fiction?' of 1988, and the 'Philologist's Stone' of 2013.

one can consider geographical texts as literary texts, as travelogues, as descriptions of far away countries and the manners and customs of their inhabitants. But geography is not just a topographical description of a region. It often contains a wealth of information on economic and other matters as well. Ibn Ḥawqal is a case in point, and the idea that he was economically active during his travels makes sense. The other story, that he probably also acted as a Fāṭimid missionary, only corroborates this, since activism for a cause and commercial activities have, in the history of Islam, always gone hand in glove.

Geographical texts are not just literary travelogues, nor just collections of commercial information, nor simply descriptions of bizarre habits of far-away peoples, nor cosmological fantasies. They contain, in fact, aspects of all of these. Geographical texts sometimes have the extra dimension of maps, and these put an extra burden on the shoulders of the editor. The earlier editions of geographical texts hardly ever had maps added to the text. On the one hand, this had to do with the limited technical possibilities of the reproduction of images at that time; on the other hand, this can be attributed to the nineteenth-century philologist's preoccupation with texts, to the exclusion of graphic images. It seems to me that De Goeje was less interested in this extra pictorial dimension as a help to the interpretation of the geographical texts he was editing. His orientation toward his subject was mostly of a literary nature.[23] As a text he wanted it to be as correct as possible, and to consider it as a collection of maps with explanatory texts was much less his approach.

Textual criticism works by the assumption that each copy of a manuscript contains the scribal errors already contained in the exemplar, plus a set of new ones. By analyzing the genealogy of errors it is possible to establish a genealogy of the manuscripts. These rules of textual criticism cannot simply be transposed to the critical analysis of a hierarchy in the tradition of maps, and their relationship with their captions, legends or accompanying texts in the manuscripts. The maps compound the matter, because their analysis is on the basis of more than just the text or legends. Art historical considerations, especially with regard to the techniques of painting and the conventions of representation, give extra information and, thereby, complicate matters. There is, of course, always the correlation with the geographical reality. In other words, the question to what extent does the map represent the reality both on the ground and in the text, is a constantly relevant one.

One example I will give here. It concerns the dilemma of the editor for the map of the Caspian Sea. In the al-Iṣṭakhrī maps two islands are drawn. If one compares the map in the Leiden manuscript with the one in the Persian

23 De Goeje never travelled in the Orient.

translation as contained in the Oxford manuscript,[24] one can see that the names, or relative positions of the two islands, have been interchanged. This is not only a problem for textual criticism, but equally for the study of cartographical representation.

De Goeje never wrote a survey of his editorial methods (as Lachmann never did, for that matter). Their philological methods are implicitly available from their editions and other works. The obvious source for the method used is the critical apparatus. When one peruses De Goeje's critical apparatus, one sees that he sometimes does use the maps, but just as additional evidence, where the text has failed him or shows a lacuna. Furthermore, De Goeje made no distinction between a critical apparatus, which serves to elucidate the selection of readings of the text, and the commentary which he needed for understanding the meaning of the text. For the editing of geographical texts this is not always feasible. Often the textual difficulties are about the spelling of names, and the correct spelling cannot always be reconstructed by textual criticism. The editor of a text must then have recourse to other works with a better reading, even though they are textually unrelated to the primary text that the editor is establishing. The numerous references in De Goeje's apparatus are eloquent witnesses to this procedure. It proves the peculiar situation with the editing of non-fictional texts, and the impossibility of complying with the requirements of editorial theory under all circumstances.

So, when father and son Fikrī, together with their two compatriots, on that quiet Sunday afternoon in August 1889 informally sat together with De Goeje in his house, smoking and talking, was it at all surprising that they posed him a question with a geographical implication, a relevant conversation piece? Where in the world, so they asked the learned editor of so many geographical texts, would he situate the legendary island of Wāq Wāq? De Goeje did not disappoint and, of course, came up with the correct answer: Wāq Wāq was Japan. He gave his visitors a full account of the numerous literary sources about Wāq Wāq and the etymology of its name, and then he produced the decisive arguments for his opinion.[25]

24 See the bibliographical references in the appendix, below, under BGA vol. 1.

25 *Irshād al-Alibbā'*, p. 527.

Appendix

A survey of the manuscripts used by M.J. de Goeje in his *Bibliotheca Geographorum Arabicorum* (BGA)

In the following De Goeje's *sigla* have been explained, according to De Goeje's own, often scanty, notes. This information has occasionally been augmented with details of the descriptions of the manuscripts, drawn from the catalogues.[26]

Before De Goeje had arrived at the idea to publish an entire library of Arab geographers, he had already edited, together with Dozy, the part on Africa and Andalusia of the *Kitāb Nuzhat al-Mushtāq fī Ikhtirāǧ al-Āfāq* by al-Sharīf al-Idrīsī.[27] De Goeje visited Oxford in order to collate the two manuscripts there. The following manuscripts form the basis of this edition:

A = MS Paris, Bibliothèque nationale de France, supplément arabe 893. Dated Almeria 744 (1343–1344).
B = MS Paris, Bibliothèque nationale de France, supplément arabe 892. The 'best' manuscript, and the basis for the edition.
C = MS Oxford, Bodleian Library, Pococke 375. Cat. Uri, No. 887. Dated Cairo 860 (1456).
D = MS Oxford, Bodleian Library, Grav. 3837–3842. The second best manuscript of the edition. An old manuscript in *Maghribi* hand, with maps.

BGA vol. 1. *Kitāb Masālik al-Mamālik* by al-Iṣṭakhrī (1870).[28]

Dedicated to Theodor Nöldeke (1836–1930).
A = MS Bologna, Biblioteca Universitaria, Cod. 3521 (Cat. Rosen, p. 94, No. 421).
B = MS Berlin, Staatsbibliothek zu Berlin, Sprenger 1 (Cat. Ahlwardt, vol. 5, p. 362, No. 6032).
C = Gotha, Forschungsbibliothek, Arab. 1521 (Cat. Pertsch, vol. 3, pp. 142–144). An extract, or shortened version, of the text.
D = the *Kitāb al-Masālik wal-Mamālik* by Ibn Ḥawqal, edited by De Goeje as BGA vol. 2 (see hereafter).

---

26 I refrain from giving the full bibliographical details of the relevant catalogues as these can all be found in Geoffrey Roper, *World Survey of Islamic Manuscripts*. London (al-Furqān Islamic Heritage Foundation) 1992–1994 (4 volumes). Libraries are referred to by their present-day names.

27 *Description de l'Afrique et de l'Espagne par Edrisi. Texte arabe publie pour la première fois d'apres les manuscrits de Paris et d'Oxford avec une traduction, des notes et un glossaire* par R. Dozy et M.J. de Goeje. Leyde (E.J. Brill) 1866.

28 *Bibliotheca Geographorum Arabicorum*. Edidit M.J. de Goeje. Pars Prima. *Viae regnorum. Descriptio ditionis Moslemicae*, auctore Abu Ishák al-Fárisí al-Istakhrí. Lugduni Batavorum (apud E.J. Brill) 1870.

E = Gotha, Forschungsbibliothek, Pers. 36 (Cat. Pertsch, pp. 61–63). A Persian translation of the work.
Ous. = The translation by William Ouseley (1767–1842) of Ibn Ḥawqal.[29]
O = MS London, India Office Library 1026. A Persian translation. See catalogue Ethé, I, col. 364–365 (No. 707). The collation from this manuscript is also given by De Goeje in his lists in *BGA* 4.

¶ After the completion of the edition, De Goeje found it necessary to collate his text with the following two manuscripts. He gives his collations in volume 4 of the *BGA* (from p. 381 onwards). He describes this procedure on p. iv of the introduction.
L = MS Leiden, University Library, Or. 3101 (Voorhoeve, *Handlist*, p. 200). An Arabic compendium, copied in 589 (1193), purchased in 1876 by Alfred von Kremer (1828–1889) in Egypt, and not yet available when *BGA* 1 was published.
F = MS Oxford, Bodleian Library, Ouseley 373. A Persian translation. Purchased in 1812 by William Ouseley (1767–1842) in Shiraz. See catalogue Sachau & Ethé, cols. 397–399 (No. 396).[30]

De Goeje also adduces textual evidence from many later historical sources. In his critical apparatus and commentary one finds references to the works of Yāqūt, Abū al-Fidā', al-Qazwīnī, al-Ya'qūbī, al-Dimashqī, al-Idrīsī and others.

BGA vol. 2. *Kitāb al-Masālik wal-Mamālik* by Ibn Ḥawqal (1873).[31]

Dedicated to William Wright (1830–1889).
L = MS Leiden, University Library, Or. 314 (Voorhoeve, *Handlist*, p. 347). Without maps, copied in Istanbul in 926 (1520).
B = MS Oxford, Bodleian Library, Hunt. 538 (Cat. Uri, p. 209, No. 963).
P = MS Paris, Bibliothèque nationale de France, Arabe 2214 (Cat. De Slane, pp. 389–390). A shortened version of the text. The photograph of this manuscript, which was used by J.H. Kramers for his second edition, is kept in Leiden University Library as Or. 8525 (Voorhoeve, *Handlist*, p. 348).
I = *Kitāb Masālik al-Mamālik* by al-Isṭakhrī, edited by De Goeje as *BGA* vol. 1.

29 *Kitāb Masālik wa-Mamālik Taṣnīf Ibn Ḥawqal. The oriental geography of Ebn Haukal, an Arabian traveller of the tenth century. Translated from a Manuscript in his own Possession, collated with one preserved in the Library of Eton College* by William Ouseley. London (Oriental Press) 1800.

30 Now edited by Iraj Afshar, *Mamālik wa-Masālik. Ta'līf-i Abū Isḥāq Ibrāhīm Isṭakhrī, Tarǧuma-yi Muḥammad b. As'ad b. 'Abdallāh Tustarī.* Tehran (Bunyād-i Mawqūfāt-i Dr. Maḥmūd Afshār) 1373/1994.

31 *Viae et regna. Descriptio ditionis Moslemicae*, auctore Abu'l-Kásim Ibn Haukal. Edidit M.J. de Goeje. Lugduni Batavorum (apud E.J. Brill) 1873.

¶ A new edition of Ibn Hawqal's work was edited in 1938–1939 under the title *Kitāb Ṣūrat al-Arḍ* (as the Istanbul manuscript has it) by that other Leiden editor of geographical texts, J.H. Kramers (1891–1951).[32] The difference between De Goeje's and Kramers' edition is, of course, the use by Kramers of the Istanbul manuscript as the basis (referred to as *al-Aṣl* in his apparatus), which had been unavailable to De Goeje.

Kramers used the following textual sources:

– MS Istanbul, Topkapı Sarayı Müzesi, No. 3346. Kramer's photographic copy of the manuscript is kept in Leiden University Library as Or. 8524 (Voorhoeve, *Handlist*, p. 347).

– The edition by M.J. de Goeje, *Kitāb al-Masālik wal-Mamālik* by Ibn Ḥawqal, in *BGA* vol. 2.

– MS Leiden, University Library, Or. 314, containing version II, the manuscript that was denoted by De Goeje as L.

– MS Oxford, Bodleian Library, Hunt. 538, containing version II, the manuscript that was denoted by De Goeje as B.

– MS Paris, Bibliothèque nationale de France, arabe 2214, containing version III, the manuscript that was denoted by De Goeje as P.

– *Kitāb Masālik al-Mamālik* by al-Istakhri, edited by De Goeje as *BGA* vol. 1.

BGA vol. 3. *Kitāb Aḥsan al-Taqāsīm fī Maʿrifat al-Aqālīm* by al-Muqaddasī (1877),[33] and the 2nd edition (1906).[34]

Dedicated to Aloys Sprenger (1813–1893).

B = MS Berlin, Staatsbibliothek zu Berlin, Sprenger 6 (Cat. Ahlwardt vol. 5, pp. 362–363, No. 6033). Sprenger had brought this manuscript from India, where he apparently had commissioned it. It is a copy ('copied for the press') of Sprenger's own manuscript, which is now MS Berlin, Staatsbibliothek zu Berlin, Sprenger 5 (Cat. Ahlwardt, vol. 5, p. 363, No. 6034).

C = MS Istanbul, Süleymaniye Kütüphanesi, Aya Sofya 2971 bis (*Defter*, p. 179). From this manuscript a copy was made at De Goeje's request and provided to him by the Dutch ambassador in Istanbul. This copy is now kept in Leiden University Library as Or. 2063 (Voorhoeve, *Handlist*, p. 7).

32 *Opus geographicum* [*auctore*] Abu al-Kasim Ibn Haukal al-Nasibi. *Secundum textum et imagines codicis Constatinopolitani conservati in bibliotheca antiqui Palatii no 3346 cui titulus est "Liber imaginis terrae" edidit collato textu primae editionis aliisque fontibus adhibitus* J.H. Kramers. Lugduni Batavorum (E.J. Brill) 2 vols. 1938–1939.

33 *Bibliotheca Geographorum Arabicorum*. Edidit M.J. de Goeje. Pars Tertia. *Descriptio imperii Moslemici*, auctore al-Mokaddasi. Lugduni Batavorum (apud E.J. Brill) 1877.

34 *Bibliotheca Geographorum Arabicorum*. Edidit M.J. de Goeje. Pars Tertia. *Descriptio imperii Moslemici*, auctore al-Mokaddasi. *Editio secunda*. Lugduni Batavorum (apud E.J. Brill) 1906.

For the 2nd, improved, edition of 1906, De Goeje did not use more manuscripts, but he gave a wider review to the existing sources, and he used the opportunity to add a considerable number of textual emendations, which he had received from a number of colleagues in the meantime.

BGA vol. 4 does not contain an edition, but comprises the indexes, glossary and the additions and corrections to the texts in vols. 1–3.[35] There is no dedication. The volume does mention, however, newly discovered materials, notably MS Leiden Or. 3101, a shortened version of al-Iṣṭakhrī, dated 589 (1193), with maps (pp. iv–vi). In the section for additions and corrections De Goeje selected material from this and other manuscripts (from p. 381 onwards).

BGA vol. 5. *Kitāb al-Buldān* by Ibn al-Faqīh al-Hamadhānī (1885).[36]

Dedicated to the memory of Otto Loth (1844–1881).

B = MS London, British Library, Add. 7496 (Rich). Cat., part 2 (1870), pp. 182–183, No. 380.

I = MS London, India Office 617 (Hastings). Cat. Loth, pp. 208–209 (No. 722).

S = MS Berlin, Staatsbibliothek zu Berlin, Sprenger 2 (Catalogue Ahlwardt vol. 5, pp. 363–364, No. 6035).

BGA vol. 6 contains two texts: *Kitāb al-Masālik wal-Mamālik* by Ibn Khordādhbeh, and excerpts of the *Kitāb al-Kharāǧ* by Qudāma b. Ǧaʿfar (1889).[37]

Dedicated to C. Barbier de Meynard (1826–1908) and Carlo de Landberg (1848–1924).

– Ibn Khordādhbeh:

A = MS Vienna, Österreichische Nationalbibliothek, Mixt. 783 (Cat. Loebenstein, p. 197, No. 248). The manuscript was originally purchased by Carlo de Landberg in the Orient and donated by him to the Vienna library, with the condition that it would be used first by De Goeje. The basis manuscript of the edition.

35 *Bibliotheca Geographorum Arabicorum*. Edidit M.J. de Goeje. Pars Quarta continens *Indices, glossarium et addenda et emendanda ad part. I–III* auctore M.J. De Goeje. Lugduni Batavorum (apud E.J. Brill) 1879.

36 *Bibliotheca Geographorum Arabicorum*. Edidit M.J. de Goeje. Pars Quinta. *Compendium libri Kitâb al-Boldân*, auctore Ibn al-Fakîh al-Hamadhâni. Lugduni-Batavorum (apud E.J. Brill) 1885.

37 *Bibliotheca Geographorum Arabicorum*. Edidit M.J. de Goeje. Pars Sexta. *Kitâb al-Masâlik wa'l-Mamâlik*, auctore Abu 'l-Kâsim Obaidallah ibn Abdallah ibn Khordâdhbeh. *Accedunt excerpta e Kitâb al-Kharâdj* auctore Kodâma ibn Djaʿfar. Lugduni-Batavorum (apud E.J. Brill) 1889.

B = MS Oxford, Bodleian Library, Hunt. 433 (Cat. Uri, p. 216, No. 993). Dated 630 (1232–1233), incomplete and lacunous. A copy of this manuscript is MS Berlin, Staatsbibliothek zu Berlin, Sprenger 4 (2), cat. Ahlwardt, vol. 5, p. 361, No. 6031.
– Qudāma b. Ǧaʿfar, the geographical and statistical parts only (*al-Manzila al-Sābiʿa*, which is all that is preserved of the text).
Cod. = Starting point for the edition was a copy of MS Istanbul, Köprülü Kütüphanesi, which was in the possession of Charles Schefer (1820–1898) in Paris. Subsequently, De Goeje's own copy of the Schefer copy was collated at De Goeje's request by the Orientalist Dr. Hermann Gies, first dragoman of the German embassy in Constantinople, with the Köprülü original. Köprülü Defter, p. 70, Cat. Şeşen, vol. 1, p. 548, No. 1076. Şeşen[38] was unaware of the manuscript having been edited in this way. De Goeje's own collated copy is now in Leiden University Library: Or. 5580 (Voorhoeve, *Handlist*, p. 158).

BGA vol. 7 contains two texts: the 7th part of *Kitāb al-Aʿlāq al-Nafīsa* by Ibn Rusta and the *Kitāb al-Buldān* by al-Yaʿqūbī (1892).[39] Already in 1860 De Goeje had published part of al-Yaʿqūbī's text, as his PhD thesis.[40]

Dedicated to Ferdinand Wüstenfeld (1808–1899).

– Ibn Rusta:
Cod. = MS London, British Library, Add. 23,378. Supplement Catalogue (1871), pp. 604–607.
– al-Yaʿqūbī:
Cod. = MS Munich, Bayerische Staatsbibliothek. According to De Goeje's preface (p. viii) at first part of the Muchlinski collection, and later purchased by the Munich library.

BGA vol. 8. *Kitāb al-Tanbīh wal-Ishrāf* by al-Masʿūdī (1894).[41]

Dedicated to Victor R. Rosen (1849–1908).

38 *Catalogue of manuscripts in the Köprülü Library*, prepared by Ramazan Şeşen, Cevat Izgi, Cemil Akpinar. Presented by Ekmeleddin Ihsanoğlu. Istanbul 1986.

39 Bibliotheca Geographorum Arabicorum. Edidit M.J. de Goeje. Pars Septima. *Kitâb al-Aʿlâk an-Nafîsa VII* auctore Abû Alî Ahmed ibn Omar Ibn Rosteh *et Kitâb al-Boldân* auctore Ahmed ibn abi Jakûb ibn Wâdhih al-Kâtib al-Jakûbî. Edit. Secunda. Lugduni Batavorum (apud E.J. Brill) 1892.

40 *Specimen literarium inaugurale, exhibens descriptionem Al-Magribi, sumtam e libro regionum Al-Jaqubii, versione et annotatione illustratam* [...] Michael Janus de Goeje. Lugduni-Batavorum (apud E.J. Brill) 1860.

41 Bibliotheca Geographorum Arabicorum. Edidit M.J. de Goeje. Pars Octava. *Kitâb al-Tanbîh wa'l-Ischrâf* auctore al-Masûdî. *Accedunt indices et glossarium ad tomos VII et VIII.* Lugduni-Batavorum (apud E.J. Brill) 1894.

L = MS London, British Library, Add. 23,270. Supplement catalogue (1871), pp. 548–549.
P = MS Paris, Bibliothèque nationale de France, arabe 1407 (Cat. De Slane, p. 284). Used by De Goeje in Leiden by special arrangement with the French National Library.

Viro Clarissimo

Th. Nöldeke

CARO ET FIDELI AMICO

HUNC LIBRUM

d. d. d.

EDITOR.

BIBLIOTHECA GEOGRAPHORUM ARABICORUM.

EDIDIT

M. J. DE GOEJE.

PARS PRIMA.

VIAE REGNORUM.

DESCRIPTIO DITIONIS MOSLEMICAE

AUCTORE

Abu Isháh al-Fárisí al-Istakhrí.

LUGDUNI BATAVORUM.

APUD E. J. BRILL,

ACADEMIAE TYPOGRAPHUM.

1870.

VIAE REGNORUM.

DESCRIPTIO DITIONIS MOSLEMICAE

AUCTORE

Abu Ishák al-Fárisí al-Istakhrí.

EDIDIT

M. J. DE GOEJE.

LUGDUNI BATAVORUM.

APUD E. J. BRILL,

ACADEMIAE TYPOGRAPHUM

1870.

PRAEFATIO.

Multa praemonenda non habeo. Commentationem de libri auctore, de titulo quo opus inscripsi, de codicibus quibus uti licuit, Germanice conscriptam, misi viris doctissimis qui acta Societatis Orientalis Germanicae curant, sperans fore ut in tomo operum Societatis vicesimo quarto etiam locum obtineret, itaque libri editioni antecederet ejusque quasi praenuncia esset. Ad quam igitur lectorem relego.

Methodus quam in edendo secutus sum pervolventi librum satis apparebit. Est unus et alter virorum doctorum, qui in notis criticis magna parsimonia utendum esse censet, quippe nullum architectum machinam quae aedificandae domui inservierit, post opus peractum conservare. Sed ut fere semper, hic comparatio claudicat. Si editor ubique pro certo scire posset veram se auctoris lectionem elegisse, nihil morarer, nec ego, nisi rarius ut codicis indoles cognosceretur, varias lectiones adnotavi quae manifesta menda sunt librariorum. Ubicunque vero haec evidentia desideratur, editoris erga lectorem officium mihi quidem esse videtur, varietatem lectionum diligentissime indicare. Quod imprimis valet de nominibus propriis. Ibi quoque ubi nomen a Jacut cum omnibus vocalibus commemoratur, nonnunquam necesse fuit codicum lectiones enumerare. Nam plus semel doctus Arabs falsam lectionem verae praetulit. Valde autem accrevit annotatio eo quod plures hujus libri in Oriente recensiones circumferebantur, quarum diversitas mihi non videbatur negligenda. Partes libri ab Anderson e codice Berolinensi, ab Arnold e codice Gothano olim editae sunt. Supervacaneum visum est memorare quoties lectionem ab iis receptam rejicere debuerim.

Tomus alter Bibliothecae Geographicae continebit librum Ibn Haucalis, quem paucis abhinc mensibus prelo committam. Tomus tertius nobile Mokaddasíi opus exhibebit. Hunc excipiet versio Germanica horum librorum, una cum indicibus et glossario.

D. G.

Dabam Lugduni Batavorum.

Nonis Decembribus anni MDCCCLXX.

ADDENDA ET EMENDANDA.

*Pag. ٣ vs. 8. l. يتعدّر.

» ٩ ann. *o*. Non necesse est legere التعاويج. **In Glossario de significatione verbi agetur.**

» ١١ vs. 9. Fortasse leg. هو لسان بحر فارس.

» ١٣ » 3 a f. l. الحدّ الغربىّ.

» ١۴ » 4 a f. l. يمتدّ.

» ١٩ ann. *m*. جمرة quoque legebatur in exemplari quo usus est Jacut, vid. IV, p. ١١٧.

» ٢٩ vs. 7. l. للبُجّة.

» ٣٠ » 9. Accuratius est كلا pro كلى.

» ٣٥ » 5 a f. l. ترتفع.

» ۴٩ » 2. l. potius من سفن.

» ٥٠ » 2 et 3. Fortasse leg. يدوم لهم فيه الماء.

* » ٥٥ » 2 a f. l. يرابطون.

* » ٥٩ » 5 a f. l. اعذآءة.

» ٥٧ ann. *e*. Non necesse est quidquam in textu mutare: باب شبيه باب شبيه بباب السرداب est pro بالسرداب.

» ٩٧ vs. 4 seqq. Cf. Jacut, II, p. ٣٠٨, 14 seqq.

» ٧٣ » 11. Fortasse legendum est كبيرة غنّاء.

» ٨٩ » 6. Lege potius cum C. et D. منقطعة.

» ٩٠ » ult. Jacut, IV, p. ٥٢٧, 18 تتخطّه.

Pag. ٩٧ vs. 4 a f. l. دارا الملك.
» ١٠٠ » 2 et ann. *d.* Cf. Jacut, II, p. ١٧٩ جيكان.
» ١٠١ » 5. l. وطرخنيشان. De nomine agetur ad textum Mokaddasíi.
» ١٠۴ ann. *p.* Dele verba: Secundum Jacut — نائين.
» ١١٠ vs. 1 جروج. Jacut eundem locum sub forma خرُون memorat II, p. ۴٢٩.
» ١٢١ » 1 et ann. *a.* In mappa Mokaddasíi سوک s. شوک prope confinium Khuzistani ponitur.
» ١٢٩ » 2 a f. Lectio quam recepi vera non est. Prior pars nominis est نـسـا. Nempe Mokaddasí disertissime dicit نسا esse nomen urbis al-Baidhá. Memoratur quoque a Jacut, II, p. ٥٩١, 12 sub دراسفيد.
* » ١٢٨, 11. l. وشيفير.
» — ann. *l.* Dele verba: Apud Mokaddasí — scribitur.
» ١٣٣ vs. 8 seq. Restituatur الخويذان. Ad textum Mokaddasíi de hoc loco sermo erit.
» — ann. *a.* Dele verba: Mokaddasí — appellat *et* igitur.
» — ann. *i.* Ad lectionem Kodámae امران, cf. supra p. ١١٠ *k* انبوران.
» ١٣٩ vs. 2 a f. l. وطرخنيشان.
* » ١۴٠ » 6 a f. l. يطلب.
» ١۴١ » 1. Fortasse legendum est يغترهم pro يغيرهم.
* » ١۴۴ » 1. l. هذه.
* » — » 5. l. على.
» ١۴٩ » 5. ابردة. Fortasse conferendum باب ايزد supra p. ١٢٥.
» — » 8. l. الماذرائيين. Cf. Jacut, IV, p. ٣٨١. Plus semel nomen corruptum legitur e. g. Ibno 'l-Athír, VII, p. ٧٨, 14, ٨٣, 4 a f., ١٠٠, 11, ١٠۴, 13, ١٣٩, 6, Abu 'l-Mahásin, II, p. ٢٥٩, 5 seqq.

Pag. ١٤٦ vs. 4 a f. l. توڤّلن.

» ١٥١ » 8. l. فيه.

* » ١٥٤ » 10. l. وبكارزين.

» ١٥٧ » 5 a f. l. السامىّ, cf. p. ١٤١.

* » ١٥٩ » 4. l. ربّما.

» ١٩١ ann. *g.* Jacut, II, p. ٦٠٠, 15 male حواشير.

» ١٩٤ vs. 5. Aliter narrat Ibno 'l-Athír, VIII, p. ٢٤٢ ult.

» ١٩٨ » 2. Verba انما هى اخواش e margine in textum irrepsisse videntur.

» ١٧١ ann. *a.* Pro *Bibliotheca Indica* l. *Journal of the Asiatic Society of Bengal.*

* » ٢٠٢ vs. 6. l. كلّها.

* » ٢٠٥ » 6. l. على.

» ٢١٥ » 4. Melius fortasse كهده (Pagus foeni).

» ٢٣٠ » 6. Accuratius est كلا pro كلى.

» ٢٣٥ » 4. Videtur legendum بالجنوبه.

» — » 11. l. تحويان.

» ٢٤٦ » 6. Videtur legendum تسمى ملاذه فقتل.

» ٢٨٠ » 5 a f. l. اليها. Verba ولا يستحق — هناك in C. desunt.

* » ٢٨٧ » 4 l. وراءه.

* » ٣٠١ » 3 a f. l. ويتشعّب.

» ٣٠٤ » 4. Pro وببن i. e. وبين potius l. ومن.

www.ingramcontent.com/pod-product-compliance
Lightning Source LLC
LaVergne TN
LVHW102226300926
849160LV00005B/35
9789004258686

كتاب

مسالك الممالك

لابى اسحاق ابراهيم بن محمد الفارسى

الاصطخرى المعروف بالكرخى

وهو معول على

كتاب صور الاقاليم

للشيخ ابى زيد احمد بن سهل البلخى

طبع

فى مدينة لَيْدَن المحروسة

بمطبع بريل

سنة ١٨٧٠ المسيحية

الحمد لله ربّ العالمين وصلّى الله على سيّدنا
محمّد النبىّ وعلى آل محمّد وسلّم ونسال
الله التوفيق وخاتمة السعادة[a]

الحمد لله مبدئ[b] النعم وولىّ الحمد وصلّى الله على محمّد وعلى آل محمّد، امّا بعد فانّى ذكرتُ فى كتابى هذا اقاليم الارض على الممالك وقصدتُ منها بلاد الاسلام بتفصيل مدنها وتقسيم ما يعود[c] بالاعمال المجموعة

---

a) Cod. Bonon. = A. Cod. Berol. = B. Epitome Gothan. cujus facsimile dedit Moeller = C. Ibn Haucal secundum textum mox edendum = D. Vers. Pers. ex MS. Gothano = E. Vers. Ouseleyi = Ous. *b*) A. مبتدى. Scripsi secundum H. Khal. V, p. 509 et Vers. Pers. MS. East India House 1026. In B. primum folium deëst. In E. initium operis hoc est (cf. H. Khal. IV, p. 112): الحمد لله الذى خلق السموات والارضين ووضع البلاد والاقاليم وبدع الاشياء العظام كما يقول فى الكتاب الكرام انّ ربكم الله الذى خلق السموات والارض فى ستّة ايام وصلّى الله على محمد سيّد الانام وآله واولاده الكرام، اما بعد چنين گويد جامع اين صور كه چون عمرى بسياحت وسفر گذرانيدم وبر احوال بسى از بلاد معموره وغيره مطلع كرديدم بدولت آن صاحبدولتى كه در ركاب او روزگار ميگذرانيدم هم بحسب لاشارهٔ او واجب ديدم ذكر بلاد مشهور وغيره از حوزهٔ اسلام وانچه قريب بدان بود مع الملحقات والمضافات نمودن تا چون بعضها عزّه بديدهٔ اعتبار دران تأمل فرمايند ومستفيد كردند كمينه‌را بدعء يا داورى فرمايند. *c*) D. يفرد.

اليها ولم اقصد الاقاليم السبعة الّتى عليها قسمة الارض بل جعلتُ كلّ قطعة افردتُها[a] مفردة مصوّرة يحكى موضع ذلك الاقليم ثمّ ذكرتُ ما يحيط به من الاماكن وما فى اضعافه من المدن والبقاع المشهورة والبحار والانهار وما يحتاج الى معرفته من جوامع ما يشتمل عليه ذلك الاقليم من غير ان استقصيتُ ذلك كراهة الاطالة الّتى تؤدّى الى ملال[b] من قرأه ولانّ الغرض فى كتابى هذا تصوير هذه الاقاليم الّتى لم يذكرها احد علمتُهُ وامّا ذكر مدنها وجبالها وانهارها وبحارها والمسافات وسائر ما انا ذاكره فقد يوجد فى[c] الاخبار ولا يتعذّر على من اراد تقصّى شىء من ذلك من اهل كلّ بلد فلذلك تجوّزنا فى ذكر المسافات والمدن وسائر ما نذكره[d]، فاتّخذتُ لجميع الارض الّتى يشتمل عليها البحر المحيط الّذى لا يُسْلَك صورةً اذا نظر اليها ناظر علم مكان كلّ اقليم ممّا ذكرناه واتّصال بعضه ببعض ومقدار كلّ اقليم من الارض حتّى اذا رأى كلّ اقليم من ذلك مفصّلًا علم موقعه من هذه الصورة ولم تتّسع هذه الصورة الّتى جمعت سائر الاقاليم لما يستحقّه كلُّ اقليم فى صورته من مقدار الطول والعرض والاستدارة والتربيع والتثليث وسائر ما يكون عليه اشكال تلك الصورة فاكتفيتُ ببيان موقع كلّ اقليم ليعرف مكانه[e] ثمّ افردتُ لكلّ اقليم من بلاد الاسلام صورة على حِدَة بيّنتُ[f] فيها شكل ذلك الاقليم وما يقع فيه[g] من المدن وسائر ما يحتاج الى علمه ممّا آتى على ذكره فى موضعه ان شآء الله تعالى ۞

ففصّلتُ بلاد الاسلام عشرين اقليمًا وابتدأتُ بديار العرب فجعلتُها اقليمًا لانّ فيها الكعبة ومكّة امّ القرى وهى واسطة هذه الاقاليم ثمّ اتبعتُ ديار العرب ببحر فارس لانّه يكتنف[h] اكثر ديار العرب ثمّ ذكرتُ المغرب حتّى انتهيتُ الى مصر فذكرتُها ثمّ ذكرتُ الشام ثمّ بحر الروم ثمّ الجزيرة ثمّ العراق ثمّ خوزستان ثمّ فارس ثمّ كرمان ثمّ المنصورة وما يتّصل بها من بلاد السند والهند والاسلام ثمّ اذربيجان وما يتّصل بها ثمّ كور الجبال ثمّ الديلم

---

a) A. اقرن بها. *b*) A. املال. *c*) A. توجد من. *d*) A. يذكره. *e*) Hic incipit B. *f*) B. تبتت. *g*) A. et B. فيها. *h*) B. يكنف.

ثمّ بحر الخزر ثمّ المفازة الّتى بين فارس وخراسان ثمّ سجستان وما يتّصل بها ثمّ خراسان ثمّ ما ورآء النهر[a] ☆

فهذه صورة الارض عامرها والخراب منها وهى مقسومة على الممالك وعماد ممالك الارض اربعة فاعمرها واكثرها خيرًا واحسنها استقامة فى السياسة وتقويم العمارات فيها[b] مملكة ايرانشهر وقصبتها اقليم بابل وهى مملكة فارس وكان حدّ هذه المملكة فى ايّام العجم معلومًا فلمّا جآء الاسلام اخذت[c] من كلّ مملكة بنصيب فأُخذ من مملكة الروم الشام ومصر والمغرب والاندلس وأُخذ من مملكة الهند ما اتّصل بارض المنصورة والمُلْتان الى كابُل[d] وطرف اعلى[e] طُخارِسْتان وأُخِذ من مملكة الصين ما ورآء النهر *وانضاف اليها هذه[f] الممالك العظيمة، فمملكة الروم تدخل فيها حدود الصّقالبة[g] ومن جاورهم من الرُّوس والسَّرِير واللّان والأرمن ومن دان بالنصرانيّة ومملكة الصين تدخل فيها سائر بلدان الاتراك وبعض التُّبّت ومن دان بدين اهل الاوثان منهم ومملكة الهند تدخل فيها السند وقَشْمِير[h] وطرف من التُّبّت ومن دان بدينهم، ولم نذكر[i] بلد السودان فى المغرب والبُجَة والزنج[k] ومن فى اعراضهم من الامم لانّ انتظام الممالك بالديانات والآداب والحكم وتقويم العمارات بالسياسة المستقيمة وهاؤلآء مهملون لهذه الخصال ولا حظّ لهم فى شىء من ذلك

---

a) Hic sequi deberet mappa mundi. E. post divisionem haec habet: واين ترتيبى است پسنديده كه عمرى بس طويل بران صرف شد وبر جميع جزئيات اين ربع اطلاع بديد شد بعضى براى العين وبرخى از ثقات مورّخين مسموع شد درين كتاب ثبت نموديم واين كتابرا مسالك الممالك نام نهاديم وبعضى از ملوك ماضيهٔ مشهوره واكثر سلاطين كه در زمان تاليف موجود بودند نيز بتقريب قسمت ذكر يافت وبالله العصمة والتوفيق. Deinde absoluta mappa haec leguntur: خداوند كتاب گويد كه اين صوره الخ. b) B. om. c) D. اخذ. d) E. تا مولتان تا كابلستان. e) A. et B. وطرفا على; E. وحدود تخارستان. f) Ex D.; A. et B. وهذه. g) E. سقلاب. h) E. وكشمير. i) B. يذكر. k) E. وزنكيان وبجّه.

فيستحقّوا به[a] افراد ممالكهم بما ذكرنا به سائر الممالك غير انّ بعض السودان المقاربين لهذه الممالك المعروفة يرجعون الى ديانة ورياضة وحكم ويقاربون اهل هذه الممالك مثل النوبة والحبشة فانّهم نصارى يرتسمون بمذاهب الروم وقد كانوا قبل الاسلام يتّصلون بمملكة الروم على المجاورة لانّ ارض النوبة متاخمة لارض مصر والحبشة على بحر القلزم وبينها وبين ارض مصر مفازة فيها معدن الذهب ويتّصلون بمصر والشام من طريق بحر القلزم، فهذه الممالك المعروفة وقد زادت مملكة الاسلام بما اجتمع اليها من اطراف هذه الممالك۞

وقسمة الارض على الجنوب والشمال فاذا اخذتَ من المشرق من الخليج الّذى ياخذ من البحر المحيط بارض الصين الى الخليج الّذى ياخذ من هذا البحر المحيط من ارض المغرب بارض الاندلس فقد قسمتَ الارض قسمين وخطّ هذه القسمة ياخذ من بحر الصين حتّى يقطع بلد الهند ووسط مملكة الاسلام حتّى يمتدّ الى ارض مصر الى المغرب فما كان فى حدّ الشمال من هٰذين القسمين فاهله بيض وكلّما تباعدوا فى الشمال ازدادوا بياضًا وهى اقاليم باردة وما كان ممّا يلى الجنوب من هٰذين القسمين فانّ اهله سود وكلّما تباعدوا فى الجنوب ازدادوا سوادًا واعدل هذه الاماكن ما كان فى الخطّ المستقيم وما قاربه۞ وسنذكر كلّ اقليم من ذلك بما يعرف قربه ومكانه من الاقاليم الّذى يصاقبه فامّا[b] مملكة الاسلام فانّ شرقيّها ارض الهند *وبحر فارس[c] وغربيّها مملكة الروم وما يتّصل بها من الارمن واللان والران والسرير والخزر والروس وبُلْغار والصقالبة وطائفة من الترك وشماليّها مملكة الصين وما اتّصل بها من بلاد الاتراك وجنوبيّها بحر فارس، *وامّا مملكة الروم فانّ شرقيّها بلاد الاسلام وغربيّها وجنوبيّها البحر المحيط وشماليّها حدود عمل الصين لانّا ضمنّا ما بين الاتراك وبلد الروم من الصقالبة وسائر الامم الى بلد الروم[c]، وامّا مملكة الصين[d] فانّ شرقيّها وشماليّها البحر المحيط وامّا جنوبيّها فمملكة الاسلام والهند وامّا غربيّها فهو البحر المحيط ان جعلنا

---

a) A. فيستحقونه, B. فيستحقون به. b) Hic incipit C. c) C. om.
d) E. جين.

ياجوج وماجوج وما ورآءهم الى البحر من هذه المملكة، وامّا ارض الهند فانّ شرقيّها بحر فارس وغربيّها وجنوبيّها بلاد الاسلام وشماليّها *مملكة الصين[a] *فهذه حدود هذه الممالك الّتى ذكرناها[b] ✿ وامّا البحار فانّ اعظمها بحر فارس وبحر الروم *وهما خليجان متقابلان ياخذان من البحر المحيط[c] واعظمهما طولًا وعرضًا بحر فارس والّذى ينتهى اليه بحر فارس *من الارض من حدّ الصين الى القلزم[d] فاذا قطعتَ من القلزم الى الصين على خطّ مستقيم كان مقداره[e] مائتى مرحلة وذلك انّك[f] اذا قطعتَ من القلزم الى ارض العراق فى البرّيّة كان نحوًا من شهر[g] ومن العراق الى نهر بلخ نحوًا من شهرين ومن نهر بلخ الى آخر الاسلام فى حدّ[h] فَرْغَانَة نيف وعشرين[i] مرحلة ومن هناك الى ان تقطع[k] ارض الخَزْلَجِيّة كلّها فتدخل فى عمل التّغَزْغز نيف وثلاثين[l] مرحلة ومن هذا المكان الى البحر من آخر عمل الصين نحوًا من شهرين، فامّا من اراد قطع هذه المسافة من القلزم الى الصين *فى البحر[m] طالت المسافة عليه[n] لكثرة المعاطف[o] والتوآء الطرق فى هذا البحر ✿ وامّا بحر الروم فانّه *ياخذ من البحر المحيط فى الخليج الّذى بين المغرب وارض الاندلس حتّى ينتهى الى الثغور الشاميّة ومقداره[p] فى المسافة نحو من سبعة اشهر[q] وهو احسن استقامة واستوآء من بحر فارس وذلك انّك[r] اذا اخذتَ من فم هذا الخليج اذّاك ريح واحدة[s] الى آخر

---

a) E. دربای چین وما چین. b) C. om. c) C. om. Deinde habet واعظمها. d) C. pro his هو بحر القلزم, et deinde plura omissa sunt quae p. 4, l. 3 seqq. alieno loco inseruntur. e) C. et D. add. نحو. f) A. et B. اقل. g) C. شهرين et sic E. et Ous. sed hic orta est lectio omissione verborum ومن العراق الى نهر بلخ نحوا من شهرين. h) C. حدود. i) A., B. et C. وعشرون. k) A. et B. يقطع. l) A., B. et C. وثلاثون. m) A. om. n) A. om.; C. فيطول عليه. o) A. المعاطب. Deinde pro والتواء الطرق C. habet والتعاريج الطريق. l. والتعاويج. p) C. haec om. q) C. add. اعنى طوله. r) A. et B. iterum اقل. s) A. et B. واحد.

هذا البحر وبين بحر[a] القلزم الّذى هو لسان بحر فارس وبين بحر الروم على سمت[b] القرَمَا اربع مراحل غير انّ بحر الروم يجاوز الفرما الى الثغور بنيف وعشرين[c] مرحلة ۞ *وقد فصّلنا فى مسافات المغرب ما يغنى عن اعادته[d] فمن مصر الى اقصى المغرب نحو من مائة وثمانين مرحلة *فكان ما[e] بين اقصى الارض من المغرب الى اقصاها من المشرق[f] نحوًا من اربع مائة مرحلة، وامّا عرضها من اقصاها فى حدّ الشمال الى اقصاها فى حدّ الجنوب فانّك تاخذ من ساحل[g] البحر المحيط حتّى تنتهى الى ارض ياجوج وماجوج[h] ثمّ تمرّ على ظهر الصقالبة وتقطع[i] ارض بلغار الداخلة والصقالبة وتمضى فى بلد الروم الى الشام *حتّى تخرج فى الشام وارض مصر والنوبة ثمّ تمتدّ فى بريّة بين بلاد السودان وبلاد الزنج حتّى تنتهى الى البحر المحيط وهذا[k] خطّ ما بين جنوبىّ الارض وشماليّها، فامّا الّذى أَعْلَمُهُ من مسافة هذا الخطّ فانّ من ناحية ياجوج الى ناحية بلغار وارض الصقالبة نحوًا[l] من اربعين مرحلة ومن ارض الصقالبة فى بلد الروم الى الشام[m] نحوًا من ستّين مرحلة ومن ارض الشام الى ارض مصر نحوًا من ثلاثين مرحلة ومنها الى اقصى النوبة نحوًا من ثمانين مرحلة *حتّى تنتهى الى هذه البريّة[n] فذلك مائتان وعشر مراحل[o] كلّها عامرة، وامّا ما بين ياجوج وماجوج[p] والبحر المحيط فى الشمال وما بين برارىّ السودان والبحر المحيط *فى الجنوب[q] *فقفر خراب ما[r] بَلَغَنَا انّ فيه عمارة ولا ادرى[s] مسافة هاتين[t] البريّتين *الى

---

a) Supplevi بحر quod in omnibus Codd. deëst. b) C. طريق ; E. راه. c) C. نيف وعشرون. d) C. om. haec. e) C. وكانما *sic*. f) C. بالمشرق. g) C. سواحل. h) A. om. i) B., C. et D. فتقطع. k) B. et D. فهذا. Deinde C. الخط. l) Hic et ter deinde A. et B. نحو pro نحوا. Pro اربعين E. et Ous. چهار صد. m) Haec omnia inde a حتى in C. desunt h. l.; mox deinde alieno loco inseruntur. n) C. om. o) C. وعشرون مرحلة. p) A. om. q) A., B. et C. om. r) C. فهو خراب وما. s) C. ندرى. t) B. et C. ما بين.

شطّ[a] البحر المحيط *كم هى وذلك انّ سلوكهما غير ممكن لفرط البرد الّذى يمنع من العمارة والحياة فى الشمال وفرط الحرّ المانع من الحياة والعمارة فى الجنوب، وامّا ما بين الصين والمغرب فمعمور كلّه، والارض كلّها مستديرة[b] والبحر المحيط محتفّ بها كالطوق وياخذ بحر الروم *وبحر فارس[c] من هذا البحر[d]✿ فامّا بحر الخزر *فليس ياخذ من هذا الخليج وانّما هو بحر[e] لو اخذ السائر على ساحله من الخزر على[f] ارض الدَّيْلَم وطَبَرِسْتان وجُرْجان والمفازة[g] على سِيَاه كُوه لرجع الى مكانه *الذى سار منه من غير ان يمنعه مانع الاّ نهر يقع فيه[h]، وامّا بحيرة خُوَارَزْم *فكذلك✿ وفى[i] اعراض بلاد الزنج[k] ومن ورآء ارض الروم خلجان وبحار لم نذكرها لقصورها عن هذه البحار وكثرتها، وياخذ من البحر المحيط خليج حتّى ينتهى فى[l] ظهر بلد الصقالبة ويقطع ارض الروم على القُسْطَنْطِينِيَّة حتّى يقع فى بحر الروم✿ وارض الروم حدّها من البحر المحيط على بلاد الجَلالِقَة وأَفْرَنْجَة ورُومِيَة وأَثِيناس الى القسطنطينيّة ثمّ الى ارض الصقالبة ويشبه ان يكون نحوًا من مائتين وسبعين مرحلة وذاك انّ من حدّ الثغور فى الشمال الى ارض الصقالبة نحوًا[m] من شهرين وقد بيّنّا انّ من الثغور الى اقصى المغرب مائتين[n] وعشر مراحل، والروم المحض من حدّ رومية الى حدّ

---

a) C. وبيين. E. كه بر كناره درياى محيط. b) E. مدوّر است. c) A. وفارس. d) Pro his omnibus inde a كم هى, C. وسلوكهما غير ممكن لفرط الحرّ والبرد. Deinde B. et C. واما. e) C. om. et habet فلو. f) C. الى. g) A. et B. والمسافة, C. فى المفازة, D. ومفازة omisso على, E. ودر بيابان بقرب. Praepositio على idem fere significat quod s. بناحية. E. بحد سياه كوه, C. habet سياكوه. h) C. من غير مانع الا الانهار. i) C. فامثالها فى. Quae sequuntur usque ad finem capitis in C. admodum abbreviata sunt et turbata. Pro خوارزم, E. habet خارزم. k) E. زنكبار. l) B. et D. الى; E. بظاهر (ut quoque supra p. ٧, l. 8 على ظهر vertit per بر ظاهر). m) A. et B. نحو. n) A. مائتان.

الصقالبة فامّا ما [a] ضممنا الى بلد الروم من الافرنجة والجلالقة وغيرهم فانّ لسانهم مختلف غير انّ الدين والمملكة واحد كما انّ فى مملكة الاسلام السنة مختلفة والمَلِكُ واحد ✿ وامّا مملكة الصين * فانّها نحو من اربعة اشهر فى [b] ثلاثة اشهر فاذا اخذتَ من فم الخليج حتّى تنتهى الى دار الاسلام ممّا [c] ورآء النهر فهو نحو من ثلاثة اشهر واذا اخذتَ من حدّ المشرق حتّى [d] تقطع الى حدّ المغرب فى ارض التُّبَّت وتمتدّ فى ارض [e] التغزغز وخِرْخِيز [f] وعلى ظهر كَيْماك الى البحر فهو نحو من اربعة اشهر، ولمملكة الصين السنة مختلفة فامّا [g] الاتراك كلّها من التغزغز وخرخيز وكيماك والغُزّيّة والخزلجيّة فالسنتهم واحدة يفهم بعضهم عن بعض فامّا ارض الصين والتُّبَّت فلهم لسان مخالف لهذه الالسنة، والمملكة [h] كلّها منسوبة الى صاحب الصين المقيم بخَمْدان * كما انّ مملكة الروم [i] منسوبة الى الملك المقيم بالقسطنطينيّة ومملكة الاسلام منسوبة الى امير المؤمنين ببغداد ومملكة الهند منسوبة الى الملك المقيم بقَنُّوج [k] ✿ وديار الاتراك متميّزة فامّا الغُزّيّة فانّ حدود ديارهم ما بين الخَزَر وكيماك وارض الخَزْلجيّة وبُلْغار وحدود دار الاسلام [l] ما بين جُرْجان الى فاراب [m] وأَسْبِيجاب، وامّا ديار الكيماكيّة فانّهم من ورآء الخزلجيّة من ناحية الشمال وهم فيما بين الغُزّيّة وخِرْخِيز وظهر الصقالبة، وياجوج هم فى ناحية الشمال اذا قطعتَ ما بين الصقالبة والكيماكيّة والله اعلم * بمكانهم وسائر بلادهم [n]، وامّا خِرْخيز فانّهم ما بين التغزغز وكيماك [o] والبحر المحيط وارض

---

a) C. وانما; E. واما انجه كُفتيم كه فرنك وجالق وغيرهم از جملهٔ رومست. b) C. فطولها اربعة اشهر وعرضها. c) C. بما. d) C. الى ان. e) A. et B. om. f) A. semper التغْزْغز et خِرخِير. g) C. add. جميع انواع, D. جميع, omisso كلها. h) C. add. الشرقية. i) C. والغربية. k) C. بقيّوج. l) Sic A., B., C. et E. (وحدود مسلمانی); D. وحدود الديلم. m) D. et E. باراب. n) C. بمقدارهم; E. مقدار زمين ايشان وعدد ايشان; D. بمقاديرهم. Pro وسائر B. habet ودسائر. o) A. والكيماك. Pro التغزغز E. habet غز et خزرّج pro الخزلجية (supra خرلخيه habuit).

الخَزْلَجِيَّة، وامّا التغزغز فانّهم ما بين التُّبَّت وارض الخزلجيّة وخرخيز ومملكة الصين، وامّا الصين فانّهم ما بين البحر والتغزغز[a] والتُّبّت والصين نفسه هو هذا الاقليم وانّما نسبنا سائر بلاد الاتراك اليها فى المملكة كما نسبنا سائر مملكة الروم الى ارض رومية والقسطنطينيّة وكما نسبنا سائر ممالك الاسلام الى ايرانشهر وهو ارض بابل ☸ وارض الصقالبة عريضة طويلة نحو[b] من شهرين فى مثلها، وبُلْغار الخارجة هى مدينة صغيرة ليس فيها اعمال كثيرة واشتهارها لانّها فرضة لهذه الممالك، والرُّوس قوم بناحية بُلْغار فيما بينها وبين الصقالبة، وقد انقطع طائفة من الاتراك[c] عن بلادهم فصاروا فيما بين الخَزَر والروم يقال لهم البَجَناكيّة[d] وليس موضعهم بدار لهم على قديم الايّام وانّما انتابوها فغلبوا عليها، وامّا الخَزَر فانّه اسم لهذا الجنس من الناس[e] وامّا البلد فانّه مصر يسمّى اِتل[f] وانّما سُمّى باسم النهر الّذى يجرى عليه الى بحر الخَزَر وليس لهذا المصر كثير رساتيق ولا سعة ملك وهو بلد بين بحر الخَزَر والسَّرير والرُّوس والغُزّيّة[g]، وامّا التُّبّت فانّها بين ارض الصين والهند وارض الخَزْلجيّة والتغزغز وبحر فارس وبعضها فى مملكة الهند وبعضها فى مملكة الصين ولهم ملك قائم بنفسه يقال انّ اصله من التبابعة والله اعلم ☸ وامّا جنوبىّ الارض من بلاد السودان * فانّ بلد السودان[h] الّذى فى اقصى المغرب على البحر المحيط بلد مكنّف[i] ليس بينه وبين شىء من الممالك اتّصال غير انّ حدّا

---

a) E. وزمين غز. Deinde A. وتُبّت. b) A. et B. نحوا. Conferatur versio Persica hujus loci quam e libro Nacíro'd-dín Tusí edidit Melgunoff, *das südliche Ufer des Kaspischen Meeres*, p. 295. c) C. et D. الترك. Hic in Ous. (p. 10) permagna incipit lacuna. d) C. البجناك, E. نُجْنَة (Melgun. بجنه). e) C. add. وليس اسما للبلد. f) A. اتك, B. ماتك, E. أتّل. g) E. ونواحى اتل بس. Melgun. همين فارس (sic) است ومتغرغر وروس وغزنه بحدود خزر ناحيتى ندارد الا چند جايكاه كه انرا فارس (sic) خوانند بحدود خزر ومتغرغر وروس وغزيه. h) C., D. et E. om. et habent التى. i) C. et D. ملتف, E. بسر خويش.

له ينتهى الى البحر المحيط وحدًّا له الى بريّة بينه وبين ارض المغرب وحدًّا له الى بريّة بينه وبين ارض مصر على ظهر الواحات وحدًّا له ينتهى الى البريّة التى قلنا انّه لا يثبت فيها عمارة لشدّة الحرّ وبلغنى انّ طول ارضهم نحو[a] من سبع مائة فرسخ فى نحوها[b] غير انّها من البحر الى ظهر الواحات اطول من عرضها، وامّا ارض النوبة فانّ حدًّا لها ينتهى[c] الى ارض مصر وحدًّا لها الى هذه البريّة التى بين ارض السودان ومصر وحدًّا لها الى ارض البُجَة[d] وبرارىّ بينها وبين القلزم وحدًّا لها الى هذه البريّة التى لا تسلك، وامّا ارض البجة فانّ ديارهم صغيرة *وهى ما[e] بين الحبشة والنوبة وهذه البريّة التى لا تسلك، وامّا الحبشة فانّها على بحر القلزم *وهو بحر فارس[f] فينتهى حدٌّ لها الى بلاد الزنج وحدٌّ لها الى البريّة التى بين النوبة وبحر القلزم وحدٌّ لها الى البُجة والبريّة التى لا تسلك، وامّا ارض الزنج *فانّها اطول من[g] ارض السودان ولا تتّصل بمملكة غير الحبشة وهى بحذاء اليمن وفارس وكَرْمان الى ان تحاذى ارض الهند، وامّا ارض الهند *فانّ طولها من عمل مُكْران فى[h] ارض المنصورة والبُدْهَة[i] وسائر بلد السند الى ان تنتهى الى[k] قَنُّوج ثمّ تجوزه الى ارض التُّبّت نحو من اربعة اشهر وعرضها من بحر فارس على ارض قنّوج نحو من ثلاثة اشهر، وامّا مملكة الاسلام فانّ طولها من حدّ فرغانة حتّى تقطع[l] خراسان والجبال[m] والعراق وديار العرب الى سواحل اليمن[n] نحو من خمسة اشهر وعرضها *من بلد الروم[o] حتّى تقطع الشام والجزيرة والعراق وفارس وكرمان الى ارض المنصورة على شطّ بحر فارس نحو من اربعة اشهر، وانّما تركنا ان نذكر فى طول الاسلام حدّ

---

a) A., B. et C. نحوا. *b*) E. نهصد فرسنك در نهصد فرسنك. *c*) A. om. *d*) Codd. interdum البُجّه. *e*) C. et D. وهم فيما. *f*) E. كه نهايت بحر فارس است. *g*) C. فان طولها اطول. Pro ارض الزنج, E. semper زنكبار habet. *h*) E. تا برابر زمين كرمان باشد. *i*) C. والبَدهه. *k*) A. et B. om. Deinde C. فيّوج. *l*) A. et B. يقطع, C. دقطع. *m*) E. ut semper كوهستان. *n*) C. et D. add. فهو. *o*) E. از كناره ولايت روم.

المغرب الى الاندلس لانّها مثل الكمّ فى الثوب وليس فى شرقىّ المغرب ولا فى غربيّها اسلام لانّك اذا جاوزتَ مصر فى ارض المغرب كان جنوبىّ المغرب بلاد السودان وشمالىّ المغرب بحر الروم ثمّ ارض الروم ولو صلح ان يجعل طول الاسلام من فرغانة الى ارض الاندلس لكان مسيرة ثلاث مائة وعشر مراحل لانّ من اقصى فرغانة الى وادى بلخ نيفًا وعشرين[a] مرحلة ومن وادى بلخ الى العراق نحوًا[b] من ستّين مرحلة ومن العراق الى مصر نحوًا من خمسين مرحلة، وقد بيّنّا فى مسافات المغرب انّ من مصر الى اقصاه مائة وثمانين مرحلة ۞ وقصدتُ فى كتابى هذا تفصيل بلاد الاسلام اقليمًا اقليمًا حتّى يعرف موقع كلّ اقليم من مكانه وما يجاوره من سائر الاقاليم[c] ولم تتّسع هذه الصورة الّتى جمعت سائر الاقاليم لما يستحقّه كلّ اقليم فى صورته من مقدار الطول والعرض والاستدارة والتربيع والتثليث وما يكون عليه اشكالها غير انّا بيّنّا لكلّ اقليم مكانًا يُعْرَف به موضعه وما يجاوره من سائر الاقاليم ثمّ افردنا لكلّ اقليم منها صورة على حِدَة فبيّنّا فيها شكل ذلك الاقليم وما يقع فيه من المدن وسائر ما يحتاج الى علمه ممّا ناتى على ذكره فى موضعه ان شآء الله ۞

ديار العرب[d]

وابتدأتُ بديار العرب لانّ القبلة بها *ومكّة فيها[e] وهى امّ القرى وبلد العرب واوطانهم الّتى لم يشركهم فى سكناها غيرهم والّذى يحيط بها بحر فارس من عَبّادان وهو مصبّ مآء دجلة فى البحر فيمتدّ على البَحْرَيْن حتّى ينتهى الى عُمَان ثمّ يعطف على سواحل مَهْرَة وحَضْرَمَوْت وعَدَن حتّى ينتهى على سواحل اليمن الى جُدَّة ثمّ يمتدّ على الجار ومَدْيَن[f] حتّى

---

a) A. et B. نيف وعشرون. *b*) A. et B. hic et deinde نحو. *c*) Quae sequuntur mera repetitio sunt eorum quae p. ٣ leguntur. *d*) Textus hujus capitis secundum C. editus est a Cl. Arnold in *Chresthomathia Arabica*, p. 76—91. *e*) A. om. In B. folium deëst. *f*) C. اليمن.

ينتهى الى أَيْلَة[a] ثمّ قد انتهى حينئذ حدّ ديار العرب من هذا البحر وهذا المكان من البحر لسان يعرف[b] ببحر القُلْزُم فينتهى الى *تاران وجُبَيْلَات[c] الى القلزم فينقطع فهذا[d] هو شرقىّ ديار العرب وجنوبيّها وشىء من غربيّها ثمّ يمتدّ عليها من ايلة الى[e] مدينة قوم لُوط والبحيرة المُنْتِنَة الّتى تعرف ببحيرة زُغَر[f] الى الشَّرَاة والبَلْقَآء وهى من عمل فِلَسْطِين وأَذْرِعَات[g] وحَوْران والبَثَنِيَّة والغُوطَة ونواحى بَعْلَبَكّ وذلك من عمل دِمَشْق وتَدْمُر وسَلَمْيَة[h] وهما من عمل حِمْص ثمّ الخُنَاصِرَة[i] وبَالِس وهما من عمل قِنَّسْرِين *وقد انتهينا الى الفُرَات[k] ثمّ يمتدّ الفُرَات على ديار العَرَب حتّى ينتهى الى الرَّقَّة وقَرْقِيسيَا والرَّحْبَة والدَّالِيَة[k] وعانة والحَدِيثَة وهيت والأَنْبَار الى الكوفة ومستفرغ مياه الفرات الى البطائح ثمّ تمتدّ[l] ديار العَرَب على نواحى الكوفة والحِيرَة على الخَوَرْنَق وعلى سواد الكوفة *الى حدّ واسِط فتصاقب ما قارب دجلة[m] عند واسط مقدار مرحلة ثمّ تمتدّ على سواد البصرة وبطائحها حتّى تنتهى الى عبّادان فهذا[n] الّذى يحيط بدِيَارِ العَرَب، فما كان من عَبّادَان الى أَيْلَة فانّه بحر فارس ويشتمل على نحو ثلاثة ارباع ديار العرب وهو الحدّ الشرقىّ والجنوبىّ وبعض الغربىّ وما بقى من حدّ الغربىّ من أَيْلَة الى بالس فمن الشام وما كان من بالس الى عبّادان فهو الحدّ الشمالىّ فمن بالس الى ان تجاوز[o] الانبار من حدّ الجزيرة ومن الانبار الى عَبّادان من

---

a) E. بابلّة et saepius ابلـة pro ايلة habet. b) A. ويعرف. c) A. مغازات وجبال, C. وحملات فاران, E. واز آنجا بتاران آيد واز آنجا بجبيلات. d) A. بهذا. e) C. على. Sequens مدينة pro quo D. مدائن habet, ab E. vertitur شارستان. f) C. زعرا et sic E. g) E. وأذرعات. Deinde C. والجوزان (Arn. والحوران). h) E. وسلمية. i) A. sine articulo. k) C. om. l) C. add. على. m) C. والى حدّ واسط دساوط (يَسَّاقَطُ .Arn) ماء الفرات ودجلة. Pro فتصاقب, A. فيصاقب. E. habet تا بحدود واسط انكاه فرات در دجله افتد. D. in suo Codice nostram lectionem habuit. n) A. فهذه; C. add. هو. o) Arn. يجاور.

حدّ العراق ۞ ويتّصل بارض العرب بناحية أَيْلَة بريّة تُعْرَف بتيه بنى إسرآئيل وهى بريّة[a] وان كانت متّصلة بديار العرب فليست من ديارهم وانّما هى[b] بريّة بين[c] ارض العَمالقة واليُونانيّة[d] وارض القِبْط وليس للعرب بها مآء ولا مرعى فلذلك لم نُدْخِلْها[e] فى ديارهم ۞ وقد سكن طوائف من العرب من ربيعة ومُضَر الجزيرة حتّى صارت لهم ديارًا ومراعِىَ فلم نذكر الجزيرة فى[f] ديار العرب لانّ نزولهم بها وهى ديار فارس والروم فى اضعاف قرى معمورة ومدن لها اعمال عريضة فنزلوا على خفارة[g] فارس والروم حتّى انّ بعضهم تنصّر ودان بدين الروم مثل تغلب من ربيعة بارض الجزيرة وغسّان وبَهْرآء وتَنُوخ من اليمن بارض الشام ۞ وديار العرب هى الحِجَاز الّذى يشتمل على مكّة والمدينة واليَمَامَة ومخاليفها ونَجْد الحجاز المتّصل بارض البَحْرَيْن وبادية العراق وبادية الجزيرة وبادية الشام واليَمَنُ المشتملة على تِهَامَة ونَجْد اليَمَن وعُمَان ومَهْرَة وحضرموت وبلاد صَنْعَآء وعَدَن وسائر مخاليف اليَمَن، فما كان من حدّ السِّرَّيْن حتّى[h] ينتهى الى[i] ناحية يَلَمْلَم ثمّ على ظهر[k] الطائف ممتدًّا على نجد اليمن الى بحر فارس مشرّقًا فمن اليمن ويكون ذلك نحو الثُّلْثَيْن من ديار العرب، وما كان من حدّ السرّين على بحر فارس الى قرب مَدْيَن راجعًا فى حدّ المشرق[l] على الحِجْر الى جبلَىْ طيّئٍ ممتدًّا على ظهر اليمامة الى بحر فارس فمن الحجاز، وما كان من حدّ اليمامة الى قرب المدينة راجعًا على بادية[m] البصرة حتّى تمتدّ على البحرين الى البحر فمن نَجْد، وما كان *من حدّ عَبّادان الى الانبار[n] مواجهًا لنجد والحجاز على أَسَد وطيّئ وتميم وسائر قبائل مُضَر فمن بادية العراق، وما كان من حدّ الانبار الى بالس مواجهًا لبادية الشام على ارض[o] تيمآء وبريّة خُسَاف[p] الى قرب[q] وادى

a) A. om. b) C. et D. كانت. c) A. من. d) C. وارض اليونان. E. habet: بلكه از عمالقه وزمين يونان بوده كه بقبطيان باز خوانند. e) C. ندخل (تدخل). f) C. مع. g) A. حُكْم. h) C. et Abulfeda, p. ٨٠. الى ان. i) A. om. k) E. ظاهر. l) A. add. راجعا. m) Abulf. ناحية. n) A. tantum من عبادان. o) C. عرض. p) C. et E. خشاف et sic Edrisi secundum omnes Codices. q) A. om.

القرى والحجر فمن بادية الجزيرة، وما كان من بالس الى ايلة مواجهًا للحجاز على بحر فارس الى ناحية مدين معارضًا لارض تَبُوك حتّى يتّصل بديار طيّئٍ فمن بادية الشام على انّ من العلمآء بتقسيم هذه الديار من زعم انّ المدينة من نَجْد لقربها منها وانّ مكّة من تِهَامَة اليَمَن لقربها منها[a] ۞

وانا[b] بمشئة الله وعونه[c] ساذكر ما انتهى اليه علمى من مدنها وما تشتمل عليه المدن ممّا يحتاج الى علمه والمشاهير من ديار العرب بها وجبالها ورمالها وجوامع من المسافات المسلوكة بها، ولا نعلم بارض العرب نهرًا ولا بحرًا يحمل سفينة لانّ البحيرة المنتنة[d] الّتى تعرف بزغر[e] وان كانت مصاقبة للبادية فليست منها ومجمع المآء الّذى بارض اليمن فى ديار سَبَأ انّما كان موضع مسيل مآء بنى فى[f] وجهه سدٌّ فكان يجتمع فيه مياه كثيرة[g] يستعملونها فى القرى والمزارع حتّى كفروا[h] بعد ان كان الله[i] جعل لهم عمارات قرى متّصلة الى الشام فسلّط *الله عزّ وجلّ[k] على ذلك المآء[l] آفة فكان[m] لا يمسك مآء وهو قوله تعالى[n] وَجَعَلْنَا بَيْنَهُمْ وَبَيْنَ ٱلْقُرَى ٱلَّتِى بَارَكْنَا فِيهَا الى قوله وَمَزَّقْنَاهُمْ كُلَّ مُمَزَّقٍ فبطل ذلك المآء الى يومنا هذا، وامّا الجداول والعيون والسوانى والآبار فانّها كثيرة ۞ ونبتدئُ من مدن ديار العرب بمكّة شرّفها الله وهى مدينة فيما بين شعاب الجبال وطول مكّة من المَعْلاة[o] الى المسفلة نحو ميلين وهو من حدّ الجنوب الى الشمال ومن اسفل جياد الى ظهر قُعَيْقِعان نحو الثُلُثَيْن من هذا وابنيتها حجارة والمسجد فى نحو الوسط منها والكعبة فى وسط المسجد[p] وباب الكعبة مرتفع عن[q] الارض نحو قامة وهو مصراع واحد وارض البيت مرتفعة عن الارض مع الباب والباب بحذآء

---

a) E. add. والله اعلم بالصّواب. Hic sequi deberet mappa Arabiae. b) A. om. c) A. وغره. d) C. الممنته (Arn. recte الميّتة). e) E. زُغرا. f) C. et D. على. g) A. om. h) C., D. et E. add. النعمة. i) C., D. et E. add. تعالى قد. k) C. et D. om. l) C. الموضع, D. المكان, E. انجايكاه. m) C. فصار. n) Qor. 34, vs. 17 seq. o) A. المَعَلّاة. p) A. add. منها: Post والكعبة D. add. المشرفة, E. مُعظّمة. q) C. et D. من.

قبّة زَمْزَم والمَقَام بقرب زمزم على خطّ[a] محاذٍ للباب، وبين يدى الكعبة ممّا يلى المغرب حائط[b] مبنىّ مدوّر وهو من البيت الّا انّه لم يدخل فيه وهو الحِجْر والطواف يحيط به وبالبيت وينتهى الى هذا الحجر من البيت ركنان احدهما يعرف بالركن العراقىّ والآخر بالركن الشامى والركنان الآخران احدهما عند الباب والحجر الاسود[c] فيه على اقلّ من قامة والركن الآخر يعرف باليمانى، وسقاية الحاجّ الّتى تعرف بسقاية[d] العبّاس على ظهر زمزم وزمزم فيما بينها وبين البيت، ودار الندوة من المسجد الحرام فى غربيّه وهى[e] خلف دار الامارة مشرعة الى المسجد وهو[f] مسجد قد جُمع الى المسجد الحرام وكان فى الجاهليّة مجتمعًا لقريش، والصَّفَا مكان مرتفع من جبل ابى قُبَيْس وبينها[g] وبين المسجد الحرام عرض الوادى الّذى هو طريق وسوق ومن وقف على الصفا كان بحذآء الحجر الاسود، والمَسْعَى[h] ما بين الصفا والمروة، والمَرْوَة حجر من جبل[i] قعيقعان ومن وقف عليها كان بحذآء الركن العراقىّ الّا انّ الابنية قد سترت ذلك الركن عن الرؤية، وابو قُبَيْس هو الجبل المشرف على الكعبة من شرقيّها، وقُعَيْقعَان هو الجبل الّذى عن غربىّ الكعبة وابو قُبَيْس اعلى واكبر منه ويقال انّ حجارة البيت من قعيقعان، ومِنًى على طريق عَرَفَة[k] من مكّة وبينها وبين مكّة ثلاثة اميال ومِنًى شعب طوله نحو ميلين وعرضه يسير وبها ابنية كثيرة لاهل كلّ بلد من بلدان الاسلام، ومسجد الخَيْف *فى اقلّ من الوسط[l] ممّا يلى مكّة، وجَمْرَة العَقَبَة فى آخر منى ممّا يلى مكّة وليست[m] العقبة الّتى تنسب[n] اليها الجمرة من منًى والجمرة الاولى والوسطى هما جميعًا فوق مسجد الخيف الى ما

---

a) C. et D. خُطّى sic legit E. (بچند گام). Deinde C. فيحاذى الباب ايضًا.
b) C. حظار, D. secundum Epit. Paris. حصار. *c*) C. add. كان. Appellatur الركن الاسود. *d*) A. سقاية. *e*) C. وهو. *f*) C. وهى. *g*) E. وميان صفا وكوه بوقبيس. *h*) A. والمَسْعَل. D. et E. والسعى. *i*) C. om.; D. et E. حد. *k*) C. et D. عرفات. *l*) E. نزديك ميانهٔ منا است. *m*) A., B., C. et E. add. جمرة; cf. *Chron. Mekk.*, II, p. ١٠٠. *n*) A. نسبت.

يلى مكّة، والمُزْدَلِفة مبيت للحاجّ ومجمع للصلاة اذا صدروا من عَرَفات وهو مكان بين بطن مُحَسِّر والمَازِمَيْن، وامّا بطن محسّر فهو وادٍ بين مِنًى والمزدلفة وليس من مِنًى ولا من المزدلفة، وامّا المازمان[a] فهو شعب بين جبلين يُفْضِى آخره الى بطن عُرَنَة[b] وهو وادٍ بين المازمين وبين عَرَفة[c] وليس من عرفة، وعرفة ما بين وادى عُرَنَة[b] الى حائط بنى عامر الى ما اقبل على الصَّخَرات الّتى يكون بها موقف الامام والى طريق حصن[d]، وحائط[e] بنى عامر نخيل[f] عند عرفة وبقربه المسجد الّذى يَجْمَعُ فيه الامام بين الصلاتين الظهر والعصر وهو حائط نخيل وبه عين وينسب الى عبد الله بن عامر بن كُرَيْز، وليس عرفات من الحرم وانّما حدُّ الحرم الى المازمين *فاذا جزتَهما[g] الى العَلَمَيْن المضروبين فما ورآء العلمين من الحلّ وكذلك التَّنْعيم الّذى يعرف بمسجد عائشة ليس من الحرم والحرم دونه وحدُّ الحرم نحو عشرة اميال فى مسيرة يوم وعلى الحرم كلّه منار مضروب يتميَّز[h] به من[i] غيره، وليس بمكّة مآء جارٍ الّا شىء بلغنى بعد خروجى عنها[k] انّه أُجرى اليها من عين كان[l] عمل فيها بعض الولاة فاستتمّ فى ايّام المُقْتَدِر امير المؤمنين ومياههم من السمآء وليست لهم آبار تشرب واطيبها بئر[m] زمزم ولا يمكن الادمان على شربه وليس بجميع مكّة فيما علمتُه شجر مثمر الّا شجر البادية فاذا جزتَ الحرم فهناك عيون وآبار وحوائط كثيرة واودية ذات خضور[n] ومزارع ونخيل وامّا الحرم فلم اره[o] ولم اسمع انّ بها شجرًا مثمرًا[p] الّا نخيلات رايتُها بفخّ ونخيلات يسيرة متفرّقة، وامّا ثَبِير فهو جبل مشرف يُرَى من مِنًى والمزدلفة وكانت الجاهليّة لا تدفع[q] من المزدلفة الّا بعد طلوع الشمس اذا اشرقت

---

a) B. et C. المازمين. *b*) A. et B. عرفه. *c*) C. عُرنه, Arn. bene عرفة. *d*) E. خصن. Est, nisi fallor, الحصن qui appellatur المَفْجَر. *e*) A. et B. وجامعا sic. *f*) C. et E. om. *g*) A. et B. فادن حداهما. *h*) A. et B. متميز. *i*) C. et D. عن. *k*) C. منها. *l*) C. et D. add. قد. *m*) A. et B. من. *n*) A. et B. حصون, C. et D. خضر. *o*) C. add. بها. *p*) A. مثمّرا. *q*) C. ترفع.

على ثبير وبالمزدلفة المَشْعَر الحرام وهو مصلّى الامام يصلّى به[a] المغرب والعشآء والصبح، والحُدَيْبِيَة بعضها فى الحلّ وبعضها فى الحرم وهو مكان صدّ فيه المشركون رسول الله صلّعم عن المسجد الحرام وهو ابعد الحلّ الى البيت وليس هو فى طول الحرم ولا فى عرضه الّا انّه فى مثل الزاوية للحرم[b] فلذلك صار بينها وبين المسجد اكثر من يوم ۞ وامّا المدينة فهى اقلّ من نصف مكّة وهى فى حرّة سبخة[c] الارض ولها نخيل كثيرة[d] ومياه نخيلهم وزروعهم من الآبار يستقون[e] منها العبيد وعليها سور والمسجد فى نحو من[f] وسطها وقبر النبىّ صلّعم من المسجد فى شرقيّه قريبًا من القبلة وهو الجدار الشرقىّ من المسجد وهو بيت مرتفع ليس بينه وبين سقف المسجد الّا[g] فرجة وهو مسدود لا باب له وفيه قبر رسول الله صلّعم وقبر ابى بكر وعمر رضّهما والمنبر الّذى كان يخطب عليه رسول الله صلّعم غُشِّىَ بمنبر آخر والروضة امام المنبر بينه وبين القبر، ومصلّى رسول الله الّذى كان يصلّى فيه الاعياد فى غربىّ المدينة داخل الباب، وبَقِيعُ الغَرْقَد خارج باب البقيع فى شرقىّ المدينة، وقُبَآء خارج المدينة على نحو من ميلين الى ما يلى القبلة وهو مجمع بيوت للانصار[h] يشبه القرية، وأُحُد جبل فى شمالىّ المدينة وهو اقرب الجبال اليها على مقدار فرسخين وبقربها مزارع فيها ضياع لاهل المدينة توازى[i] العقيق فيما بينها[k] وبين الفرع، والفُرْع من المدينة على اربعة ايّام فى جنوبيّها وبها مسجد جامع[l] غير انّ اكثر هذه الضياع خراب وكذلك حوالى[m] المدينة ضياع كثيرة واكثرها خراب، والعقيق واد من المدينة فى قبليّها[n] على اربعة اميال فى طريق مكّة واعذب مياه تلك الناحية آبار العقيق ۞ وامّا اليَمامَة فانّ مدينتها دون مدينة الرسول وهى اكثر تمرًا ونخلًا[o] من المدينة ومن

a) B. om.; C. et D. فيه. b) C. زاوية الحرم. c) C. add. من. d) C. et D. كثير. e) B. et D. يسقون. Sequens العبيد vertitur apnd E. سياهان. f) C. et D. om. g) In C. deëst. h) Secundum A.; ceteri الانصار. i) A. بوادى, B. يوازى, C., D. et E. ووادى. k) E. ميان مدينه. l) E. منبر وجامع. m) C. حول. n) C. قبلتها. o) C. et D. نخيلًا وتمرًا.

سائر الحجاز ۞ واما البَحْرَيْن فانّها فى ناحية نجد ومدينتها هَجَر وهى اكثر تمورًا [a] الّا انّها ليست من الحجاز وهى على شطّ بحر فارس وهى ديار القَرَامِطة ولها قرى كثيرة وقبائل من مُضَر ذوو عدد قد احتفّوها [b] ۞ وليس بالحجاز مدينة بعد مكّة والمدينة اكبر من اليمامة ويليها فى الكبر وادى القرى وهى ذات نخيل كثيرة وعيون ۞ والجار فرضة المدينة وهى على ثلاث مراحل من المدينة على شطّ البحر * وهى اصغر من جُدّة ۞ وجُدّة فرضة اهل مكّة على مرحلتين منها على شطّ البحر [c] وهى عامرة كثيرة التجارات والاموال ليس بالحجاز بعد مكّة اكثر مالًا وتجارة [d] منها وقوام تجاراتها بالفُرْس ۞ والطائف مدينة صغيرة نحو وادى القرى الّا انّ اكثر ثمارها الزبيب [e] وهى طيّبة الهوآء واكثر فواكه مكّة منها وهى على ظهر جبل غَزْوَان [f] ، وبغزوان ديار بنى سعد وسائر [g] قبائل هُذَيْل وليس بالحجاز فيما علمتُه مكان هو ابرد من راس هذا الجبل ولذلك اعتدل هوآء الطائف وبلغنى انّه ربّما جمد المآء فى ذروة هذا الجبل وليس بالحجاز مكان يجمد فيه المآء سوى هذا الموضع * فيما علمتُه [h] ۞ والحِجْر قرية صغيرة قليلة السُّكّان وهى [i] من وادى القرى على يوم بين جبال وبها كانت ديار [k] ثَمُود الّذين قال الله فيهم [l] وَثَمُودَ ٱلَّذِينَ جَابُوا ٱلصَّخْرَ بِٱلْوَادِ ورايتُ [m] تلك الجبال ونحتهم الّذين [n] قال الله [o] وَتَنْحِتُونَ مِنَ ٱلْجِبَالِ بُيُوتًا فَارِهِينَ ورايتُها بيوتًا مثل بيوتنا فى اضعاف جبال وتسمّى تلك الجبال الأَثَالِب [p] وهى جبال فى العيان متّصلة حتّى اذا توسّطتَها رايتَ كلّ قطعة منها قائمة بنفسها يطوف بكلّ قطعة منها الطائف

---

a) C. دهوذًا et E. وانجا جهودان بسيار باشند. *b*) D. اغتصبوها. *c*) In A. et B. haec desunt. *d*) A. تجارة ومالا. *e*) E. ميويز pro مويز (forma apud Vullers desideratur). *f*) A. عرفات. *g*) C. وديار. *h*) C. et D. om. *i*) C. et Jacut, II, p. ٢٠٨, 5 وهو. *k*) E. معدن, Jacut منازل. *l*) Qor. 89 vs. 8. *m*) C. فرايت. *n*) C. الذى. *o*) Qor. 7 vs. 72; 26 vs. 149. *p*) B. et C. الاثالث. Jacut habet الاثالث, quae lectio quoque saepius occurrit; cf. Juynboll, Ann. ad *Meràcid*, IV, p. 37 seq.

وحواليها[a] رمل لا يكاد يرتقى الى ذروة كلّ قطعة منها احد الّا بمشقّة شديدة وبها بئر ثمود الّتى قال الله فى الناقة[b] لَهَا شِرْبٌ وَلَكُمْ شِرْبُ يَوْمٍ مَعْلُومٍ ۞ وتَبُوك بين الحجر[c] وبين اوّل الشام على اربع مراحل[d] نحو نصف طريق الشام وهو حصن به[e] عين ونخيل وحائط ينسب الى رسول الله صلّعم ويقال انّ اصحاب الأَيْكَة الّذين بُعثَ اليهم شُعَيْب كانوا بها ولم يكن شُعَيْب منهم وانّما كان من مدين ۞ ومَدْيَن على بحر القلزم محاذية لتبوك على نحو من ستّ مراحل وهى اكبر من تبوك وبها البئر الّتى استقى منها موسى عَمَ لسائمة شُعَيْب ورايتُ هذه البئر مغطّاة قد بُنِىَ عليها بيت *وماء اهلها من عين تجرى لهم[f]، ومدين اسم القبيلة الّتى كان منها شُعَيْب وانّما سمّيت القرية بهم الا ترى انّ الله يقول[g] وَإِلَى مَدْيَنَ أَخَاهُمْ شُعَيْبًا ۞ وامّا الجُحْفَة فانّها منزل[h] عامر وبينها وبين البحر نحو من ميلين وهى فى الكبر ودوام العمارة نحو من فيد، *وليس بين المدينة ومكّة منزل يستقلّ بالعمارة والاهل جميع السنة الّا الجحفة ولا بين المدينة والعراق مكان يستقلّ بالعمارة والاهل جميع السنة مثل فَيْد[i] ۞ وفَيْد فى ديار طيّئ وجبلا طيّئ منها على[k] مسيرة يومين وفيها[l] نخيل وزرع قليل لطيّئ وبها ماء قليل يسكنها بادية من طيّئ ينتقلون عنها فى[m] بعض السنة للمراعى ۞ وجَبَلَة حصن فى آخر وادى ستَارَة[n] ووادى ستارة بين بطن مرّ وعُسْفان عن يسار الذاهب الى مكّة وطول هذا الوادى نحو[o] من يومين لا يكون الانسان منه[p] فى مكان من بطن هذا الوادى لا يرى فيه نخلًا، وعلى ظهر هذا الوادى واد مثل هذا يعرف

a) C. وحولها. b) Qor. 26 vs. 155. c) A. sine artic. d) Jacut, I, p. ٨٣٥, 2 add. من الحجر. e) A., B. et C. بها. f) C. om. g) Qor. 7 vs. 83. h) C. add. كبير; E. منزل بنى عامر. i) Haec omnia in C. et E. desunt. k) C. om. l) C. et D. وبها et deinde وزروع قليلة. m) A. et B. om. n) B. et E. سارة; Jacut II, p. ٣٧, 8 (ubi الستارة) addit: قرية من ناحية بتهامة. o) A. نحوا. p) C. om.; E. haec ita habet: وهر كه در ميانهٔ اين وادى باشد هيچ نخل نبيند.

بِسَايَة[a] وآخر يعرف بالسائرة، وبجبلة كانت وقعة لبنى تميم فى بكر بن وائل، وفى جُرْف[b] منها قيل ملك لَقيط بن زُرَارَة اخو حَاجِب[c] ۞ خَيْبَر حصن ذات نخيل كثيرة وزروع ۞ ويَنْبُع حصن بها نخيل ومآء وزروع[d] وبها وقوف لعلىّ[e] ابن ابى طالب عمّ يتولّاها اولاده[f] ۞ والعِيص[g] حصن صغير بين[h] ينبع والمَرْوَة[i] ۞ *والعُشَيْرة حصن صغير بين ينبع والمروة[k] تفضّل تمورها[l] على سائر تمور الحجاز الّا الصَّيْحَانىّ بخيبر والبُرْدِىّ والعَجْوَة بالمدينة[m] ۞ وبقرب ينبع جبل رَضْوَى وهو جبل منيف ذو شعاب واودية ورايتُه من ينبع اخضر واخبرنى من طاف[n] فى شعابه انّ به مياهًا كثيرة واشجارًا، وهو الجبل الّذى زعم[o] طائفة يعرفون بالكَيْسانيّة انّ محمّد بن *الحَنَفيّة ابن[p] علىّ ابن ابى طالب حىّ مقيم به، ومن رضوى يحمل حجر المسنّ[q] الى سائر الآفاق، وبقربه فيما بينه وبين ديار جُهَيْنَة *وبَلِىّ وساحل[r] البحر ديار للحسنيّين[s] حَزْرَتْ بيوت الشعر الّتى يسكنونها نحوًا من سبع مائة بيت وهم بادية مثل الاعراب ينتقلون فى المراعى والمياه انتقال الاعراب لا تميّز[t] بينهم فى خَلق ولا خُلق وتتّصل ديارهم ممّا[u] يلى المشرق بوَدّان ۞ وودّان هذه

---

a) Sic E, Jacut l. l. et III, p. ٣٦١. A. et B. بشابه, C. بسابه (cf. Jacut III, p. ٣٣٦). Nomen السائرة (omnes Codd. السايره) a Jacut non memoratur. b) E. كوشه, legens حرف ut A. et B.; sed vid. Ibno'l-Athír, I, p. ۴٣٧. c) C. add. ابن زرارة. d) A. وزروع وماء. e) C. وقف لامير المومنين على. f) C. add. عرض. g) A. العمط, B. والقيط, C. والعيص, E. يفضل تمورها على سائر التمور. h) A. بباب. i) C. المُرّ. k) C. iterum المُرّ, Jacut, III, p. ٦٨١, ult. seq. ذى المروة (ut quoque III, p. ٧٥٣, 15). A. et B. haec om. l) C. يفضل تموره, Jacut يفضل تمره. m) A. فى المدينة. n) A. add. به. Cf. quoque Abulfeda, p. ٨١ et Jacut II, p. ٧٩٠, 19. o) C. et Abulf. زعمت; D. تزعم, Jacut يزعم. p) C. et D. om. q) C. المسانّ. r) C. بلى ساحل, D. om. وبلى, E. تا ساحل, Jacut مما يلى. s) A. (ut Jacut) للحسينيين et deinde حُزِرَتْ (Jacut حزرتْ). t) C. ميز. Jacut لا يميز بينهم وبين بادية العرب. u) C. et D. فيها.

من الجحفة على مرحلة وبينها وبين الأَبْوَآء التى هى على طريق الحاجّ فى غربيّها ستّة اميال وبها كان فى ايّام مقامى بها رئيس الجعفريّين[a] اعنى اولاد جعفر بن ابى طالب ولهم بالفُرْع والسائرة[b] ضياع كثيرة وعشيرة واتباع وبينهم وبين الحسنيّين[c] حروب ودمآء حتّى استولت طائفة من اليمن يعرفون ببنى حرب[d] على ضياعهم فصاروا حزبًا[e] لهم فضعفوا ✿ وتَيْمآء حصن اعمر من تبوك وهى فى شمالىّ تبوك وبها نخيل وهى ممتار[f] البادية وبينها وبين اوّل الشام ثلاثة ايّام ✿ ولا اعلم فيما بين العراق واليمن والشام مكانًا الّا وهو فى ديار طائفة من العرب ينتجعونه فى مراعيهم ومياههم الّا ان يكون بين اليمامة والبحرين وبين عمان من[g] ورآء عبد القيس[h] برّيّة خالية عن[i] الآبار والسكّان والمراعى قفرة لا تسلك ولا تسكن، فامّا ما بين القادسيّة الى الشُّقُوق فى الطول وفى العرض[k] من قرب السَّمَاوَة الى حدّ بادية البصرة فسكّانها[l] قبائل من بنى أَسَد، فاذا جزتَ الشقوق فانت فى ديار طَيّئٍ الى ان تجاوز معدن النُّقْرة فى الطول وفى العرض من[m] ورآء جبلى طَيّئٍ محاذيًا لوادى القرى الى ان تتّصل بحدود نجد من اليمامة والبحرين، ثمّ اذا جزتَ المعدن عن يسار المدينة فانت فى سُلَيْم، واذا جزتَه عن يمين المدينة فانت فى جُهَيْنَة وفيما بين مكّة والمدينة[n] بكر بن وائل فى قبائل من مُضَر من الحسنيّين[o] والجعفريّين وقبائل من مُضَر، وامّا نواحى مكّة فانّ الغالب على نواحيها ممّا يلى المشرق بنو هِلَال[p] وبنو سعد فى قبائل من هذيل[q] وفى غربيّها مُدْلِج[r] وغيرها من قبائل مُضَر، وامّا بادية البصرة فانّها اكثر هذه البوادى[s] احيآء وقبائل واكثرها تميم حتّى يتّصلوا[t] بالبحرين واليمامة ثمّ من

a) C. للجعفريين. b) A. والسابعه, B. والسادعه. c) A. الحسينيين et sic E. h. l. d) E. حراب. e) A. حَرْبًا, B. et C. حربا. f) E. vertit خرماستان. g) D. ومن. h) C. القبيس. i) C. من. k) C. male والعرض. l) A. وسكانها. m) Male Arn. jubet legere ومن. n) C. et D. المدينة ومكة. o) A. et E. الحسينيين. p) A. هُلَايل, B. هذيل. q) C. add. ومضر. r) C. male مدحج. s) C. et D. النواحي. t) B. يتصلون.

ورآئهم عبد القيس، وامّا بادية الجزيرة فانّ بها احيآءً من ربيعة واليمن واكثرهم كلب اليمن وفى قبيلة منهم يعرفون ببنى العُلَيْص[a] خرج صاحب الشأم * الّذى فلّ[b] جيوش مصر واوقع باهل الشام حتّى قصده المُكْتَفى بنفسه الى الرّقّة فاخذه، وبادية السماوة * من دُومَة[c] الجَنْدَل الى عين التَّمْر وبرّيّة خساف[d] من بادية الجزيرة، * وبرّيّة خساف[e] فيما بين الرّقّة وبالس عن يسار الذاهب الى الشام، وصفّين ارض من هذه البادية بقرب الفرات ما بين الرّقّة وبالس * وهو الموضع الّذى كانت به حرب معاوية وعلىّ صلوات الله عليه والحرب ينسب اليه ورايتُ هذا الموضع من بعد واخبرنى من راى به قبر عمّار بن ياسر رضه وبيت المال الّذى كان يجمع فيه الفىء لعلىّ بن ابى طالب صلوات الله عليه،[f] وامّا بادية الشام فانّها ديار لفَزَارَة ولَخْم وجُذَام وبَلِىّ وقبائل مختلطة من اليمن وربيعة ومضر واكثرها يمن ۞ والرمل المذكور بالحجاز هو[g] الرمل الّذى عرضه من الشقوق الى الأَجْفُر وطوله من ورآء جبلى طَيِّئٍ الى ان يتّصل مشرّقًا بالبحر وهو رمل اصفر ليّن المسّ[h] يكاد بعضه يحكى الغبار ۞ وامّا تِهَامَة فانّها قطعة من اليمن وهى جبال مشتبكة[i] اوّلها مشرف[k] على بحر القلزم * ممّا يلى غربيّها[l] وشرقيّها بناحية صَعْدَة[m] وجُرَش ونَجْرَان وشماليّها حدود مكّة وجنوبيّها من صَنْعَآءَ على نحو من عشر مراحل[n] * وقد صوّرتُ جبال تهامة فى صورة ديار العرب[o] ۞ وبلاد خَيْوَان[p]

---

a) A. الغليظ, B. العلمط, C. القليص, E. العليظ; Ibno 'l-Athír, VII, p. ٣٥٣ habet القليص بن صمصم بن عدى بن خباب et hinc Weil, II, p. 506 Kaliss; sed vid. *Lobbo'l-lobáb*, p. ١٨٢. Pro صمصم videtur legendum ضَمْضَم; coll. ibid. Supplem. p. 172. Apud Ibno 'l-Athír, VII, p. ٣٦٧ l. antep. legitur ut edidi et ut habet D. *b*) C. الى قتال. *c*) Codd. ودومة. *d*) C. hic et mox خشاف, E. الخشاف. *e*) E. om. *f*) Haec in C. desunt. *g*) C. من. *h*) C. الملمس, D. اللمس. Deinde C. ويكاد. *i*) A. منسبكه, B. منسلة. *k*) C. يشرف. *l*) A. et B. haec om. *m*) C. et E. سعدة. *n*) E. ميل. *o*) C. om. *p*) A. جيران, B. حمران, C, D. et E. خولان; cf. Abulfeda, p. ٩٥.

تشتمل على قرى ومزارع ومياه معمورة باهلها وهى مفترشة وبها اصناف *من قبائل[a] اليمن ۞ ونَجْران وجُرَش مدينتان متقاربتان فى الكبر بهما نخيل ويشتملان على احيآء من اليمن كثيرة ، وصَعْدَة اكبر[b] واعمر منهما[c] وبها يتّخذ ما كان يتّخذ بصنعآء من الادم ويتّخذ بنجران وجرش والطائف ادم كثير غير انّ اكثر ذلك يرتفع من صعدة وبها مجتمع[d] التجّار والاموال والحسنىّ[e] المعروف بالزيدىّ[f] بها مقيم ۞ وليس بجميع[g] اليمن مدينة اكبر ولا اكثر اهلًا ومرافق من صنعآء وبلغنى انّها[h] من اعتدال الهوآء بحيث لا يتحوّل الانسان عن مكان واحد شتآء وصيفًا عمره وتتقارب بها ساعات الشتآء والصيف وبها كانت ديار ملوك اليمن فيما تقدّم وبها بنآء[i] عظيم قد خرب فهو تلّ عظيم يعرف بغُمْدان كان قصرًا لملوك اليمن وليس باليمن بنآء ارفع منه ۞ والمُذَيْخِرَة[k] جبل للجعفرىّ بلغنى انّ اعلاه نحو عشرين فرسخًا فيها[l] مزارع ومياه ونباتها الورس[m] وهو منيع لا يسلك الّا من طريق واحد *حتّى تغلّب عليه[n] القرمطىّ الّذى كان خرج باليمن يعرف بمحمّد بن الفضل ۞ وشِبَام[o] جبل

a) A. et B. om. *b*) C. وسعده اكثر. *c*) A., B. et D. منها. *d*) B. يجتمع, C. مجمع, D. كان يجتمع. *e*) Arn. male edidit والحسينى; E. habet وحسينيان بسى انجا مقام دارند. *f*) C. بالرسى. Fortasse haec est genuina lectio, nam avus primi principis Zeiditae, cognominabatur الرسى. Vid. Ibn Khaldun MS. III, f. ٢٩١ r. (caput de بَنِى الرَّسِىّ الزَّيْدِيَّة بِصَعْدَة). *g*) A. لجميع. Locum transcripsit Omára ibn abi 'l-Hasan ibn Zeidán al-Yamaní († 569) in sua *Historia* et ex hac Jacut, III, p. ٤٢١, 16 (ubi male عمران pro عمارة; cf. III, p. ٢٤٩, 17, IV, p. ٤٧٢, 15, Ibn Khallicán, n. 500 et Hádji Khal. II, p. 159 cet.). Post صنعاء hic et D. addunt وهو بلد فى خط الاستوآء. *h*) A. et B. انه, C. ان بها, Jacut وهى. *i*) C. et D. كبير. *k*) C. والمديخره. Locum ab eodem Omara transcriptum dedit Jacut, IV, p. ٤٧٢. *l*) C. et Jacut فيه. *m*) Jacut add. وفى شفيره (l. شفيرة) الزعفران. Cf. E. qui pro ونباتها الورس habet ان كوه زعفرانست. *n*) A. et B. (A. علىّ) على تغلب حيين. *o*) A. et B. وشيار, E. ستام.

منبع جدًّا فيه[a] قرى ومزارع وسكّان كثيرة[b] وهو مشهور من جبال اليمن ۞ *ويرتفع من اليمن العقيق والجزع وهما حجران اذا حُكّا خرج منهما الجزع والعقيق لانّ وجه الحجر كالغشآء وبلغنى انّهما يكونان فى صحارى فيها حصى فيلتقط من بين الحجارة[c] ، وعَدَن مدينة صغيرة وانّما شهرتها لانّها فرضة على البحر ينزلها السائرون فى البحر وبها معادن اللؤلؤ ، وباليمن مدن كثيرة هى اكبر منها ليست بمشهورة[d] ۞ وبلاد الاباضيّة بقرب خيوان وهى اعمر بلاد تلك النواحى مخاليف ومزارع واغزرها مياهًا ۞ وحَضْرَمَوْتُ فى شرقىّ عدن بقرب البحر وبها رمال كثيرة تعرف بالأَحْقَاف وحضرموت فى نفسها مدينة صغيرة ولها اعمال عريضة وبها قبر هُود النبىّ عمّ وبقربها بَلَهُوت[e] بئر عميقة لا يكاد يستطيع احد ان ينزل الى[f] قعرها ۞ وامّا بلاد مَهْرَة فانّ قصبتها تسمّى الشِحْر[g] وهى بلاد قفرة السنتهم مستعجمة جدًّا لا يكاد يوقف عليها[h] وليس ببلادهم نخيل ولا زرع وانّما اموالهم الابل وبها نجب من الابل تفضّل فى السير على سائر النجب واللُّبان[i] الّذى يحمل الى الآفاق من هناك وديارهم مفترشة وبلادهم بواد نائية ويقال انّها من عمان ۞ وعُمَان مستقلّة[k] باهلها وهى كثيرة النخيل والفواكه الجُرُوميّة من الموز والرمّان والنبق[l] ونحو ذلك وقصبتها صُحَار وهى على البحر[m] وبها متاجر البحر وقصد[n] المراكب وهى اعمر مدينة بعمان واكثرها مالًا ولا تكاد تعرف على شاطئ[o] بحر فارس بجميع بلاد الاسلام مدينة اكثر عمارة ومالًا من صُحَار ، وبها مدن كثيرة وبلغنى انّ حدود اعمالها نحو من ثلاث مائة فرسخ وكان الغالب عليها

---

a) A. et B. فيها. b) A., B. et E. om. c) Haec omnia in A. et B. desunt. E. habet وگويند ان كوه مغنيسيا است وگويند در ميان سنگ ريزه مغنيسيا يابند. Cf. Apud Vullers مغنيسا 2). d) A. مشهوره. e) C., D. et E. برهوت et sic Jacut II, p. ٢٨٥, 15. f) A. et B. om. g) A. الحِشْر, B. الحشره. h) E. addit مگر بترجمان. i) E. وكندر. k) C. مستعله. l) E. habet وناك كه به فارسى نبق خوانند. m) A. et B. البرّ. n) A. وقصب, B. وقصد. o) C. et D. شط.

الشراة[a] الى ان وقع بينهم وبين طائفة من بنى سامة[b] بن لُؤَىّ وهم[c] من كبرآء تلك النواحى حروب فخرج منهم[d] رجل يعرف بمحمّد بن القاسم السامىّ[e] الى المعتضد فاستنجده فبعث معه بابن ثور[f] ففتح عمان للمعتضد واقام بها الخطبة له وانحاز الشراة الى ناحية لهم تعرف بنَزْوة[g] والى يومنا هذا بها امامهم وبقيّة[h] مالهم وجماعتهم، وعمان بلاد حارّة جدًّا وبلغنى ان بمكان منها بعيد عن البحر ربّما وقع ثلج دقيق[i] ولم ار احدًا شاهد ذلك الّا بالابلاغ[k] ۞ وبارض سبأ[l] من اليمن طوائف من حِمْيَر وكذلك بارض حضرموت، وامّا ديار هَمْدان وأَشْعَر وكِنْدَة وخَوْلان فانّها مفترشة فى اعراض اليمن وفى اضعافها مخاليف وزروع وبها بوادٍ وقرى تشتمل على بعض تهامة وبعض نجد[m] اليمن من شرقىّ تهامة وهى قليلة الجبال مستوية البقاع، ونجد اليمن غير نجد الحجاز غير انّ جنوبىّ نجد الحجاز يتّصل بشمالىّ نجد اليمن ۞ وبين * البحرين وبين عمان[n] بريّة ممتنعة[o] ۞ وباليمن قرود كثيرة[p] بلغنى انّها تكثر حتّى لا تطاق[q] الّا بجمع عظيم واذا اجتمعوا كان لهم كبير يتبعونه مثل اليعسوب للنحل[r]، وبها دابّة تسمّى العُدار[s] بلغنى انّها تطلب الانسان فتقع عليه فان اصاب منه ذلك تدوّد[t] جوف الانسان فانشقّ[u]، ويحكى عن[v] الغيلان بها * من الاعجوبة[w] ما لا استجيز حكايته[x] ۞

واما المسافات بديار العرب فانّ الّذى يحيط بها من عبّادان الى البحرين

---

a) E. خوارجى — كه ايشان‌را شراة خوانند. b) E. أُسامه. c) A. et B. om. d) A. om. e) A., B. et E. الشامى. f) B. ثور, C. بوز, D. ابا ثور. Est محمد بن ثور, vid. Ibno 'l-Athír, VII, p. ٣٤٣. g) C. et D. بنزوى, E. ببروى. h) C. وبيت. i) C. رقيق. k) C. بالبلاغ. l) C. et E. صنعاء. m) A. add. ونجد. n) C. et D. النجدين وعمان. o) E. add. وآب وچراكاه دران يافت ميشود. p) C. add. جدا. q) Arn. male تطاف. r) E. add. وپادشاه بوزينگانرا يرعوب گويند. s) A. العذار. t) B. يدود, C. indist. u) A. وانشق. v) C. وحكى ان. w) C. et D. omittunt. x) E. add. در عجائب المخلوقات آورده. In C. folium desideratur.

نحو من ١٥ مرحلة ومن البحرين الى عمان نحو من شهر ومن عمان الى ارض مهرة نحو من شهر والى حضرموت من مهرة نحو شهر ومن اقصى حضرموت الى عدن نحو شهر ومن عدن الى جُدّة نحو شهر ومن جُدّة الى ساحل الجحفة نحو ٥ مراحل ومن ساحل الجحفة الى الجار نحو من ٣ مراحل ومن الجار الى ايلة نحو من ٢٠ مرحلة ومن ايلة الى بالس نحو من ٢٠ مرحلة ومن بالس الى الكوفة نحو من ٢٠ مرحلة ومن الكوفة الى البصرة نحو من ١٢ مرحلة ومن البصرة الى عبّادان نحو مرحلتين فهذا هو الدور الّذى يحيط بها ۞ وامّا طرقها فانّ من الكوفة الى المدينة نحو من ٢٠ مرحلة ومن المدينة الى مكّة نحو من ١٠ مراحل وطريق الجادّة من الكوفة الى مكّة اقصر من هذا الطريق بنحو من ٣ مراحل اذا انتهى الى معدن النقرة[a] عدل عن المدينة حتّى يخرج على معدن بنى سُلَيْم الى ذات عِرْق حتّى ينتهى الى مكّة، وامّا طريق البصرة فهو الى المدينة نحو من ١٨ مرحلة ويلتقى مع طريق الكوفة بقرب معدن النقرة[a]، وامّا طريق البحرين الى المدينة فنحو[b] من ١٥ مرحلة، وامّا طريق الرّقّة الى المدينة فنحو[b] من ٢٠ مرحلة وكذلك من دمشق الى المدينة نحو من ٢٠ مرحلة ومن فِلَسْطين الى المدينة نحو من ٢٠ مرحلة ومن مصر الى المدينة على الساحل نحو من ٢٠ مرحلة، ولم نفرد[c] لمصر والمغرب طريقًا لانّه يلتقى بايلة[d] مع طريق اهل فلسطين فيصير الطريقان سوى[e] وهو اوّل حدّ البادية وانّما يتفرّق قبل دخول البادية، ولاهل مصر وفِلَسْطين اذا جاوزوا مدين طريقان احدهما الى المدينة على بَدًا[f] وشَغْب قرية بالبادية كانوا بنو مروان اقطعوها الزُّهْرىّ[g] المحدّث وبها قبره حتّى ينتهى الى المدينة على المَرْوة، وطريق يمضى على ساحل البحر حتّى يخرج بالجحفة فيجتمع بها طريق

---

a) A. البقرة. *b*) A. et B. نحو. *c*) A. et B. يفرد. *d*) E. hic et deinde male ابلّه pro ايلة. *e*) B. سواء. *f*) In B. superinscribitur يعقوب et locus a Mokaddasí semper بدا يعقوب appellatur. Deinde A., B. et E. وشعب. *g*) A. للزهرى; cf. Jacut III, p. ٣٠٣, 17 seq., 21.

اهل العراق ودمشق وفلسطين ومصر۞ وامّا طريق البصرة والرَّقّة فهما لا يسلكان وقد تعطّلا وسائر الطرق مسلوكة۞ ومن عَدَن الى مَكّة نحو شهر ولها[a] طريقان احداهما على ساحل البحر وهو ابعد والآخر ياخذ على صنعاء وصَعْدَة وجُرَش ونَجْران والطائف حتّى ينتهى الى مكّة، ولهم طريق على البوادى وتهامة هو اقرب من هذين الطريقين الّا انّه[b] على احياء اليمن ومخاليفها تسلكه الخُوّاس[c] منهم، وامّا اهل حضرموت ومهرة فانّهم يقطعون عرض بلادهم حتّى يتّصلوا بالجادّة الّتى بين عَدَن وبين مَكّة والمسافة منهم الى الاتّصال بهذه الجادّة[d] ما بين ٢٠ مرحلة الى ٥٠ مرحلة، وامّا طريق عُمَان فهو طريق يصعب سلوكه فى البرّيّة لكثرة القفار بها وقلّة السُّكّان وانّما طريقهم فى البحر الى جُدّة فان سلكوا على السواحل[e] من مهرة وحضرموت الى عَدَن او الى طريق عَدَن بَعُدَ عليهم وقلّ ما يسلكونه، وكذلك ما بين عُمَان والبحرين فطريق شاقّ *يصعب سلوكه[f] لتمانع العرب فيما بينهم بها، وامّا ما بين البحرين وعبّادان فغير مسلوك وهو قفر والطريق فيها[g] على البحر، ومن البصرة الى البحرين نحو من ١٨ مرحلة فى قبائل العرب ومياههم مسلوك عامر[h] غير انّه مخوف، فهذه جوامع المسافات الّتى يحتاج الى علمها فامّا ما بين ديار العرب لقبائلها من المسافات فقلّ ما تقع الحاجة لغير اهل البادية الى معرفتها۞

بحر فارس

وسنذكر بعد ديار العرب بحر فارس فانّه يشتمل على اكثر حدودها ويتّصل بديار العرب منه وبسائر[i] بلدان الاسلام ونصوّره ثمّ نذكر *جوامع ممّا يشتمل عليه[k] هذا البحر ونبتدئ بالقلزم[l] على ساحله ممّا يلى المشرق فانّه ينتهى

---

a) A. وله. *b*) A. انها. *c*) A. الحرّاس, D. الخواص بسكنها. E. habet گذشت كسى گذرد كه با ايشان آشنائى دارد. *d*) Hic desinit lacuna in C. *e*) C. الساحل. *f*) C. om. *g*) C. et D. منها. *h*) A. et B. على برّ. *i*) A. et B. وسائر. C. om. منه. *k*) C. ساير ما على ساحل. *l*) C. add. فانها. D. deinde وساحله et sic E. ut vid.

الى أَيْلَة ثمّ يطوف بحدود ديار العرب الّتى ذكرناها وبيّنّاها قبل هذا الى عبّادان[a] ثمّ يقطع[b] عرض دجلة وينتهى على الساحل الى مَهْرُوبَان[c] ثمّ الى جَنَّابَا[d] ثمّ يمرّ[e] على سيف فارس الى سِيرَاف ثمّ يمتدّ الى سواحل هُرْمُز[f] ورآء كرمان الى الدَّيْبُل وساحل المُلْتَان[g] وهو ساحل السند *وقد انتهى حدّ بلدان الاسلام[h] ثمّ ينتهى الى سواحل الهند حتّى ينتهى الى سواحل *التّبّت فيقطعها[i] الى ارض الصين، واذا اخذتَ من القُلْزُم غربيّها على ساحل[k] البحر سرتَ فى مفاوز من حدود مصر حتّى تنتهى الى مفاوز هى للبُجّة وبها معادن الذهب الى مدينة على شطّ البحر[l] يقال لها عَيْذَاب ثمّ تمتدّ على بلد[m] الحبشة وهى محاذية لمكّة والمدينة حتّى تحاذى قرب عَدَن ثمّ ينقطع[n] الحبشة ويتّصل بظهر بلد[o] النوبة حتّى ينتهى الى بلدان الزنج وهى من اوسع تلك الممالك فتمتدّ[p] على محاذاة جميع بلدان الاسلام[q]، وقد انتهى مسافة هذا البحر ثمّ يعرض[r] فيه جزائر واقاليم مختلفة الى ان يحاذى ارض الصين[s] ۞

وقد صوّرتُ هذا البحر وذكرتُ حدوده مطلقًا وساصف ما يحيط به وما فى اضعافه جملًا يقف عليه من قرأه ان شآء الله، امّا ما كان من هذا البحر من القلزم الى ما يحاذى بطن اليمن فانّه يسمّى بحر القلزم ومقداره نحو ثلاثين مرحلة طولًا وعرضه[t] اوسع[u] ما يكون *غير مسير[v] ثلاث ليال ثمّ

---

a) C. et E. عمان. *b*) B. نقطع et deinde ننتهى, et sic vertit E. *c*) E. ماهى رويان. *d*) C. جُنّابا, E. جنايا. *e*) B. نَمُرّ et deinde نمتدّ. *f*) C., D. et E. هرموز ut infra quoque A. et B. *g*) E. مولتان. *h*) C. وينتهى حد الاسلام الى الملتان. Deinde B. bis ننتهى (= E.). *i*) B. التِّبّت فنقطعها. *k*) C. سواحل. *l*) C. النيل, E. دريای نيل. *m*) C. et E. om. *n*) C. دقطع. *o*) A. et B. om.; E. ut saepius pro النوبة habet ridicule يونانيان. *p*) C. فيمر. *q*) C. add. الماقيه (الباقية). *r*) C. et D. تعترض. Deinde A., B. et D. فيها. *s*) Hic mappa maris Erythraei sequi deberet. *t*) C. وعرضا. Jacut, IV, p. ١٥٩, 12 واوسع ما يكون عرضًا. *u*) A. ايسر, B. اكبر. *v*) C. et Jacut عبر.

لا يزال يضيق حتّى يُرَى من[a] بعض جنباته[b] الجانب الآخر[c] حتّى ينتهى الى القلزم[d] ثمّ يدور على[e] الجانب الآخر من بحر القلزم ✿ وبحر القلزم مثل الوادى به[f] جبال كثيرة قد علا المآء عليها وطرق *السفن بها[g] معروفة لا يهتدى فيها[h] الّا برُبّان يتخلّل بالسفينة فى اضعاف تلك الجبال بالنهار[i] فامّا بالليل فلا يسلك ومآؤه صاف ترى تلك الجبال *فيه ✿ وفى هذا البحر ما[k] بين القلزم وايلة مكان يعرف بتاران وهو اخبث ما فى هذا البحر من الاماكن وذلك انّه[l] دوّارة مآء *فى سفح[m] جبل اذا وقعت[n] الريح على ذروته انقطعت[o] الريح على[p] قسمين *فتنزل[q] الريح على شعبين[r] فى هذا الجبل متقابلين فتخرج الريح من كلى[s] هذين الشعبين فتتقابل[t] فيثور[u] المآء وتتبلّد[v] كلّ سفينة تقع فى *تلك الدوّارة[w] باختلاف الريحين *وتتلف فلا[x] تسلم واحدة[y] واذا كان للجنوب ادنى مهبّ فلا سبيل الى سلوكه[z] ومقدار طوله نحو *ستّة

---

a) C. et Jacut فى. *b*) Jacut جوانبه, C. المواضع. *c*) Jacut addit المحاذى له. *d*) Jacut addit وهى مدينة. *e*) C. الى. Post sequens القلزم a Jacut 8 vss. inseruntur, qui in omnibus Codd. desunt. *f*) C. et Jacut فيه. *g*) C. et Jacut السير منها. *h*) C. اليها. *i*) Jacut فى ضياء النهار. Deinde B., C. et Jacut واما. *k*) C. فى هذا البحر وما et sic fere Jacut. *l*) C. ut Jacut, I, p. ٨١١ et Kazwíní, I, p. ١١٩ ان به. Deinde Jacut دوران pro دوارة. *m*) C. بسفح. *n*) A., B., Jacut et Kazw. وقع. *o*) Jacut انقطع, C. et Kazw. انقسمت. *p*) C. et Jacut om. *q*) C. فترد (فترد). *r*) Jac. et Kazw. pro his: فيلقى (وتلقى) المركب بين شعبتين متقابلتين et deinde. *s*) A. et B. om. *t*) A. et B. فيتقابل; C. الشعبتين احداهما مقابلة للاخرى; Jacut et Kazw. المتقابلتين; E. وديگر باره هر دو باد روى; D. كل واحدة متقابلة للاخرى يكديگر آرند وبرابر يكديگر شوند. *u*) A. et B. ويثور. Deinde pro الماء ceteri omnes البحر. *v*) C. وتبارة على, Jac. et Kazw. tantum على. *w*) C., Jac. et Kazw. ذلك الدوران. *x*) C., Jac. et Kazw. فتنقلب ولا. *y*) Pro واحدة Jac. et Kazw. ابدًا; C. om. *z*) Hic in E. haec sequuntur: در حكايت

اميال[a] وهو الموضع الّذى غرق فيه فرعون[b] ۞ وبقرب تاران[c] موضع يعرف بجُبَيْلات[d] يهيج ويتلاطم امواجه باليسير من الريح وهو موضع مخوف ايضًا فلا يسلك[e] بالصبا مغرّبًا وبالدبور مشرّقًا ۞ واذا حاذى أَيْلَة ففيه سمك كثير مختلف الالوان، فاذا قابل بطن اليمن سمّى بحر عدن[f] الى ان يجاوز عدن ثمّ يسمّى بحر الزنج[g] الى ان يحاذى عُمَان عاطفًا على فارس وهذا بحر يعرض[h] حتّى يقال انّ عبره[i] الى بلد الزنج سبع[k] مائة فرسخ، وهو بحر مظلم اسود لا يرى ممّا فيه شىء وبقرب عَدَن معدن اللؤلؤ يخرج ما

آمده است كه مركبى باين جايگاه رسيد ودر گرداب افتاد وچند روزگار بماند بعد مدتى كه مردمان نوميد شده بودند وزاد كمتر گشته ماهىٔ عظيم بيامد وگرد كشتى ميگشت وسر بر آورد ويك تن‌را بربود روز ديگر هم بآن وقت باز آمد وديگرى‌را بگرفت وبر عادت هر روز مى آمد ويك تن مى برد در ميان ايشان مردى خردمند بود آن جماعت‌را گفت آگاه باشيد كه مارا ازين جايگه روى برون شدن نيست ولا محاله همه هلاك خواهيم شدن ودشمنى پديد آمده است كى هر روز از ما يكى‌را مى ربايد من شمار اجارهٔ سازم اگر فرمان من كنيد همه خرسند شدند گفت اتفاق كنيد كه يك تن هلاك شود وديگران رهائى يابند يك تن گفت من خويشتن‌را فدا كردم وبرين قرار دادند واين مرد دانا هرچه در كشتى ابريشم بود بياورد ورسنى محكم بساختند ومردرا بياوردند وبر ميان او بستند وروى او بپوشيدند وبر كرانهٔ كشتى بنهادند روز ديگر ماهى بيامد وآن شخص‌را بربود وسر رسن گردر كشتى استوار كرده بودند چون شخص‌را بربود بقوت ماهى كشتى روان شد تا از آن گرداب برون آمدند وبايمنى رسيدند رسن ببريدند واز آن گرداب باين حيلت خلاص يافتند. *a*) E. سه ميل بقياس. *b*) C. add. وجنوده. *c*) A. hic ut supra بباوان, Jac. et Kazw. الملعيين. *d*) A. et E. بجبالات. D. hic ut supra جبيلان habet, sed Jacut, IV, p. ١٩٠, 4 et 12 cum textu nostro facit. *e*) A. solus insert الا. *f*) A. العدن. *g*) E. دريا ريك وزنكبار خوانند (sic). *h*) C. يعترض. *i*) A. غبره. *k*) A. بسبع.

يرتفع منه الى عَدَن، واذا جزتَ عمان الى ان تخرج عن حدود الاسلام وتتجاوزه الى قرب سَرَنْديب يسمّى بحر فارس وهو عريض البطن جدًّا فى جنوبه[a] بلدان الزنج ✿ وفى هذا البحر هوارات[b] كثيرة ومعاطف صعبة[c] ومن اشدّها ما بين جَنَّابَة والبصرة فانّه مكان يسمّى هور جَنَّابَة وهو مكان مخوف لا تكاد تسلم منه[d] سفينة عند هيجان البحر، وبها[e] مكان يعرف بالخَشَبَات[f] من عبّادان على نحو من[g] ستّة اميال على جرى مآء دجلة[h] الى البحر ويرقّ المآء حتّى يخاف على السفن الكبار ان سلكته[i] ان تجلس على الارض الّا فى وقت المدّ وبهذا[k] الموضع خشبات منصوبة قد بُنى عليها مرقب يسكنه ناظور يوقد بالليل ليهتدى به ويعلم به المدخل الى دجلة وهو مكان مخوف اذا ضلّت السفينة فيه خيف انكسارها *لرقّة المآء[l] ✿ وبحذآء جنّابة مكان يعرف بخارك وبه معدن اللؤلؤ يخرج منه الشىء اليسير الّا انّ النادر اذا وقع من هذا المعدن فاق[m] فى القيمة غيره ويقال انّ الدُّرّة اليتيمة تقع من هذا المعدن[n]، وبعمان وبسرنديب فى هذا البحر معدن لؤلؤ[o] ولا اعلم معدنًا للؤلؤ الّا ببحر[p] فارس ✿ ولهذا البحر مدٌّ وجزر فى اليوم والليلة مرّتان من حدّ القلزم الى حدّ الصين حيث انتهى وليس لبحر المغرب ولا لبحر الروم ولا لسائر البحار مدٌّ ولا جزر غير بحر فارس وهو ان يرتفع المآء قريبًا من عشرة اذرع ثمّ ينصب حتّى يرجع الى مقداره ✿ وفى هذا البطن من البحر الّذى نسبناه خصوصًا الى فارس جزائر منها لافت[q] وخارك وأُوَال وغيرها من الجزائر المسكونة وبها[r] مياه عذبة وزرع وضرع، فهذه جوامع[s] من صفة هذا البحر *من حدود[t] الاسلام ✿

---

a) C. et D. عدوته. b) C. هورات, E. شبيهها. c) C. صعفه. d) C. فيه. e) C. ولها, D. وفيه. f) E. خشّاب. g) A. om. h) C. ut saepius الدجلة. i) A. سلكها. k) C. وهى فى هذا. l) C. لخفة الماء ورقته. m) C. جاز. n) C. supra omisso تقع hic habet وقعت ut D. o) C. اللولو. p) C. فى بحر. q) A. et B. لامُر, E. لافث. r) C. et D. فيها. s) C. et D. جملة. t) C. فى بلاد, D. et E. فى حدود.

وسأصف ما على سواحله *صفةً جامعةً[a] نبتدئ منها بالقلزم ثمّ ننتهى الى جنباته[b] ان شآء الله، وامّا القلزم فانّها مدينة[c] على شفير البحر *وينتهى هذا البحر اليها[d] وهى فى عطف هذا البحر[e] فى آخر لسانه وليس بها زرع[f] ولا شجر ولا مآءٌ وانّما يحمل لهم من آبار ومياه بعيدة منهم[g] وهى تامّة العمارة بها فرضة مصر والشام ومنها تحمل حمولات الشام ومصر الى الحجاز واليمن وسواحل هذا البحر وبينها وبين فسطاط[h] مصر مرحلتان[i] ۞ ثمّ ينتهى على شطّ البحر[k] فلا تكون بها[l] قرية ولا مدينة سوى مواضع فيها[m] ناس مقيمون على صيد[n] من هذا البحر وشىء من النخيل يسير حتّى ينتهى على تاران وجُبَيْلات[o] وما حاذى جبل الطور الى ايلة ۞ وايلةُ هذه مدينة صغيرة عامرة *بها زرع[p] يسير وهى مدينة اليهود الّذين حرّم الله[q] عليهم صيد السبت وجعل[r] منهم القردة والخنازير وبها فى يد[s] اليهود عهد لرسول الله صلعم[t] ۞ وامّا مَدْيَن وما *انتهى على هذا البحر فى[u] عطوف اليمن الى عمان والبحرين الى عبّادان فقد وصفناه فى صفة ديار العرب ۞ وامّا عَبّادان فانّها[v] حصن صغير عامر على شطّ البحر ومجمع مآء دجلة وهو رباط كان فيه محارس للقطريّة[w] وغيرهم من متلصّصة البحر وبها على دوام

---

a) C. om. et habet deinde ut D. وابتدى. b) A. et B. جنابه; D. على جنباته الى غايته et E. تا پايان. C. om. ان شاء الله. c) C. مبنية, Jacut, IV, p. ١٩٠, 6 مدينة مبنية. d) C. haec om.; A. et B. om. اليها. e) C. add. اعنى. f) C. ضرع. g) C. et D. منها et sic Jacut l.l. 8. h) C. et D. om. i) C. et E. ut Jacut l.l. 6 ثلاث مراحل; cf. Abulfeda, p. ١١٧. k) Jacut add. نحو الحجاز. l) C. فيها. m) C., D. et Jac. بها. n) C. للتصيد. Jacut add. السمك. o) A. باوان وجبالات. p) C. ذات نخل وزرع. Vid. quoque Jacut, I, p. ٤٢٣, 8. q) A. et E. om. r) C. فجعل; Jacut فخالفوا فمسخوا; E. خداى تعالى ايشانرا مسخ كرد. s) C. et D. ايدى. t) Hic desinit lacuna in Ous. u) C. ينتهى الى هذا البحر من. v) C. فانه. w) C. للقطوريه, B. للعطويه.

الايّام مرابطون ۞ ثمّ تقطع[a] عرض دجلة فتصير على ساحل هذا[b] البحر الى مهروبان[c] من حدّ فارس ويعرض[d] فيها اماكن تمنع من السلوك الّا فى المآء وذلك انّ مياه خوزستان تجتمع[e] الى دَوْرَق وحصن مهدىّ وبَاسِيَان فتتّصل بمآء البحر ۞ ومَهْروبَان مدينة صغيرة عامرة وهى فرضة ارّجان[f] وما والاها من اداني[g] فارس وبعض خوزستان ۞ ثمّ ينتهى البحر على الساحل الى شينيز[h] وهى مدينة اكبر من مهروبان ومنها يرتفع الشينيزىّ الّذى يحمل الى الآفاق ۞ ثمّ ينتهى الى جَنّابَة وجَنّابَة هذه مدينة اكبر من مهروبان وهى فرضة لسائر فارس خصبة[i] شديدة الحرّ ۞ ثمّ ينتهى على الساحل الى[k] سيف البحر الى نَجِيرَم وهذا السيف ما بين جَنّابة ونجيرم به[l] قرى ومساكن ومزارع متفرّقة[m] مفترشة شديدة الحرّ ۞ ثمّ ينتهى الى سِيرَاف[n] وهى الفرضة العظيمة لفارس وهى مدينة عظيمة *ليس بها[o] سوى الابنية شىء[p] حتّى يجاوز على جبل يُطِلّ عليه وليس بها مآء يجمد ولا زرع ولا ضرع وهى[q] اغنى بلاد فارس[r] ۞ ثمّ يتجاوز[s] على الساحل فى مواضع منقطعة تعترض بها[t] جبال ومفاوز الى ان ينتهى الى حصن ابن عُمارة[u] وهو حصن منيع على هذا البحر وليس بجميع فارس حصن *امنع منه[v] ويقال انّ صاحب هذا الحصن

---

a) B. ينقطع. b) A. om. c) E. ut solet ماهى روبان. d) C. et D. ويعترض. e) B. يجمع. Mox C. et D. الدورق. f) E. ارغان. g) C. et D. درياى, E. اراضى. h) A. et E. تُسْتَر et التسترى. Scripsi شينيز quia infra plerumque sic habent Codd. i) A. et C. حصنه, B. حصينه. k) C. على. l) C. وبها. m) C. om. n) A. et B. شيراز. o) A. et B. om.; vid. quoque Jacut, III, p. ٢١٢, 9. p) A. et B. sine punctis; C. et Jacut om. Pro iis quae sequuntur, C. tantum habet وليس بها زرع ولا ضرع; Jacut: حتى يجاوز على نظر عملها وليس بها شىء من ماكول ولا مشروب ولا ملبوس الا ما يحمل اليها من البلدان ولا بها زرع ولا ضرع ومع ذلك فهى اغنى. q) C. et D. add. من. r) B. بفارس. s) C. يحاوزه, D. ينجاوزها. Post على B. add. هذا. t) C. فيها. u) E. عمّاره. v) C. et E. اعمر منه ولا امنع.

هو الّذى قال الله فيه[a] وَكَانَ وَرَآءَهُمْ مَلِكٌ يَأْخُذُ كُلَّ سَفِينَةٍ غَصْبًا ۞ وينتهى على ساحل هذا البحر الى هُرْمُوز وهى فرضة كرمان مدينة غرّآء[b] كثيرة النخل[c] حارّة جدًّا ۞ ثمّ تسبير[d] على شطّه الى الدَّيْبُل وهى مدينة عامرة[e] وبها مجمع التجّار وهى فرضة لبلد[f] السند وبلد السند هو[g] المنصورة واراضى الزُّطّ[h] وما والاها الى المُلْتَان[i] ثمّ ينتهى على[k] ساحل بلدان الهند الى ان يتّصل بساحل تُبّت وينتهى الى ساحل الصين ثمّ الى الصين ثمّ لا يسلك بعده ۞

واذا اخذتَ من القلزم غربىّ هذا البحر فانّه ينتهى الى بريّة قفرة[l] لا شىء فيها الى ان يتّصل ببادية[m] البُجّة والبجة قوم[n] اصحاب اخبية شعر اشدّ سوادًا من الحبشة فى زىّ العرب لا قرى[o] لهم ولا مدن ولا زرع الّا ما ينقل اليهم من مدن الحبشة *واليمن ومصر والنوبة وينتهى حدّهم الى ما بين الحبشة[p] وارض النوبة وارض مصر وينتهى الى معادن الذهب، ويأخذ هذا المعدن من قرب أُسْوَان[q] مصر على نحو من عشر مراحل حتّى ينتهى الى حصن على البحر يسمّى عَيْذَاب ويسمّى مجمع الناس بهذا المعدن العَلّاقى وهو رمال وارض مبسوطة لا جبل بها واموال هذا المعدن يرتفع الى ارض مصر وهو معدن ذهب لا فضّة فيه، والبُجّة قوم يعبدون الاصنام وما استحسنوه ۞ ثمّ يتّصل ذلك بارض الحبشة وهم نصارى *وتقرب الوانهم من[r] الوان العرب بين السواد والبياض وهم متفرّقون فى ساحل هذا البحر الى ان يحاذى عَدَن وما كان من النمور والجلود الملمّعة واكثر جلود اليمن الّتى تُدْبَغ للنعال تقع

---

a) Qor. 18 vs. 78. Pro فيه C. habet عنه. *b*) A. et B. غراه, C. et D. غناء. *c*) C. et D. النخيل. *d*) C. تصير. *e*) C. كثيرة العمارة. *f*) C. بلد. Deinde A. الهند. *g*) C. om. Deinde A. et B. والمنصورة. *h*) E. وولايت زنكيان. *i*) C. et E. الصين, B. المدن. *k*) C. الى. *l*) C. قفرا. *m*) C. بناحية; A. et E. scribunt البُجّه. *n*) C. om. *o*) B. قرّ. C. tantum لا مدن لهم. *p*) Haec in A. et B. desunt. *q*) A. et B. سوان, C. et E. سواد. *r*) C. وتضرب الوانهم الى.

منها الى عدوة اليمن، وهم اهل سلم ليسوا بدار حرب ولهم على الشطّ موضع يقال له زَيْلَع فرضة للعبور الى الحجاز واليمن، ثمّ يتّصل ذلك بمفازة بلد النوبة۞ والنُّوبَة نصارى *وهى بلدان اوسع من الحبشة وبها من المدن والعمارة اكثر ممّا بالحبشة[a] ويخترق نيل مصر فيما بين مدنها وقراها حتّى يتجاوز[b] ذلك الى رملة[c] من ارض الزنج ثمّ يتجاوزه الى برارىّ *يتعذّر مسلكها[d] ۞ ثمّ ينتهى هذا البحر حتّى يتّصل بارض الزنج ممّا[e] يحاذى عدن الى ان يمتدّ على البحر وتتجاوز محاذاتها جميع حدّ الاسلام ويدخل فيما حاذى[f] بعض بلدان الهند *لسعته وكثرته[g]، وبلغنى انّ فى بعض اطراف الزنج *صرودًا فيها زنج بيض[h] وبلد الزنج هذا بلد قشف[i] قليل العمارة قليل الزروع الّا ما اتّصل بها من مستقرّ الملك[k] ۞

ديار المغرب

وامّا المغرب فهو *نصفان يمتدّان على بحر الروم[l] نصف من شرقيّه[m] ونصف من غربيّه فامّا الشرقىّ فهو برقة[n] وافريقيّة وتاهرت[o] وطنجة والسوس[p] وزويلة وما فى اضعاف هذه الاقاليم، وامّا الغربىّ فهو الاندلس وقد جمعتُهما فى التصوير۞ فامّا الجانب الشرقىّ فانّ الّذى يحيط به من شرقيّه حدّ مصر

---

a) C. وبلدهم اوسع واعمر من الحبشة. b) C. يحاوز. c) In A. superinscribitur قمار. d) C. بعيدة. e) C. التى. f) C. يحاذى. g) C. لسعتها. h) C. زنج بيض يسكنون صرودًا. i) A. نسف, B. كمف. k) C. الملوك وابن قوم زنكيانرا مدد :الذين للزنج. Pro his inde ab الا E. haec habet: وقوت وپوشش از جايها ديگر مى آرند ونه جائى باشد كه از آنجا متاع زيادت خيزد ومردم زنكباررا زيادتى هنر ودانشى نباشد وهمه مردمان يك ممتند على بحر الروم وهو. l) C. لخت باشند امّا توانا وزورناك باشند نصفان. m) C. شرقى هذا البحر. n) C. باقة. o) A. et B. ودمهر i. e. ut vid. وتيبهرت (in mappa A. تيبهوت). In Codice quo usus est Abulfeda, p. ١٣٩ تيبهرت scriptum erat. p) C. add. الاقصى. Deinde A. وزويلة.

بين الاسكندريّة وبرقة من حدّ بحر الروم حتّى يمضى على ظهر الواحات الى بريّة تنتهى الى ارض النوبة، وغربيّه[a] البحر المحيط ممتدًّا على حدّه، وشماليّه[b] بحر الروم الّذى ياخذ من البحر المحيط ياخذ من حدّ مصر على ما يحاذى برقة الى طرابلس[c] المغرب ثمّ الى المهديّة ثمّ الى تونس ثمّ الى طبرقة ثمّ الى تنس ثمّ الى جزيرة بنى مزغنّا[d] ثمّ الى ناكور[e] ثمّ الى البصرة[f] ثمّ الى ازيلة ثمّ الى السوس الاقصى ثمّ يمتدّ على بريّة ليس ورآءها عمارة، وجنوبيّه رمل من حدّ[g] البحر المحيط حتّى يمتدّ من ورآء سجلماسة الى زويلة ثمّ يمتدّ الى[h] ظهر الواحات من ارض مصر۞ وامّا الاندلس فانّه يحيط به ممّا يلى البحر المحيط من حدّ بلد الجلالقة[i] على كورة يقال لها شنترين[k] ثمّ الى اخشنبة *ثمّ الى اشبيلية[l] ثمّ الى سدونة ثمّ الى جزيرة جَبَل طارق ثمّ الى مالقة[m] ثمّ الى بجّانة[n] ثمّ الى بلاد مرسية *ثمّ الى بلاد[o] بلنسية ثمّ الى طرطوشة ثمّ يتّصل ببلاد الكفر ممّا يلى البحر ببلاد الافرنجة[p] وممّا يلى البرّ[q] ببلاد علجسكس ثمّ ببلاد بسكونس ثمّ ببلاد الجلالقة حتّى ينتهى الى البحر[r] ۞

فامّا بَرْقَة فانّها مدينة وسطة[s] ليست بكبيرة وحواليها *كورة عامرة[t] كبيرة

---

a) A. et B. وغربيها. *b*) A. et B. ممتدّ على حدها وشماليها. *c*) A., B. et C. h. l. اطرابلس. Deinde A. et B. الغرب. *d*) A. et B. h. l. رعيس, C. رعى, E. رعبى. *e*) B. باكور, C. et E. باكور. *f*) C. et E. semper البصيرة. A. et B. h. l. النصيرة, infra القُصَير. Fortasse legendum est البُصَيْرة. *g*) C. et et E. هذا. Hinc in Ous. lacuna est. Ib. p. 16—19 transponenda sunt post p. 28. *h*) C. على. *i*) E. جالقيان hic et infra. *k*) A. et B. سمرس, C. سيرس. *l*) A. et B. om. *m*) C. et E. صاليه. *n*) A., B. et C. بجايه. *o*) A. pro his مدينته. *p*) A. افرنجه. *q*) C. (ut E.) المغرب; deinde ter بلاد pro ببلاد habet. *r*) C. add. المحيط. Hic sequi deberet mappa al-Maghribi. *s*) Abulfeda, p. ١٣٩, متوسطة. *t*) Cf. Abulfeda, l. l. A. عمارة كورة, B. عمارة كثيرة, C. وكورة عامرة وغامرة وحولها كور عامرة, ut D., E. وناحيتى ابادان. Deinde B. كثيرة.

وهى فى مستوٍ من الارض خصبة وبطيف[a] بها من كلّ جانب بادية يسكنها طوائف من البربر وقد كان يخرج اليها عامل من مصر الى ان ظهر المهدىّ[b] عبيد الله المستولى على المغرب فاستولى عليها وازال عمّال مصر ✿ وامّا طَرَابُلُس المغرب فهى من عمل افريقيّة وهى مدينة مبنيّة من الصخر على ساحل بحر الروم خصبة[c] واسعة الكورة[d] حصينة جدًّا ✿ وامّا المَهْدِيّة فانّها مدينة صغيرة استحدثها[e] عبيد الله المستولى[f] على المغرب وسمّاها بهذا الاسم وهى على البحر وعبيد الله تحوّل اليها من القيروان وهى من القيروان على يومين ✿ وتُونِس مدينة كبيرة خصبة[g] واسعة المياه[h] والزروع وهى اوّل عدوة الاندلس يعبر منها ولا يعبر من دونها *الّا من المدن الّتى تلى المغرب لانّها اوّل مدينة تحاذى الاندلس وما دونها محاذ لبلاد الافرنجة[i] ✿ وطَبَرْقَة مدينة صغيرة[k] وبيّة بها عقارب قاتلة نحو عقارب *عَسْكَر مُكْرَم[l] وبها فى البحر معدن المرجان وليس يعرف فى الارض *معدن للمرجان[m] الّا بها ✿ *وامّا تَنَس فهى مدينة كبيرة وهى عدوة الى الاندلس ايضًا الّا انّها وبيّة[n]' وجزيرة بنى مَزْغَنّا[o] مدينة عامرة يحفّ[p] بها طوائف من البربر وهى من الخصب والسعة على *غاية ما تكون[q] المدن ✿ وناكور[r] على شطّ البحر مدينة كبيرة[s] يعبر منها ايضًا الى بَجّانة[t] وهى مدينة حصينة خصبة ✿ والبَصْرَة[u] مدينة كبيرة واسعة خصبة وهى بحذاء جزيرة جَبَل طارق وبينها وبين الجزيرة المذكورة[v]

---

a) C. ويطوف. A. om. بها. *b*) C. et E. om. *c*) A. et B. om. *d*) C. الكور. Deinde A. et B. خصبة. Cf. Abulfeda, p. ١٤٧. *e*) B. استجدها. *f*) C. المتغلب. *g*) E. سخت استوار. *h*) C. الثمار. *i*) A. الفرنجة, B. افرنجة. C. his pro omnibus: الى المدن التى تلى المغرب والافرنجة. *k*) C. om. *l*) E. لشكر. *m*) A. et B. المرجان; C. tantum مرجان. *n*) C. et E. haec om. *o*) C. رعى, E. رعبى hic et infra. *p*) C. تحف. *q*) C. افضل ما يكون من. *r*) B. وتاكور, C. et E. وباكور ut infra quoque A. et B. *s*) C. add. واسعة, infra omisso خصبة. *t*) A., B. et C. بجاية. *u*) A. et B. والقُصَير, C. et E. والبصيرة. *v*) C. om. H. l. in marg. A. هى سَبْتَة sine indicatione quo pertineat.

عرض البحر اثنا عشر فرسخًا ✿ وأَزِيلَة مدينة كبيرة على شطّ البحر المحيط *وهى خصبة كثيرة الخير[a] وهى اقصى المعابر[b] الى الاندلس ✿ والسوس الاقصى اسم المدينة الّا انّها كورة عظيمة ذات مدن وقرى وسعة وخصب ويحتفّ بها طوائف من البربر ✿ *وامّا البصرة[c] وأَزِيلَة فهما من اقليم طنجة، وطَنْجَة هى[d] كورة عظيمة تحيط بمدن وقرى وبوادٍ من البربر كثيرة ومدينتها العظمى الّتى هى القصبة تسمّى فاس وهى المدينة الّتى بها يحيى الفاطمىّ ولم يفتحها عبيد الله *الخارج بالمغرب الى حين تصنيف هذا الكتاب[e] ✿ وامّا ناكور وجزيرة بنى مَزْغَنَّا فى مدن وقرى كثيرة *فقريبة من[f] تاهرت الاعلى، ومدينة كورة تاهرت اسمها[g] تَاهَرْتُ وهى مدينة كبيرة خصبة[h] واسعة البرّيّة والزروع[i] والمياه وبها الاباضيّة وهم الغالبون عليها ✿ وسِجِلْمَاسَة مدينة وسطة من حدّ تاهَرْت الّا انّها منقطعة لا يسلك اليها[k] الّا فى القفار والرمال وهى[l] قريبة من معدن[m] الذهب بينها وبين ارض السودان وارض[n] زَوِيلَة، ويقال انّه لا يعرف معدن للذهب اوسع ذهبًا ولا اصفى منه الّا انّ المسلك اليه صعب[o] والاستعداد[p] شاقّ جدًّا *وهى من مملكة عبيد الله[q] ✿ ويقال انّ كورة تاهرت باسرها من افريقيّة الّا انّها مفردة بالاسم والعمل فى الدواوين ✿ وسَطِيف[r] مدينة كبيرة بين تاهرت وبين القيروان وهى حصينة[s] ولها كورة تشتمل على قرى كثيرة وعمارة متّصلة وسكّانها[t] كُتَامَة قبيلة من البربر بهم ظهر عبيد الله[u] وكان ابو عبد الله المحتسب الداعى الى عبيد الله مقيمًا بينهم حتّى *تمهّد امره بهم[v] ✿ والقَيْرَوَان هى اجلّ مدينة بارض المغرب خلا قرطبة

---

a) C. om. b) C. العلميين. c) A. et B. والقصير. C. واما البصيرة. d) C. om. e) C. الخارجى الى يومنا هذا. f) C. قريبة من; A. et B. pro his tantum ما. g) C. واسمها. h) C. om. i) C. add. والثمار. k) C. فيها. l) C. add. مدينة. m) C. المعدن. n) A. بارض. o) C. add. جدا. p) C. add. له. q) C. et E. om. r) C. et E. وشطيف. s) C. خصبة جدا et sic leg. E. et Abulfeda, p. ١٤١. t) C. وسكانه. u) C. add. المتغلب. v) C. دمهد له امره.

بالاندلس فانّها اعظم منها[a] وهى المدينة الّتى كان يقيم بها ولاة المغرب وبها كان مقام *الاغلب وبنيه[b] الى ان ازال ملكهم ابو عبد الله المحتسب وخارج القيروان[c] ابنية كانت معسكر آل الاغلب ومقامهم بها كان[d] وتسمّى الرّقّادة الى ان استحدث عبيد الله المَهْدِيّة *على شطّ البحر فاقام بها[e] وانتقل عن رقّادة ✿ وامّا زَوِيلَة فانّها من حدّ المغرب وهى مدينة وسطة لها كورة عريضة هى مناخمة لارض السودان ✿ وبلدان[f] السودان بلدان عريضة الّا انّها قفرة قشفة[g] جدّا ولهم[h] فى جبال لهم فيها[i] عامّة ما يكون فى بلاد الاسلام من الفواكه الّا انّهم لا يطعمونه ولهم اطعمة يتغذّون بها من فواكه ونبات غير[k] ذلك ممّا لا يعرف فى بلدان الاسلام، والخدم السود الّذين *يباعون فى بلدان[l] الاسلام منهم وليس هم بنوبة ولا بزنج ولا بحبشة ولا من البُجَة الّا انّهم جنس على حِدَة اشدّ سوادًا من الجميع واصفى ويقال انّه ليس فى *اقاليم السودان[m] من الحبشة والنوبة والبُجَة وغيرهم اقليم هو[n] اوسع منه ويمتدّون الى[o] قرب البحر المحيط ممّا يلى الجنوب وممّا يلى الشمال على مفازة ينتهى الى مفاوز[p] مصر من ورآء الواحات ثمّ على مفاوز بينها وبين ارض النوبة ثمّ على مفاوز بينها وبين ارض الزنج وليس[q] لها اتّصال بشىء من الممالك والعمارات الّا من وجه المغرب لصعوبة المسالك[r] بينها وبين سائر

---

a) E. add. وتعريف اين دو شهر بتفصيل در كتاب ديگر مذكور است اينجا محل ذكر آن نيست. *b*) C. آل الاغلب. *c*) C. add. على دعوة. *d*) B. om.; C. كان بها. *e*) C. om. *f*) C. وبلاد. *g*) A. et B. مشقة. Fortasse eodem sensu quo شاقة adhibitur, ut habet Mokaddasí (MS. Berol. p. 120) qui hunc locum transcripsit. Cf. Edrîsî, *Description de l'Afrique et de l'Espagne*, p. ٣٦, 1 et Gloss. p. 329. *h*) C. وهى. *i*) A. om. *k*) B. et C. وغير. Lectio A. confirmatur ab E. et Mokaddasí l. l. *l*) C. ينقلون الى بلد. In A. antea scriptum fuit يملعون; in B. superinscribitur يبلغون. *m*) C. الاقاليم التى. *n*) C. om. *o*) C. على. *p*) C. مفازة. *q*) C. فليس. *r*) C. المسلك.

الامم، وهذه جوامع ما يحتاج الى معرفته من شرقىّ البحر من المغرب ۞
واما الغربىّ من المغرب فهو الاندلس، والاندلس بلدان عريضة كثيرة المدن خصبة واسعة ومدينتها العظمى تسمّى قُرْطُبَة وهى من الاندلس فى وسطها والّذى يحيط بالاندلس البحر المحيط ثمّ يطوف بحر الروم بها الى ارض افرنجة فياخذ من مدينة شَنْتَرِين الى أُخْشُنَبَة[a] *ثمّ الى اشبيلية[b] ثمّ الى سَذُونَة[c] ثمّ الى الجزيرة ثمّ الى مالقة ثمّ الى بَجّانَة[d] ثمّ الى بلاد مُرْسِيَة[e] *على مدينة لَقَنْت[f] الى بلاد بَلَنْسِيَة[g] ثمّ الى طُرْطُوشَة وهى آخر المدن *الّتى على[h] البحر ثمّ يتّصل من جهة البحر ببلاد الافرنجة ومن جهة البرّ يتّصل[i] ببلاد عَلَنْجَسْكَس وهى بلاد حرب من النصارى ثمّ يتّصل ببلاد بَسْكُونَس *وهم ايضا نصارى[k] ثمّ يتّصل ببلاد الجَلالقة[l] وهم نصارى ايضا، فينتهى من الاندلس حدّان[m] الى دار الكفر وحدّان[m] الى البحر، وهذه المدن الّتى ذكرناها على الشطّ كلّها مدن كبار عامرة ۞ والاندلس فى ايدى[n] بنى اميّة ما افتتحت لبنى[o] العبّاس ولا قدر عليها عبيد الله ولمّا زالت دولة بنى[p] مروان عبر اليها من أَزِيلَة المغرب الى جزيرة جبل طارق بعض بنى اميّة فتغلّب عليها فهى فى ايديهم *الى وقت تصنيفنا هذا الكتاب[q] ۞
ومن مشاهير مدن الاندلس جَيّان وطُلَيْطُلَة ونَفْزة[r] وسَرَقُسْطَة[s] ولَارِدَة ووادى الحجارة وتُرْجَالة[t] وقُورِيَة ومَارِدَة وباجة وغافق ولَبْلَة وقَرْمُونَة ومَوْرُور[u] واسْتِجَة

a) C. حسىه, B. أُخْشُنَبه. b) A. et B. haec, ut supra p. ۳۷ ann. *l*, om. c) C. شذونه. d) B. et C. بجايه, e) A. et B. مرشيه. f) C. om.; B. لقمب. g) B. مَلْمنيه, C. مليشه. h) A. et B. الى. Quod Ous. p. 23 hic memorat „باريس *Baris*, a town on the sea-side" ortum est ex his verbis vers. Pers. كه باز پسين شهريست بر كنارهٔ دريا. i) B. et C. om. k) C. وهى ايضا للنصارى. l) E. جالقان. m) C. حد. n) A. يدى. o) C. لولد. p) C. om. q) C. اليوم. r) A. et B. وستره, C. om., E. سدونه. Deinde B. وسَرَقطه, C. وسرقصه. s) C. add. وبرصه. t) A., B. et C. sine punctis, E. برخاله. Deinde C. وثوره. u) A. ومَوُر, B. ومَدُور, C. et E. وموروده; Jacut موزور. Deinde C. واسمنجه.

وريّة وهى كلّها مدن عظام *وليس فيها [a] ما يقارب قرطبة فى العظم والكبر واكثر ابنيتها من حجارة وهى ابنية جاهليّة لا تعرف فيها مدينة محدثة [b] الّا بجّانة [c] فانّها محدثة فى حدّ بلاد يقال لها البِيرة [d] ✿ وشَنْتَرِين الّتى على البحر المحيط بها يقع العنبر ولم نعلم [e] ببحر *الروم والبحر المحيط [f] موضع عنبر الّا بشنترين وشى [g] وقع *فى ايّام مقامى [g] بالشام بسواحل الروم، وتقع بشنترين فى وقت من السنة من البحر دابّة تحتكّ بحجارة على شطّ البحر فيقع منها *وبر فى لين [h] الخزّ لونه لون الذهب لا يغادر منه شيئًا [i] وهو عزيز قليل فيجمع [k] وتنسج منه ثياب [l] فتتلوّن فى اليوم الوانًا ويحجر عليها ملوك بنى اميّة ولا ينقل الّا سرًّا وتزيد قيمة [m] الثوب على الف دينار لعزّته [n] وحسنه ✿ ومَالَقَة سكّانها عرب وبها الشَّفَن الّذى تتّخذ منه مقابض السيوف ✿ وجزيرة جَبَل طارق منها افتتح الاندلس فى اوّل الاسلام وجَبَل طارق جبل [o] عامر حصين بالقرى والمدن وهو آخر المعابر بالاندلس ✿ وطُلَيْطُلَة مدينة فى جبل عالٍ بناؤها من حجارة قد وُثّقت بالرصاص وحواليها [p] سبعة اجبل *كلّها عامرة منيعة [q] مسكونة وحولها نهر عظيم يقارب فى الكبر دجلة واسم هذا النهر تاجُه يخرج من بلد يقال لها شَنْتَبَرِيَّة [r] ✿ ووادى الحجارة مدينة وهى وما حواليها من المدن والقرى تعرف بمدن بنى سَالِم [s] ✿ وريّة [t] كورة *عظيمة

---

a) C. فمنها. *b*) C. add. فى الاسلام ولا قريبًا منه. *c*) C. بجايه. *d*) C. البريّة (E. بيابان), D. بجانة وهى المرية. *e*) C. ولا يعلم. *f*) A. et B. الدور والمحيط. *g*) C. منذ زمان. E. habet خداوند كتاب گويد در آن وقت كه من بشام بودم الخ. *h*) C. وبرة فى لون. Mokaddasí (MS. Berol. p. 119) quoque وبرة habet. Nomen hujus animalis est ابو قلمون. *i*) C. شى. *k*) C. add. منه. *l*) C. ثيابا. *m*) A. et B. om. *n*) B. et C. لعزه. *o*) C. om. *p*) C. ومن حولها. *q*) C. om. *r*) A. et B. شَنْتَرِينَه, C. et E. سمره. Deinde C. من وادى. *s*) A. h. l. in marg.: هى اليوم قلعة تسمى شقوره. Cf. ad h. l. Abulfeda, p. ١٧٩. *t*) C. وثوره, E. ورقه (Ous. قده).

خصيبة[a] ومدينتها ارْجذونَة[b] ومنها كان عمر بن حَفْصُون[c] الّذى خرج على بنى اميّة بها[d] ✿ وفحص البَلُّوط كورة خصبة واسعة[e] ومدينتها غافِق ✿ وقُورِيَة[f] مدينة كانت كبيرة الّا انّها خربت بعصبيّة[g] وقعت[h] بينهم فاستعان احد الفريقين بالجَلّالقة النصارى حتّى خرّبوها ✿ ومارِدَة من اعظم مدن الاندلس وكذلك طليطلة وهما ممتنعان ليس بهما عامل لبنى اميّة الّا انّه يخطب بهما لهم ✿ وشَنْتَرِين[i] كورة عظيمة[k] ومدينتها قُلُمْرِيَة[l] ✿ وثغور الجلالقة ماردة ونَفْزَة[m] ووادى الحجارة وطليطلة ومدينة الجلالقة ممّا يلى ثغور الاندلس يقال لها سَمُّورَة وعظيم الجلالقة بمدينة يقال لها أُبِيط[n] هى بعيدة عن بلدان الاسلام ✿ وليس فى اصناف الكفر الّذين يَلُون الاندلس اكثر عددًا من الافرنجة ويقال لملكهم قَارُلَه[o] غير انّ الّذين يَلُون[p] المسلمين منهم اقلّ من سائر اجناس الكفر[q] لدخولهم فى البحر والحاجز الّذى بينهم *وبين الافرنجة[r] من بلدان الشرك من غيرهم، ثمّ الجلالقة يتلونهم[s] فى الكثرة واقلّهم عددًا البسكونس وهم اشدّ شوكة، والّذين يتلون البسكونس من ثغور الاندلس سَرَقُسْطَة[t] وتُطيلَة ولاردة، ويليهم[u] قوم من النصارى يقال لهم غَلْجَسْكَس اقلّهم غائلةً[v] وهم الحاجز بينهم وبين الافرنجة ✿

---

a) C. حصىىه, E. استوار. *b*) A. et B. pro his عظيمة وارجدونه كورة عامرة وارجدونه (ش). *c*) A. et B. حَفْصُوِى, C. حفصوىه, E. حفصويه ut Jacut I, p. ١٩٥, 16. *d*) C. et D. om. *e*) C. عظيمة حصىىه, E. استوار وابادان. Post غافق C. add. وهى كورة واسعة. *f*) A. et B. وفرونه, C. وبروده, E. ىوريه. *g*) C. لعصبيّة. *h*) A. et B. ووثوب. *i*) A., B. et C. وسمرىں, E. بسبيرىں. *k*) A., B. et E. om. *l*) A. et B. لىحدادىه, C. دىحراىمه, E. دىحداىمه. Cf. infra p. ٤٧ ann. *h*. *m*) A. et B. وثغره; C., D. et E. ودىعره. *n*) A., B. et C. ادىط, E. انبط. *o*) A. et B. شارله, C. ماوله, E. فاراه. *p*) C. الذى يلى. *q*) A. et B. om. *r*) A. وبيينهم, B. om. *s*) A. يلونهم et mox deinde يلون. *t*) C. سرقىطه et deinde ومطىله. *u*) C. وىبىنهم ut quoque E. *v*) E. مردمانى بد.

والبربر الّذين هم[a] بارض الاندلس وسائر المغرب صنفان صنف يقال لهم البُتْر[b] وصنف يقال لهم البَرانِس، فنفزة ومكناسة وهوارة ومديونة من البتر[b] وهم بالاندلس وكتامة وزناتة ومصمودة ومليلة وصنهاجة من البرانس، فامّا زناتة فاوطانها بناحية[c] تاهرت وامّا كتامة فاوطانها[d] بناحية سطيف[e] وسائر البربر الّذين هم من البرانس فمفترشون فى سائر المغرب من شرقىّ بحر الروم، وامّا نفزة[f] ومكناسة فهم بالاندلس بين الجلالقة وبين مدينة قرطبة، وامّا هوارة ومديونة فهم سكّان شنتبريّة[g] ۞

وبكورة *الْبِيرَة حرير[h] كثير يفضّل ويقدّم على غيره ۞ وبالاندلس *معادن كثيرة من الذّهب[i] وبها معادن فضّة بناحية *البيرة ومرسية[k] وبقرب قرطبة بموضع يقال له كُرْتَش[l] وتفسيره بالعربيّة الدِيَار[m]، *وبناحية تُطيلة[n] سمّور كثير ۞

وزَويلة بلد فى وجه ارض السودان وهاؤلآء الخدم السود اكثرهم يقع الى زويلة ۞ وارض المغرب ما كان منها فى شرقىّ بحر الروم بقرب الساحل فتعلوهم سمرة وكلّما تباعدوا *فيما يلى[o] الجنوب والمشرق ازدادوا سوادًا *حتّى ينتهوا الى بلد السودان فيكون الناس بها اشدّ الامم سوادًا، ومن كان فى غربىّ بحر الروم بالاندلس فهم بيض زُرْق وكلّما ازدادوا وتباعدوا الى ما يلى المغرب والشمال ازدادوا بياضًا حتّى يقطع عرض الروم كلّه الى ظهر الصقالبة، فكلّما ازدادوا وتباعدوا ازدادوا بياضًا وزرقة وحمرة شعر الّا انّ طائفة منهم يرجعون الى سواد شعر وعيون وهم صنف من الروم من انجلالقة ويقال انّ

---

a) C. om. *b*) C. et E. البربر. *c*) C. hic et deinde فى ناحية. *d*) A. et B. om. *e*) C. شطيف. *f*) A. et B. نعرمرة, C. ut supra نمرة. Deinde A. et B. وكُتَامَة. *g*) A. شنتريه, B. سنتريه, C. سنيوده, E. سمنمرده. *h*) C. البمرده كثير من; E. وبكوره نمريه جزيره است بزرگ به از ديگر جزيرها. *i*) C. حرده البعادن للذهب. *k*) C. المرده ومرسه. *l*) A., B. et E. كولس, C. كولمس. Vid. Makkarí, I, p. ٩١ et ٣٩٧. Debeo emendationem viro amicissimo Cl. Dozy. *m*) A. et B. الددار, C. الريّان. *n*) C. et E. وبطليطله. *o*) C. فى ناحية.

اصلهم من الشام كما أنّ طائفة *بِخَرْشَنَة من ارض[a] الروم يرجعون الى سواد شعر وعيون يزعمون انّهم من العرب من غسّان وقعوا اليها مع جَبَلَة بن الأَيْهَم ۞ وبين المغرب وبلد السودان مفاوز منقطعة لا تسلك الّا من مواضع معروفة ۞ وكان ملوك افريقيّة وبرقة اولاد الاغلب الّذى كان قد[b] انفذ فى اوّل ايّام بنى العبّاس ليكون فى وجه ادريس بن ادريس[c] ۞ وملوك طنجة اولاد ادريس بن ادريس[d] وبينهم وبين افريقيّة[e] تاهرت الشراة وهم الغالبون عليها ۞ *وملوك الاندلس بنو اميّة ما خُطِب لبنى العبّاس بها[f] الى يومنا هذا ويخطبون لانفسهم وهم من اولاد هشام بن عبد الملك، وصاحبهم فى وقت تصنيف هذا الكتاب هو عبد الرحمان بن محمّد بن عبد الله بن محمّد ابن عبد الرحمان بن الحكم بن هشام بن عبد الرحمان بن معاوية بن هشام بن عبد الملك بن مروان بن الحكم، واوّل من عبر منهم الى الاندلس عبد الرحمان بن معاوية بن هشام بن عبد الملك بن مروان فى اوّل ولاية بنى العبّاس فتغلّب عليها وبقيت الامارة فى اولاده الى وقت تصنيفنا هذا[g] ۞ والغالب على مذاهب اهل المغرب كلّهم مذاهب[h] الحديث واغلبها عليهم فى الفتيا[i] مذهب مالك بن أَنَس ۞ والّذى يقع[k] من المغرب الخدم السود من بلاد السودان والخدم البيض من الاندلس والجوارى المثمنات *تاخذ الجارية والخادم عن غير صناعة على وجوههما الف[l] دينار واكثر وتقع منها اللبود المغربيّة[m] والبغال للسرج والمرجان والعنبر والذهب والعسل والزيت والسفن[n] والحرير والسمّور ۞

---

a) A. et B. بخرشنة omissis, habet حتى ينتهوا C. omnibus inde a .مخوسيه بارض — ويزعمون طائفة من ارض الروم. E. بخرشه بزمين روم. Cf. *Mémoire sur la conquête de la Syrie*, p. 121. b) A. et B. om. c) C. add. ابن الحسين بن الحسن بن على بن ابى طالب عم. d) C. add. الحسنى. e) C. om.; deinde E. بتاهرت. f) A. et B. add. بالاندلس. g) In C. haec omnia desunt. h) C. om. i) B. et C. الفتوى. k) C. يرتفع et post البيض insert تقع. l) C. تشترى الجارية من غير صناعة على جمالها بالف. In E. quoque tantum de كنيزك fit sermo. m) C. om., E. ونمد مغربى. n) C. والسمن.

واَمّا المسافات بالمغرب فانّ من مصر الى برقة ٢٠ مرحلة ومن برقة الى طرابلس[a] مثلها ومن طرابلس الى القيروان مثلها فذلك من مصر الى القيروان ٦٠ مرحلة، *ومن القيروان الى سطيف ١٦ مرحلة ومن سطيف الى تاهرت ٢٠ مرحلة[b] ومن تاهرت الى فاس ٥٠ مرحلة ومن فاس الى السوس الاقصى نحو ٣٠ مرحلة *فمن القيروان الى السوس الاقصى ١١٦ مرحلة[b] فجميع المسافة من مصر الى اقصى المغرب فى شرقىّ بحر الروم نحو ٦ اشهر، وحجّاج اقصى المغرب *يخرجون قرب[c] المحرّم فيذهب فى سفرهم واستراحتهم عامّة السنة حتّى يلحقوا الحجّ☆ ومن القيروان[d] الى زويلة نحو شهر ومن القيروان الى المَهْدِيّة مسيرة يومين، ومن القيروان الى تونس ٣ مراحل ومن تونس الى طَبَرْقَة نحو ١٠ مراحل ومن طبرقة الى تنس[e] نحو ١٩ مرحلة ومن تنس الى جزيرة بنى مَزْغَنّا[f] ٥ ايّام، ومن تاهرت الى ناكور ٢٠[g] مرحلة ومن تاهرت الى سجلماسة نحو ٥٠[h] مرحلة، ومن فاس الى البصرة[i] ٦ مراحل ومن فاس الى ازيلة ٨ مراحل، ومن القيروان الى سجلماسة فى البرّيّة *نحو من ٨٠[k] مرحلة *وفى العمارة ١٣٠ مرحلة فهذه جوامع المسافات فى[l] المغرب فى شرقىّ بحر الروم[m]☆

واَمّا مسافات الاندلس فانّ قصبتها قرطبة ومنها الى اشبيلية[n] ٣ مراحل والى استجة مرحلة على سمت القبلة، ومن قرطبة الى *سرقسطة ١٠[o] ايّام والى تطيلة ١٣ يومًا ومن تطيلة الى لاردة[p] ٤ مراحل، ومن قرطبة الى طُليطلة ٦ ايّام ومن طليطلة الى وادى الحجارة يومان، ومن قرطبة الى مكناسة[q]

---

a) C. h. l. اطرابلس. *b*) C. haec om. *c*) C. يرتحلون فى; E. در اول محرم. *d*) A. et B. بريّة (num برقة?). *e*) A. et B. تُنّس, C. دمس, E. سبس et دمس. *f*) C. رعى, E. رعبى. *g*) C. et E. ثلثون. *h*) C. خمسة وعشرون, E. پانزده. *i*) A. et B. القصير, C. البصيره, E. بصيره. *k*) C. et E. خمسون. *l*) A. et B. الى. *m*) C. et E. haec om. *n*) A. اِشبْلِيَة, E. اسبيليّه. *o*) C. et E. سرقصه خمسة. *p*) C. اولاذه, E. ارده (pro تطيلة, C. دطليه, E. بطيله). *q*) C. et E. مكياسه. Ous. p. 18 „to Mekiah, three (سه) or four days journey."

٤ ايّام *ثمّ الى[a] هوارة مثلها ثمّ الى نفزة[b] ١٠ ايّام ومن نفزة[c] الى مدينة
سَمُّورة ٤ ايّام، ومن قرطبة الى قورية[d] ١٢ يومًا ومن قورية الى ماردة ٤ ايّام
ومن قورية الى باجة ٦ ايّام ويباخذ في[e] طريق ماردة[f] ممّا يلي اخشنبة
فمن قرطبة الى اشبيلية الى باجة[g] الى ماردة الى قورية الى قلمرية[h] مدينة
شنترين[i] العظمى ومن باجة الى شنترين ١٢ يومًا والى[k] اقصى كور شنترين
٥ ايّام، ومن قرطبة الى فحص البلّوط يومان[l] الى مدينتها المعروفة بغافق
ومن فحص البلّوط الى لبلة ١٤ يومًا، *واشبيلية[m] على طريق سَدُونة[n] ومن
قرطبة الى قرمونة ٤ ايّام[o] ومن قرمونة الى اشبيلية ٣ ايّام[p]، *ومن استجة
الى مورور مرحلة ومن مورور الى سَدُونة يومان ومن مورور الى جبل طارق
٣ ايّام[q]، ومن استجة الى مالقة ٧ ايّام ومالقة شرقيّ قرطبة واستجة قبليّها
ومن استجة الى ارجدونة[r] ٣ مراحل، ومن قرطبة الى بجّانة[s] ٦ ايّام ومن
قرطبة الى مرسية ١٤ يومًا[t] ومن قرطبة الى مدينة بلنسية[u] ١٨ يومًا، ومن
طرطوشة الى بلنسية ٥[v] مراحل، ومن مرسية الى بجّانة ٦ ايّام ومن بجّانة
الى مالقة نحو ١٠ ايّام ومن مالقة الى جزيرة جبل طارق ٤ ايّام *ومن
الجزيرة الى سَدُونة ٣ ايّام[w] ومدينة سدونة قلسانة *ومنها الى اشبيلية ٤ ايّام[w]
والى قرمونة ٣ ايّام، فهذه جوامع المسافات بالاندلس ۞ وقد اتينا على

---

a) C. والى, E. وتنا. Deinde E. هوّاره. *b*) E. دو روزه راه. *c*) C. نقره et بقره. *d*) C. فوره. *e*) C. om. *f*) A. et B. المارده. *g*) A. hic et deinde تاجه. *h*) A. indistincte فلننيوه s. فلذاينوه, B. فليتوه, C. فلمدره, E. فليترده. Cf. supra p. ٤٣ ann. *l*. *i*) A., B. et C. سنترين, E. سبرين. *k*) A. ومن, B. ومن الى. *l*) E. يك روزه, in C. lacuna. *m*) A. hic et deinde اشلبه. *n*) C. et E. haec om. *o*) C. et E. addunt في ناحية المغرب بين باجة واشبيلية على طريق ماردة. *p*) E. دو روزه, in C. lacuna. *q*) In C. et E. haec desiderantur. *r*) C. add. قصبة (l. قصبة رية). *s*) C. ut semper بجايه. *t*) E. هفت روزه. C. add. ومن استجة الى ارحود (*sic*) ٧ ايام. *u*) C. et E. اثنا عشر. *v*) C. et E. hic quoque اثنا عشر. *w*) C. om.

جوامع ما اردناه من المغرب وتتلوه ارض مصر فى حدّ بلاد الاسلام راجعًا الى المشرق ✿

ديار مصر

وامّا مصر فانّ لها حدًّا ياخذ من بحر الروم بين[a] الاسكندريّة وبرقة فياخذ فى برارىّ حتّى ينتهى الى ظهر الواحات ويمتدّ الى بلد النوبة ثمّ يعطف على حدود النوبة فى حدّ أُسْوان الى ارض البُجَة *من ورآء اسوان[b] حتّى ينتهى الى بحر القلزم ثمّ يمتدّ على بحر القلزم ويجاوز[c] القلزم على البحر الى طُور سينًا ويعطف على تيه بنى اسرآئيل ويمتدّ حتّى ينتهى الى بحر الروم فى الجِفار خلف[d] رَفَح والعَرِيش ويمتدّ على بحر الروم الى ان ينتهى الى الاسكندريّة ويتّصل باوّل الحدّ الّذى ذكرناه[e] ✿

المسافات بمصر من ساحل بحر الروم حيث ابتدانـاه الى ان يتّصل بارض النوبة من ورآء الواحات نحو ٢٥ مرحلة ومن حدّ النوبة ممّا يلى الجنوب على[f] حدود النوبة نحو ٨ مراحل[g] ومن القلزم على ساحل البحر الى ان ينعطف[h] على التيه ٦ مراحل ومن حدّ البحر على[i] حدّ التيه الى ان يتّصل ببحر الروم نحو ٨ مراحل ويمتدّ على البحر الى اوّل الحدّ الّذى ذكرناه نحو ١٢[k] مرحلة، وطولها من اسوان الى بحر الروم نحو ٢٥ مرحلة، وبها بحيرة فيها جزائر[l] مسافتها نحو مرحلتين فى مثلها، فهذه جملة مسافاتها ✿

وامّا صفة مدنها وبقاعها فانّ مدينتها العظمى تسمّى الفُسْطَاط وهى على النيل فى شرقيّه[m] شمالىّ النيل وذلك انّ النيل يجرى[n] مُوَرَّبًا بين[o] المشرق

---

a) C. الى. b) A. et B. tantum واسوان. c) A. et B. ويجاور. d) A. et B. على. Deinde B. ربح, C. العريش ورفح. e) Hic sequi deberet mappa Aegypti. f) D. add. اخر. g) Locus corruptus videtur. Emendatio proposita est ad locum Ibn Haucalis. h) B. يتعطف, C. يعطف, D. تعطف. i) C. الى, E. تا. k) C. et E. نحو من عشر. l) In marg. E. lector addit: مسكون وغير مسكون وذكر آن دربين كتاب نكنند. m) C. et E. om., D. contra habet فى شرقى النيل. n) C. et E. add. من المشرق, D. من الجنوب. o) B. من.

والجنوب والبلد كلّه على جانب واحد الّا انّ فى عدوة النيل ابنية قليلة تعرف بالجزيرة وهى جزيرة يعبر من الفسطاط اليها على جسر فى سفن ويعبر من هذه الجزيرة *الى الجانب الآخر[a] على جسر آخر الى ابنية ومساكن على الشطّ الآخر يقال لها الجِيزَة، والفسطاط مدينة كبيرة نحو الثُلْث من بغداد ومقداره نحو ثُلْثى فرسخ، والفسطاط على غاية العمارة والخصب وبالفسطاط قبائل وخِطَط للعرب تنسب اليهم محالُّها[b] مثل ما بالكوفة والبصرة الّا انّها اقلّ من ذلك، وهى سبخة ومعظم بناآتهم بالطوب طبقات واكثر السفل بها غير مسكونة[c] وربّما بلغت طبقات الدار الواحدة ثمانى طبقات الّا فى طرف منها يسمّى المَوْقِف[d] فانّه اصلب قليلاً، وبها بنآء مفترش وذلك بالحَمْرآء[e] على شطّ النيل[f]، وبها مسجدان للجمعة بنى احدهما عمرو ابن العاص فى وسط الاسواق والآخر باعلى المَوْقِف بناه احمد بن طولون، وخارج مصر ابنية بناها احمد بن طولون تكون زيادة على ميل[g] كان يسكنها جنده تسمّى القطائع كما كان بنآء آل[h] الاغلب خارج القيروان الرقّادة، وبها نخيل وثمار[i] كثيرة وزروعهم على مآء النيل تمتدّ فتعمّ[k] المزارع من حدّ اسوان الى حدّ الاسكندريّة وسائر الريف[l] فيقيم المآء من عند ابتدآء[m] الحرّ الى الخريف ثم ينصرف[n] فيزرع ثمّ لا يسقى بعد ذلك ۞ وارض مصر لا تمطر

---

a) C. om. *b*) A., B. et D. اليها محالهم. *c*) C. مسكون. *d*) A. المرىف, B. المرفف. E. haec ita disposuit: وبناهاى ايشان همه هفت وهشت طبقه است ودر طبقه زيرين نتوان بود زيراكه زمينى شوره است وبر كنارهٔ شهر جائى است كه آنرا موقف كويند زمين محكمى است شوره نيست. Deinde de suo addidit: وآنجا شنيدم كه فسطاط نام قبيلهٔ ايست كه بانى آن بنا شدند. *e*) A. et B. بالخضراء. Pro وذلك, C. habet وكذلك. *f*) A. et B. om. *g*) E. يكفرسخ. *h*) A. om. Deinde C. اغلب. *i*) C. وكروم. *k*) A. فتعمر, B. فَيَعْمر. *l*) C. add. فعم الما *m*) A. اشمدآ, B. اشد, C. اشتداد, D. امتداد. Cf. Edrîsî, *Description de l'Afrique et de l'Espagne*, p. ١٤٣. *n*) C. et D. ينضب, E. كم شود.

ولا تثلج وليس بارض مصر مدينة يجرى فيها الماء[a] دائمًا غير الفَيُّوم، والفَيُّوم هذه مدينة وسطة يقال انّ يوسف النبىّ عمّ اتّخذ لهم مجرى *يدوم لهم الماء[b] *وقُوِّم بحجارة[c] وسمّاه اللّاهُون☆ وامّا النيل فانّ ابتداء ماته لا يعلم وذلك انّه يخرج[d] من مفازة من وراء ارض الزنج[e] لا تسلك حتّى ينتهى الى حدّ الزنج ثمّ يقطع[f] فى مفاوز *وعمارات ارض النوبة فيجرى على عمارات متّصلة الى ان يقع فى ارض مصر، وهو نهر يكون عند امتداده[g] اكبر من دجلة والفرات اذا جُمِعَا ومآؤه اشدّ عذوبة وحلاوة وبياضًا من سائر انهار الاسلام، وفى هذا النهر يكون التمساح والسقنقور وسمكة يقال لها الرعّادة لا يستطيع احد ان[h] يقبض عليها وهى حيّة حتّى يرتعش[i] وتسقط من يده فاذا ماتت فهى كسائر السمك، وامّا التمساح فانه دابّة من دوابّ الماء مستطيل[k] الراس طول راسه[l] يكون نحوًا من نصف طول بدنه وله انياب لا يعضّ[m] على دابّة ما كانت من سَبُع او جمل الّا مدّه *فى الماء[n] وربّما خرج من الماء فمشى[o] فى البرّ وليس له فى البرّ سلطان ولا يضرّ احدًا وجلده يشبه السَّفَن الّذى تتّخذ منه مقابض السيوف لا يعمل السلاح فيه الّا تحت يديه ورجليه[p] ومكان الابط، وامّا السقنقور فانّه صنف من السمك الّا انّ له يدين ورجلين ويتعالج به للجماع ولا يكون فى مكان الّا فى النيل☆ وعلى حافات النيل من حدّ اسوان[q] الى ان يقع فى البحر[r] مدن وقرى

---

a) C. et E. add. من غير النيل. *b*) C. tantum للماء. *c*) C. et D. وقومه بالحجارة. Deinde B. et C. وسماها, A. سماها. *d*) C. دمندى. *e*) A. et B. add. من مفاوز ex margine, ut videtur. D. habet hanc lectionem pro من مفازة. E. از غارى „from a cavern" (Ous.). C. om. لا تسلك. *f*) C. يقطعه. Deinde A. فى مفازته. *g*) C. ادمدايه et sic E. وانجا كه اوّل رود است. *h*) A. et B. om. Deinde B. بقبض. *i*) C. et D. ترتعش يده. *k*) A. فيستطيل. *l*) C. الراس منه. *m*) C. et D. يقبض. *n*) C. اليه. E. haec ita habet: واورا دندانها باشد كه اگر شيرى بدندان او آيد چون موشى نمايد. *o*) A. et B. ممسا. *p*) A. ونعليه, C. خاصة وتحت رجليه. *q*) C. hic ut fere semper سوان. *r*) C. البحيرة.

منظومة[a] متكاثفة وأُسْوان هذه ثغر النوبة الّا انّهم مهادنون[b] ۞ وبصعيد مصر جنوبىّ النيل معدن الزبرجد[c] فى بريّة منقطعة عن العمارة ولا يعلم فى الارض معدن له[d] غير هذا ۞ وفى شمالىّ النيل جبل بقرب الفُسْطاط يسمّى المُقطّم فيه وفى نواحيه حجر الخُماهن ويمتدّ هذا الجبل الى النوبة وعند هذا الجبل بحذاء الفُسْطاط قبر الشافعىّ فى جملة المقابر ۞ وامّا الاسكندريّة فهى مدينة على شطّ البحر كثيرة الرخام فى الفرش والابنية والعمد[e] وبها منارة قد أُسّسَتْ فى الماء من صخر رفيع السمك جدًّا تشتمل على زيادة من[f] ثلاث مائة بيت لا يصل المرتقى اليها الّا بدليل ۞ ويسمّى *ما علا من النيل عن[g] الفسطاط الصعيد وما تسفّل[h] منه الريف ۞ ومن حدّ الفُسْطاط فى جنوبىّ النيل ابنية عظيمة يكثر عددها مفترشة[i] على سائر الصعيد *تدعى الاهرام[k] وبحذاء الفسطاط على نحو من فرسخين منها[l] ابنية عظيمة *اكبرها اثنان ارتفاع كلّ واحد منهما اربع مائة ذراع وعرضه اربع مائة ذراع وطوله اربع مائة ذراع وهو فى صورة العماريّة مربّع الاسفل ثمّ لا يزال يرتفع ويضيق حتّى يصير[m] اعلاه نحو مبرك جمل ومُلئت بنيانه بكتابة يُونانيّة[n] ومن داخله[o]

---

a) C. om. *b*) C. .بهادون *c*) E. .زبرجد وزمرّد Pro بريه , C. .قريه
d) C. om. Deinde B. .بغير *e*) C. .والاعمدة *f*) C. .على *g*) C. ما على
النيل من; cf. Abulfeda, p. ١٠٤. *h*) A. يسفل , B. تسفل . C. كان اسفل ,
Abulf. .سفل Deinde B. فيه , Abulf. .عنه *i*) A., B. et C. .مفترش *k*) Ex
D. et E. restitui. *l*) A. بها , B. et C. om. *m*) A. et B. add. .الى *n*) E.
insert: گويند بعض از آن اين عبارت است بنى الهرمان والنسر طاير فى
.السرطان Locus laudatur a Makrízío ed. Bul., I, p. ١١٥, 5 a f. his verbis:
وذكر ابو زيد البلخى انه وجد مكتوبًا على الاهرام بكتابتهم خط فعرّب فاذا
هو بنى هذان الهرمان والنسر الواقع فى السرطان فحسبوا من ذلك الوقت
الى الهجرة النبوية فاذا هو ست وثلاثون الف سنة شمسية مرتين يكون
.اثنين وسبعين الف سنة شمسية *o*) C. pro his omnibus اعظمها الهرمان
وهى تاخذ فى الدقّة حتى ينتهى مقدار راسها الى مبرك جمل وقد ملئت
.بالكتابة اليونانية فى داخلها

طريق يسير فيه الناس رجّالة الى *قريب اعلاه[a] وفى هذين الهرمين طريق فى باطن الارض مخترق، واصحّ ما سمعتُ فى الاهرام انّها قبور الملوك الّذين كانوا بتلك الارض ۞ وعرض العمارة على النيل من حدّ اسوان ما بين نصف يوم *الى يوم[b] الى ان تنتهى الى الفسطاط ثمّ تعرض فيصير عرضها[c] من حدّ الاسكندريّة الى الحَوْف الّذى يتّصل بمفازة القُلْزُم نحو ثمانية ايّام، وما فى *العرض من[d] ارض مصر قفار ۞ وامّا الواحات فانّها بلاد كانت معمورة بالمياه والاشجار والقرى والناس فلم يبق فيها ديّارٌ وبها الى يومنا هذا ثمار[e] كثيرة وغنم قد توحّشت فهى تتوالد، والواحات من صعيد مصر اليها فى حدّ الجنوب نحو ثلاثة ايّام فى مفازة وتتّصل الواحات بالنوبة[f] ببريّة فتنتهى[g] الى ارض السودان ۞

وبارض مصر بحيرة يغيض فيها مآء النيل تتّصل ببحر الروم تعرف ببحيرة تنّيس اذا امتدّ النيل فى الصيف[h] عَذُبَ مآؤها واذا *نقص فى الشتآء الى[i] اوان الحرّ غلب مآء البحر عليها فملح مآؤها، وفيها مدن مثل الجزائر تطيف[k] البحيرة بها فلا طريق اليها الّا فى السفن فمن مشاهير تلك المدن تنّيس ودمْيَاط وهما مدينتان لا زرع بهما ولا ضرع وبهما يتّخذ المرتفع من ثياب مصر، وهذه البحيرة قليلة العمق[l] يسار فى اكثرها بالمَرَادىّ وبها سمكة تسمّى الدلفين فى خلقة الزقّ المنفوخ وسمكة اذا اكلها الانسان راى منامات هائلة ۞ ومن حدّ هذه البحيرة الى حدّ الشام ارض كلّها رمال متّصلة حسنة اللون تسمّى الجفار بها نخيل ومنازل ومياه مفترشة غير متّصلة ويتّصل حدّ الجفار ببحر الروم *وحدّ بالتيه وحدّ باراضى فلسطين من الشام وحدّ

---

a) C. اعلاها. b) C. om. c) A., B. et C. عرضه. Deinde C. فى. d) C. عرض. e) C. اثمار. f) C. om. g) C. دمنهى. h) In C. superinscribitur النصف. i) C. ليس يطيف بدمياط بحيرة. جزر فى الشمال k) Lector in marg. B. يسلك اليها على الارض من كلّ جهة وهى راكبة على ذنابة (ذنابة l.) النيل وعلى البحر الرومى وبها نخيل وشجر وبها زروع كثيرة فاما تنيس فكما قال محفوفة بالماء. l) C. الماء.

ببحيرة تنّيس وما اتّصل به من ريف [a] مصر الى حدود القلزم ✿ وامّا تيه بنى اسرآئيل فيقال انّ طوله نحو اربعين فرسخًا وعرضه قريب من طوله وهى ارض منها صلبة ومنها رمال وبها نخيل وعيون مفترشة قليلة *يتّصل حدٌّ له بالجفار [b] وحدٌّ [c] بجبل طور سينا وما اتّصل به وحدٌّ بازآء [d] بيت المقدس وما اتّصل به من فلسطين *وحدٌّ له ينتهى الى [e] مفازة فى ظهر ريف مصر الى حدّ القُلْزم ✿

وامّا الأشْمُونَيْن فانّها مدينة صغيرة عامرة [f] ذات نخيل وزروع ويرتفع من الاشمونين ثياب كثيرة ✿ وبحذآئها من شمالىّ النيل مدينة صغيرة يقال لها بُوصِير بها قُتل مروان بن محمّد ويقال انّ سحرة فرعون *الّذين حشرهم فى يوم موسى [g] من بوصير ✿ فامّا أُسْوَان [h] فانّ بها نخيلًا كثيرًا وزروعًا وهى اكبر مدن الصعيد ✿ واسْنَا [i] واخْميم متقاربتان فى العمارة [k] صغيرتان عامرتان بالنخيل والزروع وذو النُّون المِصْرىّ الناسك [l] من اخميم ✿ والفَرَما على شطّ البحيرة وهى مدينة صغيرة خصبة [m] وبها قبر جالينوس اليونانىّ ، ومن الفرما الى تنّيس نحو فرسخين فى البحيرة ✿ وبتنّيس تلٌّ عظيم مبنىٌّ من اموات منضّدين [n] بعضهم على بعض يسمّى هذا التلّ بُوتُون [o] ويشبه ان يكون ذلك *من قبل موسى عمّ [p] لانّ ارض مصر فى ايّام موسى كان دينهم الدفن

a) Pro his C. et E. tantum وحد [الى] التيه وريف. b) A. يتصل دميرله (?). يتصل حد من حدودها بالجار, Jacut, I, p. ٩١٣, 7 بالجار. c) C. add. له. d) C. له بارض, D. باقاصى, Jacut بارض. e) A. et B. وحد فى. Jacut om. له. f) C. add. اهلة خصبة. g) C. et E. om. et addunt كانوا. h) C. ut solet سوان. i) C. وادلمما (in mappa ابلميا). D. et E. والبلينا. k) C. et E. المفازة (در ريك بيابان). Deinde A., B. et E. om. صغيرتان. l) C. add. كان. m) C. add. واسعة الخيرات. n) A. متصدّين, D. مسندين. C. habet الاموات. o) A. et B. بوتوى, C. بركوم, E. ثركوم, Ous. تركوم. Vid. Quatremère, *Mémoires géographiques et historiques sur l'Egypte*, I, p. 331 seqq. (cf. p. 306). p) C. منذ قبل زمان موسى بن عمران عم, deinde omnibus omissis ad هذا يومنا.

ثمّ صارت للنصارى وديفهم الدفن ثمّ صارت للاسلام، ورايتُ[a] عليهم اكفانًا من جنس[b] الخيش وجماجم وعظامًا فيها صلابة الى يومنا هذا ✿ وعين شمس[c] ومَنْف هما قريتان قد خربتا *كلّ واحدة منهما من الفسطاط على نحو اربعة اميال[d] وعين شمس من شمالىّ الفسطاط ومنف من جنوبيّه ويقال انّهما كانتا[e] مسكنين لفرعون، وعلى راس جبل المقطّم فى قلّته[f] مكان يعرف بتنّور[g] فرعون يقال انّه كان اذا خرج من احد هذين الموضعين يوقد *فيه فيُعَدّ فى المكان الآخر ما يُعَدّ له[h] ✿ وفى نيل مصر مواضع لا يضرّ فيها التمساح[i] منها عند الفسطاط وبوصير وغير ذلك من اماكن معروفة ✿ وحوالى الفسطاط زرع[k] ينبت مثل القضبان يسمّى البلسان[l] يتّخذ منه دهن البلسان لا يعرف بمكان فى[m] الدنيا الّا هناك ✿ وامّا العَبّاسَة[n] وفاقوس وجُرْجِير فانّها من أرض الحَوْف ويعرف شمالىّ النيل اسفل من الفسطاط بالحوف وجنوبيّه بالريف ومعظم رساتيق مصر وقراها فى هذين الموضعين ✿ وامّا معدن الذهب فمن اسوان اليه خمسة عشر يومًا والمعدن ليس فى ارض مصر ولاكنّه فى ارض البُجَة[o] وينتهى الى عَيْذاب، ويقال انّ عيذاب ليست[p] من ارض البُجة وانّما هى من مدن الحبشة، والمعدن ارض مبسوطة لا جبل فيها وانّما هى رمال ورضراض ويسمّى ذلك المكان الّذى فيه مجمع[q] الناس العَلّاقِى، وليس للبجة[r] قرى ولا خصب[s] فيه غناء وانّما هم[t] بادية ولهم نجب[u] يقال انّ ما فى النجب[u] *اسير منها[v] ورقيقهم ونجبهم وسائر ما بارضهم يقع[w] الى مصر ✿

---

a) E. صاحب كتاب ميگويد كه من النخ. *b*) E. مانند. *c*) C. عين الشمس. *d*) C. pro his عن غربى الفسطاط. *e*) A. et B. كانا. *f*) C. قبلته. *g*) B. بقبور. *h*) C. pro his tantum فى الاخر; cf. Edrîsî, p. ١٤٥. *i*) C. add. ابدا. *k*) C. زروع. *l*) C. et D. h. l. البلسم ut Makrízi, I, p. ٢٣٠. *m*) C. من. *n*) C. العباسيه, E. قياسه. Deinde A. وقابوس, B. ودادومن, E. وقاقور. *o*) A. et B. البُنجّه, E. بنجه. *p*) Codd. ليس. *q*) B. يجمع; A. om. الناس. *r*) C. بارض البجه, D. بالبجه. *s*) C. حصن. *t*) C. هى et deinde ولها. *u*) A. et B. بحر, D. بخت. *v*) A. et B. اشقّ منه (num l. اسبق ?). *w*) C. يرتفع.

وبمصر بغال وحمير لا يعرف فى شىء من بلدان الاسلام احسن ولا اثمن منها، ولهم من ورآء اسوان حمير صغار فى مقدار الكباش[a] ملمّعة تشبه البغال الملمّعة اذا اخرجت[b] من مواضعها لم تعش[c]، ولهم[d] حمير يقال لها السملاقية[e] بارض الصعيد زعموا انّ احد ابويها من الوحشىّ والآخر من الاهلىّ فهى اسير *تلك الحمير[f]* وبالجفار حيّات فى مقدار الشبر تثب من الارض حتّى تقع فى المحامل فتلسع، واهل مصر فى اخبارهم يزعمون انّ الجفار فى ايّام فرعون كانت معمورة[g] بالقرى والمياه وانّ الّذى قال الله تعالى[h] وَدَمَّرْنَا مَا كَانَ يَصْنَعُ فِرْعَوْنُ وَقَوْمُهُ وَمَا كَانُوا يَعْرِشُونَ هو[i] الجفار ولذلك سُمّى العَرِيش عريشا*

ارض الشام

وامّا الشام فانّ غربيّها بحر الروم وشرقيّها البادية[k] من ايلة الى الفرات *ثمّ من الفرات[l] الى حدّ الروم وشماليّها بلاد الروم وجنوبيّها حدُّ مصر وتيه بنى اسرآئيل وآخر حدودها ممّا يلى مصر رَفَح وممّا يلى الروم الثغور وهى مَلَطْيَة والحَدَث ومَرْعَش والهارُونيّة والكنيسة وعين زَرْبَة والمَصِيصَة وأَذَنَة وطَرَسُوس والّذى يلى الشرقىّ والغربىّ مدن قد ذكرناها فى تصوير الشام[m]*

قد جمعتْ الثغور الى الشام وبعض الثغور تعرف بثغور الشام وبعضها تعرف بثغور الجزيرة وكلاهما من الشام وذلك انّ كلّ ما ورآء الفرات من الشام وانّما سُمّى من مَلَطْيَة الى مرعش ثغور الجزيرة لانّ اهل الجزيرة بها يرابطون وبها يغزون *لا لانّها[n] من الجزيرة، *وكور الشام انّما هى جند فِلَسْطِين

a) B. الاباس. *b*) B. et C. خرجت. *c*) C. دعشر. *d*) C. وثم. *e*) A. sine punctis, B. السملاقه, C. et E. السقلابية, D. السقلاقية. *f*) C. ما فى تلك المدن. *g*) Hinc in A. et B. lacuna est usque ad verba in capite seq. قد جمعت الثغور. Cf. Abulfeda, p. ١٠٩. *h*) Qor. 7 vs. 133. *i*) C. هى et deinde سميت. *k*) C. بالبادية et mox ببلاد الروم. Textus hujus capitis secundum C. editus est a Cl. Arnold in *Chrestomathia Arabica*. *l*) Ex D. et Ous. *m*) E. مصر. Hic sequi deberet mappa Syriae. *n*) A. et B. لا انها, C. et D. لانها omisso لا.

وجند الأُرْدُنّ وجند حِمْص وجند دِمَشْق وجند قِنَّسْرِين والعواصم والثغور، وبين ثغور الشام[a] وثغور الجزيرة جبل اللُّكام وهو الفاصل بين الثغرين وجبل اللكام هو جبل داخل فى بلد الروم ويقال انّه ينتهى فى بلد الروم الى نحو من مائتى فرسخ ويظهر فى بلد الاسلام بين[b] مَرْعَش والهارونيّة وعين زربة فيسمّى[c] اللكام الى ان يجاوز اللاذقيّة ثمّ يسمّى جبل بَهْرآء وتَنُوخ الى حِمْص ثمّ يسمّى جبل لُبْنان ثمّ يمتدّ على الشام حتّى ينتهى الى بحر القلزم ۞

وامّا جند فلسطين وهو اوّل اجناد الشام ممّا يلى المغرب[d] فانّه تكون مسافته للراكب طول يومين من رفح الى حدّ[e] اللَّجّون وعرضه من يافا[f] الى ريحا يومين، وامّا زُغَر وديار قوم لوط والجبال[g] والشراة فمضمومة اليها وهى[h] منها فى العمل الى ايلة، وديار قوم لوط والبحيرة المُنْتِنَة وزُغَر[i] الى بَيْسَان وطَبَرِيّة[k] تسمّى الغَوْر لانّها[l] بين جبلين وسائر بلاد الشام مرتفع عليها وبعضها من الاردنّ وبعضها من فلسطين فى العمل، وامّا نفس فلسطين فهو ما ذكرتُه *وفلسطين مآؤها من الامطار واشجارها وزروعها اعذآء الّا نابُلُس فانّ بها مياهًا جارية[m] وفلسطين ازكى[n] بلدان الشام ومدينتها العظيمة الرَّمْلَة وبيت المقدس يليها فى الكبر، وبيت المقدس مدينة مرتفعة على جبال[o] يصعد اليها من كلّ مكان قصد[p] من فلسطين وبها مسجد ليس فى الاسلام مسجد اكبر منه والبنآء[q] فى زاوية من غربىّ المسجد يمتدّ على نحو نصف عرض

---

a) C. pro his: وكورة الشام انما هى من حد فلسطين وحد الشام. *b*) A. et B. من ut Codd. D. *c*) A. et B. يسمى. Deinde C. اكام et mox اللبنان. *d*) A. et B. الثغور, Abulfeda, p. ٢٣٦ الغرب. *e*) C. جند. *f*) C. تلقا, E. بلقا, Ous. بلقا. Deinde C. et E. اريحا. *g*) A. et B. من الجبال; C. et E. وجبال الشراة. *h*) A. et B. وهو. *i*) A. et B. وهو, C. وعرر, sed in marg. زُغَر. *k*) C. والطبرية (l. والى طبرية, coll. Abulf.). *l*) A., B., C. et Abulf. l.l. لانه. *m*) C. haec om. *n*) Abulfeda, p. ٢٢٧ ارخى. *o*) C. جدار. *p*) A. قُصِد, C. فسط quod in praef. Chrestom. Arnold, p. XVI explicatur per فى وسط. *q*) C. add. منه.

المسجد *والباقى من المسجد فارغ الّا [a] موضع الصخرة فانّ عليه حجرًا مرتفعًا مثل الدكّة وفى وسط الحجر على الصخرة قبّة عالية جدًّا [b] وارتفاع الصخرة من الارض الى صدر القائم وطولها وعرضها متقارب يكون بضعة [c] عشر ذراعًا وينزل الى باطنها [d] بمراقٍ من باب شبيه بالسرداب الى بيت [e] يكون طوله نحو بسطة فى مثلها، وليس ببيت المقدّس ماء جارٍ سوى عيون لا تتّسع للزروع [f] وهو من اخصب بلدان فِلَسْطين، ومحراب داوُد عم بها وهو بنية مرتفعة ارتفاعها *يشبه ان يكون [g] خمسين ذراعًا من حجارة وعرضها [h] نحو ثلاثين ذراعًا، على الحزر والتخمين واعلاه بناء مثل الحجرة وهو المحراب واذا وصلتَ اليها [k] من الرَّمْلَة فهو اوّل ما يتلقّاك [l] من بناء بيت المقدّس، وفى مسجد بيت المقدّس لعامّة الانبياء المعروفين لكلّ واحد منهم محراب معروف ✿ وعلى ناحية جنوب بيت المقدّس على ستّة اميال منه قرية تعرف ببيت لَحْم [m] وهى مولد عيسى عم ويقال انّ فى كنيسة منها قطعة من النخلة الّتى اكلت منها مريم [n] هى مرفوعة عندهم يصونونها ✿ ومن بيت لحم على سمته فى الجنوب مدينة صغيرة شبيهة فى القدر [o] بقرية تعرف بمسجد ابراهيم عم وفى المسجد الّذى يجمّع [p] فيه الجمعة قبر ابراهيم واسحاق ويعقوب عم صفًّا وقبور نسائهم صفًّا *بحذاء كلّ قبر من قبورهم قبر امراة صاحبه [q] والمدينة فى وهدة بين *جبال كثيرة [r] كثيفة

---

a) C. pro his tantum الى. *b*) A. om.; E. گنبدى است بزرگ. *c*) C. دسعه (Arn. تسعة); E. دوازده, *d*) A. et B. باطنه. Abulf. تحتها. *e*) C. بمراقى وينزل من باطنها واسفلها Edrîsî, بمراقى الى بيت Abulf., p. ٢٢٧; شبيه بالسرب من باب الى سرداب. Videtur legendum شبيه بيت كالبيت المظلم بالسرداب. *f*) A. للزرع; deinde C., D. et Abulf. وهى. *g*) C. نحو. *h*) C. وعرضه. *i*) A. et B. om. Deinde C. بالتخمين. E. ex his effecit از سنگ است. *k*) C. اليه. *l*) C. et D. يلقاك. *m*) C. لحم, E. اللَّخُم. Vid. Jacut, I, p. ٧٧٩. Deinde C. et D. وهو. *n*) E. add. ودر قران ذكر ان درخت مى آيد. *o*) C. الكبر. *p*) C. يجتمع. *q*) C. om. *r*) C. الجبال.

الاشجار[a] واشجار هذه الجبال وسائر جبال فِلَسْطين وسهلها[b] زيتون وتين وجُمّيز وعنب وسائر الفواكه اقلّ من ذلك ۞ ونابُلُس مدينة السامرة يزعمون[c] انّ بيت المقدّس هو نابُلُس وليس للسامرة مكان[d] من الارض الّا بها ۞ وآخر مدن فِلَسْطين ممّا يلى جِفارَ مصر مدينة يقال لها غَزّة بها قبر هاشم بن عبد مَنَاف وبها مولد محمّد بن ادريس الشافعىّ وفيها ايسر[e] عُمَر بن الخطّاب فى الجاهليّة لانّها كانت مستطرقًا[f] لاهل الحجاز ۞ *وبفلَسْطين نحو من عشرين منبرًا على صغر رقعتها[g] وهى من اخصب بلاد الشام ۞ وامّا الجبال والشَّراة فانّهما بَلَدانِ متميّزان امّا الشراة فمدينتها تسمّى أَذْرُح وامّا الجبال فانّ مدينتها تسمّى رُوَات[h] وهما بَلَدانِ فى غاية الخصب والسعة وعامّة سكّانها العرب متغلّبون عليها ۞

وامّا الأُرْدُنّ فانّ مدينتها الكبرى طَبَرِيّة وهى على بحيرة عذبة المآء طولها اثنا عشر ميلًا[i] فى عرض فرسخين او ثلاثة وبها عيون جارية[k] حارّة مستنبطها على نحو فرسخين من المدينة فاذا انتهى المآء[l] الى المدينة على ما دخله من الفتور بطول السير اذا طُرِحت فيه الجلود انمعطت[m] ولا يمكن استعماله الّا بالمزاج ويعمّ[n] ذلك المآء حمّاماتهم ومياضىً لهم ۞ والغَوْر اوّله هذه البحيرة[o] ثمّ يمتدّ على بَيْسان حتّى ينتهى الى زُغَر وريحا[p] الى البحيرة المنتنة ،

---

a) C. add. والثمار. Deinde A. et B. واشجار هذا الجبل, C. وهذه الجبال. b) C. add. كلها et om. وتين et وعنب. Cf. quoque Abulfeda, p. ٢٤١. c) C. ويزعمون. d) C. السامرة فى مكان. e) D. استغنى, Epit. Paris. recte ايسر. A. أُسمر, B. et C. اسر. f) A. مسطرقًا, D. متجرا; cf. Abulfeda, p. ٢٣٩. g) C. om. h) D. روات, C. et E. رواد, Edrîsî دراب. i) A., B. et D. فرسخا. Secutus sum C., E., Abulfeda, p. ٢٤٣ et Edrîsî. Mokaddasí apud Jacut, III, p. ٥١٠, 16 habet طولها نحو من فرسخ بلا عرض. k) A. et B. om. جارية, C. et D. حارة. E. et Epit. Paris. utrumque habent. l) A., B. et C. om. m) C. et D. تمعّطت. n) A., B. et C. ويعمر. Deinde C. لماء. o) E. habet ونخستين حدّ غور كه ياد كرديم جنوبست ونزديك آن جايگاهى هست اوله الجنوب .كه آنرا مسجد ابرهيم گويند بحيره انجاست. In C. legitur مدينة بقربه مسجد يعرف بمسجد ابرهيم عم هى البحيرة p) C. semper اريحا.

والغَوْر ما بين جبلين غائر فى الارض جدًّا[a] وبه عيون وانهار ونخيل ولا تستقرّ به الثلوج وبعض[b] الغَوْر من حدّ الأُرْدُنّ الى ان تُجَاوِز بيسان فاذا جاوزتَهُ كان من حدّ فِلَسْطِينَ، وهذا البطن اذا امتدّ فيه السائر ادّاه الى ايلة ۞ وصُور بلد من احصن الحصون *الّتى على شطّ البحر عامرة خصبة[c] ويقال انّه اقدم بلد بالساحل وانّ عامّة حكمة اليونان منها ۞ وبالأُرْدُنّ كان مسكن يعقوب *النبىّ عم[d]، وجبّ يوسف على اثنى عشر ميلًا من طَبَرِيّة على ما يلى دمشق، ومياه طبريّة من البحيرة ۞

وامّا جند دِمَشْق فانّ قصبتها مدينة دمشق وهى اجلّ[e] مدينة بالشام كلّها وهى فى ارض واسعة بين جبال تحيط[f] بها مياه كثيرة واشجار وزروع متّصلة وتسمّى تلك البقعة الغُوطَة عرضها مرحلة فى مرحلتين ليس بالمغرب[g] مكان انزه منه، ومخرج مآئها[h] من تحت كنيسة يقال لها الفيجَة[i] وهو اوّل ما يخرج مقداره ارتفاع ذراع فى عرض باع[k] ثمّ يجرى فى شِعَب تتفجّر فيها[l] العيون فياخذ منه نهر عظيم اجراه[m] يزيد بن معاوية يعرض *فى كثير[n] ثمّ يستنبط منه نهر المِزّة ونهر القَنَوَات[o] ويظهر عند الخروج من الشِعَب بموضع يقال له النَّيْرَب ويقال انّه المكان الّذى قال الله فيه[p] وَآوَيْنَاهُمَا اِلَى رَبْوَةٍ ذَاتِ قَرَارٍ وَمَعِينٍ، ثمّ يبقى من هذا الماء عمود النهر فيسمّى بَرَدًا وعليه قنطرة فى وسط مدينة دمشق *لا يعبره الراكب غزارةً وكثرةً[q] فيفضى الى قرى الغوطة

---

a) A. et B. om. b) A. et B. وحدّ; cf. Abulf. p. ٢٢٦ ann. 7. c) In A. et B. haec desiderantur. d) A. et B. om. Pro جب E. habet جامعة. e) E. بهين ex quo Ous. (p. 41) effecit يمين "the right hand." f) C. تحتف, D. تحف. Post بها A. insert فيها. g) C. et D. بالشام et sic Ous., sed E. ut A. et B. h) A. et B. مياهها et deinde وهى. i) C. العمحه, E. فنجة. k) Abulf. p. ٢٣٠ ذراع. E. vertit واز. l) C. et ut vid. E. منها, Abulf. منه. m) A. et B. اجراها. n) A. كثيرًا. C. يعرض الدجلة, E. وبپهناى دجله. D. يغوص الرجل فيه عمقا. o) C., D. et E. القناة. p) Qor. 23 vs. 52. q) C. et D. haec om., sed exstant in Epit. Paris.

ويجرى الماء فى عامة *دورهم وسككهم وحماماتهم[a]، وبها مسجد ليس فى الاسلام مسجد *احسن ولا اكثر نفقة[b] منه واما الجدار والقبة التى فوق المحراب[c] عند المقصورة فمن بناء الصابئين *وكان مصلاهم[d] ثم صار فى ايدى اليونانيين فكانوا يعظمون فيه دينهم ثم صار[e] لليهود وملوك من[f] عبدة الاوثان فقتل فى ذلك الزمان يحيى بن زكرياء عم ونصب راسه على باب هذا المسجد *بباب يسمى[g] باب جيرون ثم تغلب عليه[h] النصارى فصار[i] فى ايديهم كنيسة *يعظمون فيها دينهم[k] حتى جاء الاسلام فصار للمسلمين واتخذوه مسجدا *وعلى باب جيرون حيث[l] نصب راس يحيى بن زكرياء نصب راس الحسين بن على عم، فلما كان فى ايام الوليد بن عبد الملك عمره فجعل ارضه رخاما مفروشا[m] وجعل وجه جدرانه[n] رخاما مجزعا واساطينها رخاما موشى[o] ومعاقد رؤوس اساطينه ذهبا ومحرابه ذهبا مرصعا بالجواهر[p] ودور السقف كله ذهبا مكتبا كما تطوف ترابيع[q] جدار المسجد، يقال انه انفق *فيه وحده[r] خراج الشام[s]، وسطحه رصاص وسقفه خشب مذهب يدور الماء على رقعة[t] المسجد حتى اذا فجر منه[u] انبسط على جميع الاركان

---

a) E. خانها وگرمابها ودكانها. *b*) C. اعمر ولا اكبر بقعة, E. بزرگتر. Cf. Abulf. p. ٢٣٠. *c*) C. add. من. *d*) C. لصلواتهم. Deinde B. وصار. *e*) Abulf. صارت. Deinde C. الى اليهود. *f*) C. om. *g*) C. الباب المعروف, D. et Abulf. tantum المسمى. *h*) A., B., C. et D. عليها. *i*) A. et B. فصارت omisso فى ايديهم. *k*) C. om. *l*) A. et B. tantum وحيث, E. وهم بآن جايگاه كه. *m*) C. om. *n*) A. جداراته. *o*) E. ملمع. Deinde pro ومعاقد, A. et B. وقواعد. Secutus sum D. et Abulf.; C. et E. tantum ورؤس. *p*) C. et D. بالجوهر. Deinde A. et B. ودون (E. وآسمانه). *q*) C. يطوف تربيع. *r*) C. بسببه. *s*) Sic quoque E. et Abulfeda, p. ٢٣٠; C. addit خمس سنين, D. سنتين quod idem exstat in Epit. Paris. et apud Edrîsî. Kazwíní, II, p. ١٢٧ habet سبع سنين et sic Mokaddasí MS. Berol. p. 80. *t*) A. رفعه, et sic Arnold in *Chrestom.*, p. 97. B. et C. رقعه ; E. بر بالاى. *u*) C. فيه.

سوآء[a] ۞ ومن جند دمشق بَعْلَبَكّ وهى مدينة على جبل عامّة ابنيتها من حجارة وبها قصور[b] من حجارة قد بُنِيت على اساطين شاهقة ليس بارض الشام ابنية[c] حجارة اعجب ولا اكبر منها ۞ وأَطْرَابُلُس مدينة على بحر الروم عامرة ذات نخل وقصب سكّر وخصب[d] ۞

وامّا جند حِمْص فانّ مدينتها حمص وهى مدينة فى مستوٍ[e] خصبة جدًّا من اصحّ بلدان الشام تربة[f] فى اهلها جمال مفرط وليس بها عقارب ولا حيّات ولها مياه واشجار وزروع كثيرة واكثر زروع رساتيقها اعذآء[g]، وبها كنيسة بعضها مسجد جامع وبعضها كنيسة[h] وهى من اعظم كنائس الشام، وعامّة طريق حمص مفروشة بالحجارة ۞ وامّا أَنْطَرْطُوس[i] فهو حصن على بحر الروم ثغر لاهل حمص وبه[k] كان مصحف عثمان بن عفّان ۞ وامّا سَلَمْيَة فهى مدينة الغالب على سكّانها بنو هاشم[l] على طرف البادية خصبة[m] ۞ وامّا شَيْزَر وحَمَاة فانّهما مدينتان[n] صغيرتان نزهتان كثيرتا المآء[o] والشجر والزرع ۞

وجند قِنَّسْرِين مدينتها حَلَب وهى عامرة بالاهل جدًّا على مدرج طريق العراق الى[p] الثغور وسائر الشامات ۞ وقِنَّسْرِين مدينة تنسب الكورة اليها وهى من اصغر المدن بها ۞ ومَعَرّة النُّعْمَان مدينة[q] هى وما حواليها من القرى اعذآء ليس بجميع نواحيها مآء جارٍ ولا عين وكذلك اكثر ما بجميع جند قنّسرين اعذآء ومياههم من السمآء ۞ وخُنَاصِرَة[r] حصن على شفير البرّيّة كان

---

a) A. et B. om.; D. بالسوية. *b*) B. om.; A. habet قصر et mox (ut quoque B.) بُنى et منه; cf. quoque Abulf. p. ٢٧١, ann. 19. *c*) A. et B. om.; C. om. حجارة. *d*) C. add. واسعة. *e*) C. et D. مستواة. Abulf. p. ٢٦١ add. من الارض. *f*) C. هواء وتربة, E. درست هوا. *g*) Abulf. عذى. E. vertit بر آب باران. *h*) E. كليسياست. *i*) C. اطرطوس, E. ازطرسوس. Deinde C. فهى; cf. quoque Abulf. p. ٢٢٩. *k*) A., B. et C. وبها. *l*) E. مردمان انجا شافعى مذهب باشند. *m*) C. add. جدا. Cf. quoque Abulf. p. ٢٦٥. *n*) C. add. خصبتان عامرتان. *o*) C. et D. المياه et deinde والزروع. *p*) C. والى; E. وثغور; cf. Abulf. p. ٢٦٧. *q*) Abulf. p. ٢٣١ add. متوسطة. *r*) C. et D. الخناصرة.

يسكنه عمر بن عبد العزيز[a] ۞ وامّا العَوَاصِمُ فاسم الناحية وليس موضع بعينه يسمّى العواصم وقصبتها انْطَاكِيَة وهى بعد دمشق انزه بلد بالشام عليها سور من صخر يحيط بها وبجبل مشرف عليها فيه[b] مزارع وارحية ومراعٍ واشجار وما *يستقلّ به[c] اهلها من مرافقها، ويقال انّ دور السور للراكب يومان[d] وتجرى مياههم فى دورهم وسككهم[e] ومسجد جامعهم وبها[f] ضياع وقرى ونواحٍ خصبة جدًّا، وامّا الصخرة *فانّها تعرف بصخرة موسى ويقال انّ موسى اجتمع مع الخضر عمّ فى هذا[g] الموضع ۞ وامّا بَالِس فهى مدينة على شطّ الفرات صغيرة وهى اوّل مدن الشام من العراق والطريق[h] اليها عامر وهى فرضة الفرات لاهل الشام ۞ وامّا مَنْبِج فهى مدينة فى بريّة الغالب على مزارعها الاعذآء وهى خصبة *ومنها البُحْتُرِىّ الشاعر *وثابت ابنه بها وسكّانها عرب[i]، وبقربها سَنْجَة[k] وهى مدينة صغيرة بقربها قنطرة حجارة[l] تعرف بقنطرة سنجة ليس فى الاسلام قنطرة اعجب منها ۞ وامّا سُمَيْسَاط فهى على الفرات وكذلك جِسْرُ مَنْبِج وهما مدينتان صغيرتان خصبتان[m] لهما زروع سقى ومباخس[n] ومآؤهما من الفرات ۞ ومَلَطْيَة مدينة كبيرة من اكبر الثغور الّتى دون جبل اللُّكَام وتحتفّ بها جبال كثيرة الجوز وسائر الثمار مباح لا مالك له وهى من قرى بلد الروم على مرحلة ۞ وحِصْنُ مَنْصُور حصن صغير فيه منبر وزروعه[o] عذى ۞ والحَدَث ومَرْعَش هما مدينتان صغيرتان عامرتان فيهما مياه وزروع

---

a) C. add. رحمه. b) A. et B. om.; C. فيها. c) C. يحتاج اليه. d) A. om.; D. يوم. B. et C. يومين. e) A. et B. ومساكنهم; cf. Abulf. p. ٢٥٧. E. سراها ومحلها. f) Abulf. ولها. g) C. pro his: التى تعرف بصخرة موسى بن عمران فانها بهذا. Alii illud saxum ponunt prope Scharwán in Armenia (vid. Jacut, III, p. ٢٨٢). h) A., B. et C. om. i) C. haec om. Verba وثابت ابنه بها in A. et B. post عرب perperam collocata sunt. Cf. Ibn Khallicán, n. 793, p. ١٤٧ (وهى (منبج وقف على ولد البحترى. k) A., B. et C. سنجه. (Male Arnold سبيخة). E. صناجة. l) In flumine cujus nomen est النهر الازرق. m) B. om. n) A. et B. حسن وبنآء (B. تسقى); cf. Abulf. p. ٢٦٧. o) A. وزروعها, Abulf. p. ٢٦٩ وزرعه.

واشجار كثيرة وهما ثغران ❀ وامّا زِبَطْرَة فانّها حصن كان من اقرب هذه الثغور الى بلد الروم خرّبه[a] الروم ❀ والهَارُونيّة من غربىّ جبل اللكام فى بعض شعابه وهى حصن صغير بناه هارون الرشيد فنسب اليه ❀ واسْكَنْدَرُونَة[b] حصن على ساحل *بحر الروم[c] صغير به نخيل ❀ وبَيَّـاس مدينةٌ صغيرة على شطّ بحر الروم ذات نخل وزروع خصبة ❀ والتِّينات[d] حصن على شطّ البحر ايضًا فيه *مجمع للخشب[e] الصنوبر الّذى ينقل الى الشامات والى مصر والثغور ❀ والكَنِيسَة حصن فيه منبر وهو ثغر فى معزل من شطّ البحر ❀ والمُثَقَّب[f] حصن صغير بناه عمر بن عبد العزيز به[g] منبر ومصحف له ❀ وعَيْن زَرْبَة بلد *يشبه مدن الغور[h] بها نخيل وهى خصبة واسعة *الثمار والزروع[i] والمرعى وهى المدينة الّتى اراد *وصيف الخادم[k] ان يدخل بلد الروم منها فادركه المعتضد هناك ❀ والمَصّيصَة مدينتان احداهما تسمّى المصّيصة والاخرى تسمّى كَفَرْبَيَّا[l] على جانبَىْ جَيْحان وبينهما قنطرة حجارة حصينة[m] جدًّا على شرف من الارض ينظر منها الجالس فى المسجد[n] الجامع الى قرب البحر نحو اربعة فراسخ، وجَيْحان يخرج من بلد الروم حتّى ينتهى الى المصّيصة ثمّ الى رستاق يعرف بالمَلُّون[o] حتّى يقع فى بحر الروم ❀ وأَذَنَة مدينة تكون مثل احد جانبى المصّيصة على نهر يسمّى سَيْحان وهى مدينة خصبة عامرة وهى

---

a) C. et Abulf. p. ٢٣٤ خربها. b) C. et D. والاسكندرية. c) C. البحر للروم. d) C. البتنيات, E. بتينات. e) A. et B. الخشب, C. بجمع خشب, D. مقطع للخشب. f) E. قبقب. Secundum Beládsorí, p. ١٦٦ Hischám ibn Abdo 'l-Melik castellum condidit. g) C. وبها. h) C. فيه الغورية. i) A. et B. tantum الزروع; vid. quoque Abulf. p. ٢٣٤, ann. 2. k) A. om.; B. per conjecturam, ut videtur, lacunam in الاصل explevit dicens اراد بابك الخُرّمى. Cf. Ibno 'l-Athír, VII, p. ٣٤٣. l) C. ridicule كغرتوثا (in mappa كفرسا), Ous. كفرتوما. B. et E. sine punctis. m) Abulf. p. ٢٥١ وهى خصبة. n) C., D. et Abulf. مسجد. o) A. et B. بالملون. Est veterum *Mallus*. (Arnold male pronunciat المُلُوّن. E. habet ملوّن).

منقطعة عن نهر سَيْحان[a] فى غربىّ النهر، وسَيْحان هو دون جَيْحان فى الكبر عليه قنطرة حجارة عجيبة البناء طويلة جدًّا ويخرج[b] هذا النهر من بلد الروم ايضًا ✿ وطَرَسُوس مدينة كبيرة عليها سوران من حجارة تشتمل على خيل ورجال وعُدّة[c] وهى فى غاية العمارة والخصب وبينها وبين حدّ[d] الروم جبال هى الحاجز بين المسلمين والروم ويقال *انّه كان بها زهآء مائة الف فارس فيما يزعم اهلها[e] وليس من مدينة عظيمة من حدّ سجستان الى كرمان وفارس والجبال وخوزستان وسائر العراق والحجاز واليمن والشامات[f] ومصر الّا وبها لاهلها دارٌ واكثرُ *ينزلها اهلُها[g] اذا وردوها ✿ وأَوْلَاس حصن على ساحل البحر بها قوم متعبّدون وهى[h] آخر ما على بحر الروم من العمارة للمسلمين[i] ✿

وامّا رَقيم[k] فانّها مدينة بقرب البَلْقَآء وهى صغيرة منحوتة بيوتها كلُّها وجدرانها من صخر كانّها حجر واحد ✿ والبُحَيْرَة المُنْتِنَة[l] من الغَوْر بقرب زُغَر وانّما تسمّى المنتنة لانّه ليس فيها شىء من الحيوان لا سمك ولا غيره وتقذف بشىء يسمّى الحُمَّر[m] منه يلقّحون كرومَ فلسطين كما يلقّح النخلُ بطَلْع *الفحّال منها[n]، وبزُغَر بسر يقال له الانْقلآء لم ار بالعراق ولا بمكان اعذب ولا احسن منظرًا منه كانّ لونه الزعفران لا يغادر منه شيئًا ويكون اربعة منه شبرًا[o]، وديار قوم لوط هى ارض تسمّى الارض المقلوبة وليس بها زرع ولا ضرع ولا حشيش وهى بقعة سودآء قد *فُرشت بحجارة[p] كلّها متقاربة

a) A. et B. om. et deinde pro فى habent من. Pro منقطعة عن, D. et Abulf. p. ٢٤٩ habent tantum على. b) A. et B. يخرج. c) C. عدة. d) A. et B. om. e) C. ان بها الوف من الفرسان. E. انه هزار هزار. f) C. والشام. g) C. (sic) اعلها دمزلوها. Male Arnold أَهْلُهَا. h) A. et B. وهم. i) Ex C. et E. addidi. k) C. et E. رقم. Vulgo الرقيم scribitur. l) C., D. et E. hic et deinde الميتة. m) A. et B. العجم, E. حمره; cf. Abulfeda, p. ٢٢٨. n) C. ذكورها. o) A. شبر, D. ويكون فى — رطل. p) C., D. et Abulf. p. ٢٢٨ فرش بها حجارة.

فى الكبير يروى انّها الحجارة المُسَوَّمَة الّتى رُمِىَ بها قوم لوط وعلى عامّة تلك الحجارة كالطابع ۞ ومُعَان[a] مدينة صغيرة سكّانها بنو اميّة ومواليهم وهو حصن من الشَرَاة ۞ وحَوْرَان والبَثَنِيّة هما رستاقان عظيمان من جند دمشق مزارعهما مباخس وهناك بُصْرَى *وعند البلقآء عَمّان[b] الّتى جآء فى الخبر فى ذكر الحَوْض[c] انّه ما بين عمّان وبُصْرَى ۞ وبَغْرَاس على طريق الثغور وبها دار ضيافة لزُبَيْدَة وليس بالشام دار ضيافة غيرها ۞ وبَيْرُوت مدينة على شطّ بحر الروم خصبة[d] من عمل دمشق بها كان مقام الأوزاعىّ[e] ۞

وامّا المسافات بالشام[f] فانّ طولها من مَلَطْيَة الى رَفَح فمن ملطية الى منبج ۴ ايّام ومن منبج الى حلب يومان ومن حلب الى حمص ٥ ايّام ومن حمص الى دمشق ٥ ايّام ومن دمشق الى طبريّة ۴ ايّام ومن طبريّة الى الرملة ٣ ايّام ومن الرملة الى رفح يومان فذلك ٢٥ مرحلة، وعرضها فى بعض المواضع اكثر من بعض فاعرضها طرفاها وأَحَدُ[g] طرفيها من الفرات من جسر منبج على منبج ثمّ على قُورُس[h] فى حدّ قنّسرين ثمّ على العواصم[i] فى حدّ انطاكية ثمّ يقطع جبل اللكام الى بيّاس[k] ثمّ الى التينات ثمّ على[l] المثقّب ثمّ على المصّيصة ثمّ على اذنة ثمّ على طرسوس وذلك نحو ١٠ مراحل، وان سلكتَ من بالس فالى حلب ثمّ الى انطاكية ثمّ الى اسكندرونة[m] ثمّ الى بيّاس حتّى تنتهى الى طرسوس فالمسافة ايضًا نحو ١٠ مراحل غير انّ السمت المستقيم هو الطريق الأوّل، وامّا الطرف[n] الآخر فهو[o] حدّ فلسطين فياخذ

---

a) A. وعمعان, B. وععمان. *b*) C. et D. عند البلقاء وعمان. Deinde C. et D. الذى. *c*) I. e. حوض الثعلب (Jacut, II, p. ٣٦١ ubi vs. 21 l. عَمّان pro عُمان). *d*) B. et fortasse A. حصينة; cf. Abulfeda, p. ٢۴٧. *e*) E. add. صاحب اخبار. *f*) A. et B. بدمشق. *g*) C. et Abulfeda, p. ٢٣٦ فاحد. *h*) A. et B. قورص وهى omisso فى. E. تا قوس تا حد قنسرين. *i*) A. العاصى, nomen fluvii Orontis, qui eodem hoc nomine in mappa notatus est. Deinde C. على. *k*) E. بانياس. *l*) B. et D. الى. *m*) C. et D. الاسكندرية. *n*) C. et D. الطريق ut in A. et B. altera manus (E. راهى ديگر); cf. Abulfeda, p. ٢٣٦. *o*) C. add. من.

من البحر من حدّ يافا حتّى ينتهى الى الرملة ثمّ الى القدس[a] ثمّ الى اريحا[b] ثمّ الى زغر ثمّ الى جبال الشراة *ثمّ الى الشراة[c] الى ان ينتهى الى معان ومقدار هذا ٩ مراحل، فامّا ما بين هذين الطرفين[d] من الشام فهو مختصّر ولا يكاد يزيد عرض موضع من الاردنّ ودمشق وحمص على اكثر من ٣ ايّام لانّ من دمشق الى اطرابلس[e] على بحر الروم يومين غربًا والى اقصى الغوطة حتّى يتّصل بالبادية شرقًا يومًا[f] ومن حمص الى انطرطوس[g] على بحر الروم يومين غربًا ومن حمص الى سلمية على البادية شرقًا[h] يومًا ومن طبريّة الى صور على البحر غربًا يومًا ومنها الى ان تجاوز فيق[i] على حدّ ديار بنى فَزارة شرقًا يومًا فهذه مسافتنا[k] طول الشام وعرضه۞ وامّا المسافة فى اضعافه *فانّا نبدأ[l] بفلسطين وهى اوّل اجناد الشام ممّا يلى المغرب وقصبتها الرملة فمن الرملة الى يافا نصف مرحلة ومن فلسطين[m] الى عسقلان مرحلة والى غزّة مرحلة ومن الرملة الى بيت المقدّس يوم ومن بيت المقدّس الى مسجد ابراهيم يوم ومن بيت المقدّس الى ريحا مرحلة ومن بيت المقدّس الى البلقآء يومان ومن الرملة الى قَيْساريّة يوم ومن الرملة الى نابلس يوم ومن ريحا[n] الى زُغَر يومان[o] ومن زُغَر الى جبال الشراة يوم ومن جبال الشراة *الى آخر الشراة يوم[p] ۞ وامّا الاردنّ فانّ قصبتها طَبَريّة فمنها الى صور يوم[q] ومنها الى عقبة فيق يوم[r] ومنها الى بيسان[s] يومان خفيفان ومنها الى عكّا

---

a) C., D., E. et Abulf. l. l. ببيت المقدس. *b*) Sic h. l. omnes Codd. *c*) D. et Abulf. om. Intelligitur مدينة الشراة i. e. اذرح. *d*) D. الطريقين, C. om. Deinde من الشام addidi ex D. et Abulf. *e*) C. et E. طرابلس. *f*) A. et B. يوم. *g*) C. طرسوس, D. ut semper انطرسوس. *h*) A. et B. om. Deinde A., B. et C. يوم et sic in seqq. *i*) C. et Abulf. اريحا واغيق. E. تا حدود فيق. *k*) C. et Abulf. مسافة. *l*) C. فانك دهدا. *m*) I. e. مدينة فلسطين s. الرملة ut habet D. *n*) C. ut semper اريحا. *o*) C. يوم, rectius ut videtur. *p*) C. الى اخر الشام ثلثة ايام, E. تا اخر ولايت ايشان سه روزه‌راه. *q*) C. et E. يومان. *r*) C. et E. نصف يوم. *s*) E. بانياس.

يوم، والاردنّ اصغر اجناد الشام واقصرها مسافة ✿ وامّا جند دمشق فانّ قصبتها دمشق ومنها الى بعلبكّ يومان والى اطرابلس يومان والى بيروت يومان والى صَيْدَا يومان والى اذرعات ۴ ايّام *والى اقصى الغوطة يوم[a] والى حوران والبثنيّة يومان ✿ وامّا جند قنّسرين فانّ[b] مدينتها قنّسرين غير انّ *دار الامارة والاسواق ومجمع الناس[c] بحلب فمن حلب الى بالس يومان ومن حلب الى قنّسرين يوم *ومن حلب الى الأَثَارِب يوم[d] ومن حلب الى قورس يوم ومن حلب الى منبج يومان ومن حلب الى الخناصرة يومان ✿ وامّا العواصم فانّ قصبتها انطاكية ومنها الى اللاذقيّة[e] ۳ مراحل ومنها الى بغراس يوم والى الاثارب[f] يومان والى حمص ۵[g] مراحل ومنها الى مرعش يومان والى الحدث ۳ ايّام ✿ وامّا الثغور فانّه لا قصبة لها وكلّ مدينة قائمة بنفسها ومنبج قريبة الى[h] الثغور ومن منبج الى الفرات مرحلة خفيفة ومن منبج الى قورس مرحلتان ومن منبج الى ملطية ۴ ايام ومن منبج *الى سميساط يومان ومن منبج الى الحَدَث يومان ومن سميساط الى شمشاط يومان ومن شمشاط الى حصن منصور يوم ومن حصن منصور[i] الى ملطية يومان *ومن حصن منصور الى زبطرة يوم ومن حصن منصور الى الحدث يوم[k] ومن الحدث الى مرعش يوم *ومن ملطية الى مرعش ۳ مراحل كبار[l] فهذه مسافات ثغور الجزيرة، وامّا الثغور الشاميّة فمن اسكندرونة[m] الى بيّاس مرحلة خفيفة[n] ومن بيّاس الى المصّيصة مرحلتان ومن المصّيصة الى عين زربة يوم

---

a) In A. et B. haec desunt. *b*) Quae sequuntur usque ad verba وقد ذكرت ininitio capitis sequentis, in A. et B. desiderantur. *c*) E. دار الملك وبازارها ومسجد اذينة وعمارت. Pro ومجمع C. habet ومجامع. *d*) Ex solo E. *e*) D. اذنة; cf. Abulfeda, p. ٢٥٧. *f*) C. الارباب. *g*) D. اربع. *h*) D. من. *i*) Ex his C. om. ومن منبج الى الحدث يومان, E. الى سميساط — ومن سميساط et الى شمشاط — ومن شمشاط, D. denique الى حصن — حصن منصور. *k*) Haec in D. desunt. *l*) Haec in C. desunt. E. يك روزه. *m*) C. et h. l. quoque E. الاسكندرية. *n*) C. om.

ومن المصّيصة الى اذنة يوم ومن اذنة الى طرسوس يوم *ومن طرسوس الى اولاس على بحر الروم يومان[a] ومن طرسوس الى الجوزات[b] يومان *ومن طرسوس الى بيّاس على بحر الروم فرسخان[c] ومن بيّاس الى الكنيسة والهارونيّة اقلّ من يوم ومن الهارونيّة الى مرعش من ثغور الجزيرة *اقلّ من[d] يوم فهذه جملة مسافات الثغور[e] ❁

بحر الروم

واما بحر الروم فانّه خليج من البحر المحيط *بين الاندلس وبين البصرة من بلاد طنجة[f] وبين طنجة وبين جزيرة جَبَل طارق من ارض الاندلس عرضه اثنا عشر ميلًا[g] ثمّ يتّسع ويعرض فيمتدّ على سواحل المغرب فيما[h] يلى شرقىّ هذا البحر حتّى ينتهى الى ارض مصر ويمتدّ على اراضى مصر حتّى ينتهى الى ارض الشام ممتدًّا عليها ثمّ يعطف بناحية الثغور فيدور على بلد الروم من[i] انطاكية وما قاربها، ثمّ يصير غربىّ البحر الى خليج القسطنطينيّة[k] ويعبره ثمّ يمتدّ على سواحل اثيناس[l] ثمّ على سواحل رومية ثمّ يمتدّ على قرب افرنجة فيصير البحر حينئذ جنوبيًّا ويكون على ساحله افرنجة الى ان يتّصل بطرطوشة[m] من بلاد الاندلس ويمتدّ على البلاد ألّتى وصفناها فى صفة الاندلس حتّى يحاذى البصرة[n] بجزيرة جَبَل[o] طارق ثمّ

---

a) Haec in D. desunt. Deinde hinc usque ad فرسخان in C. om. b) Vid. Add. ad *Merácid*, V, p. 276. E. habet حورات. c) E. om. d) D. et E. om. Deinde C. om. فهذه الخ. e) E. add. وچون اقليمهاى مغرب ومصر وشامرا شرح داديم اكنون صورت درياى روم بكاريم وشرح آن ياد كنيم ان شا الله تعالى. Cf. D. f) C. خارج من الاندلس من البصرة ومن بلاد طنجة, D. از اندلس ميان بصيره وميان ولايت طنجه. E. من ارض الاندلس وارض طنجة. Pro البصرة A. in mappa habet ut supra القصير. g) C. et E. فرسخا (ut Ous. p. 51). h) D. مما. i) C. et E. الى. Fortasse legendum est الى انطاكية. k) C. et D. h. l. sine articulo. l) C. ادماس, E. ايماس. m) E. بطرسوسة. n) C. et E. hic et deinde البحيرة. o) C. شبل.

يمتدّ على البحر المحيط الى *شنترين وهى[a] آخر بلاد[b] الاسلام على هذا البحر *من جانب بلد الروم[c]، فلو انّ رجلًا سار من البصرة على السواحل حتّى يعود الى ما يحاذيه من ارض الاندلس لا يحتاج الى ان يعبر نهرًا او خليجًا امكنه ❁

*وقد[d] ذكرتُ ما على هذا البحر من المدن والبقاع[e] من السوس الاقصى الى ان ينتهى الى ارض مصر والى آخر الشام من الثغور الى اولاس وما يحيط به من بلد الاندلس ما يغنى عن اعادته[f]، فاذا جُزْتَ أَوْلَاس دخلتَ جبالًا تنتهى الى بحر الروم[g] يقال لها قَلَمْيَة[h] وقلمية مدينة كانت للروم وبعض ابواب طَرَسُوس يسمّى باب قَلَمْيَة ينسب[i] اليها وقلمية ليست على البحر *ولا على شطّ هذا البحر[k]، واذا جُزْتَ هذا الموضع بنحو من مرحلة مكان يعرف باللّامِس قرية على شطّ البحر[l] فيه يكون الفدآء بين المسلمين والروم يكون[m] الروم فى البحر فى السفن والمسلمون فى البرّ فيتفادون[n] ❁ وأَنْطَالِيَة[o] حصن للروم على شطّ البحر منيع واسع الرستاق كثير الاهل ❁ ثمّ تنتهى الى شطّ الخليج وهو خليج مالح يعرف بخليج القُسْطَنْطينيّة وعليه سلسلة ممتدّة لا تعبر فيه سفن البحر[p] ولا غيره الّا باذن مثل المأصر[q] ويقع فى بحر الروم

---

a) C. سمردن وهو, E. دسيبرين كه. *b*) C. بلد. *c*) Ex solo D. *d*) B. قد. Post امكنه sequi deberet mappa maris Mediterranei. *e*) E. pro his وثغرهارا بهرى بشام اضافت كرديم وبهرى بجزيره وهرچه برين درياست از شهر وغير آن همه ياد كرديم. *f*) C. haec omnia inde a وقد om. Deinde C. واذا. *g*) Jacut, IV, p. ١٦٩, 19 addit وولاية. *h*) D. اقليمية ut quoque appellatur. *i*) A. et C. فنسب, D. et Jacut منسوب. *k*) C. وعلى البحر et deinde اذا, ut quoque habent A., B. et Jacut, IV, p. ٣٤٢ paen. *l*) Jacut add. من ناحية ثغر طرسوس. *m*) C. ويكون, Jacut يقدمون — فيكونون (in ed. male الغزاة bis pro الفداء). *n*) A., B. et C. فيفدون. Jac. ووقع الفداء. *o*) In B. sequitur لام quod superinscribendum erat. E. انطاكية. *p*) A. et B. om.; C. من البحر et sic E. (Ous. p. 52). *q*) D. كالمرصد; C. om. مثل. Deinde A. et B. يقع.

*من البحر المحيط من وراء الروم [a] ✿ وسواحل أثينَاس ورُومِيَة ذات قرى ومزارع ومدن كبار واثيناس ورومية مدينتان بهما مجمع النصارى [b] بقرب البحر، فامّا اثيناس فانّها دار حكمة اليونانيّين وبها تحفظ علومهم وحكمهم، وامّا رومية فانّها ركن من اركان مُلْك النصارى فانّ للنصارى كرسيًّا بانطاكية [c] وكرسيًّا بالاسكندريّة [d] وكرسيًّا برومية والكرسيُّ الّذى بالبيت [e] المقدّس محدث لم يكن فى ايّام الحواريّين انّما اتّخذوه بعد ذلك لتعظيم البيت [f] المقدّس ✿ ثمّ يتّصل [g] بالافرنجة على ساحل البحر الى ان يحاذى صقلّيّة [h] ويجاوزها حتّى يتّصل بطرطوشة [i] من ارض الاندلس ✿ وقد ذكرنا المسافة الّتى بارض المغرب ومصر والشام الى آخر الاسلام *والثغور فى كلّ مكان منه ما يغنى عن اعادته [k] ✿

وفى هذا البحر جزائر صغار وكبار وجبال، فامّا المعمور [l] بالناس فهى صقلّيّة وهى اكبرها واقْرِيطش وقُبْرُس وجبل القِلال [m]، فامّا صقلّيّة فانّها قريبة من الافرنجة *حتّى يُرَى منها ارض افرنجة وتثمر الزروع بها وهى جزيرة [n] طولها نحو سبع [o] مراحل وبصقلّيّة من الخصب والسعة والزروع والمواشى والرقيق اكثر ما *يقع منها ما [p] يفضّل على سائر ممالك الاسلام المتاخمة للبحر ✿ واقريطش دونها فى العرصة وفى العمارة *وسكّانهما جميعًا مسلمون اهل غزو وبين اظهرهم نبذ من النصارى [q] كما يكون ببلدان المسلمين ✿ وامّا قُبْرُس فانّ اهلها نصارى كلّهم ليس فيهم من المسلمين احد وهى تقارب فى الكبر

---

a) Haec in C. et E. desunt. b) A. للنصارى. c) A. بانطاليه. d) C. باسكندريه et deinde بالرومية. e) C. et D. ببيت. f) C. et D. بيت. g) Secundum E. inserendum est البحر. h) C. et E. سقلية semper. i) E. ut supra طرسوسه. k) C. pro his tantum بالثغور. l) A. et B. المعروف. m) Vid. Reinaud in Add. ad *Merâcid*, V, p. 27. n) C., D. et E. haec om.; pro وتثمر الزروع, B. habet ومثمر الزرع. o) C. et E. تسع. p) C. haec om., B. om alterum ما. q) C. وساكنها مسلمون واهل غزاة وبين ظهرانيهم نصارى.

والعمارة اقريطش[a] خصبة جدًّا افتتحها معاوية صلحًا فهادن اهلها فهى فى هدنة المسلمين وهم نصارى من الروم، وعرض هذا البحر من سواحل الشام اذا استوى الريح يومان الى قبرس ومن قبرس الى الجانب الآخر من هذا البحر نحو ذلك، ويقع بقبرس الميعة الّتى تحمل الى بلدان الاسلام من بلد الروم والمصطكى يكون بقبرس ۞ وامّا جبل القلال فانّه كان جبلًا[b] فيه مياه خرّارة[c] فوقع اليه قوم من المسلمين فعمروه وصاروا[d] فى وجوه الافرنجة[e] لا يقدر عليهم لامتناع مواضعهم ومقداره فى الطول يومان ۞ وليس فى البحار احسن[f] حاشية من هذا البحر فانّ العمارات فى الجانبين ممتدّة غير منقطعة وسائر البحار يعرض فى شطوطها المفاوز والمقاطع وتتردّد فيه[g] سفن المسلمين والروم[h] يعبر كلّ فريق الى جانب[i] الآخر *سوآء فيغنمون[k] وربّما اجتمع فيه[l] الجيوش من المسلمين والروم فى السفن فيجتمع لكلّ فريق مائة سفينة حربيّة واكثر من ذلك *فيكون حربهم[m] فى المآء، وهذه صفة هذا البحر وما يكون فيه ۞

ارض الجزيرة[n]

وامّا الجزيرة فانّها[o] ما بين دجلة والفرات وتشتمل على ديار ربيعة ومضر، ومخرج الفُرات من داخل بلد[p] الروم من ملطية على يومين *ويجرى بينها[q] وبين سُمَيْسَاط[r] ويمرّ[s] على سميساط وجسر منبج وبالس الى الرقّة[t] وقرقيسيا

---

a) C. male addit واقريطش et sic E. qui habet وابن اقريطس جزيرة سخت قديم على استوارست legens حصينة pro خصبة. b) C. et E. add. خراب, D. مر الزمان. c) C. وارض, D. واراضى, E. وزمين بسيار. d) A. فصار. e) A. et D. الافرنجيه. Deinde A. ولا. f) C., D. et E. اعمر. g) C. فيها. h) A. et B. om.; C. السفن للروم والمسلمين. E. pro والروم habet وكافران. i) A. et B. الجانب. k) In A. et B. haec desunt, lacuna indicata. l) A., B. et C. فيها. m) A. يتكون حروبهم. n) Initium hujus capitis usque ad واما مسافاتها in A. et B. desideratur. o) C. om. واما et فانها. p) C. om. q) C. وبينها. r) D. شمشاط. s) C. فيخرج. t) C. الرقتين, E. رقتين. Secundum Jacut الرقتان dicitur pro الرقة والرافقة.

والرحبة وهيت والانبار وقد انقطع حدّ الفُرات ممّا يلى الجزيرة ثمّ يعدل[a] حدّ الجزيرة فى سمت الشمال الى تكريت وهى على دجلة حتّى ينتهى عليها الى السنّ[b] ممّا يلى الجزيرة والحديثة والموصل وجزيرة ابن عمر ثمّ يتجاوز[c] آمد فينقطع[d] حدّ دجلة على بعد من حدّ ارمينية ثمّ يمتدّ مغرباً[e] الى سميساط ثمّ ينتهى الى مخرج ماء الفرات فى حدّ الاسلام[f] من حيث ابتدانا، ومخرج دجلة فوق آمد من حدّ بلد الارمن[g] وعلى شرقىّ دجلة وغربىّ الفرات مدن وقرى تنسب الى الجزيرة وهى[h] خارجة عنها لقربها منها ؛[i] ✿

وامّا مسافاتها فمن مخرج ماء الفرات فى[k] حدّ ملطية الى سميساط يومان ومن سميساط الى جسر منبج ٤ ايّام *ومن جسر منبج الى بالس ٤ ايّام ومن بالس الى[l] الرقّة يومان ومن الرقّة الى الانبار ٢٠ مرحلة ومن الانبار الى تكريت يومان ومن تكريت الى الموصل ٦ ايّام ومن الموصل الى آمد ٤[m] ايّام ومن آمد الى سميساط[n] ٣ ايّام ومن سميساط الى ملطية ٣ ايّام ✿ ومن الموصل الى بَلَد مرحلة ومن بَلَد الى نصيبين ٣ مراحل[o] ومن نصيبين الى راس عين[p] ٣ مراحل ومن راس عين الى الرقّة ٤ ايّام ومن راس عين الى حرّان ٣ ايّام ومن حرّان الى جسر منبج يومان ومن حرّان الى الرُّها يوم ومن الرُّها الى سميساط يوم ومن حرّان الى الرقّة ٣ ايّام ✿

وامّا صفة مدنها وبقاعها فانّ انزه بلد بالجزيرة واكثرها خضرة[q] بلد نَصيبين

---

a) D. يعود, Abulfeda, p. ٢٧٣ يعطف. C. et E. om. حد الجزيرة. *b*) C. واز انجا تا جزيره بر كنارهٔ شط نوان رفتن. Inde a دجلة E. habet بالس. انكاه حديثه الخ. *c*) C. دتحاوز; D. add. الى. *d*) C. ثمقطع. *e*) C. om. Deinde habet سمساط. *f*) Abulf. add. عند ملطية. *g*) Sic videtur scribendum. In C. الاندلس expunctum legitur, E. در حد مسلمانى. *h*) C. وان كانت. *i*) Hic sequi deberet mappa Mesopotamiae. *k*) C. et E. من. *l*) Pro his C. et E. tantum والى. *m*) D. اربعة عشر. *n*) Solus A. شمشاط quod fortasse praestat. Cf. quoque Abulf. p. ٢٧٣. *o*) D. خمس مراحل; Abulf. ومن الموصل الى نصيبين اربع مراحل. *p*) C., D. et E. العين. *q*) C. خيرا. Deinde A. om. بلد.

وهى مدينة كبيرة فى مستوى من الارض *ومخرج مآئها[a] من شعب جبل يعرف بيالوسا[b] وهو انزه مكان بها حتّى ينبسط فى بساتينها ومزارعها ولهم مع ذلك فيما بعد عن المدينة مباخس كثيرة وبها دير[c] عظيمة وحواليها ديارات وصوامع للنصارى كثيرة وبها عقارب كبيرة[d] قاتلة موصوفة، وبالقرب من نصيبين جبل مارِدين من الارض الى ذروته نحو من فرسخين وبه[e] قلعة منيعة لا يستطاع فتحها عنوة وبه[f] حيّات موصوفة تفوق الحيّات بسرعة القتل وهو جبل به جواهر الزجاج ۞ وامّا المَوْصِل[g] فهى مدينة على غربىّ دجلة صحيحة التربة والهوآء ليس[h] لهم سوى مآء دجلة للشفة وليس لهم *من دجلة[i] زرع ولا شجر الّا الشىء اليسير فى عدوة دجلة من شرقيّها[k] وزروعهم مباخس وفواكههم تحمل من سائر النواحى وهى مدينة عامّة ابنيتها بالجصّ والحجارة كثيرة غناآء[l] ۞ وبَلَد مدينة صغيرة على غربىّ دجلة وبها[m] مآء جارٍ سوى دجلة وشجر وزروع ومباخس كثيرة ۞ وامّا سِنْجار فانّها مدينة فى وسط بريّة[n] ديار ربيعة بقرب جبل[o] ينسب الى سنجار وبها نخيل وليس بالجزيرة بلد به نخل سوى سنجار الّا ان يكون على الفرات وبهيت[p] والانبار *وتلّ أَعْفَر[q] ۞ وامّا دارا فهى مدينة صغيرة نزهة تشتمل على مياه جارية واشجار وزروع ولها مباخس وهى فى سفح جبل ۞ وكَفَرْتُوثَا فى مستوى[r] من الارض وهى مدينة اكبر

---

a) C. ولـه نهر مخرجه. *b*) C. بالوصا. E. انرا بانوصا كوينند (Ous. p. 56). *c*) A. et B. ذكر; sed idem infra in loco de الرها habent, ubi sine dubio دير legendum est. Observandum est h. l. vocem genere feminino adhiberi. E. habet وانجا دكانى عظيم هست وكردان جايكاههاى ترسان بود (Ous. p. 56, 2 ex كردان nom. prop. finxit). C. tantum habet ومواضع للنصارى كثيرة. *d*) A. et D. om., B. كثيره. *e*) A. et B. وبها. *f*) A., B., C. et Abulf. p. ٢٧٩ وبها. *g*) A. موصل. *h*) C. وليس. Deinde A. لها. *i*) C. بالدجلة. *k*) C. شرقى الموصل. *l*) C. الغناء. E. وشهرى بزركست. *m*) D. وليس بها. *n*) A. et B. om. *o*) C. جبال; Abulf. p. ٢٨٣ بالقرب من الجبال. *p*) B. وهيت. *q*) C., D. et E. om. *r*) C. et Abulf. p. ٢٨٥ مستوٍ, ut quoque supra et infra.

من دارا ذات *نهر وشجر[a] وزروع ولها مباخس كثيرة ۞ وراس عين[b] مدينة على مستوى وارضها الغالب عليها القطن ويخرج منها زيادة على ثلاث مائة عين كلّها صافية تحكى ما تحتها على قامات فتجتمع مياهها[c] حتّى يصير منها[d] نهر الخابور الّذى يقع الى قَرْقِيسِيَا ومسافة هذا النهر[e] نحو عشرين فرسخا قرى ومزارع، وراس عين مدينة اكبر من كفرتوثا ولهم زروع واشجار مستغلة[f] عن البنيان على سنن[g] هذه المياه وهى خصبة كثيرة المباخس[h] ۞ واما

---

a) B. اشجار وانهار, C. ثمار وانهار, D. ثمر وشجر, Abulf. p. ٢٨٥ ثمار وشجر. E. ut A. b) C. et E. العين. c) C. كل ذلك. d) A., B. et C. منه. e) A. om. النهر. Pro ومسافة C. habet وعلى ; D. وكان عليه. Cf. *Merâcid* in v. راس عين. f) A. et B. habent مستقل. Deinde B. habet البيان. In C., D. et E. haec desunt. g) A. ستن. h) Deinde in C. haec adduntur: فمن مشاهير تلك القرى المَجْدَل وهى تحت راس العين بمرحلة كلّها ضياع متّصلة على جانبى الخابور، ويتلوها عَرَابَان وبينها وبين المجدل مرحلة وهى مدينة صغيرة كثيرة العلل والوباء وهى ايضًا على شاطئ الخابور، وبين المجدل وعرابان ايضًا عمارات متّصلة وضياع على جانبى الخابور فمن مشاهيرها طابان والمَطَرِيَّة والسُّخَيْمِيَّة وتُنَيْنِير ويرتفع من هذه القرى قطن كثير، وعرابان فرضة لاهل خِلَاط والموصل يجلب منها القطن الى سائر تلك النواحى، ويتلو عرابان على هذا النهر مَاكِسِين وهى مدينة صغيرة نحو من عرابان الّا انّها خصبة كثيرة الخيرات ولها جسر على الخابور ومنها ومن عرابان والمجدل يرتفع القطن الذى يحمل الى خلاط والموصل، وبينها وبين عرابان مسيرة يوم لا عمارة فيه الّا قرية يقال لها السُّكَيْر (i. e. سكير العباس) فى نحو من نصف الطريق، وبين ماكسين وبين سنجار ثلاثة ايّام فى مفازة قفرآء ليس فيها عمارة الّا الجبال (Cod. الحبال ; cf. p. ٧٣ ann. *o*.) من ضياع سنجار وهذه البرّيّة هى برّيّة سنجار ولها اتّصال بالفرات وبرّيّة البادية. Haec in A., B. et Vers. Pers. non leguntur, neque Ibn Haucal, Jacut et Abulfeda in suis Codd. habuerunt.

آمد فهى على دجلة من شرقيّها وسورها[a] فى غاية الحصانة وهى كثيرة الشجر والزروع ✿ وامّا جزيرة ابن عمر فهى مدينة صغيرة على غربىّ دجلة لها اشجار ومياه[b] ✿ وشمْشاط[c] هى ثغر الجزيرة لانّها *فى غربىّ دجلة وشرقىّ الفرات[d] ، وامّا مَلَطْيَة وما ذكرناه من[e] الثغور فى ثغر[f] الشام فانّما *نسبناها الى[g] الجزيرة لانّ اهلها *يرابطون بها[h] لقربها منهم *والّا فثغر الجزيرة على الحقيقة[i] شمشاط ✿ والحَديثة[k] *على شطّ دجلة من شرقيّه[l] وهى مدينة نزهة جدًّا[m] ذات بساتين واشجار وزروع ولها مباخس ✿ والسِّنّ على شرقىّ دجلة وهى مدينة صغيرة بقربها جبل بارمّا *على مرحلة[m] وجبل بارمّا هو جبل تشقّه[n] دجلة فتجرى *فى وسطه[o] وفى الماء منه عيون *القير والنفط[p] ، وجبل بارمّا يمتدّ الى[q] وسط الجزيرة ممّا[r] يلى المغرب ويقال انّه ممّا[r] يلى المشرق يمتدّ الى[s] حدّ كرمان وهو جبل ماسَبَذان ✿ وامّا ديار مُضَر فانّ الرقّة اكبر ما فيها من المدن[t] والرقّة والرافقة مدينتان متلاصقتان وفى كلّ واحدة منهما مسجد جامع وهما على شرقىّ الفرات كثيرتا الاشجار والمياه فى مستوى[u] ، وفى غربىّ

---

a) C. مدينة عليها سور اسود, Abulfeda, p. ٢٨٧ وهى مدينة عليها سور على غاية, D. وعليها سور اسود. Cf. Jacut, I, p. ٦٦, 21. b) C. add. وهى حصينة خصبة كثيرة الخيرات جدّا. c) A., B., C. et E. سميساط, sed vid. Abulf. p. ٢٧٧; Deinde A. et B. هو. d) C. et E. شرقى دجلة والفرات. e) A. et B. فى. f) B. ثغور. g) C. تنسب الى ثغور. h) C. ينتابونها, E. بنوبت انجا باشند. i) C. والثغر الذى هو من صلب الجزيرة. Deinde A., B. et E. سميساط, C. سمشاط. k) A. in marg. add. يعنى حديثة الموصل, quod B. in textum recepit. l) Haec in A. et B. desiderantur. C. om. من. Deinde B. فهى. m) A. et B. om. n) B. تشتقه; vid. quoque Jacut, I, p. ٤٦٤, 18. o) C. et Jacut بمحاذيته, E. بر كنارهٔ كوه. p) C. للنفط والقير, Jacut للقار والنفط. E. ودرين كوه چشمهاى قير ونفط. q) C. et Jac. على; E. بر. r) C. فيما. s) A., B. et C. على. t) Abulfeda, p. ٢٧٧ البرقة اكبر مدن ديار بكر. u) C. وهما فى مستو من الارض حصينتان.

الفرات بين الرقّة وبالس ارض صِفّين *وبها قبر عمّار بن ياسر قتيل الفئة الباغية رضه وبالرقّة موضع كان بيت مال علىّ عم ايّام صفّين[a] ✿ وحرّان تليها فى الكبر وهى مدينة الصابئين[b] وبها سدنتهم السبعة عشر وبها تلّ[c] عليه مصلًّى[d] يعظّمه الصابئون[b] وينسب الى ابراهيم خليل الله عم وهى من بين تلك المدن قليلة الماء والشجر ولها مباخس[e] ✿ والرُّهَا مدينة وسطة والغالب على اهلها النصارى وفيها زيادة على ثلاث مائة دير[f] وصوامع كثيرة ورهابين[g] ولهم بها كنيسة[h] ليس فى بلاد الاسلام كنيسة اعظم منها ولها مياه وبساتين *كثيرة وزروع وهى اصغر من كفرتوثا ' وجسر مَنْبِج وسُمَيْسَاط هما مدينتان نزهتان لهما زروع ومياه وبساتين[i] ومباخس وهما غربىّ الفرات[k] ✿ وامّا

---

a) C. haec om. b) C. الصابئة. c) C. add. عظيم, E. بلند. d) C. add. للصابئين; cf. Abulf. p. ٢٧٧. e) C. et E. add. كثيرة. f) Hic A. et B. denuo habent ذكر. Ceteri omnes (vid. quoque Abulf. p. ٢٧٧) دير. g) B. male ورقّابين. h) E. كليسبائى. i) In C. haec omnia desunt. Pro وهما habet على. k) Deinde in C. haec adduntur: وحصن كَيْفَا على شرقىّ دجلة وهى مدينة بغير سور ولها قلعة حصينة جدًّا ' نَصِيبين مدينة عظيمة واسعة كثيرة الخصب والخيرات الّا انّ بها عقارب شديدة الضرر قتّالة كبيرة جدًّا وهى ايضًا وبيّة كثيرة العلل والوباء ومنها يجلب ماء الورد الى الآفاق ' دُنَيْسَر مدينة كثيرة البيوت والعمارة الّا انّها بغير سور وهى من اعمال ماردين وبينها وبين ماردين اربعة فراسخ وبين دنيسر وبين نصيبين مرحلتان وبدنيسر بازار عظيم فى الخميس والجمعة والسبت والاحد يجتمع فيه الناس من البلاد البعيدة ليحضروا ويشترى منه كلّ ما يحتاج اليه ' مارِدِين مدينة كبيرة فى قلّة جبل شاهق (Cod. ساحق) له عقبة طويلة نحو فرسخ ' مَيّافارِقِين (Cod. ممرقمن) مدينة وسطة ذات سور دائر خصبة جدًّا كثيرة المياه الّا انّها وخمة كثيرة العلل والوباء والمياه تجرى فى منازلها واسواقها وهى احسن ديار بكر وبينها وبين ماردين ثلاثة ايّام ' أَرْزَن على نهر سَرْبَط بغير سور وغربىّ النهر لها قلعة حصينة كبيرة ' إِسْعِرْد (quoque dicitur سِعِرْت) مدينة صغيرة بغير سور خصبة ' حِينِى (quoque dicitur حانِى) مدينة وسطة لطيفة خصبة.

قَرْقِيسِيَا فانّها على الخابور ولها بساتين واشجار كثيرة وزروع نزهة ۞ ورَحْبَة مالك بن طَوْق اكبر منها وهى كثيرة الشجر والمياه *a* على غربىّ الفرات ۞ *وهيت مدينة وسطة على غربىّ الفرات وعليها حصن وهى عامرة آهلة وهى بحذآء تكريت *b* وبها قبر عبد الله بن المبارك ۞ والأَنْبَارُ مدينة وسطة وبها آثار *ابنية لابى العبّاس السفّاح أوّل خلفآء بنى العبّاس *c* *وكانت داره ٱلّتى يسكنها *d* وهى مدينة عامرة آهلة *e* ذات نخل وزرع وشجر وهى شرقىّ الفرات ۞ وبالجزيرة مفاوز يسكنها قبائل من ربيعة ومضر اهل خيل وغنم والابل عندهم اقلّ منها بالبادية واكثرهم *f* متّصلون بالقرى وباهلها *g* فهم بادية حاضرة ۞ والزابان نهران كبيران اذا جُمِعَا يكونان نحو النصف من دجلة *h* واكبرهما ممّا يلى الحديثة ومخرجهما من قرب جبال اذربيجان ۞ وتِكْرِيت بلد على غربىّ دجلة اكثرٍ *i* اهلها نصارى، *واسفل من تكريت فوهة *k* نهر دُجَيْل

---

a) In C. haec sequuntur: والخصب والغلل وهى وقرقيسيا على شرقىّ الفرات وعانة والحديثة جزيرتان حصينتان فى وسط الفرات ذات خصب ولهما سور مانع ولا سيما سور الحديثة واهلهما اعرف الناس بالسباحة والملاحة وهيت الخ. D. et E. الرحبة male collocant فى شرقى الفرات. *b*) Haec inde a ومن حدود عانة الى وهيت in A. et B. desunt. Deinde in C. haec sequuntur بعد تكريت نحو خمسين فرسخًا جميعه نخيل متّصل على حافتى دجلة اتّصالاً لا يتخلّله انفصال، وتكريت على غربى دجلة وهى مدينة وسطة ولها قلعة ليس بالعراق احصن منها. E. habet h. l. quoque تكريت بر جانب غربى دجلة نهاده است. *c*) C. ابنية القائم بالله امير المومنين. E. ابو العبّاس القايم بالله انجا مقام داشتى واثار سراى بديدست. *d*) Haec in A. et B. desunt. *e*) A. h. l. habet شرقى الفرات. *f*) A. et B. وهم. *g*) A. et B. om. et habent deinde فهى. E. habet بهرى در باديه نشينند وبهرى بنواحى جزيره در ديهها. *h*) C. add. بغداد. Deinde C. et Abulf. p. ٢٧٥ واكبر الزابين. *i*) C. et D. واكثر et post نصارى C. add.: وهى عالية على دجلة جدًّا. *k*) C. et D. ومن تكريت يشقّ, Abulf. p. ٢٨٩ وقرب تكريت يشتق نهر الدجيل.

الّذى ياخذ من دجلة فيتعمّر عليه قطعة كبيرة من سواد بغداد حتّى يقاربها [a] ✿ * وعَانَة مدينة صغيرة فى وسط الفرات يطوف بها خليج من الفرات [b] ✿ وحصن مَسْلَمَة بلغنى [c] انّه كان لمسلمة بن عبد الملك وبه طائفة من بنى اميّة ومآؤه من السمآء وبها مباخس ✿ وتلّ بنى سَيّار [d] مدينة صغيرة تسكنها عرب من غَنِىّ واكثرها كانت للعبّاس بن عمرو [e] الغَنَوىّ ✿ * وباجَرْوان منزل نزه خصب واسع [f] ✿ والدالِيَة مدينة * من غربىّ الفرات صغيرة [g] بها أُخِذ صاحب الخال [h] الّذى كان خرج بالشام ✿ والجُودىّ جبل بقرب * جزيرة ابن عمر [i] يقال انّ سفينة نوح استقرّت عليه وتحته [k] قرية تعرف بثمانين يقال انّ جميع من كان مع نوح فى السفينة ثمانون رجلًا بنوا تلك القرية فسمّيت بهم ولم يُعقب احد منهم ✿ وسَرُوج [l] مدينة خصبة [m] كثيرة الاعناب والفواكه لها رستاق من حرّان على نحو يوم ✿

العراق

وامّا العراق * فحدّها فى الطول [n] من حدّ [o] تكريت الى عبّادان على

---

a) C. مارًّا الى سواد سرّ من راى .D ; فيعمر به سواد سامرّه الى قرب بغداد. الذى يسقى سواد سامرّا الى قرب .Abulf. l. l ; فيعمره الى قريب بغداد ونماحيت سامرهرا آب دهد .E ; بغداد. b) C. om.; cf. Abulf. p. ٢٨٧. c) C. قيل. d) Secutus sum ed. Jacut in v. (I, p. ٨٩٤). In ann. ad *Merácid*, IV, p. 489 سيَّار pronunciatur. Cf. de hoc loco Ibn Khallicán, n. 745, p. ١٣٨. e) A. عَمْر, B. et C. عمر. f) Haec in A. et B. desunt. g) C. et D. صغيرة بشط الفرات من (عن) غربيه et sic habet Abulf. p. ٢٨٣. Cf. Jacut II, p. ٥٣٨. h) Abulf. add. المعروف بابى شامة القرمطى. In C. haec desiderantur. Cf. *Mémoire sur les Carmathes du Bahrain*, p. 15. i) D. الجزيرة; C. et E. نصيبين (cf. Abulf p. ٢٨٣ sub نصيبين); cf. Jacut I, p. ٩٣٤, 19, II. p. ١٤٤, 10. k) C. والى جانبه. l) A. وسَرّوج. C. et D. h. l. add. (ولها) له رستاق. m) E. بزرك. n) A. et B. فطولها. Apud C. et D. العراق masculini generis est. o) A. et B. om.

بحر فارس وفى العرض *عند بغداد من قادسيّة الكوفة[a] الى حلوان وعرضها بواسط من واسط الى قرب[b] الطّيب وعرضها بالبصرة من البصرة الى حدود جُبّى، والّذى يطوف[c] بحدودها من تكريت ممّا[d] يلى المشرق حتّى يجوز بحدود شهرزور ثمّ يطوف على حدود حلوان وحدود السيروان والصيمرة وحدود الطيب وحدود السوس حتّى ينتهى الى حدود جُبّى ثمّ الى البحر فيكون فى[e] هذا الحدّ من تكريت الى البحر تقويس ويرجع الى[f] حدّ المغرب من ورآء[g] البصرة فى البادية على سواد البصرة وبطائحها الى واسط ثمّ على سواد الكوفة *وبطائحها الى الكوفة ثمّ على ظهر الفرات[h] الى الانبار ثمّ *من الانبار الى[i] تكريت بين دجلة والفرات وفى هذا الحدّ *من البحر الى تكريت[k] تقويس ايضًا *فهذا المحيط بحدود العراق[k] ✿

وامّا المسافات فانّه من[l] تكريت الى البحر ممّا يلى المشرق مقوّس نحو شهر ومن البحر راجعًا فى حدّ المغرب[m] الى[l] تكريت مقوّس[n] نحو شهر، ومن بغداد الى سامرّا[o] ٣ مراحل ومن سامرّا الى تكريت مرحلة[p] ومن بغداد الى الكوفة ٤ مراحل ومن الكوفة الى القادسيّة مرحلة ومن بغداد الى واسط ٨ مراحل ومن بغداد الى حلوان ٦ مراحل ومن بغداد الى حدود الصيمرة والسيروان[q] نحو ذلك، ومن واسط الى البصرة ٨ مراحل ومن الكوفة الى واسط على طريق البطائح[r] ٦ مراحل ومن البصرة الى البحر مرحلتان[s]، وعرض

---

a) C. من القادسية على الكوفة وبغداد, D. ببغداد والكوفة من القادسية, E. از بغداد تا كوفه تا قادسيّه. b) C. حد; E. وقرقوب (sic) تا قيب. c) C. et D. يطيف. Deinde A. et B. ut C. et D. بحدوده. d) C. et D. فيما. e) C. من. f) C. et D. على. g) E. add. سواد. h) A. et B. وبطائحها ثم. i) C. يمر من الانبار على حد. k) A. et B. haec om. Post العراق sequi deberet mappa Iraci. l) C. et D. addunt حد. m) A. et B. المشرق. n) B. مقوّسا. o) C. et E. سامرّه, D. سر من راى. p) C., D. et E. مرحلتان. q) B. والشروان. r) C. add. نحو. s) C. h.l. habet طريق البطائح quod supra omisit.

العراق ببغداد من حلوان الى القادسيّة ١١ مرحلة وعرضه عند سامرّا [a] الى حدّ شهرزور من جهة اذربيجان نحو ٥ [b] مراحل والعامر منه اقلّ من مرحلة وعرضه بواسط نحو ۴ مراحل وعرضه بالبصرة [c] من البصرة الى حدود جبّى * نحو مرحلتين خفيفتين [d] ۞

وامّا مدنها فالبصرة مدينة عظيمة لم تكن فى ايّام العجم وانّما اختطّها المسلمون ايّام عمر بن الخطّاب [e] ومصّرها عُتْبَة بن غَزْوان وهى خطط وقبائل كلّها ويحيط بغربيّها البادية مقوّسا * وليس فيها مياه الّا انهار [f] وذكر بعض اهل الاخبار انّ انهار البصرة عُدَّتْ ايّامَ بِلَال بن ابى بُرْدَةَ فزادت على مائة الف نهر وعشرين الف نهر تجرى فيها الزواريق وقد كنتُ انكر ما ذكر من عدد هذه الانهار فى ايّام بلال بن ابى بردة حتّى رايتُ كثيرًا من تلك البقاع فربّما رايتُ فى مقدار رمية سهم عددًا من الانهار صغارًا [g] تجرى فى كلّها زواريق صغار، ولكلّ نهر اسم ينسب به الى صاحبه الّذى احتفره او الى الناحية الّتى يصبّ فيها واشباه ذلك من الاسامى [h] فجوّزتُ ان يكون ذلك فى طول هذه المسافة وعرضها [i] ، واكثر ابنيتها بالآجرّ وهى من بين سائر العراق مدينة عُشْرِيَّة ولها [k] نخيل متّصلة من عَبْدَسِى الى عبّادان نيف وخمسين فرسخًا متّصلًا [l] لا يكون الانسان * منه فى مكان الّا بحيث [m] نهر ونخيل او يكون بحيث يراهما وهى فى مستوى [n] لا جبال فيه ولا بحيث يقع البصر على جبال، وبها قبر طلحة بن عبيد الله من الصحابة * فى المدينة [o] وخارج المِرْبَد * فى البادية [o] قبر أَنَس بن مالك [p] والحسن

---

a) E. add. بر كنار دجله ; D. من دجلة. b) E. پانزده. c) A. om. d) C., D. et E. مرحلة. e) C., D. et E. add. رضه. f) C. pro his: وبشرقيها مياه ; E. بى آب ; D. وبشرقيها مياه الانهار مفترشة ; الانهار مقوسا. g) C. الصغار. h) A. الامر. Deinde C. فتحرزت. i) A., B. et C. او غيرها. k) B. وبها. l) A., B. et C. متصل. m) C. habet فيه بمكان الا فى. n) C. مستواة من الارض et deinde فيها. o) A. et B. om. Pro seq. المربد C. habet المدينة. p) E. add. وزيارتگاههاى فراوانست انجا چون كور.

البصرى[a] وابن سيرين والمشاهير من علماء البصرة وزهّادها[b] ، ولها نهر يعرف بنهر الأُبُلّة طوله اربعة فراسخ ما بين البصرة والأُبُلّة وعلى حافتى هذا النهر قصور وبساتين متّصلة كانّها بستان واحد قد مُدّت على خيط واحد ويتشعّب هذه الانهار الى[c] انهار كثيرة فمنها ما يقارب هذا النهر فى الكبر كانّ نخيلها[d] غُرست على خيط واحد وهذه الانهار كلّها مخترقة بعضها الى بعض وكذلك عامّة انهار البصرة[e] حتّى اذا جاءهم مدّ البحر تراجع الماء فى كلّ نهر حتّى يدخل نخيلهم وحيطانهم وجميع انهارهم من غير تكلّف فاذا جزر الماء انحطّ حتّى تخلو منه البساتين والنخيل ويبقى فى الانهار الّا انّ الغالب على ماءهم الملوحة وانّما يستقون[f] اذا جزر الماء الى حدّ نهر مَعْقِل ثمّ يعذب فلا يضرّه[g] ماء البحر ۞ والأُبُلّة على هذا البحر[h] وعلى ركن الأُبُلّة فى نهر الأُبُلّة خور[i] عظيم الخطر وربّما سلمت السفن من سائر الاماكن فى البحر وغرقت فى هذا الخور[k] يعرف بخور الأُبُلّة ، والأُبُلّة مدينة صغيرة خصبة عامرة[l] حدّها نهر الأُبُلّة الى البصرة وحدّ لها[m] دجلة الّتى يتشعّب منها هذا النهر عاطفًا عليها وينتهى * عمودها الى البحر بعبّادان ، وللبصرة مدن[n] فامّا عبّادان والأُبُلّة والمَفْتَح[o] والمَذار فعلى شطّ دجلة وهى مدن صغار متقاربة فى الكبر عامرة[p] الّا الأُبُلّة فانّها اكبرها ، وفى حدود البصرة بين[q] اضعاف قراها آجام كثيرة وبطائح اكثرها يسار فيها بالمدارى[r] قريبة

---

a) A, B. et E. om. *b*) A. et B. om. *c*) C. الى collocat ante هذه. E. pro his tantum habet وآبهاى بسيار باز مى پيوندد. *d*) A. نخلها. *e*) A. add. ثم. *f*) B. et C. يسقون. *g*) C. يغيره. *h*) C., D. et E. النهر. *i*) C. hic et deinde هور ut E. habet. *k*) C. add. وهو. *l*) C. جدا ; E. پر نعمت وميوها فراوان وآبادنست. *m*) A. et B. وحدها ; cf. Abulf. p. ٣٠٩. Deinde C. الى الدجلة, E. سوى دجله. *n*) C. عمنوده الى عبادان والبصرة. Post بعبادان E. insert وزمين بصره خاك سپيدست. *o*) C. et D. والمعمح, A. et B. والمصمح, E. ومصح (?المصيّح). Ous. p. 64 منتح. *p*) A. et B. om. *q*) C. et D. فى. *r*) C. بالمرادى, E. بمردى.

القعر كانّها كانت على قديم الايّام ارضًا مكشوفة ويشبه ان يكون لمّا بُنيت البصرة وشُقّت الانهار *واتّصل بعضها ببعض فى القرى والمجارى [a] فتراجعت المياه وغلبت على *ما يسفل من ارضها [b] فصارت بحارًا وبطائح ✿ وامّا واسط فانّها نصفان على شاطئ دجلة *من غربيّها وشرقيّها [c] وبين الجانبين جسر من [d] سفن وفى كلّ جانب [e] مسجد جامع وهى محدثة فى الاسلام احدثها الحجّاج بن يوسف [f] وبها خضرآء الحجّاج [g] وهى مدينة يحيط بحدّها الغربىّ البادية بعد مزارع يسيرة وهى خصبة كثيرة الشجر والنخيل والزروع وهى اصحّ هوآء من البصرة وليس لها بطائح واراضى رساتيقها [h] متّصلة معمورة ✿ وامّا الكوفة فانّها *قريبة من [i] البصرة فى الكبر وهوآؤها اصحّ ومآؤها اعذب *من البصرة [k] وهى على الفرات وبنآؤها مثل بنآء البصرة ومصّرها سعد بن ابى وقّاص وهى ايضًا خطط *لقبائل العرب [l] الّا انّها خراجيّة [m] بخلاف البصرة لانّ ضياع الكوفة جاهليّة وضياع البصرة احيآء موات فى الاسلام ✿ والقادسيّة والحيرة والخَوَرْنَق هى على طرف [n] البادية ممّا يلى الغرب ويحيط بها ممّا يلى المشرق النخيل والانهار [o] والزروع وهما [p] والكوفة فى اقلّ من مرحلة، والحيرة مدينة جاهليّة طيّبة التربة مفترشة البنآء كبيرة الّا انّها خلت [q] عن الاهل لمّا عُمرت الكوفة وهوآؤها وترابها اصحّ من الكوفة وبينها وبين الكوفة نحو فرسخ، *وقريب من الكوفة قبر علىّ [r] عم وقد اختلف فى مكانه فقيل انّه [s] فى زاوية على باب *جامع الكوفة [t] أُخْفِىَ

---

a) C. كثرت انهارها واسعاف (sic) بعضها الى بعض, D. وكثرت واستغلق بعضها على بعض فى مجاريها. Deinde D. melius تراجعت. b) C. الاراضى المستفلة. Pro يسفل, B. سفل. c) C. et D. متقابلان. d) B. et D. om. e) C. add. منها. f) C. add. الثقفى. g) E. حصن حجاج, C. om. h) A. et C. ورساتيقها. i) C. تقارب, E. از بصره كوچكترست. k) C. et D. om. l) C. وقبائل من العرب. m) C. خراجى. n) C. شرف, D. سيف, E. كرانهٔ. o) E. رودها بسيار. p) D. وهى. q) C. خوت, D. خف اهلها, E. خالى گشت. r) C., D. et E. وبالكوفة قبر امير المومنين على بن ابى طالب. s) C. addit مقبور. t) C. et E. مسجد الجامع. Deinde A. اخفيت, B. اختفيت.

من اجل بنى اميّة ورأيتُ فى هذا الموضع دكّان علّاف ومنهم من زعم انّه من الكوفة على فرسخين وعليه قنطرة [a] وآثار المقابر، والقادسيّة على شفير البادية وهى مدينة صغيرة ذات نخيل [b] ومياه وزروع ليس بالعراق بعدها ماءٌ جارٍ ولا شجرة‌ واما [c] بغداد فانّها مدينة محدثة فى الاسلام لم تكن بها عمارة فابتنى المنصور [d] المدينة فى الجانب الغربىّ وجعل حواليها قطائع لحاشيته ومواليه واتباعه مثل قطيعة الرّبيع والحَرْبيّة [e] وغيرهما ثمّ عُمرت * فلمّا كان فى ايّام المهدىّ جعل معسكره فى الجانب الشرقىّ فسمّى عسكر المهدىّ ثمّ عُمرت [f] بالناس والبنيان وانتقلت [g] الخلافة الى الجانب الشرقىّ وهى اليوم [h] اسفل هذا الجانب بالحَرِيم [i] ليس وراءها بنيان للعامّة متّصل وتفترش [k] قصور الخلافة وبساتينها من بغداد الى نهر بين [l] فرسخين على جدار واحد حتّى تتّصل من نهر بين الى شطّ [m] دجلة ثمّ يتّصل البناء [n] بدار الخلافة مرتفعًا على دجلة الى الشّمّاسيّة نحو خمسة اميال وتحاذى الشّمّاسيّة فى الجانب الغربىّ الحربيّة فيمتدّ نازلًا على دجلة الى آخر الكَرْخ ويسمّى الشرقىُّ جانب [o] الطاق وجانب الرُّصافة وعسكر المهدىّ فمن نسبه الى الطاق يعنى انّ اوّله باب [p] الطاق *وهو موضع السوق الاعظم [q] ومن نسبه الى الرصافة نسبه الى قصر كان الرشيد بناه بقرب مسجد الجامع بها ومن نسبه الى عسكر المهدىّ فانّ المهدىَّ كان عسكر [r] من هذا الجانب بحذاء

---

a) A. منظرة. *b*) C. et E. om.; in A. et B. deëst وزروع. *c*) C. et E. add. مدينة السلام. *d*) E. ابو جعفر منصور دوانيقى i. e. ابو الدوانيق, vid. Thaálibí *Latáifo 'l-ma'árif*, p. ۳۱. *e*) A. والحربية (vid. Jakubí, p. ۲۱, Beládsorí p. ۲۹۵); C. والحارثية (vid. Jakubí p. ۱۹). *f*) Haec inde a فلما in A. desunt. *g*) C. فانتقلت. *h*) In marg. B. يعنى وقت تصنيف هذا الكتاب. *i*) A. et B. بالحرم. *k*) In A. corrupte ونهر بين, B. ونهرس. *l*) C. hic et deinde نهراس. *m*) C. شد *sic*. E. تا واسط. *n*) C. et D. البنيان. *o*) C. باب, D. جانب باب, E. در باب; cf. Abulfeda, p. ۳۰۳. *p*) C. et Abulf. راس; E. ومعنى باب الطاق منسوبست بطاقى عظيم در بازار بزرگ كه آنرا سوق الاعظم كفتندى بود. *q*) A. et B. om. *r*) C. عسكره.

مدينة ابى جعفر ويسمّى الجانب الغربىّ جانب الكَرْخ، وببغداد مساجد جوامع فى ثلاثة مواضع فى مدينة المنصور وفى الرصافة[a] وفى دار الخلافة وتتّصل العمارة والبنيان بكَلْوَاذَى وبها مسجد جامع فلو عُدّ فى جملة بغداد لجاز، وقد عُقد بين الجانبين على دجلة جسران[b] من سفن ويكون من باب خراسان الى ان يبلغ[c] باب الياسريّة وذلك عرض الجانبين جميعًا نحو خمسة اميال[d] واعمر بقعة منها[e] الكَرْخ وبها اليسار[f] ومساكن معظم التجّار[g]، وامّا[h] الاشجار والانهار الّتى فى الجانب الشرقىّ ودار الخلافة فانّها[i] من ماء النهروان وتامرّا وليس يرتفع اليها من ماء دجلة الّا شىء يسير يقصر عن العمارة وينضح بالدواليب، وامّا الجانب الغربىّ فانّه قد شُقّ[k] اليه من الفرات نهر عيسى[l] لجنب قنطرة دممّا وتتحلّب من هذا النهر صُبابات تجتمع فتصير نهرًا يسمّى الصَّرَاة وينفجّر منها انهار وبها[m] عمارات الجانب الغربىّ ويقع ما يبقى من ماء الصراة الصغيرة والكبيرة فى دجلة وينتهى آخر نهر عيسى الى دجلة فى جوف مدينة بغداد وامّا نهر عيسى فانّ السفن تجرى

---

a) C. جانب الطاق, D. وفى الرصافة جامع اهل باب الطاق, E. ديگر در. *b*) E. جسرى ; C. جسر قريب من وسطها. Et auctor Epit. Paris. qui paullo post annum 560 scripsit, dicit وبين الجانب الشرقىّ والغربى جسر — وكان فى القديم جسران اثنان. Cf. Ibn Djobair, qui prope annum 580 urbem visitavit, p. ٢٢٧. *c*) C. add. الجسر ويبلغ, quod quoque D. legisse videtur. *d*) E. شش ميل (ut Ous. p. 67). *e*) C. فيها, D. بها. *f*) B. ايسار, C. om., D. اهل الياسرية. E. وكرخ ابادان‌ترست وبازرگانان گرانمايه آنجا. *g*) C. addit ثم صار من بعد ذلك الكرخ اقلّ عمارة واكثر خرابًا لانتقال العمارة الى الجانب الشرقى ويعرف اليوم بنهر معلّى. Cf. Jacut, IV, p. ٢٥٥, 10. Auctor Epit. Paris. dicit وانتقلت العمارة الى نهر معلى وقد سُوّر فى زماننا هذا وهو عشر الستين وخمس مائة. *h*) C. et D. فاما. *i*) C. فقد كانت. *k*) C. سيق. *l*) C. et D. add. من قرب الانبار. Deinde pro لجنب ut B. habet, A. بحيث, C. et D. تحت. *m*) C. فيها.

فيه من الفرات الى ان يقع فى دجلة وأمّا الصراة فانّ فيها حواجز تمنع من جرى السفن فتنتهى السفن منها الى قنطرة الصراة ثمّ يُحَوَّلُ ما فيها [a] ويُتجاوَزُ به ذلك الحاجز الى سفن غيرها ۞ وبين بغداد والكوفة سواد مشتبك غير متميّز تخترق اليه [b] انهار من الفرات فاوّلها ممّا يلى بغداد نهر صَرْصَر عليه [c] مدينة صرصر تجرى فيه السفن وعليه جسر من سفن تعبر عليه القوافل ومدينة صَرْصَر صغيرة عامرة بالنخيل والزروع وسائر الثمار من بغداد على ثلاثة فراسخ، ثمّ ينتهى على فرسخين الى نهر المَلِك وهو نهر كبير اضعاف نهر صرصر وعليه جسر يُعْبَر من سفن * وينتهى نهر الملك الى قصر عمر بن هُبَيْرة الفَزارىّ باحدى شعبتيه والاخرى ترمى فى دجلة عند كُوثَى نحو ضيعة تعرف بالكيل ثمّ يمتدّ عمود الفرات حتّى يخرج منه نهر سُورًا وهو نهر [d] كثير المآء ليس يخرج من الفرات شعبة [e] اكبر منه حتّى ينتهى الى سورا ثمّ الى سائر سواد الكوفة ويقع الفاضل فى البطائح ۞ وكَرْبَلَآء من غربىّ الفرات فيما يحاذى قصر ابن هبيرة [f] ۞ وأمّا سَامَرّا [g] فانّها كلّها فى شرقىّ دجلة وليس معها فى الجانب الشرقىّ مآء جارٍ * لاكنّ عماراتها وزروعها واشجارها فيما يقابلها من غربىّ دجلة وسامرّا [h] مدينة اسلاميّة ابتدأها المعتصم

---

a) E. add. .بر چهار پا. *b*) C. اليها. *c*) A. et B. عليها. *d*) Pro his C. D. et E. habent: ونهر الملك مدينة اكبر من صرصر عامرة بالاهل كثيرة النخيل والزروع والثمار ثم ينتهى الى قصر ابن هبيرة وليس بين بغداد وبين الكوفة مدينة اكبر منها وهى بقرب نهر الفرات الذى هو العمود ويطلع اليها انهار متفرّقة ليست بكبار الّا انها تعمّهم وتنفصل (C. والبها; vid. Abulfeda, p. ٣٠٥) فى قربه. وهى اعمر هذا السواد ثم ينتهى الى نهر سورا. *e*) C. شعب. *f*) C. add. وبها قبر الحسين بن على عم. et deinde C. et D. *g*) C. et D. سر من راى, E. سامره hic et infra. *h*) C., D. et E. pro his habent: الا نهر القاطول الذى ينصبّ الى السواد ببعد عنها فاما ما يحيط بها فبرّيّة وعماراتها واشجارها ومياهها فى الجانب الغربىّ بحذآئها وهى ممتدّة معما يتّصل بها من الكرخ والدّور نحو مرحلة لا ينقطع بنآؤها وهى. Cf. Jakubí, p. ٣٢.

وتمّمها[a] المتوكّل * ومكثت[b] بُرْهَةً دارَ خلافة[c] وهوآؤها وثمارها اصحّ من بغداد ✿ وامّا النَّهْرَوان فانّها مدينة يشقّ نهر النهروان وسطها صغيرة عامرة من بغداد على اربعة فراسخ ونهر النهروان يفضى الى سواد بغداد فيما يسفل[d] عن دار الخلافة الى اسْكاف[e] وغيرها من المدن والقرى، فاذا جزتَ النهروان الى الدَّسْكَرَة خفّت المياه والنخيل ثمّ يصير من الدسكرة الى حدّ حُلْوَان كالبادية المنقطعة[f] العمارة مفترشة منفردة[g] المنازل والقرى حتّى تدور على تامرّا وحدود شهرزور الى حدّ تكريت ✿ وامّا المداين فمدينة *صغيرة جاهليّة[h] قد كانت عظيمة فنُقل عامّة ابنيتها الى بغداد وهى من بغداد على مرحلة وكانت مسكن الاكاسرة وبها ايوان كسرى الى يومنا هذا وهو ايوان عظيم معقود * من آجُرّ[i] وجصّ ليس للاكاسرة ايوان اكبر منه، ولم نكثر من وصف بغداد لاشتهار وصفها عند الخاصّ والعوامّ فاكتفينا من وصف بغداد بالجملة يسيرة ذكرناها لئلّا يطول به الكتاب ✿ وبَابِل قرية صغيرة الّا انّها * اقدم ابنية العراق وينسب ذلك الاقاليم اليها لقدمها[k] وكانت ملوك الكنعانيّين وغيرهم يقيمون بها وبها آثار ابنية تشبه[l] ان تكون فى قديم الايّام[m] مصرًا عظيمًا ويقال انّ الضحّاك[n] اوّل من بنى بابل ✿ وكُوثَى ربّا يقال انّ ابراهيم الخليل عم بها طُرِح فى النار، وكوثى اثنان احدهما كوثى الطريق والآخر كوثى ربّا وبكوثى ربّا الى هذه الغاية تلال عظيمة من رماد يزعمون انّها نار نمرود بن كنعان الّتى طُرح فيها ابراهيم[o] عم ✿ والجامعان[p] منبر صغير حواليها رستاق عامر

---

a) C. et D. واستتم بناءها. E. وبر دست متوكل تمام شد. b) A. ملئت. c) C. et E. pro his وهى خراب كلها (addidi sec. E. وهمه) ربّما يسير الرجل فى مقدار فرسخ منها لا يجد بها دارًا معمورة. d) B. يستقل l. يستغل. e) C. et E. addunt بنى جُنَيد. f) B. المنقطع, C. et D. منقطعة. g) A. منفردة. h) A. et B. om. i) C. باجر, E. بسنك. k) C. ولذلك اقدم الاقليم ابنية; Abulfeda, p. ٣٠٣ quoque نسب habet. l) B. يشبه; Abulf. نسب اليها الاقليم احسبها et sic fere E. كه بروم كمان است مانده. m) C. et D. الزمان (C. وكانت فى قديم الزمان). n) C. et E. add. بيوراسب. o) C. add. خليل الرحمان. p) C. الجامعين ut Jacut in v. Loco hujus oppiduli anno 495

خصب جدًّا ۞ * والمداين من شرقىّ دجلة من بغداد على مرحلة [a] ويقال انّ ذا القَرْنَيْن اقام بالمداين الى ان مات والاخبار عنه تكذب [b] فانّ الاكثرين على انّه سُمّ فى منصرفه من ارض الصين وحُمل تابوته الى امّه باسكندريّة ، ويقال انّ جانبى المداين المكتنفين لدجلة كانت على عهد الفرس موصولًا بينهما بجسر على دجلة مبنىّ بالآجُرّ وليس لذلك اثر [c] ۞ وامّا عُكْبَرآء [d] والبَرَدان والنُّعْمانيّة ودير العاقُول وجَبُّل وجَرْجَرايا [e] وفَم الصِّلْح ونهر سابُس وسائر ما ذكرنا على شطّ دجلة من المدن فهى متقاربة فى الكبر ليس بها مدينة كبيرة وهى مشتبكة العمارة وكذلك [f] لكلّ مدينة من ذلك كورة ۞ وامّا حُلْوان فهى [g] مدينة عامرة ليس * فى ارض [h] العراق بعد البصرة والكوفة وبغداد وواسط وسامرّا والحيرة مدينة اكبر منها واكثر ثمارها التين وهى بقرب الجبل وليس للعراق مدينة بقرب الجبل غيرها وربّما سقط بها الثلج فامّا اعلى جبلها فانّ الثلج يسقط به دائمًا ۞ والدَّسْكَرَة مدينة بها نخيل وزروع عامرة وخارجها حصن من طين داخله فارغ وانّما هو مزرعة يقال انّ الملك كان يقيم هناك فى بعض فصول السنة فسمّيت دسكرة الملك لذلك ۞ وامّا من تكريت الى ان تجاوز سامرّا الى قرب العَلْث فكانّك [i] تطوف على مثال القوس الى الدسكرة ثمّ تطوف على مثال القوس الى حدّ عمل واسط من حدّ العراق الى حدّ الجبل فانّه قليل العمارة فيها قرى مفترشة والغالب عليها الاكراد والاعراب وهى مراعٍ [k] لهم ، وكذلك من تكريت عن غربيّها الى ان تنتهى الى الانبار بين [l] دجلة والفرات قليل العمارة وانّما العمارة منه ما

Hilla est condita, vid. Abulfeda, p. ٢٩٩, Jacut II, p. ٣٢٢ seq. Deinde A. بلد منبر cum signo delendi in voce منبر; B. منبر بلد. E locum de الجامعين om.

a) Haec in A. et B. desunt. b) E. وگمان برم كى اين خبر درست نيست. c) C. ويقال انه كان قد عُقد بها جسر من آجرّ فى ايام الفرس ولم يبق له اثر. E. وما اثرا اثر نديديم. d) A. et B. وعكبرا. C. add. واوانا. e) A. et B. om. f) C. et D. om. g) C. et Jacut II, p. ٣١٧, 10 فانها. h) C. et Jac. بارض. i) C. كما. k) A. مزارع. l) A. et B. فمن.

يحاذى سامرّا امياًلا [a] يسيرة والباقى بادية ✿ ولم ابالغ فى وصف العراق لاكثار الناس فيها واشتهار عامّة ما يذكر منها فهذه صفة جامعة لها وجيزة اذ كان قصدى فيها وفى غيرها الى * تخطيط هيآتها [b] ✿

خوزستان

وامّا حدود خوزستان فانّ شرقيّها حدّ فارس واصبهان [c] وبينها وبين حدّ فارس من حدّ اصبهان نهر طاب وهو الحدّ الى قرب مَهْرُوبَان [d] ثمّ يصير الحدّ بين الدَّوْرَق [e] ومهروبان على الظهر الى البحر وغربيّها حدّ رستاق واسط ودور الراسبىّ وشماليّها حدّ الصَّيْمَرة وكَرْخا [f] واللُّور حتّى يتّصل على حدود الجبال الى اصبهان على انّه يقال انّ اللُّور كانت من خوزستان فحُوّلت الى الجبال وحدّ خوزستان ممّا يلى فارس واصبهان وحدود الجبال [g] وواسط على خطّ مستقيم فى التربيع الّا انّ الحدّ الجنوبىّ من عبّادان [h] الى رستاق واسط يصير مخروطًا فيضيق فى التربيع عمّا قابله وفى [i] حدّ الجنوب ايضًا من حدّ عبّادان على البحر الى حدّ فارس تقويس يسير فى الزاوية فينتهى هذا [k] الحدّ الجنوبىّ الى شىء من البحر ثمّ الى دجلة * حتّى يجاوز بَيَان [l] ثمّ ينعطف من ورآء المَفْتَح والمَذَار الى ان يتّصل برستاق واسط من حيث ابتدأنا [m] ✿

وامّا ما يقع فيها من المدن فانّها كُوَر منها الأَهْوَاز واسمها [n] هُرْمُزْشَهْر وهى

---

a) A., B. et C. امیال. b) A. et B. ىحطط اهياتها, D. هیاتها, E. اثبات هیاتها که مجهولات معلوم گردد. c) C. semper habet اصفهان. d) E. ut semper ماهی رویان. e) C. دورق. f) C. والكرحه, D. والكرج, E. وكرخه (ut Ous. p. 73). Jacut habet كرجة, sed Abulfeda p. ٣١١ الكرخة, Edrîsî, I, p. 379, كرخة et *Kâmus* كَرْخ et كَرْخَة. Mokaddasi (Ms. Berol. p. 197) habet quoque كرخة. g) C. الحجاز! h) A. عند بيان (عند بَيَان? cf. D.). i) C. وفيه. k) A. et B. الى. Deinde C. البحر. l) B. ثم ىحاور ىىان, C. حين تجاوز بيان. A. pro حتّى habet حين ut B. m) Hic sequi deberet mappa Khuzistani. n) C. وهو اسم.

الكورة العظيمة الّتى ينسب اليها سائر الكور وعَسْكَرُ مُكْرَم وتُسْتَر a وجُنْدَى سَابُور والسُّوس ورامَهُرمُز وسُرّق b وكلّ ما ذكرنا من * كورة فهى c اسم المدينة غير سُرّق d فانّ مدينتها الدَّوْرق وهى المعروفة بدورق الفرس وايذَج ونهر تيرى وحَوْمة الزُّط والخابَران e وهما واحد وحَوْمة البُنْيَان f وسوق سَنْبيل g ومَنَاذِر الكبرى ومناذر الصغرى وجُبّى والطّيب وكَلِيوَان h فهذه مدن لكلّ مدينة كورة، ومن مدنها المعروفة المشهورة بَصِنّى i وأَزم وسوق الاربعاء وحصن مهدىّ وبَاسِيَان وبَيَان وسُلَيْمانَان k وقُرْقوب ومَتّوث وبِرذَوْن l وكَرْخا ✿

وخوزستان فى مستوى m وارض سهلة ومياه جارية فمن اكبر انهارها نهر تستر وهو النهر الّذى بنى عليه سابور الملك شاذروان بباب تستر حتّى ارتفع ماؤه الى ارض n المدينة لانّ تستر على مكان مرتفع من الارض فيجرى هذا النهر من وراء عسكر مكرم على الاهواز حتّى ينتهى على نهر السِّدْرة الى حصن مهدىّ ويقع فى البحر، ويجرى من ناحية تستر نهر المَسْرُقان حتّى ينتهى الى عسكر مكرم * سفلى الاهواز وآخره بالاهواز لا يتجاوزها فاذا انتهى الى عسكر مكرم o فعليه جسر كبير نحو عشرين سفينة تجرى فيه p السفن العظام وقد ركبتُهُ انا من عسكر مكرم الى الاهواز والمسافة ثمانية

---

a) E. شوشتر. *b*) D. cum articulo. C. والسوق i. e. fortasse سوق الدورق, nam E. بازار دورق habet. *c*) C. فهو كورة اسم. *d*) C. سوق. *e*) A. et B. hic et infra sine punctis. D. الحايران, E. حايزان et جايران (ut Epit. Paris.), C. h. l. الجايران, infra sine punctis. Vid. Jacut, I, p. ٣٨٣ et II, p. ٧٩١ (ubi male السرط). *f*) A. et C. المماں, B. h. l. الممار, D. h. l. الثنيان, deinde البنيان, Epit. Paris. الثننات et الثنيات, E. المان et دمماں. Cf. Belâdsori, p. ٣٨٣ et ann. *c.*, ubi الثيمان. Vid. Jacut I, p. ٧٤٩, 9. *g*) B. سلمل, C. سنبل. *h*) Vocales apud Jacut I, p. ٩٥٩, 5. C. وكليون. *i*) A. et B. نصنا; deinde A., B. et C. وازم. *k*) A. et B. وسلحانان. *l*) A. et B. وبرودن. Deinde C. et D. وكرجه. *m*) C. مستو. *n*) C., D. et Jacut, I, p. ٨٤٨, 10 om. *o*) In A. et B. haec desiderantur. (C. دمحاوزه). *p*) C. عليه.

فراسخ فسرنا فى الماء ستّة فراسخ ثمّ خرجنا وسرنا فى وسط النهر وكان الباقى من[a] هذا النهر * الى الاهواز[b] طريقًا يابسًا ولا يصبغ من هذا الماء شىء وانّما تسقى به اراضى قصب السكّر وما فى اضعافه من النخيل والزروع، وما بخوزستان كلّها على كمال عمارتها بقعة هى اعمر وارخى[c] من المسرقان، ومياه خوزستان من الاهواز والدورق[d] وتستر وغير ذلك ممّا يصاقب هذه المواضع كلّها تجتمع عند حصن مهدىّ فتصير هناك نهرًا[e] كبيرًا ويغزر[f] ويصير له عرض ثمّ ينتهى الى البحر، وليس بها بحر الّا ما تنتهى اليه زاوية من مهروبان[g] الى قرب سليمانان[h] بحذاء عبّادان فانّه شىء يسير[i] وهو من بحر فارس ☆ وليس بجميع خوزستان جبال ولا رمال الّا شىء يسير يتاخم نواحى تستر وجندى سابور وبناحية ايذج واصبهان والباقى من خوزستان كانّه ارض العراق ☆ وامّا هواؤها ومآؤها وتربتها وصحّة اهلها فانّ مياهها طيّبة عذبة جارية ولا اعرف بجميع خوزستان بلدًا مآؤهم[k] من البئر لكثرة المياه الجارية بها[l]، وامّا ترابها[m] فانّ ما بَعُدَ عن دجلة الى ناحية الشمال ايبس واصحّ وما كان الى دجلة اقرب فهو من جنس ارض البصرة فى التسبّخ[n] وكذلك[o] الصحّة ونقاء البشرة فى الناس فيما بعد عن دجلة ☆ وامّا المَسْرُقان[p] خاصّةً فانّ بها رطبًا يسمّى الطّنّ[q] يقال انّ ذلك الرطب اذا اكله الانسان وشرب عليه ماء[r] المسرقان لم تُخطّه[s] الحُمّى، وليس بخوزستان موضع

---

a) C. فى. *b*) A. et B. om. *c*) C. et D. وازكى. *d*) A. et B. h. l. دورق.
e) A. et B. om. *f*) A. et B. ويغور, Abulfeda, p. ٣١٧ om. Deinde A. او يصير.
g) C. add. ودمان i. e. وبيان. *h*) A. et B. سلمانان. *i*) C. add. يتاخم.
k) A. مياههم, C. et D. ماؤه; cf. Jacut, II, p. ٤٩٩, 16. *l*) A. et B. om. بها.
m) D. et Jacut تربتها. *n*) B. التسبّح; apud Jacut l. l. corrupte السنج l. السبخ ut D. habet. *o*) C. et Jac. add. فى. *p*) E. مشرقان ut h. l. D. et ut Cod. Berol. Mokaddasii. *q*) A. et B. الصلق, E. طم. *r*) C. الماء من. *s*) A., B. et C. تخطه. Cogitari posset de تَخْطُه (a وخط), sed E. habet حالى تنب لم تخطى (l. يتخطاه) منه رائحة النخر. D. habet محرق ثيردش.

يجمد فيه الماء * ولا يقع[a] فيه الثلج ولا يخلو من النخل[b]، والعِلل بها كثيرة وخاصّة لمن انتابها، وامّا ثمارهم وزرعهم[c] فانّ الغالب على بلاد خوزستان من الاشجار النخل ولهم عامّة الحبوب من الحنطة والشعير والباقلاء واكثر حبوبها بعد الحنطة والشعير الارزّ فيخبزونه وهو لهم قوت * وكذلك فى رستاق العراق[d] وليس من بلد ليس به قصب سكّر[e] من هذه الكور الكبار ولكنّ[f] اكثر ما بها من السكّر بالمسرقان ويقع[g] جميعه الى عسكر مكرم وليس بعسكر مكرم فى القصبة كثير سكّر وكذلك بتستر والسوس * فانّه يتّخذ منه السكّر والقصب فى سائر المواضع انّما هو للاكل دون ان يتّخذ منه السكّر[h]، وعندهم عامّة الثمار لا يكاد يخطئهم الّا الجوز وما لا يكون الّا ببلاد الصرود ۞ *وامّا لسانهم فانّ عامّتهم يتكلّمون[i] بالفارسيّة والعربيّة غير انّ لهم لسانًا آخر خُوزِيًّا ليس بعبرانىّ ولا سريانىّ ولا فارسىّ، وزيّهم زىّ اهل العراق فى الملابس من القمص والطيالسة والعمائم وفى اضعافهم من يلبس الأُزُر والميازر، والغالب على اخلاقهم سوء الخُلُق[k] والمنافسة فيما بينهم فى اليسير من الامور وشدّة الامساك والغالب على خَلْقهم[l] صفرة اللون والنحافة وخفّة اللحى والضخامة ووفور الشعر فيهم اقلّ ممّا فى غيرهم من المدن وهذه صفة عامّة الجروم[m]، وامّا ما ينتحلونه من الديانات فانّ الغالب بخوزستان الاعتزال والغلبة عليهم[n] دون سائر النِحل وفى[o] سائر كورهم من اهل الملل نحو ما فى سائر الامصار ۞

---

a) C. ويقع, Jacut l. l. ويروح. *b*) C. et D. النخيل. *c*) C., D. et Jacut وزروعهم. *d*) Jacut: كرستاق كسكر من واسط. *e*) C. et Jacut السكر. *f*) C. et Jacut الا ان. *g*) Jacut ويرفع. *h*) Haec corrupta esse patet. C. habet: وانما يتخذ به السكر والقصب من سائر المواضع وما لهم من قصب السكر فانما هو للاكل فقط دون ان يعملوا منه سكرًا et Jacut, II, p. ۴۹۷, 2 وانما يحمل عليها القصب من نواحى اخر والذى فى هذه الثلاثة بلاد انما يكون بحسب الاكل لا ان يستعصر منه سكرٌ. *i*) Ridicule C. واما نسوانهم فان عامتهن يتكلمن. *k*) Jac. add. والبخل المفرط. Deinde C. والمناقشة. *l*) B. خِلقهم, C. خلقتهم. *m*) C. الخوز. *n*) Sic A. et B.; D. لهم (بهم), Epit. Paris. لاهله. In C. et Jac. desunt haec duo verba. *o*) A. et B. sine copula.

واما الخاصّيات بها فانّ عندهم بتُسْتَر الشاذروان الّذى بناه سابور وهو من اعجب البناء واحكمه بلغنى انّ امتداده يقرب من ميل قد بُنى بالحجارة [a] كلّه حتّى تراجع الماء فيه وارتفع الى باب تستر ۞ *ولهم بالسُّوس بلغنى والله اعلم انّ [b] تابوتًا وُجد فى ايّام ابى موسى الاشعرىّ فذكروا انّ فيه عظام دانيال النبىّ عم وكان اهل الكتاب يديرونه فى مجامعهم ويتبرّكون به ويستسقون المطر به اذا اجدبوا فاخذه ابو موسى وعمد الى نهر على باب [c] السوس فشقّ منه خليجًا وجعل فيه ثلاثة قبور مطويّة بالآجُرّ ودفن ذلك التابوت فى احد القبور ثمّ استوثق منها كلّها وعمّاها ثمّ فتح الماء حتّى *غلب زيد [d] النهر الكبير على ظهر تلك القبور والنهر [e] يجرى عليها الى يومنا هذا ومن نزل الى قعر الماء وجد تلك القبور۞ ولهم بناحية آسَك [f] متاخمًا لارض فارس جبل يتّقد منه نارٌ [g] ابدًا لا ينطفىُٔ ويُرَى منها الضوء بالليل والدخان بالنهار وهو فى حدّ خوزستان ويشبه [h] فيما اظنّ انّه عين نفط او زفت او غيره ممّا تعمل فيه النار فوقع فيه على قديم الايّام نار فعلى قدر ما تخرج يحترق ابدًا فيما احسبه من غير ان رايتُ علامة لذلك ولا سمعتُ به * وانا اقوله [i] ظنًّا۞ ولهم بعَسْكَرِ مُكْرَم [k] صنف من العقارب صغار على قدر ورق الأنْجُذان تسمّى الجرّارة [l] قلّ ما يسلم من لسعة [m] وهى ابلغ فى القتل من بعض الحيّات۞ واما تُسْتَر فانّ بها يتّخذ الديباج الّذى يحمل الى الدنيا وكسوة مكّة من الديباج يتّخذ بها [n] وبها للسلطان

---

a) E. بكج وسنكٔ. Abulfeda, p. ٣١٥ شاذروان vertit per سكر. *b*) C. ولهم بالسوس بلغنى ان امتداده قريب من ميلين قد رايت به, quae verba variam lectionem continent ad ea quae praecedunt. De seqq. cf. Beládsorí, p. ٣٧٨. *c*) A. et B. بلد. *d*) A. ربد, C. et D. قلب ذلك. *e*) A. فالنهر. *f*) A., B., C. et E. اسل. *g*) C. et D. النار. *h*) B. وبسه, D. وشببه; E. وكمان چنانست. *i*) C. وانما اقول. *k*) E. semper لشكر pro عسكر مكرم habet. *l*) D. الحرارة, C. الكروره l. الكزودة. E. habet كزوره. *m*) C. لسعها احد, D. لسعتها (لسعها). *n*) E. وكسوت وطرازخانهٔ كعبه سازند آنجا.

طراز ✿ وامّا السُّوس فانّه تعمل بها الخزوز[a] ومنها تحمل الى الآفاق ، وبالسوس صنف من الاترجّ شمّامات ذكيّة[b] كالاكفّ باصابعها لم ار مثلها فى * بلدان الاترجّ[c] ✿ وبقُرْقوب[d] السُّوسَنْجِرْد الّذى يحمل الى الآفاق وبها[e] وبالسوس طراز للسلطان ✿ وببَصِنّى تعمل الستور الّتى تحمل الى الآفاق المكتوب عليها عمل بصنّى * وقد تعمل ببِرْذَون[f] وكَلِيوَان وغيرهما من تلك المدن ستور يكتب عليها بصنّى[g] وتدلّس فى ستور بصنّى الّا انّ المعدن بصنّى ✿ وبرَامْهُرمُز من ثياب ابريسم[h] ما يحمل منها الى كثير من المواضع ، ويقال انّ مانى بها قُتِل وصُلب ويقال انّه مات فى محبس[i] بهرام حنف انفه فقطع رأسه واظهر قتله[k] ✿ وجُنْدَىْ سَابُور[l] مدينة خصبة واسعة الخير وبها نخل وزروع كثيرة[m] ومياه ونزلها يعقوب بن اللَّيْث الصَّفّار لخصبها واتّصالها بالمِيَر[n] الكثيرة فمات بها وبها قبره ✿ ونهر تِيرَى تكون بها ثياب تشبه ثياب بغداد وتحمل الى بغداد فتدلّس بالبغدادىّ[o] وتُقْصَر ببغداد ✿ وجُبّى مدينة ورستاق عريض مشتبك[p] العمارة من النخيل وقصب السكّر[q] ومنها ابو علىّ

---

a) A., B. et C. الخزور. E. جامهاى خز مرتفع. b) C. مركبة et sic Epit. Paris. (B. زكية). E. ترنج كه آنرا بنج انكشت خوانند لغايت خوش بوى. c) C. فى سائر البلدان الا ببغداد. d) A. et B. وبها ثوب et sic E. qui habet آنجا جامهاى خز خيزد. e) A. et B. om., E. وبقرقوب, C. وبالسوس وبها. f) Jacut, I, p. ٩٥٩, 5 ببيرون, sed vid. p. ٥٩١, 21. Deinde C. om. وكليوان et habet وغيرها. g) Haec inde a وقد in A. et B. desunt. h) C. et D. الابريسم, E. ابريشم. Deinde A., B. et C. om. ما. i) C. مجلس. k) C. add. ورامهرمز ايضا من بين مدن خوزستان يجتمع فيها من النخيل والجوز والثلج (Cod. والبلح) ما ليس بغيرها وكذلك الاترج; cf. Jacut, II, p. ٧٣٨, 10 et 11. l) E. كنديشابور. Deinde pro خصبة Jacut, II, p. ١٣٠, 14 حصينة. C. om. واسعة الخير et Jac. الخير. E. habet بزرك واباداں. m) C. add. وثمار. n) C. بالمدن et sic Jacut l. l. 18, qui pro لخصبها habet لحصانتها. o) C. فى البغدادى. p) B. مشتمل. q) C. et D. add. وغيرهما.

الجُبّائىّ امام المعتزلة فى عصره ۞ وتتّصل زاوية من خوزستان بالبحر فيكون له خور[a] * يخاف على سفن[b] البحر اذا انتهت[c] اليه فانّه يعرض، واستجمع[d] مياه خوزستان بحصن مهدىّ فيتّصل[e] بالبحر ويعرض هناك حتّى ينتهى فى طرفه المدّ والجزر ويتّسع حتّى[f] كانّه من البحر ۞ وتتّخذ بالطّيب تِكَكٌ[g] تشبه الارمنىّ[h] قلّ ما تتّخذ فى مكان من الاسلام بعد ارمينية احسن منها فيما علمتُهُ[i] ۞ واللُّور بلد خصب[k] الغالب عليه هوآء الجبل وكان من خوزستان الّا انّه أُفرد فى اعمال الجبل[l] ۞ وامّا سَنْبيل فانّها كورة متاخمة لفارس وقد كانت مضمومة[m] الى فارس من ايّام محمّد بن واصل الى آخر ايّام السّجْزيّة ثمّ حُوّلت الى خوزستان ۞ والزُّطُّ والخابَرَان[n] هما كورتان عامرتان * على نهرين جاريين[o] ۞ والبُنْبَان[p] متاخمة للسَّرْدَان[q] من ارض فارس ولاصبهان وهوآؤها هوآء الصرود وليس بخوزستان رستاق يقارب الصرود غير البنبيان ۞ وامّا آسَك فانّها قرية ليس فيها[r] منبر وحواليها نخيل كثير وبها كانت وقعة الازارقة الّتى يقال انّ اربعين من الشُّرَاة قتلوا نحوًا من الفى

---

a) C. semper هور ut Jacut, II, p. ۴۹۷, 10. *b*) C. على شغير, Jacut تدخله سفن. *c*) C. انتهى. *d*) C. وستجمع, Jacut وتجتمع. *e*) B. ويتصل. Jacut وتنفصل منه الى البحر فيتصل به. *f*) A. et B. om. *g*) E. شلواربندها. *h*) A. الارمينى, E. رومى. *i*) C. add.: الا ما قد استحدث بطوس فانهم يعملون بها ما هو احسن من الطيبى. *k*) C. et Jacut IV, p. ۳۹۹, 16 خصيب. Deinde Jacut الجبال omisso هواء. *l*) Jacut add. لاتصاله بها. *m*) C. كان مضموما et mox حول (ut Jacut, III, p. ۱۵۷, 8). *n*) Vid. supra p. ۸۹ ann. *e*. *o*) Pro his A. et B. tantum جاردمان, D. متجاورتان. E. بر رودى ذهاده. Cf. Jacut, II, p. ۷۹۱, 5. *p*) Vid. supra p. ۸۹ ann. *f*. In C. locus deëst. *q*) A. للسيروان, B. للسوادن, Jacut, I, p. ۷۴۹, 10 السَّرْدن, quae est lectio D. et B. infra. Bekrí praescribit السَّرْدَان (non السَّرْدان ut in ann. 1 ad *Merácid*, II, p. ۲۴). Apud Mobarrad, Ms. p. 754 scribitur الشَّرْدَان. *r*) C. بها, D. لها. E. دهيست كوجك.

رجل تبعتهم من اهل[a] البصرة، والدوشاب الأَرَّجانيّ الّذى يحمل الى الآفاق منها[b] ۞ وامّا مَنَاذِر الكبرى والصغرى فانّهما كورتان عامرتان بالنخل[c] والزروع ولهما ارتفاع كثير ۞

واما المسافات بها فانّ من فارس الى العراق *طريقين شارعين[d] احدهما الى البصرة ثمّ الى بغداد والآخر الى واسط ثمّ الى بغداد فامّا طريق البصرة فانّك تاخذ من ارّجان الى آسك قرية[e] مرحلتين خفيفتين ثمّ الى زَيْدَان[f] مرحلة وزيدان قرية ثمّ منها الى الدَّوْرَق مرحلة والدورق مدينة كبيرة وهى مدينة سُرَّق ثمّ من الدورق الى خان مَرْدَوَيْه[g] وهو خان تنزله السابلة مرحلة ومن خان مردويه الى بَاسِيَان مدينة وسطة فى الكبر عامرة يشقّ النهر فيها فتصير[h] نصفين مرحلة ومن باسيان الى حصن مهدىّ مرحلتين فيها منبر ويُسْلَك بينهما[i] فى الماء وكذلك من الدورق الى باسيان[k] يسلك فى الماء وهو ايسر من البرّ ومن حصن مهدىّ الى بَيَان مرحلة على الظهر وبيان فيها منبر وقد انتهيتَ الى آخر حدود خوزستان، وبَيَان على دجلة فاْركب منها الماء ان شئتَ الى الأُبُلّة وان شئتَ على الظهر الى ان تحاذى الأُبُلّة ثمّ تعبر ۞ وامّا الطريق الى واسط ثمّ الى بغداد فانّ من أرّجان الى سوق سَنْبِيل مرحلة ثمّ الى رامَهُرْمُز[l] مرحلتين ثمّ من رامهرمز الى عسكر مكرم ٣ مراحل ومن عسكر مكرم الى تستر مرحلة ومن

---

a) E. لشكر. *b*) A. et B. منه. E. pro his: ارغان شهريست از آنجا دوشاب خيزد ودر افاق برند. *c*) C. et D. بالنخيل. *d*) A. et B. طريقان شارعان. *e*) C. قريب. *f*) A., B. et D. ديرا, sed in mappa A. زيرا. C. ددران et Jacut I, p. ٤٩٨, 1 دَبَرَان. E. ديران, Ous. ديدان, Edrîsî, I, p. 389 ذيرا. Vid. Jacut II, p. ٩٩٥, 17. Deinde A. om. مرحلة. *g*) D. مندونه et من دونه, C. مردوه, B. semel مزروده; cf. Abulfeda, p. ٣١٢ ann. 1. et vid. Jacut, I, p. ٤٩٨. In E. recte scribitur nomen. *h*) A. ويصير. *i*) Ex Abulfeda. A., B., C. et D. منها. *k*) A. et B. h. l. الماسان. *l*) E. رامز hic et deinde, quae est pronunciatio vulgaris, vid. Jacut in v.

تستر الى جندى سابور مرحلة ومن جندى سابور الى السوس مرحلة ومن السوس الى قرقوب مرحلة ومن قرقوب الى الطيب مرحلة ويتّصل بعمل واسط ، ومن[a] العسكر الى واسط طريق اخصر من هذا فلا يدخل تستر[b] ولكنّا ذكرنا هذا المسلك لانّ قصدنا ذكر مسافة ما بين المدن ولم نرد نفس الطُرق[c] الى بغداد فكان هذا اجمع لما اردنا ان نذكره ✿ ومن العسكر الى ايذج ٤ مراحل ومن العسكر الى الاهواز مرحلة ومن الاهواز الى أَزَمَ[d] مرحلة ومن الاهواز الى الدورق ٤ مراحل ومن عسكر مكرم الى الدورق نحو هذا ، ومن الاهواز الى رامهرمز نحو ٣ مراحل ايضًا لانّ الاهواز وعسكر مكرم فى سمت واحد ورامهرمز منهما كاحدى زوايا المثلّث ، ومن عسكر مكرم الى سوق الاربعآء[e] مرحلة وجُبّى تحاذى[f] سوق الاربعآء ومن سوق الاربعآء الى حصن مهدىّ مرحلة ، ومن الاهواز الى نهر تيرى يوم ، ومن السوس الى بصنّى اقلّ من مرحلة ومن[g] السوس الى بردون مرحلة خفيفة ومن السوس الى متّوث مرحلة ، فذلك جوامع المسافات بها ✿

بلاد فارس

وامّا فارس فالّذى يحيط بها ممّا يلى الشرق حدود كرمان وممّا يلى الغرب كور خوزستان واصبهان وممّا يلى الشمال المفازة الّتى بين فارس وخراسان وبعض حدود اصبهان وممّا يلى الجنوب بحر فارس ، وصورة فارس على التربيع الّا من الزاوية الّتى تلى اصبهان والزاوية الّتى تلى كرمان ممّا يلى المفازة وفى الحدّ الّذى يلاصق[h] البحر تقويس قليل من اوّله الى آخره * وانّما وقع فى زاويتيها[i] ممّا يلى كرمان واصبهان زنقة لانّ من شيراز وهى وسط فارس اليهما من المسافة نحو نصف ما بين شيراز وخوزستان وبين شيراز وجروم كرمان[k] ✿

a) A. et B. ومن. b) A. et B. h. l. التستر. c) Abulfeda الطَريق. d) A. B. et C. ارم. e) E. (sic) بازار چهار شنبه. f) C. ut Abulf., p. ٣١٢, بحذاء. g) Hinc incipit lacuna in A. et B. usque ad verba قد صورت فارس. h) Jacut III, p. ٨٣٦, 15 يلى. i) Jacut وانما قلنا ان فى زاويتها. k) Hic sequi deberet mappa Persidis.

قد صوّرتُ بلاد فارس بحدودها ولم اصوّر فيها رستاقًا لانتشار ذلك وكثرته ولا الجبال لانّه ليس بفارس بلد الّا وبه جبل او يكون الجبل * منه بحيث [a] تراه الّا اليسير، وما صوّرتُ فيها الّا مدينة لها منبر * مذكور مشهور [b] وقد ذكرتُ فى الرسالة ما يُعْلَمُ من قرأها موضعَ كلّ كورة برساتيقها ومواضع المدن بها ان شآء الله تعالى ۞

ذكر ما بفارس من الكور والمدن والزموم [c] والاحيآء والحصون وبيوت النيران والانهار والبحار كُوَرُ فارس خمس فاوسعها عرضًا [d] واكثرها مدنًا ونواحى كورة اصْطَخْر ومدينتها اصطخر وهى اكبر مدينة بهذه الكورة [e]، وتليها فى الكبر كورة أرْدَشير خُرّه [f] ومدينتها جُور ويدخل فى هذه الكورة [e] قُبَاذ خُرّه [g] وبكورة اردشير خُرّه مدن هى اكبر من جور مثل شِيرَاز وسِيرَاف وانّما صارت جور هى [h] مدينة اردشير خُرّه لانّها بنآء اردشير ودار ملكه وشيراز وان كانت قصبة فارس [i] كلّها وبها الدواوين ودار الامارة فهى مدينة محدثة فى الاسلام، وتليها فى الكبر كورة دارابْجِرْد ومدينتها دارابْجِرْد وفَسَا [k] هى اكبر مدنها واعمر غير انّ الكورة منسوبة الى دار الملك ومدينته الّتى ابتناها لهذه الكورة [e] دَارَابْجِرْد، وتليها فى الكبر كورة أرَّجان ومدينتها العظمى ارّجان وليس بهذه الكورة [e] مدينة اكبر من ارّجان، وتليها فى الكبر كورة سابُور وهى اصغر كور فارس ومدينتها سابور وبهذه الكورة [e] مدن هى [l] اكبر

---

a) Male Jacut l. l. 19 لا بحيث. *b*) A. om., E. منبرى مساجد ادينه. *c*) A., B., C., D., E. et Edrîsî semper زم pl. زموم habent, vid. Jacut II, p. ٨٢١ (et *Moschtarik*, p. ٢١.). Deinde A. والاجناد. *d*) A. عرصة. *e*) C. الكور. *f*) C. et D. pro خره habent جرّه, E. دره. *g*) A. et B. قباذ كره, C. قباذ جرّه, Jacut, I, p. ١٩٩, 17 كورة قَنَاخُرّه (III, p. ٩١٨, 4 قَنَاخُرّه). Cf. Jacut IV, p. ٣٩, 1 cum III, p. ٨٣٩, 20. *h*) C., D. et Jac. (I, p. ١٩٩) om. Jacut كانت habet pro صارت. *i*) C. et D. لفارس. *k*) Codd. interdum habent فسّا, ut semper Cod. Berol. Mokaddasii, Cod. Mobarrad. Leid. p. 775, vs. 7 a. f., Cod. Belâdsorii ad edit. p. ٣٩١. ann. *c.*; et sic pronunciat Chardin, *Voyages*, ed. 1811, t. VIII, p. 415 „Phessa ou Pessa." Deinde A. et B. وهى, et pro مدنها, C. et D. منها. *l*) A. om.

منها مثل النُّوبَنْدَجان *a* وكَازَرُون ولكنّ هذه الكورة تنسب الى سابور لانّ سابور الملك هو الّذى بنى مدينة سابور ۞ وامّا رمومها *b* فهى خمسة واكبرها رمّ جيلَوَيْه *c* ويعرف برمّ الزميجان *d* ، ثمّ الّذى يلى هذا الرمّ فى الكبر رمّ احمد بن الليث ويعرف باللّوالجان *e* ، ويلى ذلك فى الكبر رمّ الحسين *f* بن صالح ويعرف برمّ الديوان *g* ، ثمّ رمّ شهريار *h* ويعرف برمّ البازنجان *i* والبازنجان

---

a) C. النوبندجان. *b*) E. h. l. habet جند جايكاهست در بارس كه آنرا بزم خوانند ومراد از ان قبيله باشد. Jacut III, p. ٨٣٦ ult. dicit فالرم منزل الاكراد ومحلتهم. *c*) A. et B. حلويه et حلونه, D. semel خَبْلَوَيْه. Jacut l. l. جيلويه, sed II, p. ٨٢١ (*Moschtarik*, p. ٢١.) الحسن بن جيلويه. Ibn Khordádbeh, ed. Barbier de Meynard, p. 55 الحسين بن جيلويه, quae lectio confirmatur loco ex *Kitábo 'l-boldán* (Mus. Britt. *Rich.* n. 7496) f. 55 v., ubi hic locus laudatur. Edrîsî, I, p. 406 الحسن بن خالويه. Forma كيلويه occurrit apud Quatremère, *Histoire des Mongols*, p. 382, 384 et 385 ann. *d*) A. الدممحان et الرممحان, B. الرممجان, C. الرمبيجان, Codd. D. الزمنجان, الزمنجان, الزمنجان et الدمجان, E. رممحان, ميجان et درسحان (Ous. p. 82 et 90 الرمجان, p. 122 مسجان) Edrîsî الرميجان, Jacut الزنيجان, Ibn Khordádbeh الميانجان. *e*) A., B. et C. h. l. باللوانجان. E. semper لوابحان. *f*) A., B., C., D. et E. h. l. male احمد et sic quoque Mokaddasí, Ms. Berol. p. 215 et Jacut III, p. ٨٣٩, 22 (et *Moschtarik*, p. ٢١١), sed II, p. ٨٢١ habet الحسن. Vid. infra et Ibn Khordádbeh. *g*) Ibn Khord. الريزان, Jacut الزيزان, C. الدوان ut semel E., Mokaddasí الزيراز. *h*) A. سهربزار, B. شهريزار, infra uterque شهرى راز. E. شهرسار. *i*) A. et B. sine punctis ut C. h. l. Infra C. المادنجان; D. المازنجان. E. بادبحان, مادبحان et داربحان (Ous. p. 83 et 123 بادنجان, p. 91 بادانجان). Mokaddasí البادنجان. Vid. Jacut et cf. Quatremère in *Notices et Extraits*, XIII, p. 304, ubi ex Masudii operibus duae formae المادنجان et البازنجان memorantur. In ed. Masudii Paris., III, p. 254 eodem modo المادنجان legitur, pro quo Codd. Leid. habent الماذنجان.

ٱلّذين فى حدود اصبهان ناقلة من هذا الرمّ، ورمّ احمد بن الحسن [a]
ويعرف برمّ الكاريان [b] وهو رمّ اردشير ✿ وامّا احياء الاكراد فانّها تكثر فى [c]
الاحصاء غير انّهم بجميع فارس يقال انّهم يزيدون على خمس مائة الف بيت
شعر ينتجعون المراعى فى * المشتى والمصيف [d] على مذاهب العرب ويخرج
من بيت واحد من الارباب والاجراء والرعاة واتباعهم ما بين واحد الى
عشرة من الرجال ونحو ذلك [e] ، وسأذكر من اسامى احيائهم ما يحضرنى
ذكره على انّهم لا يتقصّون [f] فى العدد الّا من الدواوين للصدقات ✿ وامّا
انهارها الكبار الّتى تحمل السفن اذا اجريت فيها فانّها نهر طاب ونهر شيرين
ونهر الشاذكان ونهر درخيد [g] ونهر الخوبذان ونهر رتين [h] ونهر سكّان [i] ونهر
جرشيق [k] ونهر الاخشين ونهر كرّ ونهر فرواب [l] ونهر تيرزة [m] ✿ وامّا بحارها فانّها

---

a) Ex C., E. et Jacut, III, p. ٨٣٦, 22; ceteri h.l. الحسين. *b*) A. h.l. الكاديان, deinde ut B. sine punctis; C. et D. الكاربان, E. كادمان h.l., infra كاردان et كاريان ut Ous p. 91 et 123. *c*) C. et D. عن et sic E. بيش از آنست كى در شمار آيد. *d*) C. et D. الشتاء والصيف et sic Jacut, III, p. ٨٣٨, 9. *e*) C. tantum جماعة; E. وكس باشد ازيشان كه دويست مرد پيوسته دارد. از چوبان ومزدور وشاگرد وغلام وانچ بايد ماند. *f*) A. ينقصون. *g*) D. h.l. دوخيد, E. درجمد et دوحمد, Ous. p. 84 درجند, p. 96 دوجند, Edrîsî, I, p. 410 درجند, C. infra دوحمد, Mokaddasí Ms. Berol. p. 214 ult. درجيد. Cf. Jacut, III, p. ٨٣٨, 10. *h*) Sic A. et B. h.l., infra sine punctis. C. h.l. ردمر, infra ردبن; D. h.l. رس, infra اوس; E. h. l. درودن (Ous. p. 84 روبيين), infra ردن (Ous. p. 97 رميين); Edrîsî l. l. ورش, Mokaddasí l. l. رس. *i*) A. et B. h.l. سكاب. *k*) B. حرسق, infra A., B., C. et D. جرسيق ut semper E. — Jacut l. l. جَرْسَق, Mokaddasí p. 215, vs. 1 حرسس. *l*) Sic recte C.; A. h.l. قروان, B. et D. فروان, E. فروات ut infra D. — Mokaddasí et Jacut l. l. recte فرواب habent; cf. Ritter, *Erdkunde*, VIII, p. 863 Polwar, Farwar, Parvab. *m*) Lectio incertissima est. A. et B. دورة, C. دمردة, E. دمورة, D.

بحر فارس وبحيرة البَخْتَكان [a] وبحيرة دَشْت أَرْزَن وبحيرة التوز [b] وبحيرة الجوبانان [c] وبحيرة جنكان [d] ۞ وامّا بيوت نيرانها فانّها لا تخلو ناحية ولا مدينة بفارس الّا القليل من بيوت النيران والمجوس اكثر ملل اهل الكتاب بها ولهم * من هذه البيوت بيوت [e] يفضّلونها فى التعظيم وسنذكر ذلك ۞ وامّا حصونها فانّ فى عامّة نواحى فارس حصونًا بعضها امنع من بعض واكثرها بناحية سيف * بنى الصّفّار [f] ۞

وسأفصّل كلّ ما ذكرتُه مجملًا فابتدئ بذكر ما فى كلّ كورة من النواحى الّتى تشتمل على القرى وشهرت [g] فى الدواوين باعمال مفردة ورساتيق مستقلّة بضياعها فمنها ما يخلو من المنابر ومنها ما بها منابر وربّ كورة هى اكبر واعرض ومدنها ونواحيها فى التسمية اقلّ ممّا هو اصغر منها ثمّ اتبع ذلك بتفصيل كلّ ما ذكرتُه مجملًا ان شآء الله ۞ نواحى كورة اصطخر ناحية يزْد وهى اكبر ناحية منها [h] وبها من المدن كَثَه [i] وهى القصبة ومَيْبُذ ونائين والفَهْرَج [k] وليس فى هذه النواحى كلّها ناحية بها اربع منابر غير هذه

بوره ، يوره ، دمره et تيبرزه , Jacut l.l. بيرده , Mokaddasí بيرزه , Edrîsî, I, p. 411 نيره , *Djihân-Numa*, p. ٢٧۴, 11 a f. نيرزه. Fortasse intelligitur ille fluvius de quo agit Quatremère, *Hist. des Mongols*, p. 443 et inter sese confusa sunt duo nomina نيريز s. تيبريز et خوره. *a*) A., B. et C. sine punctis; D. البجكان ut Abulfeda, p. ٩٠ et Jacut l.l.; E. دبخكان, Ous. et *Djihân-Numa*, p. ٢٧۴ recte بختكان. Vid. Ritter, *Erdkunde*, VIII, p. 766. Mokaddasí المحكان. *b*) D. البور , E. التور. Infra D. بون , E. بوف. Secutus sum C. *c*) A. et B. حردايان et الحردادان , الجردادان. E. الخوباذان , C. h. l. om.; D. الحوناذان (cf. Ous. p. 84 et 107). Vid. infra ad itinerarium a Schiráz ad Sirdján. Jacut l. l. الجَوْذان. *d*) Abulfeda, p. ۴٣ الجمكان (الجُكمان). Mokaddasí الحمكان , Jacut l. l. جنكان. *e*) C. (p. 58) مع هذه البيوت دمرانا. *f*) A. et B. بين الصفاق et sic semper الصفاق habent pro الصفار. *g*) A. ويسهرت , B. وشهرت , D. وتنحرف. *h*) C. et D فيها. *i*) A. كثّه. *k*) E. وبهرة. Vocales secundum *Kámus*. In ed. Jacut الفهْرج exstat.

الناحية، وناحية الرُّوذان كانت من كرمان فحُوّلت الى فارس ويكون امتداد a
هذه الناحية فى الطول نحوًا من ستّين فرسخًا، وأَبَرْقُوه ومدينتها ابرقوه،
واقْلِيدُ ومدينتها اقليد، والسَّرْمَق ومدينتها السرمق b، والجُوْبَرْقان c ومدينتها
مُشْكان، والأَرْخُمان d ومدينتها الارخمان، وجارين e ليس بها منبر، *وقوين
ليس f بها منبر، وطرخيشان g ليس بها منبر، وبَوّان h ومدينتها المُرَيْزِجان i،
والرُّهْنان k ليس بها منبر، وبرم l مدينتان اباذه m وهى *قرية عبد الرحمن
ومهرزنجان n، وَخورستان o وليس بها منبر، والبودنجان p ومدينتها البودنجان

---

a) C. et E. مقدار. Hinc incipit magna lacuna (7 s. 8 foliorum) in optimo Codice A. *b*) E. بپارسی آن هردورا كليد وسرمه خوانند. *Djihán-Numa*, p. ٢٦٦ quoque formas جرمق et سورمه habet. *c*) B. والحوىرىان, C. والخوبرنان, D. والخويبرقان. Epit. Par. والجوبرقان. *d*) B., C. et D. sine punctis. E. ارجمان. *e*) B. وهاردون. Secutus sum E. (cf. Ous. p. 86), qui etiam infra idem habet, nempe حارين, ubi B. حاربر, D. جبان. C. habet وجام. *f*) Sic edidi secundum E.; B. العس وفرىد; C. et D. om. *g*) B. وطرحمسان, C. وطرحسان, E. طرخينسان, D. infra طرحشيان, ubi E. طرحسان (Ous. p. 113 طرجنسان), Edrîsî, I, p. 414 الطرخشان, Mokaddasí MS. p. 215 وطرحسان. *h*) Secundum E.; B. حوان. Vid. infra et cf. Jacut in *Moschtarik*. *i*) Recte sic, ut videtur, E., vid. Jacut in v.; B. sine punctis, D. المزيرجان, Epit. Par. المويرجان. Infra idem nomen in alia provincia occurrit. *k*) B. h. l. sine punctis. E. hic وهبان, infra رهبان ut Ous. p. 113. D. infra sine punctis. Edrîsî رهنان, Mokaddasí l. l. وهنان. Cf. Jacut in v. *l*) D. h. l. بزم, Mokaddasí l. l. دروم sed p. 211 درم. *m*) B. et D. h. l. sine punctis. Jacet secundum mappas inter Iklîd et Jezdakhast. Sub nomine قرية عبد الرحمن memoratur a Kazwínio. *n*) B. haec om.; E. مهرويان. Edrîsî, I, p. 416 مهريرجان quocum conferre jubet Juynboll مِهْرِبجان (*Merâcid*, III, p. ١٨٠), quod vero secundum *Moschtarik* est من قرى كازرون. Eadem autem nomina in diversis provinciis interdum recurrunt. *o*) B. وحوسا, E. حوستان. Secutus sum Sprenger, *Post- und Reiserouten*, p. 72. Cod. Mokaddasii, p. 218, minus recte خوزستان. Vid. inter alios Chardin, *Voyages*, ed. 1811, t. VIII, p. 495: "il y a — une sorte de prunes sauvages, que les gens du pays nomment *kauré*, mot qui se prend aussi pour toute sorte d'arbre épineux; et c'est d'où est venu le nom de *Courestoon*, que porte ce village, qui veut dire *amas d'arbres épineux*." Langlès pronunciat Khârehstâun et Khâurehstâun. Urbs ejusdem nominis in provincia Ardaschír Khorra exstat. *p*) B. h. l. المرودىحان, infra البرودىحان, C. h. l. المردىحان,

*وهى قرية الآس، وصاهك الكبرى ولها منبر [a]، وصاهك الصغرى ليس بها منبر، ومروسف [b] ليس بها منبر، وشهر فاتك ومدينتها شهر فاتك، وقَرَاةُ ومدينتها هراة، والرُّوذان ومدينتها الروذان وبها [c] أَبَان وكمس [d] وخَبْر [e]، والأَذْكان ومدينتها الاذكان، وسرشك [f] وليس بها منبر، والراذان [g] ليس بها منبر، والبَيْضَآء ومدينتها البيضآء، وهزار ومدينتها هزار، ومائيين [h] ومدينتها مائيين، وايرج [i] ومدينتها ايرج، ونوده [k] ليس بها منبر، ورَامَجِرْد وليس بها منبر، والطَّسُّوج ومدينتها خُرَّمَة [l]، والجيبرة وبها منبر، والكاشْكان [m] ليس به منبر، والمهرجاسقان [n] ليس بها منبر، وجفوز [o] ليس بها منبر، وحمر [p] ليس بها منبر، *والفَارُوق

---

infra الموددحان, D. h. l. om.; E. h. l. بوددحان, infra sine punctis, sed Ous. p. 107 ut D. بودنجان. *a*) Haec in B. desunt, excepto nomine corrupto صاحل. Mokaddasí apud Sprenger, p. 70 et *Djihán-Numa*, p. ٣٦٧ صاهه habent pro صاهك. *b*) Secundum E.; D. مروست. *c*) B. add. هو. Deinde B. ادار, C. ادان, D. اثار, E. om. Infra in itinerario B. ادار, C. et E. انار, D. recte ابان. Sprenger l. l. p. 76 dat انار et اباد. Vid. Jacut in v. et sub روذان. *d*) Sic B., C. et D.; E. كلسن i. e. fortasse كُلشن *rosetum* (Ous. p. 86 كلس). Videtur legendum أَنَاس. *e*) B., C. et D. وحمر, E. وحمره (Ous. دحيره). Lectio constat loco mox secuturo ubi de خَبْر in provincia Ardaschír Khorra mentio fit (cf. Jacut, II, p. ٣٩٩, 2 seqq.) et additur: وهى غير خبر كورة اصطخر. *f*) Secundum E.; B. et C. وسرسك. *g*) B, والداوان. *h*) B., C. et D. ومادمن, Epit. Paris. ومابَيْن, E. مابس, Ous. ماهين. *Djihán-Numa*, p. ٣٦٧ مابين. *i*) B. ادلح et mox ايدج, C. وادلج, Epit. Par. وايذخ, D. et E. ابرج. Vid. Jacut in v. et infra. *k*) *Sic*. A solo B. memoratur. *l*) B., C. et D. sine punctis, E. جرمه. Infra bis recurrit nomen. Vid. Jacut in v. Mokaddasí p. 206 جرمه, p. 215 الجربه. Deinde B. et C. والحدره, E. حمره. Veram lectionem praebet D., vid. *Moschtarik*. *m*) E. كاسان, C. et D. om., vid. Edrîsî, I, p. 392. Locum ejusdem nominis prope Kázerun memorat Jacut. *n*) Secundum E.; B. المهرحاسمان; C. et D. om. Mokaddasí MS. p. 211 memorat regionem مهرجاناواذ in hac provincia. *o*) A solo B. hic memoratur et scribitur حغور. In hac regione est lacus Bakhtegán, in cujus descriptione infra nomen recurrit. Scribitur ibi حفوز in B., حعون in C., جغرو in D. *p*) Secundum E.; B. (ليس) حنرلس, C. et D. om.

ليس بها منبر[a]، والسرواب[b] وبها منبر، وكمين[c] وبها منبر، والرون[d] ليس بها منبر، والأُرْد[e] ومدينتها بَنَجّه، وكُرْد[f] ومدينتها كرد، وكَلّار[g] ليس بها منبر، وسروستان[h] ليس بها منبر، والاوسبنجان[i] ليس بها منبر، والسُّرْدَن[k] ومدينتها اللُّورجان[l]، واسلان ليس بها منبر، والبامان[m] ليس بها منبر، والخُمَايْجَان[n]

---

a) E. فاروق اين هر چهار منبر دارند. Vid. Jacut et Mokaddasí apud Sprenger l. l. p. 69 ult. B. العارون. *Djihán-Numa*, p. ٣٩٧, 7 فارون. b) B. et C. والسرواب, Ous. p. 87 سرواب, E. سروات; D. والسرداب. c) Pro كمين Ous. habet مكى, Epit. Par. كمير. *Djihán-Numa* l. l. كمين. E. de hac et superiore regione dicit هر دو منبر ندارد. d) Secundum E. et Mokaddasí apud Sprenger p. 70. B. الروق, C. et D. om., Ous. habet رادن. Apud Quatremère, ***Histoire des Mongols***, p. 445 occurrit رون, sed ipsi lectio suspecta est. Infra in catalogo urbium regionis frigidae B. الروق, D. الرون, E. رور (Ous. روز), Edrîsî الزوز. e) C. والادر, D. والازد, E. لاذن, Ous. لاون. Jacut habet: أُرْد كورة بفارس قصبتها تيمارستان. Nihilominus lectio satis certa est. Infra in catalogo urbium regionis frigidae B. اللارد, D. الارد, sed E. يزد. Mokaddasí habet الازد et ازد. Pro بنجه B. habet دنحر. f) E. كورد. g) B. وولار, C. et D. om. Recte scribit E. h) E. سراوسين, Ous. سرادسين, B. سروس, C. et D. om. Sprenger l. l. p. 69 et p. 73 Sarwistán (Cypressenhain) secundum Mokaddasí (qui bis سروستان, semel شروستان, semel سروشير habet). Infra in catalogo urbium regionis frigidae B. سروسنى, E. سروسين, D. سروشتن. *Djihán-Numa*, p. ٣٦٤, 7 سروستان. i) E. اسپندان, Ous. اسدان, B. والاوسنجار. Sprenger p. 70 ex Mokaddasí اسبانجان. Infra B. الاوشجان, E. اوليخان, D. الاستجان et ibi Mokaddasí habet الاوسخان. *Djihán-Numa*, p. ٣٦٦ اسفندان. k) B. السرون, C. والسردر. l) E. لوردكان, Epit. Paris. اللوردجان. Utraque forma a Jacut memoratur. Deinde B. اسار. Secutus sum E. m) Secundum E.; B. والبادنان, C. et D. om. n) B. sine punctis, E. خمايكان.

السفلى ليس بها منبر ۞ وامّا نواحى كورة أَرْدَشِير خُرّه فانّ شِيرَاز هى مستقرّ العمّال ولها ثلاثة عشر[a] طسّوجًا فى كلّ طسّوج قرى وعمارات متّصلة ينفرد كلّ طسّوج بعمل فى الديوان مفرد منها[b] طسّوج كفرهٔ العليا طسّوج كفرهٔ السفلى[c] طسّوج كبير[d] طسّوج جُوَيم[e] طسّوج الدسكان[f] طسّوج تَنْبُوك[g] طسّوج الكارنيان[h] طسّوج الاشاربانان[i] طسّوج انبديان[k] طسّوج شاهمرنك[l] طسّوج شَهْرَسْتان[m] طسّوج الطيريان[n] طسّوج خان[o]، وبهذه الطساسيج منبران احدهما شيراز وهى محدثة فى الاسلام من هذه الطساسيج من الشاهمرنك ومن الطيريان ومن * اشاربانان ومن التنبوك ومن الكارنيان وموضع مسجد الجامع والاسواق من شاهمرنك وموضع دار الامارة من الاشاربانان، وجُوَيم بها منبر ۞ وسائر نواحى اردشير خُرّه جُور ومدينتها جور، مِيمَنْد ومدينتها مائين[p]، والصِّيمكان ومدينتها الصيمكان، وخُوَار[q] ليس بها منبر،

---

a) E. دوازده. Enumeratio eorum in C. et D. deëst. b) B. فيها. c) B. المعلام. Recte E. (Pro كفره B. habet لفره). d) B. كمير, E. كمير; vid. Ous. p. 87. e) B. جرمر et infra جرمر, E. جرم, Ous. حديم. f) Sic B., E. دسكان, Ous. ديركان. g) B. et E. دمول, Ous. تيبول. Vid. Naçr apud Jacut. h) B. hic et mox الكاردمان, E. الكادنيمان, Ous. الكاديان. Infra templum ignis Schirazense appellatur a D. الكارنيان, ubi B. الكارسان. In E. hoc nomen cum praecedente conjungitur, tamquam unum nomen unius regionis. i) B. bis sine punctis, semel الاشاردادان, E. انباردانان, Ous. انباربانان. k) Sic E.; Ous. ابنديان, B. امريان. l) B. h. l. سهر صودد, mox bis ساهمودد. E. شاهمرنك, Ous. شاهرنك. m) B. سهرسان. n) B. sine punctis. E. incertum utrum طيران an طيريان habeat. Ous. طيرار. o) B. حاد. p) B. مادر, C. ماير, D. نابين, E. مادين, Epit. Par. مابين. Mokaddasí apud Sprenger, p. 69 مادن. Secundum Jacut, I, p. ۱۹۹, 15 ذائين. Si recte edidi, duo loca ejusdem nominis esse apparet, alter in provincia Istakhr, alter in provincia Ardaschír Khorra. q) B. حوان, C. h. l. والصرّار. D. om., E. male repetit صمكان. Vid. Jacut, I, p. ۱۹۹, 11 et in v.

الفرجان [a] وليس بها منبر، الباسكجان [b] وليس بها منبر، الخُنَيْفغان [c] ليس بها منبر، خَبْر [d] وبها منبر وهى غير خبر كورة اصطخر، الباذوان [e] ليس بها منبر، خورستان [f] ومدينتها خورستان، الفوسكجان [g] ومدينتها الفوسكجان، همند [h] ليس بها منبر، *جيمرين ليس بها منبر، [i] هُرْمُز [k] ليس بها منبر، النشكانات [l] ليس بها منبر، الحسكان [m] ليس بها منبر، همكجان [n] ليس بها منبر، الكوهكان [o]، كيزرين [p] ليس بها منبر، سِيف * بنى الصفّار [q] ليس بها منبر وفيها باسكوت [r] وباورم، سِيف آل ابى زُهَير ليس بها منبر، سِيف عُمارة ليس بها منبر ويعرف

---

a) Jacut l. l. et Edrîsî, I, p. 392 البرجان. *b*) B. الباسكان, E. sine punctis; ceteri om. *c*) B. الحسهمار, E. حمعان, Ous. p. 88 خنيفان; vid. Jacut, II, p. ٤٧٩, 6. *d*) B. حمر et mox خمير; vid. Jacut in v. et I, p. ١٩٩, 11 et 15 (Sprenger p. 69 جبير, ut semel in Cod. Mokaddasii). *e*) Sic C.; B. الماروان, E. داذوان, Ous. نادوان, D. om. *f*) B. primum حورسان, deinde حورسان ut C.; E. حورستان, Ous. خورسان. Mokaddasí bis خوزستان ut D. et Jacut, I, p. ١٩٩, 15, sed recte Sprenger, p. 69 pronunciat *Chawarustán*, vide supra ad p. ١٠١, *o*. *g*) D. الغوثنجان, E. فرسكان, Ous. دريكجان, C. om. Fortasse legendum الفوشجان. *h*) E. هميد ut Ous.; C. et D. om. *i*) Addidi ex E., qui habet حمرس. B., C. et D. om. Sed infra in catalogo urbium regionis calidae nomen recurrit et scribitur in B. حمرير, in D. uti in textum recepi. In E. ibi deëst. *k*) C. وهرمزا, E. هرمر. D. om. *l*) B. sine punctis, C. والنشكانات, E. سكانات, Ous. سكانات. Ab hac regione non videtur differre ea quae in *Djihán-Numa*, p. ٢٦٣, ult. بوشكانات appellatur. *m*) Sic B. et C. ut Ous.; E. حمكان, D. om. *n*) B. حمهان, C. وهكجمان, E. هكان, Ous. هبكجان, D. om. *o*) C. وكوكهان, E. لومكان, Ous. كومكان. *p*) B. كمر, C. كيرى, E. كمرى, Ous. كسرى. Infra in catalogo urbium regionis calidae recurrit, ibique B. habet كمررين, D. كيرريق, E. om., Mokaddasí, p. 215 كيزرين. *q*) B. ببين الصفاق. *r*) ? B. داسكور, C. داسكوت, ceteri om. Deinde B. ودررم. Secutus sum C.

بالجُلَنْدَى [a]، كُرَان [b] ومدينتها كران، وسيراف وبها ثلاثة منابر سيراف وهى القصبة ونَجيرَم وجَمّ [c]، دَشْتُ بَارِين [d] وقصبتها الغُنْدِكَان وبها الفَهْلَق [e] مدينة، دشت الدستقان [f] ومدينتها صفارة [g]، تَوَّج ومدينتها توّج، الاغرِسْتان [h] ومدينتها الخربق، كبير [i] ومدينتها كبير، وكَارَزِين [k] ومدينتها كارزين، ايرز [l] ومدينتها ايرز، سَمِيرَان [m] ومدينتها سميران، كُوَار ومدينتها كوار، الكَهْرجان [n] ليس لها مدينة ۞ ومما فى البحر من الجزائر المنسوبة الى كورة اردشير خُرّه جزيرة

---

a) B. h. l. male بالجلودى. *b*) B. كسوان, C. وكراز. Vid. Jacut in v. et I, p. ١٩٩, 15, et sic restitue III, p. ٢١٧, 11 pro كوان (eadem pagina bis corrigatur نجيرم pro بباحيبرم). *c*) B. حمه, infra جم; D. حم, infra خم; Epit. Paris. et Edrîsî, I, p. 397 خم. E. h. l. خم, infra جم; C. h. l. om., infra جم. *d*) E. دشت وارين. Urbem ut D. et Jacut appellat غندجان. *e*) Sic edidi secundum D.; B. العهان, ceteri om. *f*) B. الرسعان, D. الرستغان, E. omisso دشت habet دستيقان ut Ous. Infra in descriptione fluvii نهر الشاذكان, B. habet الرستقان, D. الرستغان, E. رشيقان, et in catalogo urbium regionis calidae B. et D. الدستقان ut quoque Jacut III, p. ٢١٧, 16 (E. ibi habet ورشاق). *g*) B. h. l. صعاده ut D.; infra صعاره. Secutus sum E. Quod *Merácid* in v. habet صفارة male est pro صفاوة. *h*) B. الاعرسان. Vocales ex Epit. Paris.; E. fortasse melius ut Ous. habet لاغرستان. Nomen urbis scripsi secundum E. (Ous. خريق); D. الحرنق, B. الحوس, Epit. Par. الجَرْمق. Infra in catalogo urbium regionis calidae B. الحريق, E. خرمى, D. om. *i*) Secundum B., D. et Edrîsî, I, p. 392. Epit. Par. كبز. E. h. l. شكبير, sed infra كبير. C. infra كمر. *k*) B. et E. كازرين (vid. Ous.), D. male كازرون. *l*) ? B. h. l. ايزر, D. ابور et اتور, E. ادرن, Ous. امدن. Infra E. ابور et ادكر, D. ابرد, B. ادرو; Mokaddasí p. 215 ادور. *m*) Mokaddasí شميران (locum transscripsit Jacut, I, p. ١٩٢, 15). Vid. *Djihán-Numa*, p. ٢٦٨, 3 a f. (Pro ابراهستان ibi lege ابراهستان). *n*) B. الكمرحان. Recte C. et E. Infra B., D. et E. كهركان habent.

* بنى كاوان [a] وهى لافت وبها مدينة [b] ، وأُوَال وبها مدينة ، وخارك بها منبر [c] ✿
وامّا نواحى كورة دارابَجِرْد كرم [d] وبها منبران احدهما اباذه [e] والآخر كردبجرد [f] ،
المصّ [g] ومدينتها المصّ ، فَسَا [h] ومدينتها فسا ، طَمِسْتان [i] وبها منبر ، المُحوّلة [k]
ليس بها منبر ، الكُردبان [l] وبها منبر ، ازبراه [m] ومدينتها ازبراه ، سنان [n] ومدينتها
سنان ، جُوَيم ومدينتها جويم ، جَهْرَم ومدينتها جهرم ، الفستجان [o] وبها منبر ،

---

a) D. et E. بركاوان. Deinde E. لافث. *b*) C. منبر. *c*) C. مدينة. *d*) Sic B., Mokaddasí MS. p. 28 et 205 et ex eo Jacut, II, p. ٥٩٠, 12, Edrîsî, I, p. 395. Sprenger, p. 69, cf. p. 75 pronunciat Kurem. E. كويم, Ous. كوم, D. لاوم, Epit. Par. كُوم, C. كروم. Infra in itinerario a Schiráz ad Tárim appellatur a B. كور, C. كورم, D. كر (Edrîsî كرا), E. (Ous. p. 109) كروم. Vid. porro Ritter, *Erdkunde*, VIII, p. 759. Chardin, *Voyages*, ed. 1811, t. VIII, p. 488 pronunciat Gormouth. *e*) Sic B.; D. اياده, E. et Epit. Par. sine punctis, Ous. اماده. *f*) B. كردنجرد, E. كردىحره, Ous. كرد, D. كربجرد. *g*) Sic B., C., D.; E. فص (Ous. قصر). *h*) E. بسا. *i*) B. طمسيان, C. et E. طمسان, Ous. طبيسان. D. recte طمستان, Epit. Par. طمِسْتان. Jacut praescribit طَمَسْتان (III, p. ٥٤٧). In itinerario B. habet طمسان, C. طمسان et طمسان, D. طبستان, E. طميسان. Mokaddasí p. 28 طمبستان, p. 205 طستان et ex eo Sprenger p. 69 (cf. p. 75) et Jacut II, p. ٥٩٠, 12 طبستان. *k*) B. المُحوّلة, E. بحولة ut Ous., ceteri om. *l*) B. الكردار, C. الكردمان, E. et Ous. كردمان. D. الكودبان, Epit. Par. الكُوذيان. Mokaddasí p. 28 الكوديان, p. 205 الكُرديان et hinc Sprenger l. l. كوديان. Jacut, II, p. ٥٩٠ ut edidi. *m*) Sic D., qui hanc regionem et sequentem inter regiones provinciae Arradján enumerat, et Mokaddasí, p. 28. P. 205 habet hic ازيراه ut Epit. Par.; B. اربراه, E. sine punctis, Ous. اربيراه. *n*) D. h. l. et E. sine punctis. Ous. سان ut B. infra in itinerario. E. ibi habet سبان. Edrîsî, I, p. 405 سيان s. سبان; Mokaddasí p. 28 سيباد. *o*) Scriptio hujus nominis incertissima est. B. hic et infra in itinerario العسكان; D. h. l. الفشيكان, infra الفيشكان; C. h. l. الفشيكان, infra المستجان

الداركان [a] وبها منبر، ايج [b] وبها منبر، الاصطهبانان [c] وبها منبر، نَيْرِيز [d] ومدينتها خيار [e]، المُرَيْزَجان [f] وبها منبر، المارْوَان [g] وبها منبر، حشوا [h] ومدينتها رُونْنج [i]،

(cf. Sprenger p. 75); Epit. Par. الفسْتجان; E. مسىكان et مسيكان (Ous. مسيكان); Edrîsî, I, p. 405 الفسيكان. Apud Beládsorí, p. ٣٨٩, habentur formae فشكاتي et الفيشكان. Denique apud Jacut فُسْتَجان (pro quo Juynboll ad *Merácid*, II, p. ٣٥٣, ann. 9 restituit فستكان), فشتجان, فسنْجان (?) et المُسْتَجار (? Nomen المستجان infra occurret in provincia Sabur). *a*) B. hic et infra الدراكان, C. الدركان, E. h. l. وركان, infra داواكان. Vid. Quatremère, *Histoire des Mongols*, p. 442. *Djihán-Numa*, p. ٢٦٩ زركان. *b*) B. الج, D. ابج. Quoque dicitur ايك, vid. Jacut, I, p. ٤١٥ et II, p. ٥٩., 12, *Djihán-Numa*, p. ٢٦٩. Apud D. haec et sequens regio denuo in provincia Arradján collocatae sunt. *c*) B. الاصطهاذات, E. اصطهادات, Ous. اصطهغايان, C. الاصفهانان, D. الاصبهنذات; infra B. الاصطهمادات, D. الاصبهبذات, E. اصطهباجان. Secundum Mokaddasí scribendum est الاصبهانات, nam sic habet non tantum p. 28 et 205, sed quoque p. 14 in capite ذكر الاسامي واختلافها, et hinc Sprenger p. 69 et Jacut I, p. ٢٩٢, 5. Vid. *Hist. des Mongols*, p. 441 et *Djihán-Numa*, p. ٢٦٩. *d*) B., C., D. et E. دمردن, Ous. عريز; Mokaddasí p. 205 نيردن, p. 208 دبودن, p. 215 النبردز; Edrîsî, I, p. 395 نبرين. Vid. *Hist. des Mongols*, p. 441, 443, 449, Jacut in v. et *Djihán-Numa*, p. ٢٦٩. Dsahabí, ed. de Jong, p. ٦٨ نيريز. A Jacut, I, p. ٣٠٠, 4 inter urbes provinciae Istakhr enumeratur. *e*) B. حيار, C. خان, D. حمار, Mokaddasí p. 205 خبار. Idem locus, ut videtur, in *Hist. des Mongols*, p. 441, 446 appellatur خيره. *f*) B. المريوحان, E. دردحان, C. المزيرجان, D. المزيرجان. Infra in itinerario B. المريزحان, D. برجان, E. مزيزجان (Ous. p. 110 مرزندخان). Supra ejusdem nominis urbs in provincia Istakhr memorata est. *g*) B. et C. المادوان; vid. Jacut. *h*) B., C. et Ous. حسوا, E. حسو, Epit. Par. حشوا; Mokaddasí p. 205 حسوا. *Hist. des Mongols*, p. 442 حشو (Haschou). *i*) B. رودج, Ous. رويج, E. رودمج, C. رودمج.

رستاق الرستاق وبها منبر، قنطرة[a] ليس بها منبر، سوانجان[b] ليس بها منبر، فرّج[c] وبها منبر، تارم[d] وبها منبر، الماسكانات[e] وبها منبر، *شقّ الرستاق[f] ليس بها منبر، *شق الروذ ليس بها منبر، تالات ليس بها منبر، شقّ الماسَنان ليس بها منبر[g]، رمّ شهريار[h] مدينتها الرمّ ✩ وامّا نواحى كورة سابور[i] سابور ومدينتها سابور، كازرون ومدينتها النّجنّجان[k]، الماءال[l] ليس بها منبر، وجَفْتة[m] ليس بها منبر، دزبز[n] ليس بها منبر،

---

a) Secundum E. et Ous.; B. فطره, ceteri om. *b*) Secundum Ous.; C. et E. سوادىحان, B. سولىجان. *c*) B. ومدح, E. فرح, Ous. مدح. Vid. Sprenger, p. 69 et Jacut in v. Diversus ab hoc locus est پرک s. برک quem in vicinia memorant Mokaddasí et *Djihán-Numa*, p. ٢٦٩. *d*) B. بادر, Epit. Par. بازر, E. et Ous. بازم. Edrîsî et Mokaddasí apud Sprenger بارم, sed p. 207 تارم; vid. Jacut in v. et cf. infra in itinerario. C. et D. om. *e*) B. الباسكابات, D. الماسكنان, Epit. Par. المَسْكَنَات, E. مساكىات, Ous. مساكنات. Infra B. المشيكادات, E. مشكانات. Jacut habet ماسكَنَات. Mokaddasí p. 205 et 215 المسكانات. *Histoire des Mongols*, p. 442 ممسكانات, quod p. 446 pronunciatur ميشكانات. *f*) Secundum E.; B. سوق المرساق, C. sine punctis. *g*) Haec omnia in B. desiderantur. Supplevi ex E. qui habet شق السروذ، بالاب، شق الماسادان, ex C., ubi سف الدولاب سف الماسان, et Ous. p. 89. *h*) B. سهيران ut Mokaddasí p. 205. D. شهريران, E. سهيراز, Ous. شهرازو, C. سهوىار. Pro رم omnes, exceptis B. et C., scribunt زم. Infra in itinerario a Schiráz ad Tárim eadem urbs significatur nomine رمّ المهدى et الرمّ, ubi omnes Codd. زم. *i*) E. semper شابور. Nomen urbis in *Djihán-Numa*, p. ٢٧٠, 1 est بشاور. *k*) B. et E. sine punctis, Epit. Par. الخنىجان; vid. Jacut in v. *l*) Sic E. et C. in marg. cum صح (in textu السامات); B. الماسال, Ous. باسان. *m*) E. حفمه, Ous. حفيه, C. وحمعه, B. وحمه. Infra B. حصه aut حمه, C. et D. حمعه, Epit. Par. حفنه. Scripsi جفتة, quum, ut infra patebit, ibi sit pyraeum, quod eodem jure quo جنبذ appellatur, quoque جفتة dici potest. *n*) Conjec-

جروج [a] ليس بها منبر، خَشْت [b] ليس بها منبر، كُمارِج [c] وبها منبر، هنديجان سابور وليس بها منبر، التيرَمَرْدان [d] ليس بها منبر، الزاماجان [e] ومدينتها الزاماجان، الخُوبَذان ومدينتها الخوبذان، النُّوبَنْجان [f] ومدينتها النوبنجان، شعب بَوّان ليس بها منبر، *تَنْبُوك المورستان [g] بها منبر، الجُويخان [h] ليس بها منبر، درخيد [i] ليس بها منبر، انبوران [k] ليس بها منبر، *جُنْبَذ المَلّجان [l]

---

tura sic edidi دز پير „vetus castellum" coll. nomine simili apud Jacut, II, p. ٥٧٣, 15. B. ودير s. دردر, infra in itinerario درين; E. h. l. دريف (Ous. (درتق), infra دوين. D. infra دير, Ous. دوبين, Edrîsî, I, p. 401 زرين s. رزيق. Mokaddasí MS. p. 28 دردن s. درق, p. 205 دران, p. 210 دردن s. دردف. Cf. Sprenger p. 69 et 74. *a*) B. حورح, C. حروع, E. خورج (Ous. (خواج). Edidi جروج locum non diversum putans a جروز (Jacut, II, p. ٩٩). *b*) Sic E. et Ous.; B. حشت. Jacut in v. خَشْت, sed III, p. ٥٠, ult. خَشْت. Mokaddasí habet حشت, جشت et خشت. *Djihán-Numa*, p. ٢٧٠ خشت. *c*) B. كمادح, C. كمارخ, E. كهاوم, Ous. كياوم. Est *Komaredsch*, vid. e. g. Petermann, *Reisen*, II, p. 162 seq. *d*) Sic recte E. et Ous., vid. Jacut in v.; B. السرمروان. *e*) B. السرامحان, C. الزامحان, D. الزايجان, Epit. Par. الزَنْجان. E. om. *f*) E. نوبندكان. *g*) B. دمول الموريان, E. دمول المورسيان, Ous. بينول المور, C. تمبوك, D. عمل المورستان, Edrîsî, I, p. 392 تنفوك. Locus ejusdem nominis in ditione Schiráz exstat. Vid. supra p. ١٠٤ ann. *g*. *h*) B. sine punctis, E. درنحان, Ous. درنجان. Vid. Jacut in v. *i*) B. درخند, E. ذرحيد, Ous. درجند, C. درحيد. Ab hoc loco fluvius nomen habet, vid. supra p. ٩٩, paen. et infra. *k*) Secundum E. et Ous.; B. sine punctis; Edrîsî, I, p. 397 انبودان. Infra, ubi describitur fluvius نهر الخوبذان, B. habet امووران, D. ابيوران, E. دموران, Ous. p. 97 هوران. *l*) B. حسد الملحان, C. حيد الملحان, E. et Ous. كنبذ ملغان. Vid. Jacut sub ملجان. Idem III, p. ٥, ult. ex Mokaddasio habet جنبذ, ut hic p. 28 et 205 habet, sed p. 210 legimus: وجنبذ ملغان مدينة — وملغان قرية من حد ارجان خربة. Videtur legendum ملغان s. ملقان.

ليس بها منبر، المامغان[a] ليس بها منبر، * آسك ليس بها منبر[b]، فرطاست[c] ليس بها منبر، بين[d] ليس بها منبر، دمر[e] ليس بها منبر، كرو[f] ليس بها منبر، بادست[g] ليس بها منبر، بَهْلُو[h] وليس بها منبر، البهيسكان[i] ليس بها منبر، ازادجرد[k] ليس بها منبر، الروديجان[l] ليس بها منبر، كام فَيْرُوز ليس بها منبر ولها خمسة رساتيق ارز[m] وبازر واشتادان[n] وكاكان وآتشْجاه[o]، المستجان[p] ليس بها منبر، الزنجان[q] ليس بها منبر، بندرهبان[r] ليس بها منبر، خُمَايْجان[s] العليا ليس بها منبر، السيسكان[t] ليس بها منبر، مُورَق[u] ليس بها منبر،

---

a) ? B. الماوعان, C. المافعان, E. دامغان. Fortasse Edrîsî, I, p. 392 المامان. *b*) B. et C. om. Supplevi ex E. qui اسد, et Ous. qui اسل habet. *c*) Secundum E. et Ous.; B. داست ومطا *sic*. *d*) B. دنى, ceteri om. Pagum hujus nominis prope Fairuzabád i. e. Djur memoratur in *Kamus*. *e*) *Sic* B., et quidem verba دمر ليس بها منبر bis habet. E. شكره, Ous. شكير (fortasse ortum conjunctione hujus et sequentis nominis). *f*) Secundum B. et C.; E. om. *g*) Ex solo B. *h*) B. دهلو, C. نساهو, E. دهلوق, Ous. بهلوق. *i*) B., C. et E. sine punctis. Ous. دهلسكان. *k*) Solus B. habet et quidem sine punctis. *l*) Sine punctis a solo B. memoratur. *m*) B. et E. ارر, Ous. ارزو. Deinde E. وبادر, Ous. نادر. *n*) Secundum E. (Ous. استادان); B. واساوان. *o*) B. والسجاه, E. ادشكاه; vid. Ous. Pro praecedente كاكان fortasse legendum est كاسكان (Jacut, IV, p. ٢٢٧). *p*) Secundum C.; B. السجار, E. مسجان, Ous. مسيحان, Jacut المستجار. Cf. supra ad p. ١٠٧ ann. *o*. *q*) ? B. الدبحان, C. الردحان, E. ردحان, Ous. سجان. *r*) B. sine punctis. E. بندد هسان, Ous. مدر هيباز. *s*) B. et C. sine punctis. E. خمايكان. Apud Jacut, III, p. ٥, 22 quoque السغلى e provincia Istakhr huc trahit. *t*) B. et C. sine punctis. *u*) B. مور, C. et E. صور. Infra in descriptione locus بحيرة تور; D. habet موروق, C. دمور, E. مور. In enumeratione urbium regionis calidae infra D. habet برز et يوز, B. موز, E. موز. Cf. Jacut in v.

دانيين [a] ليس بها منبر، دَوّان [b] ليس بها منبر، خُرّة [c] ومدينتها خُرّه، صِرّام [d]
ليس بها منبر ۞ وامّا نواحى كورة أرّجان ارّجان ومدينتها ارّجان، بازرنج [e]
ليس بها منبر، بلاد سابور [f] بها منبر، رِيشَهْر [g] بها منبر، بنوليس [h] بها منبر،
كهكاب ليس بها منبر، دير ايّوب [i] ليس بها منبر، المَلّجان [k] ليس بها منبر،

---

a) B. داودن, infra semper دارين, E. h. l. دارسن, Ous. دادى. Vid. Abulfeda, p. ٣٢٣. In *Djihán-Numa*, p. ٢٦٤, 6 دادين scribitur. *b*) A solo B. memoratur. Vid. Jacut. *c*) B. خُدّه, C. حره, D. جره, Epit. Par. خَره; E. om. Infra in descriptione fluvii نهر جرشيق, B. خُره, D. et E. جره, et in catalogo urbium regionis calidae B. حره, C. et E. جره. Mokaddasí MS. p. 28 et 210 جره, sed p. 205 خره. Beládsorii Codd. (p. ٣٨٨ seq.) خره et خُره. Jacut (II, p. ٩٧) habet جِرّه s. كُرّه tamquam صقع فارس, non vero dicens in qua provincia jacet; III, p. ٥ ex Mokaddasio memorat جَره inter urbes provinciae Sábur. Vullers habet جُرّه „vicus Schirazensis." In *Djihán-Numa*, p. ٢٧٠ جره s. كره inter urbes provinciae Sábur (خوره شابور s. كره شابور). Vid. porro Jaubert ad Edrîsî, I, p. 396 et cf. quoque Langlès ad „*Voyages du chevalier Chardin*," ed. 1811, t. VIII, p. 211. *d*) C. صوام. Vid. Jacut sub چرام. Infra E. corrupte حمران (Ous. جران), Mokaddasí MS. p. 215 ضرار. *e*) B. h. l. دادردنج, infra بازرنج; E. h. l. بازريج, infra بازرنك; C. sine punctis, D. infra تازرْنج et بازرنج. Edrîsî, I, p. 409 بازيج, Abulfeda, p. ٥٨ نازرنج. Mokaddasí MS. p. 215 بارردنج. *f*) Haec regio, cujus urbs est جُومَه. a Mokaddasio appellatur بلا سابور, quod pro بالا سابور „Sabur superior" esse videtur, nam provincia montosa est et a rege Sabur nomen habet. Nostri Codd. hic et infra بلاد habent. *g*) B. ديشهر, C. ودسهر, E. ريسهن. *h*) Secundum E. (Ous. نبوليس); B. habet بيو ليس, C. دولس. *i*) Sic recte E.; B. ابور, C. sine punctis. Male Juynboll (Add. ad *Merácid*, V, p. 529) secundum Mordtmann, p. 64, pron. دير ابونا. *k*) B. et C. sine punctis. In E. deëst. Vid. supra ad p. ١١٠ ann. *l*.

السلجان[a] ليس بها منبر، الجلادجان[b] ليس بها منبر، دير العُمْر[c] ليس بها منبر، فرزك[d] بها منبر، هنديجان ارّجان ليس بها منبر، مَهْرُوبَان بها منبر، جَنَّابَة *بها منبر[e]، شينيز[f] بها منبر، صوان النجس[g] ليس بها منبر ✿

واما رمومها فانّ لكلّ رمّ منها مدنًا[h] وقرى مجتمعة قد ضُمّن[i] خراجَ كلّ ناحية منها رئيس من الاكراد وأُلْزِموا اقامة رجال لبذرقة القوافل وحفظ الطرق[k] ونوائب السلطان اذا عرضت وهى كالممالك، فامّا رمّ جِيلَوَيْه المعروف بالرميجان فانّ مكانه فى الناحية الّتى تلى اصبهان وهى تاخذ طرفًا من كورة اصطخر وطرفًا من كورة سابور وطرفًا من كورة ارّجان فحدّ منه ينتهى الى البيضآء وحدّ منه ينتهى الى حدود اصبهان وحدّ منه ينتهى الى حدود خوزستان وحدّ منه ينتهى الى ناحية سابور وكلّ ما وقع فى هذه[l] من المدن والقرى فمن[m] هذا الرمّ، وسجاختهم[n] فى عمل اصبهان البازنجان *وهم صنف من البازنجان الّذين هم برمّ[o] شهريار وليس من هاؤلآء البازنجان احد فى عمل

a) ? B. الممليحان, C. ادملحان, E. اسلجا, ultima littera vitio Codicis deleta, Ous. اسلجار. *b*) B. الحلاديحان, C. الحلادحسار, E. om. Infra B. الحلادكان et الحلادحان, C. الحلادحان, D. ut recepi, E. خلايكان et جلادحان, Edrîsî, I, p. 409 خلاجان. *c*) Vocales conjectura apposui. E. دير عمر. *d*) ? B. et E. فردك, C. دورك. Infra B. عدرل, C. ىورك, D. قرزل s. قرزل, E. دروك; Edrîsî, I, p. 409 بدرك. Denique Ibn Khordâdbeh, p. 55 بزرك. *e*) Supplevi ex D.; B. om. *f*) E. سيس, C. سنكىر. *g*) Secutus sum E.; B. et C. sine punctis; Ous. النخس. *h*) B. مدن. *i*) C. habet واضمن, Jacut, II, p. ٨٢١, 10 تضمن. E. كه خراج ومعاملت بضمان دارد. *k*) C. et Jacut الطريق et deinde ولنوائب. *l*) B. هذا; vid. Jacut l. l. 15. *m*) B. ومن. Vid. Jacut et E. وهر چه درين حدود باشد ديه وشهر جمله در شمار اين زم آيد. *n*) C. omnibus de رم جيلويه omissis, habet ويتاخم احد الزموم; Jacut l. l. 16 locum sic exhibet ويتاخمهم فى عمل اصبهان رمّ شهريار وهو رمّ البازنجان وهو رمّ جبيل من الاكراد وهم من البازنجان رهط شهريار وليس الخ. *o*) C. et E. quoque قوم habet pro رم. وهم من قوم

فارس الّا انّ لهم بها قرى وضياعًا كثيرة ، وامّا رمّ الديوان المعروف للحسين[a]
ابن صالح وهو من كورة سابور فانّ حدًّا منه يلى اردشير خرّه وثلاثة حدود
يحيط[b] بها كورة سابور وكلّ ما كان من المدن والقرى فى اضعافها فهى
منها ، وامّا رمّ اللّوالجان[c] لاحمد بن الليث وهو[d] فى كورة اردشير خرّه فحدّ
منه يلى البحر ويحيط بثلاثة حدود له كورة اردشير خرّه وما وقع فى اضعافه
من القرى والمدن فهو منه ، وامّا رمّ الكاريان فانّ حدًّا منه الى سيف بنى
الصّفّار[e] وحدًّا منه الى رمّ البازنجان[f] وحدًّا منه الى[g] حدود كرمان وحدًّا
منه الى[g] اردشير خرّه وهى كلّها فى اردشير خرّه ۞ وامّا احياء الاكراد بفارس
فهم الكرمانيّة والرامانيّة[h] ومدثر[i] وحىّ محمّد بن بشر والبقيليّة[k] والبندادمهريّة[l]
وحىّ محمّد بن اسحاق والصباحيّة والاسحاقيّة والاذركانيّة[m] والشهركيّة[n]
والطهمادهنيّة[o] والزباديّة[p] والشهرويّة والبندادكيّة[q] والخسرويّة[r] والزنجيّة

---

a) D. بالحسين , Jacut للحسن. *b*) Jacut تطيف. *c*) Jacut الريحان, sine dubio falso. *d*) B. om.; Jacut habet وهى. *e*) B. الصفاق. *f*) Jacut, II, p. ٨٢٢, 1 الريحان et bis addit ينتهى ante الى. *g*) Addidi الى, quod in B. et D. deëst, coll. E. et Jacut. *h*) Secutus sum E. (رامانيان) et Mokaddasí, MS. p. 215, coll. Jacut sub رامان. B., C. et D. الزامانية. *i*) Secundum Mokaddasí. B. sine punctis, E. وزم مدبر (Ous. p. 92 (زم بروحى), C. مدبر, D. مدين. *k*) B. sine punctis, C. البقلية, D. والقبيلة, E. دقبيلة, Ous. قبليه. Mokaddasí autem habet الثعلبية. *l*) B. والمنداذمهرده, C. والبندادمهريه, E. بندامهرنان, D. والبندامهريه et sic Mokaddasí. Deinde pro حىّ E. habet قبيلة. *m*) C. والاردكانيّة, D. والاذركانيه, E. اذركامان, Ous. اذركانيان, Mokaddasí الاذركانية. Fortasse leg. الارزكانية, coll. Jacut sub ارزكان. *n*) B., D., E. et Mokaddasí والسهركية. *o*) Sic B. sine punctis; C. والطبادهمه, D. والطارهنيه, E. طهاديان, Ous. طهاريان, Mokaddasí الطهادهنية. *p*) Secundum C., D. et Ous.; B. والزباديه, E. زباديان; Mokaddasí الزدادية. Deinde B. والسهرويه. *q*) B. والممداذليه, E. بنداذ كمان; C. والبنداقيه, D. والبنداوفيه; Mokaddasí البندافية. *r*) Secundum Mokaddasí; B. والحشرويه, C. et D. والجسرويه, E. حسرويان, Ous. خرويان. Deinde E. زنكيان.

والصفريّة[a] والشهياريّة[b] والمهركيّة والمباركيّة والاشتامهريّة[c] والشاهونيّة[d] والفراتيّة[e] والسلمونيّة والصيمريّة[f] والازاددختيّة[g] وانبرازدختيّة[h] والمُطّلبيّة والممالبّة[i] والشاهاكانيّة[k] والكجنيّة[l] والجليليّة، فهاؤلآء الّذين حضرنى ذكرهم من اسمآء هذه الاحيآء ولا يتهيّأ تقصّيهم الّا من ديوان الصدقات ويقال انّهم يزيدون على خمس مائة الف بيت ويخرج من الحىّ الواحد الف فارس الى مائة فارس واقلّ من ذلك واكثر وينتجعون فى المشتى والمصيف على المرعى الّا القليل منهم على حدود الصرود والجروم فلا ينتقلون ولهم من العدّة والباس والقوّة بالرجال والدوابّ والكراع ما يستصعب على السلطان امرهم اذا اراد تخيّفهم ويزعمون انّهم من العرب وهم اصحاب اغنام ورماك والابل فيهم قليل وليس للاكراد خيل الّا للبازنجان الّذين انتقلوا الى حدّ[m] اصبهان وانّما دوابّهم براذين وهم على حسن حال ويسار ومذاهبهم فى القنية والنجعة مذاهب قبائل العرب وقبائل الاتراك[n] وهم فيما يقال يزيدون على مائة حىّ وانّما حضرنى نيّف وثلاثون حيًّا ۞

---

a) C. et Mokaddasí والصقرية. b) C. والشهانبوده, E. سهماريان, Ous. شهماريان; Mokaddasí om. c) C. والاتسامهريه, D. والاسامهريه, E. اسامهريان, Mokaddasí اسبامهريّة. d) B. والساهويه, E. ساهويان, Mokaddasí السهاوبية. e) C. والمرانيّه, Mokaddasí القراثيّة. Deinde hi duo habent والسلموية. f) Secundum D. et Mokaddasí; B. et E. sine punctis, C. والصبريه. g) B. والاردادحمه, C. والسلاذحمه, D. والارادوجبيه, E. ازاددرختيان, Mokaddasí الارادذحيّة. h) B. والمرازدختيه, C. et D. والبرازدخنيه, E. بران دحتيان, Mokaddasí om. Deinde C. المطلبيه, D. والملطيمه, Mokaddasí المطبية. i) B. والهماليه. k) B. والساهلكانيه, C. et D. والشاهكانيه, E. سهاكانيان (Ous. سماكامان), Mokaddasí الساكانية. l) Ex solo C. addidi, nam Mokaddasí in suo Codice Istakhrii nomen habuisse videtur, quia dicit وبه من احياء الاكراد ثلاثة وثلاثون. Deinde E. et Ous. خليليان, Mokaddasí الحليلية. m) C. حدود. n) C. البربر.

واما حصون فارس فانّ منها مدنًا محصّنة بحصن ومنها حصون داخل المدينة وحواليها ارباض ومنها قهندزات [a] فى مدن ومنها حصون فى جبال منيعة مفردة عن البنيان قائمة بانفسها   واما المدن المحصّنة فانّها [b] اصْطَخر بها [c] حصن حواليه ربض، مدينة كَثَه بها [c] حصن وربض، البيضاء بها [c] حصن وربض، الشَّرْمَق بها [c] حصن وربض وقهندز، اقْليد [d] لها قهندز وربض، قرية الآس [e] لها قهندز وربض، شيراز لها قهندز * يسمّى قلعة شهموبذ [f]، جُور عليها حصن * وليس بها ربض [g]، كازَرين لها قهندز وربض، * كير لها قهندز وربض [h]، * ايرز لها قهندز وربض [i]، * سَميرَان لها قهندز وربض [k]، وفَسَا لها حصن وربض، دارابَجِرْد لها حصن وربض، رُوَيَنْج لها حصن وربض، وسابور لها سور وليس لها ربض، الجَنَّجان [l] لها حصن وليس لها ربض، جفتة [m] لها حصن ✿ واما القلاع فانّه يقال فيما بلغنى انّ لفارس زيادة على خمسة الآف قلعة مفردة [n] فى الجبال وبقرب المدن وفى المدن ولا يتهيّأ تقصّيها الّا من الدواوين وكذلك ما ذكرناه من المدن المحصّنة فانّى لا اقدر على تقصّيها وانّما اذكر جوامع ما اعرفه من ذلك الّا انّ فى هذه القلاع ما لم يذكره [o] لاحد من الجبابرة انّه قدر على فتحها عنوة منها قلعة ابن عُمَارة وتسمّى قلعة الدِّيكْدان [p] وتنسب الى الجُلَنْدَى ولا يقدر احد ان يرتقى اليها

---

a) Lector in marg. B. قهندز يعنى القلعة القديمة واصله كهن دز i. e. كُهَن دز. *b*) Addidi فانها. *c*) C. et D. لها. *d*) C. وقليد, E. كليد. *e*) E. addit ديه مورد. *f*) C. et E. pro his ولها ربض. D. شهموند, Epit. Par. سهمُوند. *g*) C. وربض. Deinde B. et C. كازرين, D. كازرون. *h*) Ex C. et E.; B. et D. om. Vid. supra ad p. ١٠٩ *i*. *i*) Ex solo E., qui habet ادور; vid. supra ad p. ١٠٩ *l*. *k*) Ex D. et E.; B. et C. om. *l*) B., C. et D. sine punctis, E. حيحان; vid. p. ١٠٩ *k*. *m*) Vid. supra ad p. ١٠٩ *m*. *n*) C. et D. منفردة; vid. quoque Jacut, III. p. ٨٣٨, 14. *o*) B. نذكر. *p*) B. الروادان, C. الديكان, E. دايفان, D. داكباياه بيره, *Djihân-Numa*, p. ٢٧١, antep. دانبان. Vid. Jacut l.l. et in v.

بنفسه الّا ان يرقى به فى شىء من البُنْجُر a وهى مرصد لآل عمارة فى البحر يعشرون b منها المراكب، وقلعة الكاريّان c على جبل طين * قصدها محمّد ابن واصل d فى جيشه فتحصّن بها احمد بن الحسن الازدىّ e فلم يقدر عليها، وقلعة سعيداباذ برامَجِرْد من كورة اصطخر وهى على جبل شاهق يرتقى f اليها فرسخًا وكانت فى الشرك تعرف بقلعة أسفندباذ g فلمّا كان فى h الاسلام تحصّن فيها زياد بن ابيه أيّام امير المؤمنين علىّ بن ابى طالب عمّ ونسبت i الى زياد ثمّ تحصّن بها آخر أيّام بنى اميّة منصور بن جعفر k وكان واليًا على فارس فنسبت القلعة اليه l فعرفت بقلعة منصور فتعطّلت m مدّة ثمّ بناها محمّد بن واصل الحنظلىّ فنسبت القلعة اليه وكان واليًا على فارس فلمّا اخذه n يعقوب بن الليث لم يقدر على فتحها الّا بامر محمّد بن واصل فخرّبها ثمّ احتاج اليها فاعاد بناءها وجعلها محبسًا لمن سخط عليه، وقلعة اِشْكَنْوان o من رستاق ماثين p المرتقى اليها صعب وهى

---

a) C. et Jacut, II, p. ٧١١ المحامل. *b*) B. يغسرون, D. يعرفون. Vid. C., Jacut et E. آنجا از كشتيها ده يك ستانند. *c*) C. et D. الكاربان, E. كاذيان, B. sine punctis, *Djihân-Numa*, p. ٢٧٢, 5 كاويان. Vid. Jacut in v. *d*) E. h. l. فاضل. Jacut, IV, p. ٢٢٥, 3 habet كان عمرو بن الليث الصفار قصدها. *e*) B. الارذى. *f*) Jacut, III, p. ٩٣, 21 يسير المرتقى. *g*) E. اسفنديار, Jacut اسْفيدباذ (B. om. بقلعة), *Djihân-Numa*, p. ٢٧١ ult. قلعهٔ سپيد. *h*) C. et D. om. *i*) C. et Jacut فنسب. Jacut post زياد addit مدة. *k*) Sic omnes. Legendum est جمهور, vid. e. g. *Kitâbo 'l-Oyun*, p. ١٥٣. Ceterum cf. Ibno 'l-Athír, III, p. ٣٢١. *l*) Jacut denuo addit مدة. *m*) B. فتعطل. Jacut ثم تعطلت مدة وخربت ثم استجدّ عمارتها. *n*) Jacut ملك — فارس. *o*) Sic B., D. et E.; C. اشكنون; Epit. Paris. اسكنواز. Vocales adscripsi secundum Jacut, qui habet اشكنوار. Idem locum quoque memorat sub nomine أُسْكُبُون. Vid. porro Defrémery, *Mémoires d'Histoire orientale*, p. 122 seq. *p*) Secun- B., C. et E.; D. et Jacut, I, p. ٢٥٣ ناثين.

منيعة جدًّا *وفيها عين ماء جارٍ a، وقلعة جُوذَرْز b صاحب كَيْخُسْرُو بموضع يسمّى السويقة c من كام فيروز وهى منيعة جدًّا، وقلعة الجِصّ d بناحية أرّجان فيها مجوس وباذْكُذارات e الفرس وأيّامهم تتدارس فيها وهى منيعة جدًّا، وقلعة إيرَج f وهى منيعة جدًّا، وامّا القلاع المنيعة التى g يقدر على الاحتيال لفتحها h فهى اكثر من ان يبلغها i حفظى ❀

وامّا بيوت نيران فارس فتكثر k عن احصائى وحفظى اذ ليس من بلد ولا رستاق ولا ناحية الّا وبها عدد كثير من بيوت النيران الّا القليل غير انّ المشاهير التى تفضّل على غيرها فى التعظيم منها بيت نار الكارِيان l ويعرف ببارْنوا m، وبيت نار بخُرّه n ينسب الى دارا بن دارا وبه يحلف المجوس فى المبالغة بايمانهم، وبيت نار عند بِرْكة جُور ويسمّى بارين o وحدّثنى من رأى به قد كتب عليه بالفهلويّة انّه أُنفق عليه ثلاثون الف p درهم، وبيت نار على باب سابور يعرف *بشَبَر خشين q، وبيت نار بباب سابور ايضًا على باب ساسان يعرف بجُنْبَذ كاوس r، وبكازُرون بيت نار يعرف بجفته s وبكازرون

a) Jacut وبها عين من الماء حارة. b) E. خودرت. c) Secundum D.; B. السويقه, C. الشريعه, Jacut, II, p. ١٤٥, 5 الشريعة. d) Secundum C., E. et Jacut sub قلعة (IV, p. ١٩٤). B., D. et Epit. Par. habent الحصن et sic Codd. *Merâcid*, II, p. ٤٤٢, 6 (ubi repone آثار pro آبار). Jacut, III, p. ٨٣٨, 17 الحَضْر habet. e) B. واياذكارات, C. واباذكارات. f) B. المرح. g) B. et C. om. h) C. بفتحها. i) B. et C. يبلغه. k) B. تكثر. l) B. et E. sine punctis. Vid. Jacut in v. et cf. Mas'udi, IV, p. 76. m) B. بنار فرا, E. om.; in Codd. D. tantum fragmentum nominis superest. Cf. Mas'udi l. l. p. 80. n) B. et D. sine punctis. Est خره سابور, vid. Mas'udi, p. 78. o) B. et E. نارين. p) D. add. الف. E. quoque سى هزار دينار. q) Conjectura sic scripsi. E. دسمر حسمن, D. بشر خشين, E. سوى حسين, Edrîsî, I, p. 413 سير حسين. r) Sic proponit legere Ous. p. 95. B. كلوسين, D. كلوشن, E. كلوش. s) Cf. supra ad p. ١٠٩, m. B. et D. بجفنه, E. حفمه, Ous. خيفه.

ايضا بيت نار يعرف بكلازن [a] ، وبشيراز ايضا بيت نار يعرف بالكارنيان [b]
وبشيراز بيت نار آخر يعرف بهرمز وعلى باب شيراز بقرية تعرف بالبَرْكان [c] بيت
نار يعرف بالمسوبان [d] ومن دين المجوس انّ المرأة اذا زنت فى حملها او
حيضتها لم تطهر الّا بان تاتى هذه النار [e] فتتعرّى لبعض الهرابذة فتطهر ببول
البقر ۞

وامّا انهار فارس فانّها نهر طاب يخرج من جبال اصبهان بقرب البُرْج [f]
فيصبّ الى نهر مَسن [g] وهو نهر يخرج من حدود اصبهان فيظهر بناحية
السُّرْدَن فيجتمعان عند قرية تدعى مسن ثم * يجرى الى [h] باب أَرّجان
تحت قنطرة تكان [i] وهى قنطرة بين فارس وخوزستان فيسقى رستاق رِيشَهْر [k]
ثمّ يقع فى البحر عند حدّ تُسْتَر [l] ۞ وامّا نهر شيرين فمخرجه من جبل
دينار [m] الّذى بناحية بازرنج [n] فيسقى فرزك [o] والجلادكان ثمّ يخترق [p] حتّى
يقع فى البحر نحو جنّابة ۞ وامّا نهر الشاذكان فانّه يخرج من [q] بازرنج
وجبالها حتّى يدخل *تَنْبُوك مورستان [r] وخان حمّاد فيسقى رستاق

---

a) B. بكلاول, E. كلاذن. *b*) B. بالكارسان, E. om. Vid. supra ad p. ١٠۴ *h*. *c*) D. بالنبوكان, Edrîsî, I, p. 413 نزكان; vid. Vullers sub برگان. *d*) Secutus sum E.; B. بالحسوبان, D. indistincte بالميسريان. *e*) B. الـدار, sed quoque E. بآتشگاه. *f*) C. et Abulfeda, p. ٥٨, المرج. Deinde C. فينصب فى, Jacut, III, p. ۴٨٥, 18 حتى ينصب فى. *g*) C. مسرقان. *h*) C. يخرج من. *i*) B. h. l. مكان, infra دكان; Codd. D. تكان et تمكان; E. دكان et infra نكار; Jacut l. l. ركان, (*Merácid*, II, p. ١٨٩ كتان). Vid. porro Defrémery, *Mémoires*, p. 142. *k*) B. ريشبر, C. ورشهر, E. روستاى زم. *l*) B. شبر, C. دستر, E. شوشتر, vid. Jacut l. l.; D. et Abulfeda, p. ٥٨, شينبز, *Djihán-Numa*, p. ٢٧۴, 3 شتر habet. Pro حد Jacut habet نهر. *m*) Secundum Abulfedam l. l. et *Djihán-Numa*. B. et C. sine punctis, D. دينمان, E. ددمان. *n*) Vid. supra ad p. ١١٢ *e*. *o*) De hoc et seq. nomine vid. supra ad p. ١١٣ *d* et *b*. *p*) Abulfeda addit بلاد فارس. *q*) C. add. ناحية. *r*) B. دممول سرومان, C. tantum دمبوك, D. corrupte بتميزل وهو رستاق; E. (Ous. p. 96) دمول مره. Vid. supra ad p. ١٠۴ *g* et ١١٠ *g*.

زيزانَرَد[a] ونائين والكَهْركان ثمّ يمتدّ الى دشت الدستقان[b] فيدخل البحر ۞ وامّا نهر درخيد فانّه يدخل من جبال الجُويخان[c] فيقع فى بحيرة درخيد ۞ وامّا نهر الخُوبَذان فانّه يخرج من الخوبذان فيسقى الخوبذان وانبوران[d] ثمّ ينصبّ الى الجلادجان متعرّجا فيقع فى البحر ۞ وامّا نهر رتين[e] فيخرج من خُمَايْجان العليا حتّى يصير بالزِّبريان[f] فيقع فى نهر سابور ثمّ ينحدر من نهر سابور فيمضى الى توّج فيمرّ ببابها ومنها الى البحر ۞ ونهر اِخْشين يخرج من خلال جبال داذين[g] فاذا بلغ الجُنْقان[h] وقع فى نهر توّج ۞ وامّا نهر سَكّان فانّه يخرج من رستاق الرُّويْجان[i] من قرية تدعى شاذفرى[k] فيسقى زروعها ثمّ ينحدر الى رستاق سياه فيسقيها ومنها الى كُوَار فيسقيها * ومنها الى خَبْر فيسقيها[l] ثمّ الى الصِّيمكان[m] فيسقيها ثمّ الى كَازَرين[n] فيسقيها ثمّ الى قرية تسمّى سكّ[o] وينسب هذا الوادى الى سكّ ثمّ يقع فى البحر وليس فى انهار فارس نهر اكثر عمارة من هذا النهر ۞ وامّا نهر جرشيق[p] فانّه يخرج من رستاق ماصرم[q] ويخترق رستاق المُشْجان[r] حتّى يجرى تحت

---

a) B. درابرد, E. دسرار. Secutus sum D. *b*) Vid. supra ad p. ١٠٩ *f*. *c*) B. sine punctis. D. الخوبجان et الحوبجان, E. جوىىدان. Vid. supra ad p. ١١٠ *h*. *d*) Vid. supra ad p. ١١٠ *k*. *e*) Vid. supra ad p. ٩٩ *h*. *f*) B. بالددرسان, C. بالراىان, D. بالزيران, E. ددسران, Edrîsî, I, p. 401 ببرزيان. Cf. Jacut in v. Deinde C. et D. فيسقط. *g*) B. داربن hic et infra; vid. ad p. ١١٢ *a*. *h*) B. الحعمان, C. الحمفان, حمفان, Ous. حيفان, D. الخيفان, Edrîsî الجيفان. Cf. Jacut in v. *i*) B. sine punctis, C. الودىحان, D. الروبجان, E. ردمحان (Ous. ديحان), Edrîsî الرويجان. Vid. Abulfeda, p. ٥٩, coll. Jacut in v. *k*) B. سادقرى, D. سارفرى, Abulfeda ساذفرى. Secutus sum E.; *Djihán-Numa*, p. ٢٧٤ شاد آفرين. *l*) B. haec om.; D. habet حمر, E. خمر. *m*) B. الحمكان, E. جمكان. *n*) B. et E. كازرين et sic *Djihán-Numa* l. l.; D. كازرون. *o*) D. شك et sic E. apud quem autem fluvius appellatur شكان. Edrîsî habet آسك. *p*) Codd. h. l. جرسيق. *q*) C. et D. ماصوم, E. ماسرم, Edrîsî ماشرم. *r*) B.

قنطرة حجارة عادية تعرف بقنطرة سَبُوك[a] حتّى يدخل رستاق خُرّه فيسقيها ثمّ الى رستاق دادين ويقع فى نهر اخشين ۞ وامّا نهر * الكُرّ فانّه يخرج[b] من كروان[c] من حدود الأرُدّ[d] وينسب الى كروان[e] هذا النهر * فيخرج من شعب بوّان ثمّ يسقى رستاق كام فَيْروز[e] وينحدر فيسقى قرية رامَجِرْد وكاسْكان والطسّوج وينتهى الى بحيرة بجفوز[f] وتسمّى بحيرة البَخْتَكان[g] ويقال انّ له منبعًا يخرج من بعض كور دارابجرد فينتهى الى البحر ۞ وامّا نهر فَرَوَاب[h] فانّه يخرج من الجَوْبَرْقان[i] من قرية تعرف بفرواب فيجرى على باب اصطخر تحت قنطرة خراسان حتّى يسقط الى نهر * الكرّ ۞ ومنها نهر[k] يعرف بتيرزة[l] يخرج من ناحية دارجان سياه[m] فيسقى رستاق الخُنَيْفغان[n] وجور حتّى يخترق رساتيق اردشير خُرّه ثمّ يقع فى البحر، وامّا الانهار الّتى تقصر عن هذا المقدار فى العظم فانّها تكثر عن احصائى ۞

---

et E. sine punctis. D. المسكان, Ous. مسيحار, Edrîsî مستجار. Cf. *Kámus* sub مشكن. *a*) B. سوك, C. شوك (fortasse confudit cum قنطرة الشوك in urbe Bagdad), D. سيبول, E. سبول et sic Edrîsî. Cf. Jacut in v. *b*) B. الكروانة فيخرج et sic quoque in E. appellatur رود كروانه. Cf. Chardin, *Voyages*, ed. 1811, VIII, p. 235. *c*) B. كروار; cf. Ritter, *Erdkunde*, VIII, p. 767 et 868. (*Djihán-Numa* l. l. كلار). * Legendum autem videtur in textu: وينسب كروان الى هذا النهر. *d*) B., C. et D. الازد. *e*) Haec verba suo loco esse non videntur. Ponantur post sequens والطسوج. *f*) B. بحفوز, E. بحيرة عمرو (Ous. p. 98 et sic *Djihán-Numa* l. l.); D. بحفر ورد سيم, C. om. Vid. supra ad p. ١٠٣ *o*. *g*) C. sine punctis, B. المكان, E. بحكان; vid. supra ad p. ١٠٠ *a*. *h*) B. دوراق, D. et E. فروات. Vid. supra ad p. ٩٩ *l*. *Djihán-Numa* l. l. amnem فراوان, pagum فروات appellat. *i*) B. الجوسرقان, C. الجودان, D. الجويزقان. Vid. supra ad p. ١٠١ *c*. *k*) B. الكور منها. *l*) Vid. supra ad p. ٩٩ *m*. *m*) B. وارجان سياه, D. داراجار سياه, E. دارجان شاذ. Cf. Edrîsî, I, p. 411. *Djihán-Numa* habet ارجان. *n*) B. الجمستان, E. جمعقان, *Djihán-Numa* جيغان; vid. supra ad p. ١٠٥ *c*.

واما بحار فارس فانّ منها بحر فارس وهو خليج من البحر المحيط فى حدّ الصين وبلد الواق واق حتّى يجرى على حدود بلدان الهند والسند[a] وكرمان الى فارس وينسب هذا البحر من بين سائر الممالك الّتى عليه الى فارس لانّه ليس عليه مملكة اعمر[b] منها ولانّ ملوك الفرس كانوا على قديم الزمان اقوى سلطانًا *وهم المستولون الى يومنا هذا على ما بعد وقرب من شطوط هذا البحر[c] ✿ ومن بحيراتها الّتى تحيط بها القرى والعمارات بحيرة البختكان[d] الّتى يقع فيها نهر *الكرّ وهى من ناحية جفوز[e] الى قرب كرمان فيكون طولها نحو عشرين فرسخًا ومآؤها مالح وينعقد فيها الملح وحواليها مسبع[f] وتحيط بها رساتيق وقرى وهى فى كورة اصطخر✿ وبحيرة بدَشْت أرزن من كورة شابور طولها نحو عشرة فراسخ ومآؤها عذب وربّما تجفّ حتّى لا يبقى فيها من المآء الّا القليل وربّما امتلأت نحو عشرة فراسخ وتحتفّ بها القرى والعمارات وعامّة سمك شيراز منها ✿ وبحيرة توّز[g] من كورة سابور بقرب كازرون[h] وطولها نحو عشرة فراسخ الى قرب مُورَق[i] ومآؤها مالح وفيها صيد كثير✿ وبحيرة الجنكان[k] مالحة طولها نحو اثنى عشر فرسخًا ويرتفع من اطرافها الملح وحواليها قرى الكَهْرجان وهى من اردشير خرّه اوّلها من شيراز على فرسخين وآخرها حدّ[l] خوزستان✿ وبحيرة الباسفهوية[m] طولها نحو ثمانية فراسخ ومآؤها مالح وصيدها كثير وفى اطرافها آجام كثيرة فيها قصب وبَرْدىّ وحلفآء وغير ذلك ممّا ينتفع به اهل شيراز وهى فى كورة

---

a) Hinc incipit lacuna in B. ad صفة معظم المدن. b) C. اعم. c) Haec inde a وهم quoque in C. desunt. Supplevi ex D. et E. (Pro وهم hic habet (مردمان بارس). d) C. sine punctis, E. دنحكان. e) C. كور متاحمه حعون. Vid. supra p. ۱۳۱ *f*. f) C. مُسبّع. g) Vid. supra ad p. ۱۰۰ *b*. h) Secundum C., D. et Mokaddasí MS. p. 215, qui lacum بحيرة كازرون appellat. E. (Ous. p. 99) habet كارزين. i) Vid. supra ad p. ۱۱۱ *u*. k) C. المحنكان, E. ut supra p. ۱۰۰ حمكان. l) Sic D. et E.; C. حذا. m) C. sine punctis, E. ما سفهونه, Ous. سفهويه, D. الباسفرية, Mokaddasí l. l. الباشفرية.

اصطخر متاخمة للزرقان[a] من رستاق هَرَاة[b] ✿

صفة معظم المدن فى مقاديرها وابنيتها ونحو ذلك   امّا اِصْطَخْرُ فهى مدينة وسطة وسعتها مقدار ميل وهى من اقدم مدن فارس واشهرها وبها كان يكون[c] ملك فارس حتّى حوّل[d] اردشير الملك الى جُور، ويروى فى الاخبار انّ سليمان بن داوود عم كان يسير من طَبَرِيَّة اليها من غدوة الى عشيّة وبها مسجد يعرف بمسجد سليمان ويزعم قوم من عوامّ الفرس الّذين لا يرجعون الى تحقيق[e] انّ جَمَ الّذى كان قبل الضحّاك هو سليمان، وكان فى قديم الايّام على اصطخر سور قد تهدّم وبنآؤهم من الطين والحجارة والجصّ على قدر يسار البانى وقنطرة خراسان خارج من[f] المدينة على بابها ممّا يلى خراسان الّا انّ ورآء القنطرة ابنية ومساكن *ليست بقديمة[g] ✿ وامّا سابور[h] *فانّها مدينة بناها سابور الملك وهى فى السعة نحو من[i] اصطخر الّا انّها اعمر واجمع للغنآء[k] وايسر اهلًا وبنآؤها نحو بنآء اصطخر وبها وباصطخر وبآء الّا انّ خارج المدينة صحيح الهوآء ✿ وامّا دَارَابْجِرْد فانّها من بنآء دارا ولذلك سمّيت دارابجرد وتفسيرها عمل دارا وعليها سور عامر جديد مثل سور جور وعليها خندق يتولّد[l] المياه فيه من النزّ والعيون وفى هذا المآء حشائش[m] ان دخله انسان او دابّة التفّت عليه فلا يتهيّأ له عبوره ولا يكاد يسلم الّا على[n] شدّة، ولها اربعة ابواب وفى وسط المدينة *جبل حجارة كانّه قبّة[o] ليس له اتّصال بشىء من الجبال وبنيانهم من طين وليس بها فى زماننا *كثير اثر[p] العجم ✿ وامّا جُورُ فانّها من بنآء اردشير ويقال انّ

---

a) C. للزرمان, D. للمرزقان. Hodie appellatur Serqûn, vid. e. g. Petermann, *Reisen*, II, p. 185 seq.; Brugsch, II, p. 163. *b*) C. هزار, quae lectio quoque bona esse potest. Hic finis lacunae in B. est. *c*) Jacut, I, p. ٢٩٩, 15 مسكن, C. om. *d*) Jacut تحول. *e*) D. تحصيل. E. وغلط مى گويند. *f*) Jacut خارجة عن. *g*) C. محدثة. *h*) E. بشاوور. *i*) C. فهى بعد. *k*) C. للبناء ut Jacut, III, p. ٥, 20. *l*) C. ودغرل i. e. ويغزل. Cf. quoque Abulfeda, p. ٣٣١. *m*) E. ذياهها وخارها. *n*) C. بعد. *o*) C. جبل كانه قبة من حجارة. *p*) B. كنفا اى, C. كثير من. In E. parva lacuna est.

مكانها كان مآء واقفًا كالبحيرة فنذر اردشير ان يبنى مدينة على المكان الّذى يظفر فيه بعدوّه ويبتنى فيها بيت نار فظفر[a] هناك فاحتال فى ازالة مآء[b] ذلك المكان بما فتح من مجاريه فبنى بذلك المكان جور[c]، وهى قريبة فى السعة من اصطخر وسابور ودارابجرد وعليها[d] سور عامر من طين وخندق ولها اربعة ابواب باب[e] ممّا يلى المشرق يسمّى[f] باب مِهْر[g] وممّا يلى المغرب باب بهرام وممّا يلى الشمال باب هرمز وممّا يلى الجنوب باب اردشير، وفى وسط المدينة بنآء مثل الدكّة يسمّى الطِّرْبال ويعرف بلسان الفرس *بأَيْوان وكيَاخُرّه[h] وهو بنآء بناه اردشير ويقال انّه كان من الارتفاع بحيث يشرف منه الانسان على المدينة جميعها[i] ورساتيقها وبنى اعلاه بيت نار واستنبط بحذآئه من جبل مآء حتّى *اصعده الى[k] اعلى هذا الطربال كالفوارة ثمّ ينزل فى مجرى آخر وهو بنآء من جصّ وحجارة وقد *استعمل الناس اكثره وخرب[l] حتّى لم يبق منه الّا اليسير، وفى المدينة مياه جارية وهى مدينة نزهة جدًّا يسير الرجل منها من كلّ باب نحوًا[m] من فرسخ فى بساتين وقصور ۞ فامّا مدينة شيراز فانّها مدينة اسلاميّة ليست بقديمة وانّما بنيت فى الاسلام بناها محمّد بن *القاسم بن ابى[n] عَقيل ابن عمّ الحجّاج ابن يوسف[o] وسمّيت بشيراز تشبيهًا بجوف الاسد وذلك انّ عامّة *المير

---

a) Jacut, II, p. ١۴٧, 3 addit بـه. *b*) C. et Jacut مـيـاه. *c*) B. جـورًا. *d*) B. عليها. *e*) B. et C. om. *f*) C. ويسمى. *g*) B. بِهْر. Urbi Schahrastán i. e. Sábur quoque quatuor portae sunt, quarum tres Báb Mihr, Báb Behrám et Báb Hormoz appellantur. Vid. Mokaddasí MS Berol. p. 209 (Jacut, III, p. ٣۴٢, 18 et 19). *h*) E. شابران وكاجره, In D. utrumque, in B. posterius nomen corruptum legitur, nempe وكيابيوه, deinde correctum in وكيماكره. *i*) B. C. et D. om.; addidi ex E. et Jacut, II, p. ١۴٧, 7. *k*) B. صعد. Deinde pro اعلى C. et Jacut habent راس. *l*) C. et Jacut خرب واستعمل الناس اكثره. *m*) B. نحو. *n*) B. القايم بن. C., E. et Jacut, III, p. ٣۴٩, 2 om. ابى. *o*) C. et Abulfeda, p. ٣٢٩, add. الثقفى.

بتلك النواحى تحمل[a] الى شيراز ولا تحمل منها الى مكان وكانت معسكرًا للمسلمين لمّا اناخوا على فتح اصطخر فلمّا فتحوا اصطخر نزل[b] بهذا المكان فجُعل معسكر فارس وبناها مدينة وهى نحو من فرسخ فى السعة وليس عليها سور وهى مشتبكة البناء[c] كثيرة الاهل بها شحنة الجيش لفارس ابدًا ودواوين فارس وعمّالها[d] وولاة الحرب فيها ۞ وامّا كَازَرِينُ[e] فانّها مدينة صغيرة نحو الثُّلث من اصطخر ولها[f] قلعة وليست من الكبر وقوّة الاسباب بحيث يجب ذكرها الّا انّا ذكرناها لانّها قصبة كورة قباذ خرّه[g] ۞ ومن اجلّ المدن الّتى بكورة اصطخر ممّا يلى خراسان كَثَه[h] وهى حومة يَزْد وأَبَرْقُوه وبناحية كرمان الرُّوذان وهريبة[i] من شقّ كرمان ومن ناحية اصبهان كُرْد والسَّرْدَن ٬ وامّا كثه فهى حومة يزد فانّها مدينة على طرف المفازة ولها طيب هوآء *البرّيّة وصحّته وخصب المدن الجليلة[k] ولها رساتيق تشتمل على خصب ورخص والغالب على ابنيتها آزاج الطين ولها مدينة محصّنة بحصن وللحصن بابان من حديد يسمّى احدهما باب ايزد[l] والآخر باب المسجد لقربه من الجامع وجامعها فى الربض ومياههم من القنىّ الّا نهرًا[m] لهم يخرج من ناحية القلعة من قرب قرية فيها معدن الآنك وهى نزهة جدًّا ولها[n] رساتيق عريضة خصبة وهى ورساتيقها كثيرة الثمار[o] يفضل لكثرتها ما يحمل الى

---

a) C. المدن بتلك النواحى تحمل اموالها. *b*) C. نزلوا, D. تبرك. *c*) C. العمارة والبناء. *d*) C. واعمالها, D. وعملها, E. وعاملان از انجا روند. *e*) D. male كازرون. *f*) B. وبها. Vid. quoque Jacut, IV, p. ٣٢٤, 15. *g*) B. et D. قبـاذ جـرد; cf. supra p. ٩٧ *g*. *h*) E. hic مـكـه, mox كـمـيـن et sic infra (p. ١٢٨, vs. 6). *i*) B. ووهـوده, C. وهـومـز; E. وهربة, sed Ous. p. 102 ut D. هريه. Est altera forma pro هراة et infra B. هراة habet ubi D. هريبة. Cf. Vullers sub هرات. *k*) Ex D. addidi الجليلة. C. وتربة وصحتها, et in Codd. D. quoque التربة pro البـريـة ut recte in Epit Par., et sic in suo Codice habuit Jacut, IV, p. ٣٣٩, 13 et 14. *l*) Secutus sum Jacut l. l. 16. B. ادكر, C. et E. ابرد, D. et Edrîsî, I, p. 418 انـدور. *m*) B., C. et Jacut نـبـر. *n*) B. وبـهـا. *o*) C.

اصبهان وغيرها وجبالهم[a] كثيرة الشجر والنبات الّذى يحمل منها الى الآفاق وخارج المدينة ربض يشتمل على ابنية واسواق تامّة[b] العمارة والغالب على اهلها الادب والكتابة[c] ✿ وامّا أَبَرْقُوه فانّها مدينة محصّنة كثيرة الزحمة تكون نحو الثُّلث من اصطخر * وهى مشتبكة البناء[d] والغالب على بناءها * وبناء يزد[e] الآزاج وهى قرعاء ليس حواليها شجر ولا بساتين الّا فيما بَعُدَ عنها[f] وهى خصبة رخيصة الاسعار[g] ✿ وامّا الرُّوذان فانّها قريبة من ابرقوه فى الشبه فيما وصفنا ✿ وامّا هَرَاة[h] فهى اكبر من ابرقوه وهى فى الابنية وسائر ما وصفنا مقاربة لابرقوه الّا انّ لها مياهًا وثمارًا كثيرة تفضل عن اهلها فتحمل الى النواحى ✿ وامّا كُرْد فانّها اكبر من ابرقوه وارخص سعرًا واخصب وبناؤهم من طين وهى كثيرة القصور ✿ والسَّرْدَن[i] اخصب منها وارخص سعرًا وهى كثيرة الاشجار ✿ والبَيْضاء اكبر مدينة فى كورة اصطخر وانّما سمّيت البيضاء لانّ لها قلعة تبصّ[k] من بُعْد ويرى بياضها فكان بها[l] معسكر المسلمين يقصدونها فى فتح اصطخر فامّا اسمها بالفارسيّة فهو نشانك[m] وهى مدينة تقارب فى الكبر اصطخر وبناؤهم من طين وهى تامّة العمارة خصبة جدًّا يتّسع[n] اهل

تفضل عن اهلها حتى تحمل الى المياه والثمار. Deinde Abulfeda, p. ٣٢٣

a) D. et Jacut وجبالها. *b*) C. et Jacut add. فى. *c*) C. et Jacut والكتبة. *d*) C. العمارة. *e*) B. et Jacut, I, p. ٨٦, 5 om. *f*) B. et D. منها. *g*) Pro his E. habet ودرخت بسيار وميوه فراوان باشد وانار انجا بجانب كرمان وبارس وعراق واذربايكان ببرند وديگر ميوها هم نيكو باشد واز انجا متاع ابريشميه وكرباس بسيار خيزد. *h*) D. هرية. In C. et E. verba واما — لابرقوه desiderantur, ut quoque in exemplari quo usus est Jacut (II, p. ٨٣٠). *i*) B., C. et D. والسرمق et sic quoque Jacut, III, p. ٨٢ et Abulf. p. ٣٢١. Secutus sum E. (coll. p. ١٢٥, vs. 9), qui addit شهرى كوجكست, dum Jacut habet وهى اكبر من ابرقوه, D. اكبر منها. *k*) B. دبيص, C. نمبص, Jacut, I, p. ٧٩١ paen. تبيّن. *l*) B. بها. *m*) Sic Abulfeda, p. ٣٢٩. Edrîsî, I, p. 420 نشابك, B. سابل, D. نسابل, Jacut, I, p. ٧٩٢, 1 نسابك, *Djihân-Numa*, I, p. ٢٦٦, 5 a f. شانك. *n*) C. et Jacut ينتفع.

شيراز بميرتهم ☼ وامّا كورة سابور فانّ معظم مدنها كازرون وخُرّه [a] والنُّوبَنْدجان وابنيتها وابنية سائر هذه المدن من طين ويستعمل فيها الجصّ والحجارة ايضًا وهى من العمران والسعة والخصب واشتباك الابنية على التمام، ☼ وامّا كازرون والنوبندجان فهما متقاربتان فى الكبر الّا انّ بناء كازرون اوثق واكثر [b] قصورًا واصحّ تربة وهواء [c] وليس بجميع فارس اصحّ هواء وتربة من كازرون ومياههم من الآبار وهى مدينة خصبة [d] واسعة الثمار، واخصب مدن كورة سابور كازرون والنوبندجان اكبر منها ☼ وامّا كورة دارابجِرْد فانّ اكبر مدنها فَسَا وهى مدينة مفترشة البناء واسعة الشوارع تقارب فى الكبر شيراز الّا انّها اصحّ هواء من شيراز واوسع ابنية منها وبناؤهم من طين واكثر الخشب فى ابنيتهم السرو وهى مدينة قديمة ولها [e] مدينة عليها حصن وخندق ولها ربض واسواقها فى ربضها وهى مدينة يجتمع فيها ما يكون فى بلاد الصرود والجروم من الثلج [f] والرطب والجوز والاترجّ وغير ذلك * وامّا سائر [g] المدن من كورة دارابجرد فانّها كلّها * عامرة خصبة [h] ☼ وامّا مدن اردشير خُرّه فانّا قد ذكرنا جور وشيراز واكبر مدينة بها بعد شيراز سِيرَاف وهى تقارب شيراز فى الكبر وبناؤهم بالساج وخشب يحمل من بلاد الزنج وابنيتهم طبقات وهى على شفير البحر مشتبكة البناء كثيرة الاهل يبالغون فى نفقات الابنية حتّى انّ الرجل من التجّار لينفق على داره زيادة عن [i] ثلاثين الف دينار * وليس حواليها [k] بساتين واشجار وانّما فواكههم [l] واطيب مياههم من جبل مشرف

---

a) Vid. supra ad p. ١١٢ *c*. C. et E. وجره, B. وحره. E. habet ut D. نوبندجان. *b*) B. واكبر; cf. quoque Jacut, IV, p. ٢٢٩, 3. *c*) B. et Jacut om. Vid. quoque Abulfeda, p. ٣٢٥. *d*) Jacut minus recte حصينة; C. add. جدًّا. *e*) D. وبها; Jacut, III, p. ٨٩١ ult. ولها حصن. *f*) Sic B. et Abulfeda, p. ٣٣١. D. et Jacut, III, p. ٨٩٢, 2 البلح. C. et E. om. *g*) B. وسائر, Jacut وباقى. *h*) E. et Jacut متقاربة, D. متقاربة فى عماراتها وخصبها. *i*) D. et Jacut, III, p. ٢١٢, 18 على. *k*) Jacut corrupte ويعملون فيها; vid. quoque Abulfeda, p. ٣٢٧. *l*) Jacut سقيها وفواكههم.

عليهم يسمّى جَمّ[a] وهو اعلى جبل بقربها يشبه الصرود وسيراف اشدّ تلك المدن حرّا ۞ وامّا ارّجان فانّها مدينة كبيرة كثيرة الخير بها نخيل كثيرة وزيتون وفواكه الجروم[b] وهى برّيّة بحريّة سهليّة جبليّة ومآؤها سيح[c] وبينها وبين البحر مرحلة ۞ واكبر مدن فارس شيراز ثمّ تليها فى الكبر فسا ثمّ تليها فى الكبر سيراف وتلى سيراف فى الكبر ارّجان وتلى ذلك توّج وسابور واصطخر وكثّه ودارابجرد وجور وجنّابة والنوبندجان[d] والغُنْدجان وهى متقاربة فى الكبر، وتوّج مدينة شديدة الحرّ فى وهدة بناؤها[e] طين وهى كثيرة النخيل، والنُّوبَنْجان مدينة حارّة فيها نخيل قليلة وبقربها شعب بوّان مقدار فرسخين قرى ومياه متّصلة قد غطّت الاشجار تلك القرى حتّى لا يراها الانسان الّا[f] ان يدخلها وهى انزه شعب بفارس وهى من الصرود، وجَنّابَة وسينيز[g] ومَهْرُوبَان على البحر شديدة الحرّ وبها نخيل وما يكون فى الجروم من الفواكه ۞

وسنذكر المسافات بفارس فالطريق من شيراز الى سيراف من شيراز الى كفره[h] قرية ٥ فراسخ ومن كفره الى بخرة[i] قرية ٥ فراسخ ومن بخر الى *كُوار غلوة وهى مقسم مآء مدينة كوار[k] ومن بخرا[l] الى البنجمان[m] قرية ۴ فراسخ

a) C. et Jacut حم, D. et Edrîsî, I, p. 397 خم. Secutus sum B., E. et *Djihân-Numa*, p. ٢٦۴, 16, coll. Jacut in v. Deinde B. om. وهو. b) Jacut, I, p. ١٩۴, 6 minus recte addit والصرود. c) Jacut يسيح. d) B. et C. hic et mox والمودىحان. Deinde C. والعىدىحان, E. هندجان. e) C. et Abulf., p. ٣٢٧ add. من. f) C. الى. Aliter E. جنانكه آفتاب دشوار درو جهد. g) B. شير, C. مسمر. h) D. et E. كفر. Vid. supra p. ١٠۴, vs. 3. Viatores hodierni locum appellant *Kafer*. Deinde B. قريب habet pro قرية. i) Videtur esse Behry s. Bihri (e. g. Chardin, VIII, p. 474). B. دىخر, C. دىحر, E. داجر (Ous. p. 105 بحر et Sic Sprenger, p. 75). D. et Edrîsî, I, p. 400 نجد. k) C. كلوان علوده وهى مسمعم دىها مدينة ستة فراسخ. E. haec om. l) C. كلوان i. e. كوار. Apud Mokaddasi MS. p. 207 male idem nomen كول scribitur. m) B. النمحان, C. المىحمان, E. السجمان, D. الميمجان, Edrîsî الينجمان.

ومن البنجمان الى جور[a] مدينة ٦ فراسخ ومن جور الى دَشْت شُوراب[b] ٥ فراسخ ومنها الى خان آزادْمَرْد ٦ فراسخ *وهو خان فى صحرآء قدرها ٣ فراسخ كلُّها نرجس مضعّف[c] ومن خان آزادمرد الى كيرند[d] قرية ٦ فراسخ ومن كيرند الى مَى قرية ٦ فراسخ ومن مى الى راس العقبة[e] بادركان[f] خان ٦ فراسخ ومن بادركان خان الى بركانة[g] خان ٤ فراسخ ومن بركانة الى سيراف مدينة نحو ٧ فراسخ فذلك ٦٠ فرسخا[h] ☆

والطريق من شيراز الى *كَثّه حومة يزد[i] وهو طريق خراسان فمن شيراز الى الزرقان[k] قرية ٦ فراسخ ومن الزرقان الى اصطخر مدينة ٦ فراسخ ومن اصطخر الى بيبر[l] قرية ٤ فراسخ ومن بير الى كهمند[m] قرية ٨ فراسخ ومن كهمند الى قرية بِيْد[n] ٨ فراسخ ومن قرية بيد الى ابرقوه مدينة ١٢ فرسخا ومن ابرقوه الى قرية الاسد[o] ١٣ فرسخا ومن قرية الاسد الى قرية الجوز[p]

Mokaddasí دوماهان. Secutus sum Ous. p. 105. Pro قرية B. habet rursus قريب et pro اربعة E. dat بنج. Male, nam distantia Schiráz inter et Djur 20 Par. est; vid. Ibn Khordádbeh, p. 55, Jacut, II, p. ١٤٧, 10, Abulf. p. ٣٢٥, Mokaddasí, MS. p. 218. *a*) C. كوار جور, E. كوراب, Ous. كوار. *b*) B., C. et D. سوراب. *c*) B. om. Supplevi ex C., D., E. et Edrîsî inter sese collatis. *d*) B. كمويد, C. كمريد, D. كيربد, Edrîsî كبرند. Secutus sum E. *e*) C. add. سنة ومن راس العقبة الى. *f*) D. اذركان, Edrîsî ادركان. *g*) B. بركاده, C. دركانه, E. دركاده. *h*) C. et E. ٦٨ فرسخا. Vid. quoque Jacut, III, p. ٢١٢, 14, Abulf. p. ٣٢٢. *i*) B. et D. tantum يزد, C. tantum كمه, E. كثّه. *k*) B. sine punctis; C. الدوقان et sic Ous. p. 108. E. دوفان. Edrîsî, I, p. 403 زركان. Vid. supra ad p. ١٢٣ *a*. *l*) C. et Mokaddasí l. l. بمر, D. بيين, E. پمر s. بير, Ous. پيبر, Edrîsî تبيز. *m*) C. et E. كهند, Mokaddasí كيبهنده, Ous. كهندز. *n*) B. بند ut Ous. et Edrîsî; C. دند, D. نيد, Mokaddasí دند, E. بيد; cf. Jacut in v. Videtur significare nomen „pagus salicis." Hodie appellatur *Dehebîd*, vid. e. g. Brugsch, *Reisen*, II, p. 129. *o*) E. ديه شير. Petermann l. l. p. 199 *Dehschire*. *p*) D. الخور, E. ديه خوار.

٦ فراسخ ومن قرية الجوز الى قلعة المجوس قرية ٦ فراسخ ومن قلعة المجوس الى مدينة كثه[a] حومة يزد ٥ فراسخ ومن يزد الى آبخيزه[b] ٦ فراسخ وآبخيزه مكان ليس بقرية وانّما هى صحرآء فيها اصول تين *وهو آخر عمل فارس فذلك ٨٠ فرسخًا[c] ☆

والطريق من شيراز الى جنّابة فمن شيراز الى خان الاسد *وهو على نهر السكّان[d] ٦ فراسخ ومن الخان الى دشت ارزن خان ٤ فراسخ ومن دشت ارزن الى تيره قرية ٤ فراسخ ومن تيره الى كازرون مدينة ٦ فراسخ ومن كازرون الى قرية دريز[e] ٤ فراسخ ومن قرية دريز الى راس العقبة خان ٤ فراسخ ومن راس العقبة[f] الى توّج مدينة ٤ فراسخ ومن توّج الى جنّابة مدينة ١٢ فرسخًا فذلك ٤٤ فرسخًا ☆

والطريق من شيراز الى الشيرّجان[g] فمن شيراز الى اصطخر ١٢ فرسخًا ومن اصطخر الى زيادآباذ قرية[h] ٨ فراسخ ومن زيادآباذ الى كلوذر[i] قرية *وهو

---

a) E. كثّه. *b*) Conjectura sic edidi. B. الحمره, C. الحمر, D. ابخيره, E. حيره, Mokaddasí الحيرة (cf. Sprenger, p. 73, vs. 3). Infra in descriptione deserti Khorásánici A. et B. habent ادحمره s. الحمره, D. ادخيره et ادحمره, Edrîsî, I, p. 436 حيره, E. حبر, *Djihán-Numa*, p. ٣٢٨, 12 جيز. De urbe الحيرة autem provinciae Istakhr cogitari nequit, nam, ut hic et infra docemur, آبخيزه ليس بها قرية ولا ساكن وبها عين ماء وحوض يجتمع فيه ماء المطر. *c*) Secundum C. et E. pro his legendum foret: ومن آبخيزه (الحيره) الى كثه ٧٠ فراسخ وهو آخر عمل فارس فذلك ٨٧ فرسخًا. *d*) In B. desunt. E. habet خان الاسد pro خان شير. *e*) Vid. supra ad p. ١٠٩ ann. *n*. Loco hujus et stationis sequentis Mokaddasí unam stationem قرية الحطب memorat. *f*) E. add. تا كريوه چهار فرسنك واز كريوه تا توج. *g*) Sic B. ut infra et D. h. l.; C. sine punctis, E. سيركان. *h*) B. om. Deinde D. add. من رستاق جور, E. از شمار خوزستان. Locus a Jacut memoratur. *i*) Sic B.; E. كلودر ut *Djihán-Numa*, p. ٢٧٦, 11; C. كاودر, D. كلون s. كلوان ut Edrîsî. Apud Qodáma

مرصد[a] ٨ فراسخ ومن كلوذر الى الجوبانان[b] قرية وبها بحيرة ٦ فراسخ ومن الجوبانان الى قرية عبد الرحمان ٦ فراسخ وهى مدينة تسمّى اباذه[c] ومن قرية عبد الرحمان الى قرية الآس[d] مدينة وتسمّى البودنجان ٦ فراسخ ومن قرية الآس الى صاهك الكبرى مدينة ٨ فراسخ ومن صاهك الى رباط السَّرْمَقان رباط ٨ فراسخ ومن رباط السرمقان الى پُشْت خَم[e] رباط ايضًا ٩ فراسخ ومن پشت خم[f] الى الشيرجان *مدينة كرمان[g] ٩ فراسخ، ورباط السرمقان من فارس وما بعده من كرمان فذلك من شيراز الى حدّ السرمقان ٦٢[h] فرسخًا☆

والطريق من شيراز *الى جروم كرمان[i] فمن شيراز الى خان ميم[k] قرية من رستاق الكَهْرجان[l] ٧ فراسخ ومنه الى خورستان[m] مدينة ٧[n] فراسخ ومن

(MS. Schefer) haec statio deëst. Ibn Khordádbeh (p. 55 seq.) pro hac et praecedenti statione habet tantum خضر (حفر sec. Sprenger, p. 71). *a*) B. et D. om. Addidi ex C. et E. (رصدگاه). *b*) Sic C. et D.; B. حوبادان, E. حرنانان et حربايان. Qodáma حوحيابان et حوحيان, Edrîsî, I, p. 404 خوبايان, *Djihán-Numa* l. l. چرمابان. Vid. porro supra ad p. ١٠٠ *c*. Ibn Khordádbeh hanc stationem appellat البحيرة. *c*) Ibn Khordádbeh sine dubio corrupte ارسينجان (sec. Sprenger ارسنحان). *d*) E. ديه مورد. Nomen hujus loci apud Ibn Khordádbeh corruptum legitur الاسس sec. Sprenger, الاستف sec. Barbier de Meynard. Pro بودنجان *Djihán-Numa* habet بودى جان. *e*) Sic E. et infra in itineraria Kermáni A., B. et D. H. l. C. نست جم, B. دسد حم, D. et Edrîsî سيف خم, Qodáma يسحم, *Djihán-Numa* بشت هم. *f*) Qodáma hic insert الى تميد ١٠ فراسخ ومن تميد, quae verba recepit Sprenger, p. 71 ult. seq. *g*) B. om. *h*) B. et C. male ٦۴. Secundum Qodáma distantia inter Schiráz et Sírdján est 76 Par., inter Sarmakán et Sírdján 24 Par., inter Sarmakán et Istakhr 40 Par. *i*) C. الى تارم; E. videtur legisse الى تارم من حد كرمان. *k*) Sic D. et Edrîsî, I, p. 404. E. مدم, B. et C. مبهم. *l*) D. et E. الكهركان. *m*) D. et E. خوزستان, B. خورسان, C. حورسان. Vid. supra ad p. ١١١ *o*. *n*) C. et E. ٩.

خورستان الى منزل يعرف بالرباط ۴[a] فراسخ ومن الرباط الى كرم[b] مدينة
۴ فراسخ ومن كرم الى فسا مدينة ٥ فراسخ ومن فسا الى طمستان[c] مدينة
۴ فراسخ ومن طمستان الى جومة الفستجان[d] مدينة ٦ فراسخ ومن[e]
الفستجان الى الداركان ۴ فراسخ ومن الداركان الى المريزجان مدينة
۴ فراسخ ومن المريزجان الى سنان[f] مدينة ۴ فراسخ ومن سنان الى دارابجرد
مدينة فرسخ ومن دارابجرد الى رمّ المهدى[g] مدينة ٥ فراسخ ومن الرمّ الى
رستاق الرستاق مدينة[h] ٥ فراسخ ومن رستاق الرستاق الى فرّج[i] مدينة
٨ فراسخ ومن فرج الى تارم[k] مدينة ١۴ فرسخا فذلك من شيراز الى تارم
٨٢ فرسخا ✿

الطريق من شيراز الى اصبهان من شيراز الى هزار[l] مدينة ٧[m] فراسخ ومن
هزار الى مائين[n] مدينة ٦ فراسخ ومن مائين الى كنسا[o] مرصد ٦ فراسخ
ومن كنسا الى كنار[p] قرية ۴ فراسخ ومن كنار الى قصر اعين قرية ٧ فراسخ
ومن قصر اعين الى اصطخران قرية ٧ فراسخ ومن اصطخران الى خان اويس[q]

---

a) C. ٧. *b*) Vid. supra ad p. ١٠٧ *d*. *c*) Vid. supra ad p. ١٠٧ *i*. *d*) Vid. supra ad p. ١٠٧ *o*. Scripsi جومة secundum E., coll. Jacut in v. (Ous. p. 110). B., C. et D. habent حومة. *e*) E. add. جومة. *f*) Vid. supra ad p. ١٠٧ *n*. *g*) B. زمّ مهدى et omnes hic habent زم. Cf. supra ad p. ١٠٩ *h*. *h*) B. om. *i*) B. دسرج, C. فرج, D. تىرج, E. فرج. Vid. supra ad p. ١٠٩ *c*. *k*) E. بتارم; vid. supra ad p. ١٠٩ *d*. *l*) B. هواز, C. ut semel E. هزار. *m*) C. et E. ٩, Ous. ١٩, Edrîsî ۴. Videtur legendum esse ٦. Numerus ٩ recte sese habere potest, si cum Qodáma distantiam inter Máïn et Kansá 3 (non 6) Par. esse accipimus. *n*) B., C. et E. مادين, D. بادين, Qodáma (vid. Sprenger p. 73) ماير. *o*) Sic E. (Ous. p. 109 كيسا); B. كسنا, C. كسا, D. كيسا, Qodáma عقبة كنا. Sequens مرصد ab E. vertitur رصدگاه. *p*) E. كمار, C. كمان et كمار. Pro hac statione Qodáma memorat aliam stationi sequenti 2 Par. propriorem, nempe حوسكان. Pro قصر اعين idem scribet قصر اين. *q*) B. روش, D. دوشن, Qodáma حوارس.

قرية ٧ فراسخ ومن خان اويس الى كوز[a] قرية ٧ فراسخ ومن كوز الى كره ٨ فراسخ[b] ومن كره الى خان لنجان قرية ٧ فراسخ ومن خان لنجان الى اصبهان ٧[c] فراسخ، وحدّ فارس الى خان اويس[d] من شيراز اليها ٤٣ فرسخًا فذلك من شيراز الى اصبهان ٧٢ فرسخًا ۞

الطريق من شيراز الى خوزستان[e] فمن شيراز الى جويم ٥ فراسخ ومن جويم الى خلان[f] قرية ٤ فراسخ ومن خلان الى الخرارة[g] قرية كبيرة قليلة الماء ٥ فراسخ ومن الخرارة الى الكركان[h] ٥ فراسخ ومن الكركان الى النوبنجان مدينة كبيرة ٦ فراسخ ومن النوبنجان الى الخوروان[i] قرية ٤ فراسخ ومن الخوروان الى درخيد[k] قرية ٤ فراسخ ومن درخيد الى خان حمّاد قرية ٤ فراسخ ومن خان حمّاد الى بندك[l] قرية ٨[m] فراسخ ومن بندك الى قرية

---

a) B. كرو, D. كرد, C. om.; Mokaddasi stationem كرو memorat et stationem sequentem وكره لا appellat. Fortasse igitur lectio كرو praeferenda est. *b*) Qodáma ومن حوارس (خان اويس) الى سرال (سراى Sprenger) ماس ومروة ٤ فراسخ ومن ماس ومروة الى كرو ٧ فراسخ. Pro خان لنجان idem habet النجار. *c*) C. et E. ٩. Qodáma quoque ٧ habet. *d*) B. اوش, C. روسمى, D. دوشن. E. h. l. خان روس — كى پيش ازين خان اويس گفتيم. Mokaddasí habet خان رش. *e*) C. ارجان, E. ارغان. *f*) B. حلار, C. حلان, D. et Qodáma جلان, E. صلان; Ous., Mokaddasí et Edrîsî خلان. Fortasse est Jacuti خُلّار. Quatremère, *Histoire des Mongols*, p. 384 proponit legere خلاو. *g*) Sic D. et Mokaddasí. B., C. et E. sine punctis, Ous. الخواره (et B. semel الحواره). Edrîsî الجراره, Ibn Khordádbeh, p. 54 الحرارة, Qodáma الحزاره. *h*) B. كركان. Ibn Khordádbeh الكرخان, Mokaddasí جركان. *i*) C. et E. الحوروان, B. الحويدان, D. الجويدان, confuso hoc pago cum urbe الخوبذان, sive cum الجويخان. Mokaddasí خواذان. Loco hujus et sequentis stationis Qodáma habet امران, 9 P. a Khán Hammád, 6 P. a Nubindján. *k*) C. et E. sine punctis, Ous. درجند, Mokaddasí حوند ده s. دحوند. *l*) C. et E. بندل, B. et D. بيدل; Edrîsî بندل s. بيدل. Ous. بندل, Mokaddasí بندق, Qodáma بيدق. *m*) D. ٣, Qodáma ٩.

العقارب وتعرف بهِير[a] ٤ فراسخ ومن هير الى راسين[b] ٤ فراسخ ومن راسين الى ارّجان ٧ فراسخ ومن ارّجان الى سوق سنبيل ٦ فراسخ والحدّ بينهما[c] قنطرة تكان[d] تكون من ارّجان على غلوة، فذلك من شيراز الى ارّجان ٦٠ فرسخا ۞

فامّا المسافات بين المدن الكبار بفارس *فمن فسا الى كارزين ١٨ فرسخا ومنها الى جهرم ١٠ فراسخ والى كارزين ٨ فراسخ[e]، وقد مرّ انّ من شيراز الى اصطخر ١٢ فرسخا ومن شيراز الى كوار ١٠ فراسخ ومن شيراز الى جور ٢٠ فرسخا ومن شيراز الى فسا ٢٧ فرسخا ومن شيراز الى البيضاء ٨ فراسخ ومن شيراز الى دارابجرد ٥٠ فرسخا وقد مرّ انّ من شيراز الى سيراف ٦٠ فرسخا ومن شيراز الى النوبنجان ٢٥ فرسخا ومن شيراز الى يزد ٧٤ فرسخا ومن شيراز الى توّج ٣٢ فرسخا[f] ومن شيراز الى جنّابة ٤٤[g] فرسخا ومنها[h] الى ارّجان ٦٠ فرسخا وقد مرّ ذلك ومنها[h] الى سابور ٢٥[i] فرسخا *ومن شيراز الى كازرون ٢٠ فرسخا ومن شيراز الى خرّه ٢٥ فرسخا[k] ومن شيراز الى

---

a) C. هيبن, D. هبر et نهر. b) B. راستن, D. راشين. c) ? B. بمتمه; in D. non legitur. d) Vid. supra ad p. ١١٩ *i*. e) E. pro his از بسا تا كازرون هژده فرسنك واز بسا تا جهرم ده فرسنك. D. pro كارزين bis habet كازرون, quod necessario falsum esse debet. Jacut etiam pejus habet وبين فسا وكازرون ثمانية فراسخ (III, p. ٨٩٢, 3 et IV, p. ٢٢٩, 6 ubi pro نسا l. فسا). Mokaddasí inter فسا et كارزين distantiam unius diei esse dicit. Sprenger, p. 76 sic locum explicat: „von Fasá nach Kárzín 18 F.; nämlich von dort (Fasá) nach Djahram 10 F., und (von Djahram) nach Kárzín 8 F." et sic habet in mappa N°. 10. f) Sprenger, coll. Ous. p. 111, insert: ومن شيراز الى ابرقوه ٣٢ فرسخا minus recte. Nam supra vidimus distantiam esse 44 P. g) Sic recte B. et Abulfeda, p. ٣٢٢. C., D., E. et Abulfeda, p. ٣٢٥ habent ٥٤, quae secundum Jacut, II, p. ١٢٢, 14 distantia est inter Síráf et Djannába. C. et E. scribunt جنابا (جبابا, جنايا). h) I. e. ومن شيراز. i) Secundum Mokaddasí ١٨. k) Haec in solo B. habentur (pro خره Cod. حره). Distantia inter Schiráz et Kázerun secundum Jacut

خُرّمَة[a] ١۴ فرسخا ومن شيراز الى جهرم ٣٠ فرسخا، ومن جور الى كازرون[b] ١٦ فرسخا ومن سيراف[c] الى نجيرم ١٢ فرسخا ومن مهروبان[d] الى حصن ابن عمارة وهو *طول فارس[e] على البحر نحو ١٦٠ فرسخا ۞ والّذى يحيط بالمفازة[f] من حدّ كرمان الى حدّ اصبهان من الرُّوذان الى أبان[g] ١٨ فرسخا ومن ابان الى فَهْرَج ٢٥ فرسخا ومن فهرج الى كَثَه ٥ فراسخ ومن كثه الى مَيْبُذ ١٠ فراسخ ومن ميبذ الى عُقْدَة[h] ١٠ فراسخ ومن عقدة الى نائين ١٥ فرسخا ومن نائين الى اصبهان ۴٥[i] فرسخا فمن روذان الى نائين ٨٣ فرسخا ۞ ومساحة الحدّ الّذى يلى كرمان من حدّ السيف[k] من لدن حصن ابن عمارة الى ان ينتهى الى تارم[l] ثمّ يمتدّ الى الروذان حتّى ينتهى الى برّيّة[m] خراسان مثل ما من[n] البحر على خطّ شيراز الى ان ينتهى الى مفازة خراسان وهو ١٢٠ فرسخا، والحدّ الّذى يلى خوزستان ومهروبان حتّى ينتهى الى ارّجان وبلاد سابور والسردن الى أوّل حدّ اصبهان نحو ٦٠ فرسخا ۞

ذكر الماء[o] والهواء والتربة بفارس   ارض فارس مقسومة على خطّ من لدن أرّجان الى النوبنجان *الى كازرون[p] الى خُرّه[q] تمرّ على حدود السيف الى

(IV, p. ٣٣٦, 1) et Mokaddasí est 3 dierum s. 18 Par., secundum itinerarium praecedens 20 Par.; Abulf. p. ٣٢٢ habet نحو عشرين فرسخا. *a*) Secundum E.; B. et C. حومه, D. جرمه. Vid. supra ad p. ١٠٣ *l*. *b*) B. et C. كاردين. Deinde C. ٣٦; cf. Abulfeda, p. ٣٢٥. *c*) C. شيراز. *d*) E. ut solet ماهى رويان. *e*) C. اطول المسافات بفارس sed in marg. cum نسخه ut recepi. *f*) C. et E. بالعمارة sive ex var. l. in textu C. به العمارة. *g*) B. ادار, C. et E. اثار; vid. supra ad p. ١٠٣ *c*. Lectionem اثار hic quoque in suo Codice habuit Jacut, III, p. ٩٢٥, 4. *h*) E. عقدا. *i*) B. et D. ٢٥. Pro نائين ut B., C., D. et E. perspicue habent, Ous. et hinc Sprenger, p. 76 male مائين, quem locum inter et Ispahán 60 Par. distantia est. *k*) E. سيف ابن عمّاره (semper sic cum *teschdíd*). *l*) E. بمازم. *m*) E. حد. *n*) B. بمن, E. ميمان. *o*) C. et D. المياه. Deinde C. والاهوية. *p*) C. om. *q*) B. خُره, D. جره, C. حره.

كازرين [a] حتّى تمتدّ الى [b] الرموم ودارابجرد الى فُرْج [c] وتارم فما كان من ناحية الجنوب فجروم وما كان يلى الشمال فصرود، ويقع فى جرومها ارّجان والنوبنجان ومهروبان وشينيز وجنّابة وتوّج [d] ودشت الدستقان وخرّه وداذين [e] ومُورَق [f] وكازرون ودشت بارين وجيبرين [g] ودشت البوسقان [h] ورمّ اللوّالجان وكيزرين [i] وايرز [k] وسميران وخُمايجان والخربق [l] وكُرّان وسيراف ونجيرم وحصن ابن عمارة وما فى اضعاف ذلك، ويقع فى الصرود اصطخر والبيضاء ومائين [m] وايرج [n] وكام فيروز وكُرْد [o] وكلّار وسروستان [p] والاوسبنجان والأرّد [q] والرون وصرام وبازرنج والسردن وخُرّمة [r] والحبيرة والنّيريز [s] والماسكانات والايج والاصطهبانان وبرم ورهنان وبوّان [t] وطرخيشان والجوبرقان واقليد والسّرمق [u] وأبرقوه ويزد وجارين [v] ونائين وما فى اضعاف ذلك، وعلى الحدّ مدن فيها ما فى الصرود

---

a) Sic restitui sec. E. et Mokaddasí MS. p. 215. B., C. et D. كازرون. b) C. et D. على. c) B. برج, D. توج. Deinde E. ut solet بازم (D. بارم). d) B. وادرج, E. ونوج. e) Codd. h. l. وداربن. f) B. وموز, E. ومور. Vid. supra ad p. ١١١ *u*. Mokaddasí habet وموركان. g) B. وحمرير. Vid. supra ad p. ١٠٥ *i*. h) Secundum D.; B. المرسغان, Mokaddasí الموستاقان. i) Vid. supra ad p. ١٠٥ *p*. Apud Mokaddasí praecedit كبير de qua urbe vid. p. ١٠٩ *i*. k) Vid. supra ad p. ١٠٩ *l*. l) B. والحريق, E. وخرمن, Mokaddasí والجرق, D. om.; vid. ad p. ١٠٩ *h*. m) D. ونائين. n) B., E. et Mokaddasí sine punctis, D. وايدج. o) E. كورد. Deinde B. وكلاع, D. وجلان, Mokaddasí وكلان. p) Vid. supra ad p. ١٠٣ *h* et de nomine seq. ib. ann. *i*. q) B. واللارد, E. يزد, Mokaddasí وازد; vid. supra ad p. ١٠٣ *e*. Deinde B. والروق, E. رور; vid. ad p. ١٠٣ *d*. D. et Mokaddasí recte الرون habent. r) E. حرمه (Ous. حومه), B. et D. الحومه, Mokaddasí الجبرة. Vid. supra ad p. ١٠٢ *l*. s) B. sine punctis, D. والسرين, E. ستودر. Vid. supra ad p. ١٠٨ *d*. t) E. h. l. بوار. u) Mokaddasí والجرمق. v) Vid. supra ad p. ١٠١ *e*. Deinde B. وبادمن, D. باس, Edrîsî, I, p. 414 غانيين. E. et Mokaddasí om.

والجروم من النخيل والجوز مثل فسا وجور وشيراز وسابور والنوبندجان وكازرون ، فامّا الصرود فانّ فيها اماكن يبلغ من شدّة البرد فيها ان لا ينبت عندهم شىء من الفواكه سوى الزرع كالأُرْد والرون وكرد والرساتيق الاصطخريّة والرهنان ، وامّا الجروم فانّ بها ما يبلغ من شدّة الحرّ فى الصيف الصائف ألّا ينبت[a] عندهم شىء من الطيور من شدّة الحرّ مثل الاغرسْتان[b] وهى رستاق ولقد خبّرنى بعض الناس انّه كان فى بيت يشرف على وادٍ فيه حجارة فراى نصف النهار يتفلّق فيه الحجارة كما يتفلّق فى النار[c] ، والصرود كلّها صحيحة الهوآء والجروم الغالب عليها فساد الهوآء وتغيير الالوان وليس فيها اكثر وبآء من مدينة دارابجرد ثمّ توّج واصحّ الهوآء فى الجروم ارّجان وسيراف وجنّابة وشينيز واعدل هذه المدن ما كان فى هذين الحدّين مثل شيراز وفسا وكازرون وجور وغير ذلك وليس بجميع فارس هوآء اصحّ من هوآء كازرون ولا اصلح ابدانًا وبشرة من اهلها وامّا المياه فانّ اصحّ المياه بها مآء نهر كرّ واردأُ المياه مآء دارابجرد ٭

ذكر صور اهل فارس وزيّهم ولسانهم واديانهم   امّا صورهم فانّ اهل الجروم الغالب على خلقتهم نحافة الخلق وخفّة الشعر وسمرة اللون واهل الصرود اعبل[d] اجسامًا واكثر شعورًا واشدّ بياضًا ٭ ولهم ثلاثة السنة الفارسيّة الّتى يتكلّمون بها وجميع اهل فارس يتكلّمون بلغة واحدة يفهم بعضهم عن بعض الّا الفاظًا[e] تختلف لا تستعجم على عامّتهم ، ولسانهم الّذى به كتب العجم[f] وايّامهم ومكاتبات المجوس فيما بينهم هو الفهلويّة الّتى تحتاج الى تفسير حتّى يعرفها الفرس ، ولسان العربيّة الّتى بها[g] مكاتبات السلطان والدواوين وعامّة الناس[h] ٭ وامّا زيّهم فانّ زىّ السلطان بها الاقبية وربّما

a) B. ينبت. b) B. الاغرسان ; vid. supra ad p. ١٠٩ h. c) E. pro his: وجائى هست كه بتابستان گوشت بر سنگ بريان شود. d) C. اكبر. e) B. et C. الفاظ. f) C. الذى كتب به العجم الكتب. E. وزبان پهلوى كه. g) E. وزبان تازى كه اكنون ببروزگار قديم مكاتبات بآن بودى. h) C. add. والامراء.

لبسوا الدراريع الّتى هى اوسع فرجة واعرض جُرُبّانًا وجيوبًا من دراريع الكُتّاب *والعمائم الّتى تحتها قلانس مرتفعة[a] ويلبسون السيوف بحمائل وفى اوساطهم المناطق وخفافهم تصغر عن خفاف اهل خراسان[b]، وامّا قضاتهم فانّهم يلبسون الدّنّيّات وما اشبهها من القلانس المشمّرة عن الاذنيّن مع الطيالسة والقُمُص والجباب ولا يلبسون درّاعة ولا خُفًّا بكسر[c] ولا قلنسوة تغطى الاذنين[d]، وامّا زىّ الكتّاب فانّهم يلبسون الدراريع والعمائم فان لبسوا تحت العمائم قلانس جعلوها خَفِيّة[e] تُوَقّى الوسخ ولا تظهر ويلبسون الخُفّ المكسّر الطف من خفّ السلطان ولا يلبسون قباء ولا طيالسة، وامّا التُنّاء والتجّار والملوك فلباسهم شىء واحد من الطيالسة والعمائم والخفاف الّتى لا كسر فيها والقمص والجباب والمبطّنات وانّما يتفاضلون فى الجودة فى الملابس، فامّا الزىّ فواحد وزيّهم زىّ اهل العراق✿ وامّا اخلاق ملوكهم والتنّاء منهم والمخالطين للسلطان من عُمّال الدواوين فالغالب عليهم استعمال المروّة فى احوالهم والنزاهة عمّا يقبح به الحديث من الاخلاق الدنيّة والمبالغة فى تحسين دورهم ولباسهم واطعمتهم والمنافسة فيما بينهم فى ذلك[f] والآداب الظاهرة فيهم، وامّا تجّارهم فالغالب عليهم محبّة جمع المال فامّا اهل سيراف والسواحل فانّهم يسيرون فى البحر حتّى[g] ربّما غاب احدهم عامّة عمره فى البحر ولقد بلغنى انّ رجلا من سيراف أَلِفَ البحر حتّى ذكر انّه لم يخرج من السفينة نحو من اربعين سنة وكان[h] اذا قارب البرّ اخرج صاحبه بقضاء حوائجه فى كلّ مدينة يتحوّل من سفينته الى اخرى اذا انكسرت وتشعّثت

---

a) E. وموزهاى ايشان تنگ ساق باشد. b) E. ودستارهاى كوچك بندند. c) B. يُكْسر, C. بكسر, D. بكسره. d) E. وقاضيان كلاهها بر سر نهند چنانك گوش بپوشد وطيلسان بر افكنند وپيراهنهاى باريك پوشند ودراعه وموزه ندارند دبيران فارس درّاعه ودستار ندارند واكر كلاه يا دستار دارند. e) B. خَفِيّة. E. وزى ايشان بزى اهل العراق: et ceteris omissis statim addit پوشيده باشد. f) E. وسراها وعمارتها بغايت نيكو سازند وميزبانها كنند نزديكست. g) C. add. انّه. h) B. om.

فاحتيج الى اصلاحها، وقد أُعْطوا من ذلك حظًّا جزيلًا حتّى انّ احدهم يبلغ ملكه *اربعة آلاف[a] الف دينار وفى عصرنا قد بلغنى ما هو اكثر من ذلك فتراه فى لباسه لا يتميّز من اجيره، وامّا اهل كازرون وفسا وغير ذلك فهم اهل تجارات فى البرّ وقد أُعْطوا من ذلك حظًّا جزيلًا[b] وهم اهل صبر على الغربة[c] وحرص على الجمع وفيهم اليسار الظاهر حيث ما كانوا[d] وما علمتُ مدينة فى برّ ولا بحر فيها قوم من الفرس مقيمون الّا وهم عيون تلك المدينة والغالب عليهم اليسار واستقامة الحال والعفّة ۞ وامّا اديانهم فانّ السواحل من سيراف الى مهروبان الى ارّجان واكثر الجروم الغالب عليهم مذاهب اهل البصرة فى القدر[e] واقلّهم المعتزلة واهل جهرم الغالب عليهم الاعتزال واهل خُرّه هم[f] شيعيّة، وامّا الصرود فانّ شيراز واصطخر وفسا الغالب عليهم مذاهب اهل الجماعة[g] على مذاهب اهل بغداد والغالب على اهل فارس فى الفتيا مذهب اهل الحديث، فامّا اهل الملل منهم فانّ فيهم اليهود والنصارى والمجوس وليس بهم صابئٌ ولا سامرىٌّ ولا من سائر النحل احد ظاهر واكثر هذه الملل المجوس وهم الغالبون على سائر الملل فى الكثرة ثمّ النصارى ثمّ اليهود اقلّهم فامّا كتب المجوس وبيوت نيرانهم واديانهم[h] وما كانوا عليه فى ايّام ملوكهم فانّهم يتوارثونه وذلك فى ايديهم ويتديّنون به وليس المجوس ببلد اكثر منهم بفارس لانّ بها دار ملكهم واديانهم وكتبهم ۞

---

a) C. اربعين الف. E. خداوند كتاب كويد من انجايكه كسانى ديدم كى هريكرا چهار بار هزار دينار بود. b) C. addit: حتى ان احدهم ليبلغ ملكه المال العظيم. c) B. غُرْبة. d) C. addit: واخبر بعض الثقات انه حضر البصرة سنة ٣٢٤ فوردت كتب التجار من عمان انه وقع بها حريق والريح جنوبىٌّ فاحرق ثم تحوّلت شماليًّا فاحرقت لرجل يعرف بابن مروان من العبيد السود سوى البيض اثنى عشر الف نسمة واحترق له من الامتعة والعطر ما لا يحدّ الّا الكافور فانه احترق له اربع مائة بركة والبركة خمسون وقرًا. e) E. بقدر كويند. f) B. هى. Deinde C. qui habet جره, scribit شيعة. E. pro شيعية legisse videtur سبعية, nam vertit كرمسيريان هفت كونه اند. g) E. سنت جماعت. h) E. واداب كبردى.

ذكر طبقات الناس بفارس   امّا طبقات الناس بفارس فانّ لهم فى قديم الايّام على ما يذكره الفرس فى كتبهم ملوكًا ملكوا الدنيا مثل الضحّاك وجَمّ وافريدون فى آخرين كانوا ملوك الارض حتّى قسم افريدون الارض بين بنيه فصار ملوك الفرس سُكّان ايرانشهر الى ان قتل ذو القرنين دارا بن دارا الملك فصارت الممالك طوائف حتّى كان ايّام اردشير فعادت المملكة الى واحد فما زالت فيهم يتولّاها مثل سابور وبهرام وقباذ وفيروز وهرمز وسائر الاكاسرة حتّى جآء الاسلام فزال الملك عنهم وانّما سكن بابل الاكاسرة فى آخر ايّامهم ونقلوا عن ديارهم عن فارس *الى قرب[a] من الروم والعرب كما انتقل التبابعة من اليمن لمّا ملكوا الآفاق وكما انتقل ملوك الاسلام من العرب عن ديار العرب الى بابل لتوسّط الممالك والاشراف على كلّ ناحية، ولسنا نكثر فى ذكر ملوك الفرس لانتشار اخبارهم وعلم الناس بايّامهم* فامّا فى الاسلام فانّ لهم ملوكًا منهم فى تقليد الامارات ومنهم من قعد عنها على استقلاله بها وكفايته من الفرس والعرب الّذين توطّنوا فارس فصاروا من اهلها والّذين تغرّبوا عنها، فمنهم الهُرْمُزان من الاساورة أُسر فى ايّام عمر فقدم عليه فاطلقه وآمنه فاسلم وله الى آل ابى طالب صهر فاتّهم بقتل عمر بن الخطّاب مع ابى لؤلؤة[b] عبد للمغيرة بن شعبة فقتله عبد[c] الله بن عمر بعد موت عمر، ويقال انّ سلمان الفارسىّ من ولد الاساورة وانّه تزهّد وخرج بطلب الدين ويتصفّح الملل حتّى وقع الى المدينة فاسلم عند ورود النبىّ صلعم المدينة، ومنهم[d] آل عُمارة ويعرفون باولاد الجُلَنْدى ولهم مملكة عريضة وضياع كثيرة وقلاع[e] على سيف البحر بفارس متاخمة لحدّ كرمان ويزعمون انّ ملكهم هناك قبل موسى[f] وانّ الّذى قال الله[g] وَكَانَ وَرَآءَهُمْ مَلِكٌ يَأْخُذُ كُلَّ سَفِينَةٍ غَصْبًا هو الجلندى وهم قوم من ازد اليمن ولهم الى يومنا هذا منعة وعُدّة[h] وباس وعدد لا

---

a) B. المقرب et الغرب pro العرب. E. مملكة بوسيد روم وبحدود تا عرب ايشان وعلم ولوا از مملكة فارس بيرون آمد چنانچه. b) B. لؤلؤ. c) B. عبيد. d) Seqq. transcripsit Jacut, II, p. ٧١١. e) E. add. استوار. f) Jacut add. بن عمران عم. g) Qor. 18 vs. 78. h) Jacut وجدّ l. وحدّ.

يستطيع السلطان ان يغيّرهم[a] واليهم أرصاد البحر وعشور السفن، وقد كان عمرو بن اللّيث ناصب *حمدانَ بن عبد الله[b] الحرب نحو سنتين فما قدر عليه حتّى استعان عليه بابن عمّه العبّاس بن احمد بن الحسن، *واحمد ابن الحسن[c] الّذى نسبنا اليه رمّ[d] الكاريان وهو من آل الجلندى ازدىّ وابنه حجر بن احمد هو على الرمّ فى منعة وقوّة الى يومنا هذا ✿ وآل الصّفّار[e] الّذين نسب اليهم سِيفُ بنى الصفّار هم آل الجلندى وهاؤلآء اقدم من ملوك الاسلام بفارس وامنعهم جانبًا ✿ ومنهم[f] آل ابى زُهَير المدينىّ ينسب اليهم سِيفُ بنى زهير وهم من سامة بن لُوَىّ[g] ملوك ذلك السيف ولهم منعة وعدد فمنهم ابو سارة[h] الّذى خرج متغلّبًا على فارس يدعو الى نفسه حتّى بعث المامون من خراسان محمّد بن الاشعث فواقعه فى صحرآء كسّ[i] من شيراز وفرّق جيشه وقتله وكان الوالى بفارس حينئذ يزيد بن عقال، وجعفر ابن ابى زهير الّذى قال فيه[k] الرشيد وقد وفد عليه فى ملوك فارس لولا *طرش به[l] لاستوزرتُه، والمظفّر بن جعفر الّذى كان يملك عامّة الدستقان[m] وله مملكة السِّيف[n] من حدّ جَنّابا[o] الى حدّ نجيرم وسائر آل ابى زهير من حدّ نجيرم الى حدّ بنى عُمارة ومسكن آل ابى زهير كُرَان[p] ومسكن المظفّر

---

a) Jacut قهرهم. E. pro his habet وهنور آثار ايشان واولاد ايشان باقى اند وخراج دريا ستانند وهيچ ملكى ايشانرا ازان باز نتوانند داشت. Cf. porro Jacut, IV, p. ٩٧٤ sub هنزو. b) E. عبد الله بن احمد الجلندى; Jacut addit بن الحارث omisso الحرب. c) Jacut om. d) B. semper زمّ. e) B. والى الصعاق, ut semper A. et B. الصفاق pro الصغار habent. f) Seqq. transscripsit Jacut, III, p. ٢١٧. g) Jacut add. وهم بنو غالب. h) Male Jacut ابو سامة بن لوى. i) Jacut كش. k) Addidi ex Jacut. l) Jacut شربه. m) Secundum Jacut l. l. 16. B. الرستاق. Ous. رشاق ut quoque p. 112 pro الدستقان habet. n) E. perperam legit السَّيْف, vertens خداوند شمشير بودند (Ous. p. 119). o) E. حمى (Ous. حى), Jacut جى. Pro نجيرم bis male بنجيرم habet Jacut. p) B. et Jacut كوان.

على ساحل البحر بصفارة[a] ٭ ومنهم آل حنظلة بن تميم من ولد عُرْوَة بن أُدَيَّة[b] الّذين عبروا من البحرين الى فارس فى ايّام بنى اميّة بعد قتل عروة ابن اديّة فسكنوا اصطخر ونواحيها وملكوا الاموال الكثيرة والقرى النفيسة، وكان منهم عمرو بن عُيَينة وبلغ من يساره انّه ابتاع بالف الف[c] درهم مَصاحف فوقفها فى مدن الاسلام وكان مبلغ خراج هذا البيت فى ضياعهم نحو عشرة آلاف الف درهم، وكان المامون ولّى عُمَر[d] بن ابراهيم غزو[e] البحر لقتال القطريّة وابنه مرداس بن عمر المكنى بابى بلال بلغ من ماله ان كان خراجه نحو ثلاثة آلاف الف[c] درهم، وكان ابن عمّه محمد بن واصل[f] ملكه مثل ملك هذا وخراجه مثل خراجه لا يتفاوت بكبير شىء وكان اجلّ اهل هذا البيت عمرو بن عيينة وكانت من قوّة هذا البيت انّ الاتراك لمّا استولوا على الخلافة فلم يطقهم الخلفآء فُرّقوا فى اقطاعات عريضة وولّوا فارس ليبعدوا عن البلد وكان منهم من عظمآء الاتراك نحو من اربعين اميرًا ورئيسهم المولّد[g] وكان يمنعهم الظلم فتشغّبوا عليه وهمّوا عليه حتّى استجار بمرداس بن عُمَر[h] فاجاره واخرجه الى بغداد وولّوا على انفسهم ابراهيم بن سيما وكتب عبيد[i] الله بن يحيى عن المعتمد الى مرداس فى قتلهم فاستعفى وكتب الى محمّد بن واصل فجمع حاشيته واهل طاعته حتّى قُتِل هاؤلآء الامرآء عن آخرهم الّا ابراهيم بن سيما واربعة نفر[k] وكان رئيس الاتراك بعد المولّد بفارس واستولى محمّد بن واصل على فارس فبُعث اليه من بغداد عبد الرحمان بن مُفْلح وكان على جيشه طاشم[l] فى جيش عظيم

---

a) B. بصعارة. Cf. supra p. ١٠٦ *g*. b) B. اذنه; vid. e. g. Ibn Kotaiba, p. ٢٠٩. c) E. om. الف. d) E. عمرو. In B. عمرو semper scribitur عُمْر. e) B. وغزو. f) De eo vid. Ibno 'l-Athír, VII, p. ١٦٦ et deinde. g) احمد المولّد, vid. e. g. Ibno 'l-Athír, VII, p. ١١٠, ١٧٠, sive محمد المولد ib. p. ٣٣٧. h) E. عمرو. i) E. عبد. k) Cf. Ibno 'l-Athír, VII, p. ١٦٦ et ١٨٩, sed ibi الحارث بن سيما vocatur عامل فارس post المولد. l) Hic طاشتمر appellatur ab Ibno 'l-Athír, p. ١٨٩.

فهزم جيش عبد الرحمان وقتل طاشم واستأسر[a] عبد الرحمان وقتله فصفت له فارس حتّى قصد ابن عمّه مرداس بالخنق مخافة على نفسه فاستدعى[b] يعقوب بن الّليث فدخل يعقوب بن الّليث فارس لمعاضدة مرداس حتّى حارب محمّد بن واصل بمروسدان[c] بناحية البيضاء راجعًا من محاربة عبد الرحمان بن مفلح فهزمه وفرّق جيشه واستوسر بسيراف فى البحر فسُلّم الى يعقوب وانفذه الى قلعة ثمّ فحبسه بها سنتين حتّى كان يعقوب بجنديسابور[d] فتغلّب هو والمحبّسين[e] على القلعة فبعث يعقوب من قتلهم الّا القليل ✿ ومن ملوك الفرس ممّن ملك بغير فارس آل سامان فانّهم من ولد بهرام وكان بهرام من اهل خَبْر[f] من اردشير خُرّه فسكن الرىّ ثمّ ولى محاربة الاتراك فقصد بلخ[g] وفرّق جمع الاتراك واثّر فيهم فاستفحل امره وقويت شوكته حتّى خافه كسرى ذلك العصر على نفسه وملكه فهمّ بمحاربته وازالة ملكه فاضطرّه بهرام الى ان استجار بملك الروم واخلى مملكته الى ان يقصد ملك الروم فرجع وكان من حديثه ما قد ذُكر فى الكتب، وآل سامان من ولده فكانوا ملوك ما وراء نهر بلخ المعروف بجيحون وامراءَهُ يتوارثونه بينهم الى ان انتهت الامارة الى اسماعيل بن احمد بن اسد فبلغ من سلطانه وتمكّن امره ان ازال ما كان استصعب على المعتضد فى شهامته وصولته وبأسه من ملك عمرو بن الليث وتفريق جمعه حين ملك خراسان كلّها وما وراء النهر وجرجان وطبرستان وقومس والرىّ[h] وابهر وزنجان وهذه مملكة ما علمتُ انّ الاكاسرة جمعتها لرجل واحد وقمع مع هذه المملكة الاتراك وذلّلهم حتّى بلغت صولته وهيبته حدود الصين وهابته ملوك الترك حتّى صار ممّا يلى مملكة الاسلام من بلدان الاتراك من الامن مثل دار الاسلام[i] ، ثمّ ملك بعده ابنه

---

a) Hic desinit magna lacuna in A. b) A. et B. واستدعى. c) A. بمروسمدان, E. corrupte بمرو شد از جانب بيضا. d) E. در حدود نشابور. e) B. والمحمنين. E. با كروهى باز داشتكان. Cf. Ibno 'l-Athír, VII, p. ٢٠١, 6 a f. seqq. f) A. et B. حمر, E. خير. g) E. هرات. Vid. de بهرام جوبين Ibno 'l-Athír, I, p. ٣٤٢—٣٤٩, VII, p. ١٩٣. h) E. add. وقزوين. i) E. vertit تا تركستان وحدود چين همچنان ايمن شد كه ولايت اسلام.

احمد بن اسماعيل فازداد الى هذه المملكة فتح سجستان واذلال بقايا السجزيّة وبسط من حسن النظر للرعيّة ما انتشر به ذكره، ثمّ ملك بعده نصر بن احمد فبلغ من باسه وقمع مَنْ عارضه فى ملكه وقوّة دولته انّه ما اعترض[a] فى ملكه احد الّا قمعه وكانت الغلبة له ۞ وامّا من ملك من فارس من غير الفرس فغلب عليه فانّ منهم علىّ بن الحسين بن بِشْر[b] من الازد المقيمين كانوا ببخارا فانتقل الى فارس وكان من الشحنة[c] فقوى فى ايّام المعتزّ والمستعين فغلب على فارس وكان له باس ومنعة حتّى حاربه يعقوب ابن اللّيث بقنطرة سكّان[d] بقرب شيراز فهزمه واستاسره فاقام فى حبسه مدّة ثمّ قتله ۞ وامّا ملوك الزموم الّذين على ابوابهم الجيوش الدائمة من الف رجل الى ثلاثة آلاف رجل فانّ منهم فى رمّ[e] الزميجان[f] المعروف برمّ جيلَوَيْه المهرجان بن *روزبه وهو[g] اقدم من جيلويه واعظم شوكةً ومنزلةً واخوه سلمة ابن روزبه بعده وكان جيلويه ناقلة اليهم من خمايجان السفلى من كورة اصطخر وكان يخدم سلمة فلمّا مات تغلّب جيلويه على هذا الرمّ واستفحل امره حتّى نُسب الرمّ اليه الى يومنا هذا وبلغ من شوكته ان اوقع بآل ابى دُلَف وقتل معقل بن عيسى اخا ابى دلف ثمّ قصده ابو دلف فقتله وحمل راسه فكان لآل ابى دلف الى ان انقضت[h] ايّامهم يقيمون براسه فى الحروب يحمل بين ايديهم على رمح وقد ضُبِّب القِحْف بالفضّة حين وقع فى يد عمرو بن الليث لمّا هزم احمد بن عبد العزيز بالزرقان[i] فكسره، ورياسة[k] هذا

---

a) B. اعرض. *b*) Ibno 'l-Athír, VII, p. ١٢٨ et ١٣٠ شبل pro بشر habet. *c*) A. السخمه, B. السجنه, E. شحمكى i. e. شحنكى. *d*) E. مشكان. *e*) A., B. et E. semper fere زمّ. *f*) A. et B. sine punctis, E. ميجان. *g*) A. et B. زرونه وهى. Secutus sum E. *h*) A. اقتضت. Post ايامهم E. addit داشتندى بفال سراورا. *i*) A. et B. بالرزمان, E. برزوان. Vid. supra ad p. ١٢٣ *a*. Ahmed hic anno 266 Ispaháni praefectus factus est (Ibno 'l-Athír, VII, p. ٢٢٧), et hoc eodem anno pugna collocanda videtur (ib. p. ٢٣١), nam de pugna anno 271 (ib. p. ٢٩١) hic cogitari nequit. Deinde A. et B. habent وكسره. *k*) B. وراسة.

الرمّ فى اولاد جيلويه الى يومنا هذا ✿ وامّا رمّ الديوان فكان رئيسهم ازادمرد ابن كوشهاذ[a] من الاكراد فملكها دهرًا ثمّ استعصى فقصده السلطان وهرب الى عُمان وبها مات وصار الامر بعده الى الحسين بن صالح من الاكراد فصار الرمّ فى يده ويد اولاده الى ايّام عمرو بن الليث فنقله عنهم الى ساسان بن غزوان[b] من الاكراد فهو فى اهل بيته الى يومنا هذا ✿ وامّا رمّ اللّوالجان فكان فى ايدى آل الصفّار[c] الى ان ولى محمّد بن ابراهيم[d] الطاهرىّ فجعله فى يدى احمد بن الليث[e] رجل من الاكراد فهو فى يدى اهل بيته الى يومنا هذا، ومحمّد بن ابراهيم هو الّذى اوقع بازادمرد بن كوشهاذ حتّى هرب ✿ وامّا رمّ الكاريان فهو فى يدى آل الصفّار الى يومنا هذا على قديم الايّام ورئيسهم اليوم حجر[f] بن احمد بن الحسن ✿ وامّا رمّ البازنجان[g] فانّ رئيسهم كان يسمّى شهريار[h] من الاكراد والرمّ منسوب اليه وكان مصاهرًا لجيلويه وصار بعده للقاسم[i] بن شهريار ثمّ انتقل الى موسى ابن القاسم[k]، والبازنجان الّذين هم فى حدّ اصبهان هم من هذا الرمّ فانتقلوا عن فارس الّا انّ لهم فى حدود فارس ضياعًا كثيرة وكان رئيسهم موسى بن عبد الرحمان ثمّ صارت لموسى بن مهراب وصارت بعده لابنه ابى مسلم[l] محمّد بن موسى ومن بعده لاخيه فارس بن موسى ومن بعده لاحمد ابن موسى والرئاسة فيهم الى يومنا هذا ✿

وامّا من يصلح من الفرس للدواوين من الكتّاب والعمّال والادباء فانّ منهم عبد الحميد بن يحيى وكان له فى بنى اميّة ولاء ينسب اليهم وكان من كتابته[m] واستقلاله ما اغنى عن ذكره لاشتهاره، ومنهم عبد الله بن المقفّع[n]

---

a) E. بن كوهستان, Ous. كوهستانى. *b*) A. عزوان, B. عروان. *c*) A. et B. ut solent الصفاق et sic E. *d*) Anno 232. Male Ibno 'l-Athír, VII, p. ٢٣, 2 a f. ابراهيم بن محمد بن مصعب; vid. p. ٣٥ seq. *e*) Sermo de eo est apud Ibno 'l-Athír, VII, p. ١٩٦, ٢١٣ et, ut videtur, p. ٢٥٩ ubi محمد بن الليث appellatur. *f*) A. et B. حجون. *g*) Codd. sine punctis. *h*) A. et B. شهرى راز hic et deinde. *i*) A. et B. القاسم. *k*) E. ابرهيم. *l*) B. مسلمه. *m*) A. كتمنه, B. كبتمه. *n*) A. المقفّع.

كان فارسيًّا اقام بالبصرة وقُتل فى ايّام المنصور بالبصرة وكان كتب امانًا لعبد الله بن علىّ من[a] المنصور فشرط فيه برآءة المسلمين من بيعته لو[b] خان فى امانته فوجد المنصور عليه فامر عامل البصرة بقتله سرًّا فقتله سرًّا، ومنهم سِيبَوَيْه[c] وكان مقيمًا بالبصرة ويقال انّه من اهل اصطخر فاقام بالبصرة الّا انّه مات بفارس[d] وقبره بشيراز بباب يعرف بباب ابردة[e] فى مفترشة يعرف بالمزدكان وله الكتاب المنسوب اليه فى النحو، والفرس هم شحنة دواوين الخلافة والعمّال الّذين بهم قوام السياسة فمن الوزرآء وسائر عمّال[f] الدواوين فمنهم البرامكة وآل ذى الرئاستين والى يومنا هذا من * المادَرَائيّين والفيرَيابيّين[g] وسائر شحنة الخلافة من اولاد الفرس الّذين انتقلوا الى السواد فى ايّام الاكاسرة فاقاموا فى ارض النبط وليس فى سائر دواوين الاسلام ديوان هو اصعب عملًا واكثر انواعًا من ديوان فارس لاختلاف ربوعها وتقارب الاخرجة على اصناف زروعها واختلاف ابواب اموالها وتشعّب الاعمال على المتقلّدين لها حتّى لا يكاد يبلغ الرجل الواحد الاستقلال بتلك الاعمال كلّها الّا فى الفرد وما علمنا احدًا منهم جمع من العلم ما يولد الدواوين الّا نفرًا يسيرًا منهم معلّى[h] بن النضر كاتب الحسن[i] بن رجآء وكان من اهل العراق يوطن شيراز فمات بها[k] وكذلك الحسن بن رجآء جمع له الحرب واعمال الدواوين مات بشيراز وقبره عند دار الامارة يعرف بدار هَدّاب[l] بن ضِرار المازنىّ الّتى كان المامون ابتناها لمّا ارجف باختياره بفارس ويكنى المعلّى بابى علىّ

---

a) A. et B. om., E. habet بن. Cf. Ibn Khallicán n. 186, p. ١٣٩ ed. Wüstenfeld. b) A. et B. او. c) E. add. النحوى الكاتب. d) E. بشيراز. Cf. Ibn Khallicán, n. 515, p. ١١٩. e) ? A. ادرده, B. ارده. f) A. et B. اعمال. g) In B. et E. haec duo nomina relativa desunt. h) A. et B. h. l. يعلى et deinde النضر. i) A. الحسن. Obiit hic anno 244, vid. Abu'l-Mahásin, I, p. ٧٥١, 4 ubi male de Jong proposuit (ann. p. 163) legere شجاع pro رجاء. Imo l. l. 4 a f. رجاء reponendum est pro شجاع. k) B. فيها. l) E. هدّاف, vid. Ibn Doraid, p. ١٣٩. Deinde E. ضرّار, A. et B. ضرار.

فكان يتقلّب فى اعمال الدواوين نحو خمسين سنة *وعاش بعد الحسن ابن رجآء نحوًا من ستّ سنين[a]، وماهان[b] بن بهرام من اهل سيراف[c] كتب لعلىّ بن الحسين بن بشر[d] ومحمّد بن واصل وجمع له الدواوين فاستقلّ بها واخوه كامل بن بهرام ويكنى بابى الليث كان لا يوصف فى الاستقلال الّا بديوان الرسائل فقط، ومنهم الحسن[e] بن عبد الله ويكنى بابى سعد *واسم عبد الله بزرجمهر[f] بن خدايداد[g] بن المرزبان وبلده فسا توطّن شيراز[h] *وهو من جانب امّه منسوب الى بنى مروان، ومنهم محمّد بن يعقوب من اهل يزد استقلّ بدواوين فارس وتوطّن ببخارى[i] ✿

وبفارس قوم يقال لهم اهل البيوتات يتوارثون فيما بينهم اعمال الدواوين منهم آل حبيب وكان مشايخهم مُدْرك واحمد والفضل بنو حبيب واصلهم من كام فيروز ومنشاهم شيراز قطنوها وتقلّدوا الاعمال الجليلة الشريفة، وكان المامون الخليفة استدعى مدرك بن حبيب الى بغداد للحساب وغيره من وجوه الخدمة وحظى عنده وقرأ عليه فمات ببغداد ايّام المعتصم *واتّهم يحيى بن اكثم به[k] ✿ وآل *ابى صفيّة[l] من موالى[m] باهلة منهم يحيى وعبد الرحمان وعبد الله بنو محمّد بن اسماعيل ناقلة توطّنوا بها *فى زمان المامون وتقلّدوا الديوان[n] ✿ *وامّا آل المرزبان بن زادية فانّهم كانوا من اهل

---

a) E. وچون در اخر عمر نابينا شد شش سال ديگر بزيست. b) A. et B. وماهار. c) E. شيراز. d) A. et B. وبرديس. e) E. h. l. حسين. f) E. ونام او عبد الله بن بزرجمهر. g) A. et B. خوابداد, E. فرانداد. h) Hic in A. et B. aliquantum desideratur, quod supplevi ex E. et D. i) Haec Arabice verti ex E., apud ceteros desunt. Verba Persica sunt واز جانب مادر نسبت بآل مروان باز برد محمد بن يعقوب از اهل يزد در ديوانهاى بارس بر آمد وببخارا وطن ساخت. Quae sequuntur quoque in D. exstant. k) E. گويند يحيى اكثم كسىرا فراز كرد تا اورا در نهان هلاك كردند. l) E. بنى صغار, Ous. p. 126 بنى صفر. m) E. قبيله. n) Ex solo E. addidi, ubi بروزگار مامون در بارس وطن ساختند وعمل ديوان كردند.

شيراز[a]، وكان الحسن بن المرزبان بندارًا لمحمّد بن واصل ومن بعده ليعقوب بن الليث، وكان جعفر بن سهل بن المرزبان كاتب ابى[b] الحارث ابن فريغون[c]، وخدم علىّ بن المرزبان عمرو بن الليث على ديوان الاستدراك ۞ آل المرزبان بن خدايداد[d] الّذين يقال انّ اصلهم من فسا وهم اقدم اهل هذه البيوتات واكثرهم عددًا[e] ومنهم ابو سعد[f] الحسن بن عبد الله ونصر بن منصور * بن المرزبان وعبد الرحمان بن الحسين بن المرزبان وخدايداد[g] بن مردشاد[h] بن المرزبان واحمد بن خدايداد[i] فى جماعة تركنا تقصّى عددهم يتولّون طرفًا من اعمال الديوان الى يومنا هذا ۞ وآل مردشاد ابن نسبة[k] منهم علىّ بن مردشاد واولاده الحسن والحسين واحمد والى يومنا هذا منهم عُمّال[l] العمالات، فهاؤلآء مع آخرين لم نذكرهم اهل بيوت يتوارثون هذه الاعمال ۞

وقد انتحل قوم من الفرس ديانات خرجوا بها من المذاهب فدعوا اليها وانتصبوا لها لولا انّ اهمال امرهم ضرب من العصبيّة وباب من التحامل فنذكر المحاسن ولا نذكر غيرها لكان من الواجب اهمال ذكرهم لشناعة امرهم وفظاعة اخبارهم ولاكنّ الوقوف على ما امكن من اخبار الناس وسيرهم من محمود ومذموم غير مكروه، فممّن[m] عُرف من هاؤلآء واشتهر ذكره الحسين ابن منصور المعروف بالحلّاج من اهل البيضاء وكان رجلًا حلّاجًا ينتحل النسك فما زال يرتقى به طبقًا عن طبق حتّى انتهى به الحال الى ان زعم انّ من هذّب فى الطاعة جسمه واشغل بالاعمال الصالحة قلبه وصبر على

---

a) D. corrupte واما ال مرزبان .E ;والى المرزبان بن زادية البيهم فى الاعمال بن زاديه از اهل شيراز بود *b*) E. om. *c*) Secundum E.; D. افريقون. Vid. de hoc nomine infra in capite de Khorasán ubi de الغور sermo est. *d*) E. فرادداد (Ous. فراسداد), D. فرابند اذْ *sic*. *e*) Hic desinit lacuna in A. et B. *f*) D. سعيد; E. add. بن. *g*) A. et B. وحوادسداد, D. وحوىمدار. *h*) Haec in E. desunt. *i*) E. حداداذ, A. et B. خوابيداد, D. خوادمده. *k*) *Sic* A. et B.; E. ىسه. *l*) A. et B om.; E. عمال ديوان. *m*) A. et B. فمن.

مفارقة اللذّات وملك نفسه فى منع الشهوات ارتقى به الى مقام المقرّبين ثمّ لا يزال يتنزّل فى درج المصافاة[a] حتّى يصفو عن البشريّة طبعه فاذا لم يبق فيه من البشريّة نصيب حلّ فيه روح الله الّذى كان منه عيسى بن مريم فيصير مطاعًا فلا يريد شيًا الّا كان من كلّ ما ينفذ فيه امر الله وانّ جميع فعله حينئذ فعل الله وجميع امره امر الله، فكان يتعاطى هذا ويدعو الى نفسه بتحقيق ذلك كلّه حتّى استمال جماعة من الوزرآء وطبقات من حاشية السلطان وامرآء الامصار وملوك العراق والجزيرة والجبال وما والاها وكان لا يمكنه الرجوع الى فارس ولا يطمع فى قبولهم ايّاه فخاف على نفسه منهم لو ظهر لهم فأُخذ وما زال فى دار السلطان ببغداد الى ان خيف من قبله ان يستغوى كثيرًا من اهل دار الخلافة من الحجّاب والخدم وغيرهم فصُلب حيًّا الى ان مات ۞ ومنهم الحسن الجَنّابىّ[b] ويكنى بابى سعيد من اهل جَنّابة كان دقّاقًا اظهر مذهب القرامطة فنُفى عن جنّابة فخرج منها الى البحرين فاقام بها تاجرًا[c] يستميل العرب بها ويدعوهم الى نحلته حتّى * استجابوا له وملك[d] البحرين وما والاها فكان من كسره عساكر السلطان وعيثه[e] وعدوانه على اهل عمان وسائر ما يصاقبه من بلدان العرب ما قد انتشر ذكره حتّى قُتل[f] وكفى الله امره، ثمّ قام ابنه سليمان بن الحسن فكان من قتله الحاجّ[g] وانقطاع طريق مكّة فى ايّامه[h] والتعدّى فى الحرم وانتهاب كنوز الكعبة[i] وقتل المعتكفين بمكّة ما قد اشتهر ذكره، ولمّا

---

a) Codd. المصافات. *b*) A. الجُنّابى ut saepius quoque habet جُنّابة; B. الجُبّادى. Locum transcripsit Jacut, II, p. ١٢٢. *c*) Jacut l. l. 17 addit وجعل. *d*) Jacut استجاب له اهل. *e*) A. وعَبْثه. Deinde A. et B. وعدوه انه, restitui ex marg. A. Jacut ورعيته وعداوته et habet من pro على. *f*) Jacut addit على فراشه. *g*) Jacut حُجّاج بيت الله الحرام. *h*) Jacut addit بسببه. *i*) Jacut addit: ونقله الحجر الاسود الى القَطيف والأَحساء من ارض البحرين et deinde habet وبقى عندهم احدى وعشرين سنة ثم ردّ ببذول بذلت لهم وقتله.

اعترض[a] الحجّاج بما كان منه[b] أُخذ عمّه اخو ابى سعيد وقراباته[c] فحُبسوا بشيراز مدّة وكانوا مخالفين له[d] فى الطريقة يرجعون الى صلاح وسداد وشُهد لهم بالنزاهة[e] من * القرمطة فخُلّىَ عنهم[f] والله الحافظ للاسلام واهله والشرّ لمن حادّ الله فى امره ۞

وسنذكر الخاصّيّات[g] بها   بناحية اصطخر ابنية حجارة عظيمة الشان من تصاوير واساطين وآثار ابنية عاديّة[h] يذكر الفرس انّه مسجد سليمان بن داوود عم وانّ ذلك من عمل الجنّ وهى تشبه ابنية رايتُها ببعلبكّ وارض الشام ومصر فى العظم وممّا يعجز عن مثله اهل هذا العصر، وبناحية اصطخر تفّاح تكون التفّاحة الواحدة منه بعضها حامض وبعضها حلو حدّث مرداس بن عُمَر به[i] الحسنَ بن رجآء فراى فى وجهه انكارًا لذلك فاحضره حتّى رآه، وبقرية عبد الرحمان بئر عمقها قامات كثيرة جافّة القعر عامّة السنة حتّى اذا كان الوقت المعروف من السنة[k] ينبع منها مآء يرتفع الى وجه الارض ويجرى منه ما يدير الرحى حتّى ينتفع به فى سقى الزروع وغير ذلك ثمّ يغور، وبناحية سابور جبل قد صُوّر فيه صُوَر كلّ ملك وكلّ مرزبان معروف للعجم[l] وكلّ مذكور من سدنة النيران وعظيم من موبذ وغيره وتتابع[m] صُوَر هاؤلآء وايّامهم وقصصهم فى ادراج وقد خُصّ بحفظ ذلك قوم سكّان بموضع بناحية أرّجان يعرف بحصن الجصّ، وبجور بركة[n] على باب البلد ممّا يلى شيراز تعرف بنزّه[o] قد أُكِبّ على قعرها قِدْر نحاس عظيمة يخرج من ثقبة فى اعلى تلك القدر ضيّقة جدًّا مآء عظيم ليس فى تقدير[p] راى

---

a) B. اعرض.   b) Jacut الحجّاج وكان منه ما كان.   c) Jacut وقرابيه. Deinde A. et B. فحبس.   d) A. et B. لهم.   e) Jacut بالبَرَآءة.   f) Jacut القرامطة فانطلقوا.   g) E. بناهاى.   h) E. از سنك صورتها كرده وبر نبشته ونكاشته. i) A. et B. om.   k) E. وقت كشاورزى.   l) C. فى العجم.   m) A. et B. sine punctis. C. ودمماسح. Verba E. videantur apud Ous. p. 129.   n) C. add. عظيمة.   o) A. et B. بمر; C. et E. om. ut quoque Mokaddasi MS. p. 214 et Kazwíní, II, p. ١٢١ (qui habet اكبّوا — قدرًا).   p) E solo C.

العين انّ مثل ذلك الماء على كثرته يخرج من *ذلك الثقب على ضيقه[a]، وبقرب ابرقوه تلال عظيمة من رماد يزعم قوم انّها نار نمرود بن كنعان الّتى اوقدها لاحراق ابراهيم وهذا خطأ لانّ الصحيح فى الاخبار انّ نمرود كان مقيمًا[b] ببابل وكذلك ملوك الكنعانيّين قبل ملوك الفرس، وقد ذكرنا المومياى فى جملة ما يرتفع من دارابجرد، وبكورة ارّجان بقرية يقال لها صاهك[c] الغرب بئر يذكر اهلها انّهم امتحنوا قعرها بالمثقلات والارسان فلم يقفوا منها على عمق يفور منها الدهر كلّه ماء بقدر ما يدير رحى *ويسقى تلك[d] القرية، *وبكورة سابور رستاق[e] يعرف بالهنديجان فيها[f] بئر بين جبلين يخرج منها دخان فيعلو حتّى[g] لا يتهيّأ لاحد ان يقربها واذا طار فوقها طائر سقط فيها واحترق، وبدَشْت بَارِين[h] قرية تعرف بجور[i] هى *نحيسة لا شجر[k] فيها فيها اهل بيت ينسبون الى السحر ويسألون عن الاخبار ويحكى عنهم ما استفظع حكايته فى كتابى، وبكورة اردشير خُرّه على باب شيراز عين ماء يشربه الناس لتنقية الجوف فمن شرب منه قدحًا اقامه مجلسًا ومن زاد[l] فلكلّ قدح مجلس، وبناحية كام فيروز بقرية تعرف بالمورجان[m] بين جبال شاهقة كهف فيه جرن وفى سقف هذا الكهف ماء

---

a) C. تلك الثقبة على ضيقها. b) E solo C. c) A. et B. صاهل. Deinde C. et E. العرب. d) A. et B. فسمى بذلك, quod fortasse spectat epitheton الغرب sensu *molae* accipiendum ut e. g. Jacut, I, p. ٨٥٧, 19. Recepi lectionem C., E., Mokaddasí l. l. et Kazwíní, I, p. ٢٠٠, II, p. ٩٥. e) C. et E. وبكورة رستاق الرستاق قرية, Abulfeda, p. ٣٢٣ فى كورة رستاق. Mokaddasí male habet بالغندجان. Kazwíní, II, p. ١٨٩ locum هنديان appellat. f) A. et D. om. g) C. حرها. Secundum E. qui habet وحرارت جنانست كى اگر مرغى الخ legendum foret حرها حتى. Deinde A. legit تهيا. h) E. وبدشت وارين, C. وبدشت بمردن. i) A. et B. بسحور. E. جروهيبن (Ous. p. 130 خسروهيبن). k) A. et B. بحيسه لا ساحر. E. آنجا هيچ درخت نبود. l) C. ازداد. m) C. بمورجان et sic Mokaddasí.

ينقطر الى [a] الحجر فيزعم الناس انّ عليه طلسمًا فان دخل ذلك الكهف رجل خرج ما [b] يكفى رجلًا وان دخله الف رجل خرج بقدر حاجتهم، وعلى باب ارّجان ممّا يلى خوزستان قنطرة على نهر طاب تنسب الى الديلمىّ طبيب الحجّاج وهى [c] طاق واحد سعة الطاق على الارض ما بين العمودين نحو ثمانين خطوة [d] وارتفاعه مقدار ما يجوز فيه راكب الجمل بيده علم من اكبر ما يكون، وبناحية كُرّان [e] طين اخضر كالسلق [f] يؤكل ليس فيما علمتُه فى بلد مثله، وبناحية جنّابا فى البحر مكان يعرف بخارك معدن اللؤلؤ يقال [g] انّ النادر منه لا يفوقه شىء فانّ [h] الدرّة اليتيمة منه ان صحّ ذلك، وبناحية شيراز ريحان يعرف بسوسن نرجس ورقه مثل ورق السوسن وداخله مثل عين النرجس سوآء، وبناحية دانين [i] نهر مآء عذب يعرف بنهر اخْشين [k] يشرب ويسقى الاراضى واذا *غُسل به ثياب [l] خرجت خُضْرًا، وبدشت بارين فى جبالها بقرية تسمّى بر [m] عين مآء قليل [n] يعرف بمآء نوح يتداوى به من العلل والعين ويقال [o] انّه ربّما حُمل منه الى حدود الصين لاشتهاره واستعمال الناس ايّاه فينتابه [p] الناس من خراسان والبلدان النائية ✿

فامّا ما يرتفع من بلدان فارس ممّا ينقل الى الامصار وما يفضّل فى جنسه على سائر ما يرتفع فى البلدان فمن ذلك *مآء الورد [q] الّذى يرتفع من جور فانّه يفضّل فى جنسه وينقل الى البحر فيفرّق [r] فى الحجاز واليمن

---

a) Secundum C. et D.; A. من, B. عن. E. habet از سقف آن غار آبى مى يقطر من جكد et sic D. يقطر اليه من سقف هذا الكهف ماء, Mokaddasí وفى سقفه ماء, Kazwíní, II, p. ١٨٣ يقطر الماء من سقفه. C. quoque habet وفى سقف.
b) A. مآء. c) A. et B. om. d) E. دويست كام. e) A. et B. كوان, C. الكوان, D. كوار, E. كوران (Ous. كوان). f) E. چون برك چغندر. g) B. ويقال. h) B. وان. i) B. دارين, C. دادن. k) A. احسنتين, C. احسن, E. حسين. Post يشرب D. et E. add. منه. l) C. et D. غسلت به الثياب. m) B. et C. بو, E. خوانند (l. ماذوج) دادوج انرا هست دهى. n) C. قيل انه. o) A. et B. ينقال. p) A. ينتابه. q) A. et B. المولود, i. e الماورد. r) C. فمغدوا sic.

والشام ومصر والمغرب وخوزستان وخراسان ويرتفع من غير جور ما هو اجود الّا انّ معظم الجهاز منه، ويرتفع بجور مآء الطلع ومآء القيصوم الّذى لا نعرفه[a] فى بلد غير جور ومآء الزعفران المسوسن[b] ومآء الخلاف الّذى يفضّل على جنسه فى سائر البلدان ✿ ويرتفع من سابور الادهان من كلّ جنس ما يفضّل على ادهان سائر المدن الّا الخيرىّ والبنفسج فانّ الّذى بالكوفة *منهما خير[c]، والانبجات[d] الّتى تحمل الى الآفاق منها ✿ ويرتفع من شينيز[e] وجنّابا وكازرون وتوّج ثياب كتّان وللسلطان فى كلّ بلد منها طراز غير كازرون وتحمل هذه الثياب الى الآفاق من بلدان الاسلام كلّها ✿ ويرتفع من فسا انواع من الثياب الّتى تجلب الى الآفاق وبها طراز الوشى والشعر[f] والسوسنجرد للسلطان، فامّا الوشى فانّ المذهّب المرتفع منه اجود ممّا[g] يكون لغيره من الامصار وامّا غير المذهّب فانّ الّذى بجهرم اجود واكثر منه، وامّا الشعر فانّه يعمل للسلطان ثياب مثقاليّة تاخذ قيمة كثيرة وكللٌ[h] مرتفعة وسائر اصناف الشعر، ويتّخذ من القزّ للسلطان ستور مُعْلَمَة ويرتفع من ثياب القزّ والشعر ما يحمل الى كثير من امصار الاسلام، والسوسنجرد الّذى يكون بها ارفع ممّا يكون بقرقوب وتوّج وتارم[i]، وبها اكسية القزّ الّتى تبلغ قيمة كثيرة ✿ ويرتفع من جهرم ثياب الوشى المرتفع[k] والبسط والنخاخ والمصلّيات والزلالىّ المعروفة بالجهرميّ ✿ ويرتفع من يزد وابرقوه[l] ثياب قطن تحمل الى الآفاق ✿ ويرتفع من الغُنْدِجان[m] قصبة دشت بارين من البسط والستور والمقاعد واشباه ذلك ما يوازى به عمل الارمينىّ وبها طراز للسلطان وتحمل منها الى الآفاق ✿ وانّما فضل سوسنجرد فسا على سوسنجرد قرقوب[n] لانّ القرقوبىّ ابريسم وهذا صوف[o] والصوف اجود من الابريسم فى الصنعة ✿

---

a) C. يعرفه احد. b) B. المسوس, C. المرشوس. c) C. يعمل اجود.
d) A. et B. sine punctis et deinde الذى. e) A. et B. سمرس. f) C. والشعرى. E. طراز ديبا et pro السوسنجرد habet سوزن كرد. g) A. et B. ما. h) A. et B. وكللا. i) A. et B. sine punctis. k) A. والمرتفع. l) A. et B. h. l. وبرقويه. m) A. الغُنْدَجان. n) A. قَرْقوب. o) E. ودر بسا از ريسمان وپشم.

ويحمل من سيراف ما يقع اليها من امتعة البحر من العود والعنبر والكافور والجواهر والخيزران والعاج والآبنوس والفلفل والصندل وسائر الطيب والادوية والتوابل الّتى يكثر تقصّيها الى جميع فارس والدنيا كلّها وهى فرضة لهذه المواضع واهلها ايسر اهل فارس ومنهم[a] من يجوز ماله ستّين الف الف درهم ما اكتسبه الّا من تجارة البحر وهم الغالبون على مدن تلك[b] السواحل وعلى البحر كلّه ✿ ويرتفع من أرّجان دوشاب يكون بآسك[c] وآسك هذه الّتى كان بها وقعة الازارقة وكانوا اربعين رجلًا فقصدهم نحو الفى رجل من اصحاب البصرة فقتلوا الالفين عن آخرهم، ويفضّل هذا الدوشاب على ما يكون بالعراق وسائر المدن *الّا السيلان الّذى يكون بالاحساء وهجر فانّه يفوقه[d]، وبارّجان زيت يحمل الى الآفاق منه فيفضّل على غيره ✿ وبكارَزين[e] تمر يقال له الجيلاندار[f] يتفرّد[g] به ذلك الموضع ولا يكون بالعراق والحجاز وكرمان وسائر مواضع التمور ويحمل منها[h] الى العراق[i] على كثرة تمورها ✿ وبدارابجرد سمك بالخندق الّذى يحيط بالبلد لا شوك فيه ولا عظم ولا فقار وهو من الذّ السموك، ويرتفع من دارابجرد مثل العمل الطّبَرىّ الّذى يكون بطبرستان ✿ ويرتفع من كازرون ثياب كتّان تنقل الى الآفاق ✿ ومن قرية من دارابجرد المومياى الّذى يحمل * الى السلطان[k] وهو غار فى جبل

---

a) E. وخداوند كتاب كويد من بازركانان ديدم درين شهر. b) B. بلد.
c) Vid. supra p. ٩٥. E. از دهى كه آنرا اشك خوانند. d) A. et B. om. Addidi ex C. et E. e) E. وبكازرون et sic legit Jacut IV, p. ٢٢٥, 22. B. وبكازرين. f) ? A. الحملاراب, B. الحملادار, E. حيلان; Ous. p. 133 خيلان, Jacut l. l. جيلان et sic Kazwíní, II, p. ١٦٢. g) B. ينفرد. h) Jacut منه. i) E. بسباهان وعراق. Jacut addit فى الهدايا, Kazwíní للهدايا. k) D. الى الافاق وهو ملك للسلطان. Ibno 'l-Fakíh speluncam in Arradján collocat, vid. Jacut, I, p. ١٩٤ seq. et II, p. ٥٩.. Mokaddasí, qui tum Ibno 'l-Fakíhi, tum al-Istakhríi libro usus est, utrumque recepit, dicens وبه معادن المومياى بدرابجرد وبارجان ايضا.

قد وُكل به من يحفظه فيفتح فى كلّ سنة فى وقت معروف وقد استجمع فى نقر[a] حجر هناك مآء قد اجتمع الموميا‌ى فى اسفله فاذا جُمع يكون[b] مثل الرمّانة فيختم ويشهده[c] ثقات السلطان من الحكّام واصحاب البرد والمعدّلين ويرضخ[d] للّذى يحضره بالشىء اليسير[e] وهو الموميا‌ى الصحيح وما عدا الموميا‌ى الّذى يحمل الى السلطان فشىء مزوّر يشبه الموميا‌ى وليس بالصحيح، وبقرب هذا الغار قرية تسمّى آبين[f] فينسب هذا اليها[g] ويسمّى موم قرية آبين، وبناحية دارابجرد جبال من الملح الابيض والاصفر والاخضر والاسود والاحمر تنحت من هذه الجبال موائد[h] وغير ذلك وتحمل الى سائر المدن والملح الّذى فى سائر المدن انّما هو من[i] باطن الارض او مآء يجمد وهو ملح[k] جبل ظاهر، وبدارابجرد دهن رازقىّ[l] يقال انّه ليس فى مكان مثله ۞ ويكون بارض فارس عامّة المعادن من الفضّة والحديد والآنك والكبريت والنفط واشباه ذلك ممّا يستقلّ * به اعلها[m] ممّا يكون فى سائر الاقطار الّا انّ الفضّة بها قليلة بناحية يزد بموضع[n] يعرف بنائبين[o] ولا اعرف بها معدن ذهب ومعدن الصفر بالسّردن يحمل منها الى البصرة والى سائر النواحى والحديد يرتفع من جبال اصطخر وبقرية من كورة اصطخر تعرف بدارابجرد[p] معدن للزيبق ۞ ويعمل بفارس مداد اسود للمدواة والصبغ يفضّل على غيره، وبشيراز ابراد تحمل الى الآفاق، وبجانات[q] من كورة اصطخر ثياب قطن مستحسنة تعرف بالادادى[r] رقيقة ۞

---

a) C. نقرة. *b*) C. يجمع كان. *c*) C. et D. بمشهد, B. ويشهد. E. ومعتمدان سلطان حاضر شوند. *d*) A. et B. ويوضح, C. ويُرخّص. *e*) C. للذى عصره فى الشىء اليسير. *f*) A. om., B. آس, D. ادمن; vid. Vullers sub آبين et آئين. *g*) A. et B. اليه. *h*) Jacut, II, p. ٥٩٠, 21 et 22 addit والصحون والغضائر وغيرها من الظروف, Kazwíní, II, p. ١٣٩ وصحون وزبادى. *i*) C. فى, Jacut om. *k*) A. et B. املح. Jacut جبل ملح. C. om. جبل. *l*) A. et B. رازقى. *m*) A. et B. باهله. *n*) A. et B. موضع. *o*) A. et B. دماس, E. مابين. *p*) Jacut, I, p. ٣٩٩ ult. et seq. *q*) A. دحادات. *r*) *Sic* A. et B. sine punctis. C., D. et E. om. Deinde C. رفيعة جدّا.

فامّا نقودهم واوزانهم ومكاييلهم   فالبيع [a] والشراء بجميع فارس بالدراهم وانّما الدنانير عندهم بالعرض وليس على سكّة الدراهم والدنانير الّتى تعرف بفارس الّا اسم امير المؤمنين من ايّام السجزيّة [b] الى يومنا هذا، فامّا اوزانهم فانّ وزن الدرهم كلّ عشرة دراهم سبعة مثاقيل وليس مثل اليمن وغيرها من المواضع الّتى تختلف مقادير اوزان الدرهم بها، وأمّا ما توزن به الامتعة فانّ المنا بشيراز اثنان صغير وكبير فالكبير الف درهم واربعون درهمًا وما رايتُ ولا بلغنى انّ فى موضع من المواضع المنا على هذا الوزن الّا باردبيل *والآخر هو منا بغداد [c] وزن مائتين وستّين درهمًا وهذا المنّ مستعمل بجميع فارس وعامّة ما دخلتُهُ من امصار المسلمين وان كان لهم اوزان غير هذا، والمنا بالبيضاء وزن ثمان مائة درهم وباصطخر وزن اربع مائة درهم وبخرّه المنا مائتان وثمانون درهمًا وبسابور المنا ثلاثمائة درهم وببعض نواحى اردشير خُرّه المنا بها مائتان واربعون درهمًا٭ وامّا الكيل فانّ بشيراز الجريب عشرة اقفزة والقفيز ستّة عشر رطلًا فى التقدير يزيد وينقص القليل اذا كان المكيل حنطة والرطل وزن مائة وثلاثين درهمًا ولهذا القفيز كيل على حدَة ولهذا القفيز نصف [d] وربع كلّ واحد منهما كيل قائم بنفسه وكيل صغير هو جزء من اربعة وعشرين من هذا القفيز وجريب اصطخر وقفيزها على النصف من جريب شيراز ومكاييل البيضاء تزيد على مكاييل اصطخر بنحو العشر *ونصف العشر [e] ومكاييل كام فيروز وما يتّصل بها على الخُمْسَيْن [f] من مكاييل البيضاء ومكاييل ارّجان تزيد على مكاييل شيراز الربع ومكاييل سابور وكازرون تزيد على مكاييل شيراز العشرة ستّة ومكاييل فسا تنقص عن مكاييل شيراز العشر٭

ابواب المال   لبيت المال على الناس والرسوم ابواب المال الّتى تطبق [g] عليها الدواوين *من خراج [h] الارضين والصدقات واعشار السفن واخماس المعادن

a) A. et B. البيع. b) E. سكزيان, quod pessime Ous. p. 134 vertit "of the Kesris." c) E. وسنك كوچك سنك بزرگ بغداد بود. d) D. et E. addunt وثلث. e) A. ونصف, B. والنصف. Restitui العشر ex E. f) E. عشرى. g) A. sine punctis, B. تطيق. E. كه در دواوين مثبت بود. h) Secundum E.; A., B. et D. وخراج.

والمراعى والجزية وغلّة دار الضرب والمراصد والضياع والمستغلّات واثمان الماء وضرائب الملّاحات والآجام، فامّا خراج الارضين فعلى ثلاثة اصناف على المساحة والمقاسمة والقوانين الّتى هى مقاطعات معروفة لا تزيد ولا تنقص زُرع او لم يزرع وامّا[a] المساحة والمقاسمة فان زُرع أُخذ خراجه وان لم يزرع لم يؤخذ وعامّة فارس مساحة الّا الرموم فانّها مقاطعات الّا شيئًا يسيرًا[b] من المقاسمات، وتختلف الاخرجة فى البلدان على المساحة فاثقلها بشيراز وعلى كلّ صنف من الزرع شىء مقدّر فعلى الجريب الكبير من الارض يزرع فيه الحنطة والشعير السيح مائة وتسعون[c] درهمًا والجرجر[d] السيح مائة واثنان وتسعون درهمًا *والرطاب والمقاثى السيح للجريب الكبير[e] مائتان وسبعة وثلاثون درهمًا ونصف ومن الجريب الكبير من القطن السيح مائتان وستّة وخمسون درهمًا واربعة دوانيق *وعلى الجريب الكبير من الكرم[f] الف واربع مائة وخمسة وعشرون درهمًا، والجريب الكبير ثلاثة اجربة وثلثان[g] بالجريب الصغير والجريب الصغير ستّون ذراعًا فى ستّين ذراعًا بذراع الملك وذراع الملك تسع قبْضات[h] هذا خراج شيراز للسيح، وخراج جور[i] على الثُّلُثَيْن[k] من هذا لانّ جعفر بن ابى زهير الشامى كلّم الرشيد فردّه الى ثلثى الربع[l]، وخراج اصطخر ينقص من خراج شيراز فى الزرع بشىء يسير، هذا خراج السيح والبخوس[m] خراجه على ثلث السيح والطّوِىّ والنضح والمندّى على ثلثى الخراج والسقى ما نُدّى وسُقى سقيةً[n] فيقبض الربع من الخراج واذا نُدّى وسقى سقيتين فهو السيح وقد استتمّ الخراج، وكورة

---

a) A. et B. وانما. b) A. et B. شى يسير. c) D. وسبعين, sed lectio A. et B. non tantum confirmatur ab E., sed quoque a Mokaddasí MS. p. 216. d) A. et B. الجرجر, D. الشجر. E. habet pro his چون گندم وجو باشد آبى صد ونود درم خراج جريب بود وبى آبرا صد ونود ودو درم. e) E. جريب پاليزرا. f) E. جريب باغهارا. g) E. ودو چهاريک. h) D. et Mokaddasí قصبات. i) D., E. et Mokaddasí كوار. k) E. دو چهاريک. l) E. بفرمود تا ثلث بربع باز آوردند. m) Codd. والنخوس. n) B. بسقية et mox بسقيتين.

دارابجرد وأرّجان وسابور زراعتهم ومقادير الخراج على ارضيهم بخلاف هذا يزيد وينقص ✿ وامّا المقاسمة فانّها على وجهين ضياع فى ايدى قوم من اهل الرموم وغيرهم معهم عهود من علىّ بن ابى طالب عم ومن عمر بن الخطّاب رضه وغيرهما من الخلفآء فيقاسمون على العشر والثلث والربع وغير ذلك والوجه الآخر[a] مقاسمات على قرى صارت لبيت المال[b] فيزارع الناس عليها ✿ وامّا ابواب اموال الضياع فانّ الضياع السلطانيّة خارجة عن المساحة وانّما تؤخذ من السلطان بالمقاسمة او المقاطعة[c] وعلى الاكرة فيها ضرائب من الدراهم يؤدّونها[d]، وامّا الصدقات واعشار السفن واخماس المعادن والجزية ودار الضرب والمراصد وضرائب الملّاحات والآجام واثمان المآء والمراعى فانّها تقرب[e] فى الرسم ممّا فى سائر الامصار، وليس بفارس دار ضرب الّا بشيراز، وامّا المستغلّات فانّها تربة اسواق بشيراز وغير شيراز ابنيتها للناس ويؤدّون اجرة الارض والطواحين[f] للسلطان واجرة الدور الّتى يعمل فيها مآء الورد، وكان الرسم القديم بفارس انّ كلّ حومة بفارس لا خراج على الكروم فيها ولا على الاشجار بجميع فارس الى ان ولى علىّ بن عيسى الوزارة سنة ٣٠٢ فالزمهم فيها كلّها الخراج، وبفارس ضياع قد الجأها اربابها الى الكبرآء من حاشية السلطان بالعراق فهى تجرى باسمآئهم وخُفّف عنهم الربع فهى فى ايدى اهلها باسمآء هاؤلآء يتبايعونها ويتوارثونها ✿

بلاد كرمان

وامّا كرمان فانّ شرقيّها ارض مكران ومفازة ما بين مكران والبحر[g] من ورآء البلوص[h] وغربيّها ارض فارس وشماليّها مفازة خراسان وسجستان وجنوبيّها بحر

---

a) A. et B. الاول. b) E. وديگر ضياع جور باشد. c) E. مقاطعه ومسامحه ومقاسمه. d) A. دونها. e) A. et B. تقرّر. f) A. et B. وطواحين. Apud D. et E. haec paullo aliter exstant, praesertim apud E. qui habet مستغلات زمين سلطانرا باشد اجرت زمين ندهند از دكان وآسياب وكارگاه ما الورد. g) A. et B. والبحار, E. وبحرين. Vid. quoque Jacut, IV, p. ٣٦٤, 2. h) E. بلوج.

فارس ولها فى حدّ الشيرجان[a] دخلة فى حدّ فارس مثل الكمّ وفيما يلى البحر لها تقويس[b] ۞

وكرمان لها صرود وجروم وصرودها تقصر عن صرود فارس * فى البرد[c] وليس فى جرومها شئ من الصرود وفى صرودها ربّما عرض بعض الجروم[d]، وامّا ما يقع فيها من المدن الّتى اعرفها فهى الشيرجان وجيرفْت وبَمّ وهُرْمُوز وفى اضعاف ذلك ما بين فارس وجيرفت مدينة روبين[e] وبعضهم يزعم انّها مدينة ليست من كرمان وبعضهم يقول انّها من كرمان ومدينة كشيستان[f] وجيروقان[g]

a) Sic semper A. et B.; cf. Jacut, III, p. ٣٥. seq. E. سيركان. b) Hic sequi deberet mappa Kirmáni. c) A. et B. om.; D. فى البر. Secutus sum C. d) E. ودر هم آميخته بود deinde addens: واز ميوها همه گونه باشد وغالب گرمسير باشد وبعضى از كنارهٔ دريا بر حد كرمان افتد وآنجا فرضهٔ ايشان باشد وبغايت گرمسير مثل هرمز وجرون وغيره وهواى تا خوش عفن باشد. De urbe جَرُون vid. Ibn Batuta, II, p. 230 seqq., *Djihán-Numa*, p. ٢٥٧ et Juynboll ad *Merácid*, V, p. 58. Abulfeda, p. ٣٣٩, scribit زَرُون. e) A. et B. h. l. ووسى (A. in mappa ررمى), infra in itinerario A. رودن, B. دودن. E. h. l. دوبين (Ous. p. 139 دوهين), infra روبين (Ous. p. 145 روبين). C. in itinerario دردن, in mappa رومين, h. l. om. D. h. l. السرودان, in itinerario lacunam habet. Cf. Sprenger p. 79. f) A. h. l. كشيسبان, B. كشيبان, infra in itiner. A. et B. كَشْسُبَان; C. h. l. كمسمان, infra كسنمان et in mappa كيسان, D. كسبستان et ut Epit. Paris. كشتشتان; E. h. l. كيسار, infra كسينان (Ous. p. 145 كسيسان); Edrîsî, I, p. 427 كشستان. Eadem urbs appellatur a Mokaddasí MS. p. 28 et 224 كستان. g) Sic cum punctis E. h. l. (Ous. p. 139), C. in itinerario, Edrîsî et Mokaddasí. A. h. l. حمروقان, infra sine punctis, ut C. hic et in mappa, et B.; D. خيروقان et خبزوقان. E. infra حمروان (Ous. p. 145 جيروان). Epitom. Paris. جُوبرقان.

ومرزقان [a] والسورقان [b] وولاشكرد [c] ومغون [d]، ومما يلى جيرفت الى الشيرجان ناجت [e] وخير [f]، وما بين الشيرجان وبمّ الشامات [g] وبهار وخنّاب [h] وغبيرا [i]

---

a) Sic h. l. E. (Ous. مزرقان) et D.; A. h. l. مررقان, infra مرزوان et المرزوان, B. h. l. مرزفان, infra مرزوان, C. h. l. مورفان, infra المرفان et مزرفان, in mappa مورقان; E. infra موروان. Epitom. Paris. سرزقان, sed in mappa مزرقان. Mokaddasí p. 28 موزقان, p. 224 مـورقـان. *b*) Sic h. l. A., B., D. et E.; C. h. l. الورفان, infra et in mappa السوروان, E. infra سوروان, A. et B. infra السرمقان. Mokaddasí habet ut recepi. *c*) A. et B. h. l. والاسكرد, D. والاشكرد, E. h. l. الاس كرى, A. et B. infra اولاشكرد, C. in mappa لاشكرد. Ceteris locis ولاشكرد. Mokaddasí habet ولاشجرد. *d*) A., B., C. et E. ubique معون (A. semel معور), D. infra ut Mokaddasí (apud Sprenger, p. 77) مغون, Jakubí, p. ٩٤ معور (vocales ab editore sunt additae), *Merácid* معورا s. معور, Jacut مَعُوز. *e*) A. et B. h. l. داحت, infra ناحب ut C.; E. h. l. ناحيت, D. باخته, Edrîsî ناجبة. Sprenger, p. 78, ex vers. Pers. dat lectionem تاحت quod in meo Cod. ut in Ous. deëst. Mokaddasí hoc et sequens nomen habet sine punctis ناحب et حمر (p. 224), sed p. 28 واحت وحمر. *f*) A. et B. h. l. حتر, infra حى, C. in mappa حمر, E. et Edrîsî حير, D. h. l. خير, infra جبر. *g*) H. l. A. et B. ut C. السامات. E. in itinerario habet سيماب. Deinde A. et B. دهار hic et infra, C. بهان, دهار et دهاز, D. h. l. كهان, infra دهار et نهار. Secutus sum E., Edrîsî et Mokaddasí, p. 222. *h*) B. حباب et خماب et C. in mappa خباب. Recte A. et D. Vid. Jakubí l. l. et Jacut in v. *i*) A. h. l. عمرا, B. غيرا, infra uterque عُبيدا; C. عمرا, E. غُمرا (Ous. غيرا), D. غُبيرا. Edrîsî عبيدا s. عبيرا. In mappa Epitom. Paris. عبيرا. Modaddasí habet semel غبترا (unde Sprenger p. 78 male غبتّرا), alibi غبيرا.

وكوغون ^{a} ورائين ^{b} وسَرْوِسْتان ^{c} ودارجين ^{d}، وما بين جيرفت وبمّ مدينة هرمز ^{e} تعرف بقرية الجُوز، وما بين الشيرجان وفارس أُنَاس وكردكان ^{f} وبيمنْد، وبين الشيرجان وبين فارس ايضًا الى حدود دارابجرد خَسّتاباذ وكاهون، ومن الشيرجان الى ما يلى المفازة بَرْدَشِير ^{g} وجَنْزَرُوذ ^{h} وزرنْد وفِرْزِين ^{i} وماهان وخَبيص،

---

a) A. et B. h. l. كوعرن, infra A. كوغرن, B. كوغون, C. et D. كرعون et sic Ous. p. 144 (E. كرعان); Edrîsî ut D. in itiner. جوين. Mokaddasí كوغون, semel كرغون. *b*) A. et B. h. l. رادين, infra رادمى; C. in mappa رانين, in itiner. راس; E. رادين (Ous. رابين), D. رايين (Edrîsî بـابين), Mokaddasí p. 28 ut recepi, p. 221 رادن p. 224 رابـين (apud Sprenger, p. 77, رايـن). *c*) A. h. l. سمروسمان, B. سيروسار, infra A. سرواستان, B. سرواسان et سـرواسـتـان, C. in mappa سروستان, E. in itiner. شارستان, D. سروشتان et سروستان. Mappa Epitom. Paris. سروستان, Edrîsî شورسيان. Jakubí, p. ٦۴ سُروسان pro سـروسـتـان uti شارسان pro شارستان dicitur (vid. Vullers in v.). *d*) Sic A. et B. h. l., D. et Edrîsî. Infra A. دار جمى, دار حنى et دار جمى, B. دار حنى et دار حينى. C. in mappa دار خير, in itiner. دار حمن; cf. Jacut, II, p. ٥١٨, 11 et ٥٢٠, 11. Ultiore loco videtur legendum دار زين, quae lectio fortasse exstat apud Sprenger p. 77, vs. 22. (Mokaddasí p. 221). Mokaddasí p. 28 et 224 داز خين, p. 226 دارحمن. *e*) A. h. l. هردن, B. هرنر ut C., infra A. et B. هردن, هوبر et هربر, D. h. l. recte هرمز الملك, infra هريز et هرار, Mokaddasí p. 226 هوبر. Vid. Edrîsî, I, p. 423. Oppon. هرمز (هرموز) الساحلية. Pro قرية الجوز (A. et B. النخور) E. habet كوز ده et nomen جُوز memoratur a Jacut, II, p. ١٥١. *f*) A. h. l. كردمان, B. كـردمان, C. كـركـان. *g*) Sic saepius in Codd. scribitur quam بردسير. Cf. quoque *Djihân-Numa*, p. ٢٥٦ et Abulfeda, p. ٣٣٧. Docetur ibi ut apud Mokaddasí, p. 221, Jacut in v. et Dimaschkí, ed. Mehren, p. ١٧٩, locum appellari quoque كواشير. E. (Ous. p. 145) واشير habet. Forma جواسير legitur apud Jacut, II, p. ٩٢٧, 7, جواشير ib. IV, p. ٣٦٥, 13. A. et B. h. l. corrupte بردشير, D. يزدشـيـر. *h*) A. h. l. حيروز, B. حيروذ, infra A. et B. جيروذ, C. in mappa خبروذ, in textu sine punctis, D. جنزوذ, Mokaddasí جيروذ et حنزوذ, E. h. l. حمروه, infra خـمـروذ. In mappa Epitom. Paris. خيرور. Vid. Jacut II, p. ١٣٢. *i*) A., B. et C. قردين, E. فروين (Ous. p. 139 فرديين), D. infra فرديين,

ومّما يلى المفازة بناحية بمّ *a* نَرْماشهر وفهرج *b* وسَنبيج *c* الّا انّ سنبج فى وسط المفازة منقطعة عن حدود كرمان وان كانت مضمومة اليها وصوّرناها فى مفازة فارس وخراسان، والخاش *d* على انّ منهم من زعم انّ خاش من عمل سجستان فصوّرناها على حدود كرمان، وحوالى جبل بارز *e* الريقان *f* ومدينة قفير *g* وحومة قُوهِسْتان ابى غانم، وفيما يلى هرموز وجيرفت مدينة كُومين *h*

---

h. l. جوين; Mokaddasí قوذين et فردذن; in mappa Epit. Paris. فزرذن. Vid. Jacut in v. et *Qámus*. *a*) In A. et B. deëst بم. Nomen sequens in A. et B. scribitur بيرماشهر (interdum درماشهر); C. habet برماشير (quoque درماشير) et in mappa نرماشير (sive فرماشير). E. برماشير (et برماسير), D. نرماسير et ترماشير (s. نرماشير). Mokaddasí habet نرماسين (cf. Sprenger, p. 77 seq.) et sic videtur legendum apud Jakubí, p. ٩٤, vs. ult. In mappa exstat نرماسير uti habet Jacut. *Djihán-Numa*, p. ٢٥٧, 9 ترماشير. *b*) Sic A., B., D. et C. in mappa, sed plerumque cum articulo. C. in itinerario دهرج et الفرج, E. مهرج et دهرج (Ous. بهرج). Mappa Epit. Paris. بهرج. Sprenger, p. 79, confusionem nominum suspicatur. Cf. in C. mappam XVI ubi دهرج cum دهرج ح invenitur. Jakubí, p. ٩٥, habet الفهرج quod ab incolis فهره appellari dicit. Infra in medio capite C. de hac urbe loquitur eamque دهرج (l. بهرج) appellat. Jacut, I, p. ٧٦٩, nomine بَهْرَه hunc locum intelligere videtur. Vide porro Beládsorí p. ٣٩٣, vs. 2. Mokaddasí, p. 226, et Jacut, IV, p. ٧٧٥, 6 الفهرج. *c*) A. سمج, B. سمج, C. سنج et in mappa سنج, E. سنج, D. مستنج; *Djihán-Numa*, p. ٢٥٧ سپنج. Vid. infra in descript. deserti Khorasánensis. *d*) A. et B. والخاش, in ceteris lectio الخُوَاش recte aut corrupte legitur. De utraque vid. Jacut, II, p. ٤٨٦. Infra appellantur الخواش et الاخواش. Dimaschkí ed. Mehren, p. ١٧٦, habet الاحواش. *e*) A. et B. قارب, C. نارزن (sed expunctum est), E. كوه قارن. *f*) A. الرهان, B. الرهار, C. in textu الرودهان, sed in mappa الريقان. Secutus sum E. et D. *g*) A. et B. sine punctis, C. دهمر, sed in mappa دهمر; E. قفمر, D. وقفير دهج. In mappa Epit. Paris. قعيز. *h*) A. et B. كرمير, infra كرمهر et كرمهمر, C. infra كردمر.

ونهرزنجان[a] ومَنُوجان[b]، فامّا شهرو[c] على البحر فليس بها منبر فهذا ما علمتُهُ ☆ ومن مشاهير جبالها المنيعة جبال القُفْص وجبال البارز[d] وجبال معدن الفضّة، وليس ببلاد كرمان نهر عظيم ولا بحور[e] الّا بحر فارس الّا انّ خليجًا من بحر فارس يخترق الى هرموز *يسمّى الجير[f] فيدخل فيه السفن من البحر وهو مالح، وفى اضعاف مدن كرمان *مفاوز كثيرة[g] وليس اتّصال عماراتها مثل اتّصال عمارات فارس ☆ وجبال القُفْص هى جبال جنوبيّها

---

a) A. h. l. مهرنحان, B. بهردحان, C. بهر راعان et in mappa بهر راغان (Jacut, IV, p. ٣٣٠, 15 نهر راغان), D. ابهردنحان, E. هرزنكان (Ous. مرزنكان). In itinerario A. et C. دهرزنكان, B. نهرزنكان, D. تهرززكان, E. نهرزنكان (Ous. بهرزنكان). Mokaddasí, p. 222 urbem زنكان memorat, cujus vero nomen p. 28 et p. 221 scribitur ريكان. *b*) A. et B. hic ميوىحان, infra المموىحان, C. h. l. ميوجان deinde المموجان, D. semper المنوجان, E. منوحان (Ous. منوجان), منوكان (Ous. ميوكان) et منونجان (Ous. منونجان). Jakubí habet متوجان, sed Jacut et Mokaddasí منوقان et *Djihán-Numa*, p. ٢٥٨ منوكان. *c*) A. et B. شهروا, E. شهرو, C. سهروا, D. سورو et in mappa Epit. Paris. سوروا. Mokaddasí MS. p. 226 شهروا et سوروا tamquam duos locos diversos memorat. *d*) A. et B. h. l. sine punctis, infra A. المازر, B. النارز et نارز; C. ut supra المارن et infra sic: واما جبال العارن وفى نسخة اخرى واما جبال المارزن; loco tertio فارن et in mappa جبال القاران; E. h. l. جارن, infra مازن (supra habuit قارن) et deinde دارم. In D. et in mappa Epit. Paris. semper recte البارز. Vid. quoque Dimaschkí, ed. Mehren, p. ١٧٩ et Mokaddasí MS. p. 225. Lectio prava القارن quoque exstat apud Jacut, IV, p. ١٤٨, 22. Edrîsî habet semper البارذة. *e*) C. بحر, D. بحيرة. *f*) B. فيسمى الحبر. Nomen incertae orthographiae est. A. الحير, C. الحمر, D. الخبر, Edrîsî, I, p. 424 الحيز, E. جير quod recepi. *g*) A. et B. مفازة كبيرة; vid. quoque Abulfeda, p. ٣٣٤. Apud E. معدنها بسيار legitur.

البحر وشماليّها حدود جيرفت والرّوذبار[a] وقوهستان ابى غانم وشرقيّها الاخواش[b] ومفازة بين القفص ومكران وغربيّها البُلُوص[c] وحدود المنوجان ونواحى هرموز ويقال انّها سبعة اجبل وبها نخيل كثيرة وخصب من زرع وضرع وهى جبال منيعة ولكلّ جبل رئيس وهم ممتنعون وللسلطان[d] عليهم جراية يستكفّهم[e] بها وهم مع ذلك يقطعون الطريق فى عامّة كرمان والى مفازة سجستان والى حدّ فارس وهم رجّالة[f] لا دوابّ لهم والغالب على خلقتهم النحافة والسمرة وتمام الخلق ويزعمون انّهم من العرب ويوصف من[g] بلادهم انّ بها من الاموال المجموعة والذخائر ما يكثر عن الوصف ۞ وامّا البُلُوص فهم فى سفح جبل القفص[h] لا يخاف القفصُ من[i] احد الّا من البلوص *وهم اصحاب نعم وبيوت شعر مثل البادية ولا يقطعون الطريق ولا يتاذّى بهم احد[k] ۞ وامّا جبال البارز فانّها جبال خصبة فيها اشجار بلد الصرود وتقع فيها الثلوج وهى جبال منيعة واهلها لا يتاذّى بهم احد[l] ولم يزل اهلها على المجوسيّة ايّام بنى اميّة كلّها لا يقدر عليهم وكانوا شرًّا من القفص فلمّا ولى الامر بنو العبّاس اسلموا وكانوا مع ذلك فى منعة شديدة الى ايّام السجزيّة فاخذ يعقوب وعمرو ابنا الليث رؤوسهم[m] وملوكهم واخلوا تلك الجبال

---

a) A. et B. والروددال , E. روذان. *b*) A. et B. h. l. الاحراس, C. et D. الاحواس et E. احواس, sed Ous. p. 140 خواس. Mokaddasí l. l. الاخواس et sic inde Jacut, IV, p. ١٤٨, ult. *c*) A. et B. البلوط, E. البلوج. *d*) In C. antea على السلطان nunc من السلطان. E. واز سلطان جامگى دارند. *e*) C. يستكفمهم. *f*) A. رجال. *g*) C. عن. *h*) E. add. (v. Ous. p. 140): وقفص ببارسى كوچ باشد واين دو قومرا كوج وبلوج خوانند; vid. Ous. Appendix, p. 288 seqq. et Vullers in v. بلوج. Perperam in ed. Jacut scribitur بَلُوص. *i*) A. et B. om. من, quod supplevi ex E., Epit. Paris. et Jacut, I, p. ٧٣٢, 22; C. et D. احدًا. *k*) Vers. Pers. habet: ومردمانى صحرانشين باشند راه نزنند وكسى را رنجه ندارند, in quibus legendum esse بزنند recte observavit Ouseley l. l. *l*) A. et B. om. *m*) C. et D. رؤساءهم.

من عيالهم[a] وهى اخصب من جبال القفص وبها معادن حديد ۞ وامّا جبال المعادن[b] فهى جبال بها فضّة وتمتدّ من ظهر[c] جيرفت على شعب يعرف بدرباى[d] الى جبل الفضّة مرحلتين ودرباى هذه شعب خصب عامر بالبساتين والقرى نزه جدّا ۞ وجروم كرمان اكثر من صرودها ولعلّ صرودها نحو الربع وهى ممّا يلى الشيرجان فيما حواليها الى جهة فارس والمفازة والى ما يلى بمّ، والجروم فيها من حدّ هرموز الى حدّ مكران وحدّ فارس وحدّ الشيرجان فيقع فى اضعافها[e] هرموز والمنوجان[f] وجيرفت وجبال القفص *ودشت رويست وبشت خم[g] وما فى اضعاف ذلك من المدن والرساتيق وكذلك بمّ وما فى اضعافها الى المفازة والى حدّ[h] مكران والى خبيص، والغالب على اهل كرمان نحافة الجسم والسمرة لغلبة الحرّ وليس بعد جيرفت وبمّ ممّا يلى المشرق

---

a) C. et D. عتاتهم. Vers. Pers. سران ايشانرا نيست كَرْدند وايشان آرام كُرفتند. *b*) A. et B. المعدن. Secutus sum C., D., E., Mokaddasí et Abulfeda, p. ٣٣٥. *c*) Abulfeda طرف, E. حدود. *d*) A., B. et C. دردساى (C. semel ردنارى, A. et B. semel دربای), E. دريان (Ous. دربان et درمان), Abulfeda, p. ٣٣٥ دُرباى, D. درفارد (Edrîsî دارفارد), Mokaddasí, p. 28 ردرقانى, p. 224 دررفانى, p. 225 الدريانى et دريانى. *e*) A. et B. اضعافه. *f*) A. et B. والمسردجان, vid. supra p. ١٦٣ *b*. *g*) A. ودشت دروربست, B. ودسب درو وده وبارست, C. ودسب درون ودسب, E. ودشت ويران وبست, D. وبست خم, Mokaddasí, p. 225 وحسب بروربست: Sunt duo loci quorum alter infra appellatur in A. روبست, B. دوبست, C. et E. رويست, D. ده بارست (Epit. Paris. دهنا رُسْت, sed in mappa ده بارست) et in itinerario in A. روبسب, B. زوسب, C. رردسب et in mappa رويست, D. زويست, E. رودست, Edrîsî رويست. Jacet inter تارم et شهرو. Videtur locus tria habere nomina رويست (روبست, زويست), ويران et بارست. De altero loco jam egimus ad itinerar. supra p. ١٣١ *e*, et infra in itinerario memoratur. Jacet inter Schîrdjân et Ribát as-Sarmakán. *h*) A. et B. om.

شىء من الصرود ومما يلى المغرب من جيرفت صرود تقع فيها الثلوج ما بين جبل الفضّة الى دربای الى ان تشرف على جيرفت وكذلك فى وجه جبل بارز، وبقرب جيرفت موضع يعرف بالميجان [a] وعامّة فواكه جيرفت والحطب [b] والثلوج تحمل اليها من ميجان ودربای ويخترقها [c] نهر [d] يعرف بديورود [e] شديد الجرى له وجبة وخرير شديد يجرى بالصخور لا يستطيع احد ان ينزله الّا متوقّيا على رجليه من تلك الحجارة وهو مقدار ما يدير عشرين رحى ۞ وهُرْمُوز انّما هى مجمع تجارة كرمان وهى فرضة البحر وموضع السوق وبها مسجد جامع [f] وليس بها كثير مساكن وانّما مساكن التجّار فى رستاقها [g] متفرّقين فى القرى نحو فرسخين وبلدهم كثير النخل والغالب على زرعهم [h] الذرة ۞ وامّا جِيرُفْت فانّ طولها نحو ميلين وهى متجر [i] خراسان وسجستان ويجتمع فيها ما يكون بالصرود والجروم من الثلج والرطب والجوز والاترج [k] وماؤهم من نهر ديورود وهى خصبة جدّا وزروعهم سقى ۞ وامّا بَم فانّ فيها نخيلا [l] ولها قرى كثيرة وهى اصحّ هوآء من جيرفت ولها قلعة منيعة [m] وهى فى المدينة وبمدينة بَم ثلاثة مساجد يجتمعون فيها الجمعات فمنها مسجد للخوارج فى السوق عند دار منصور بن خردين [n] ومسجد جامع فى البزّازين [o] لاهل [p] الجماعة ومسجد جامع فى القلعة [q] وفى مسجد

---

a) A. بالمماحاب, B. بالمحاب, deinde A. ممحان, B. ممحار, C. بالمجار, D. بالميزان, E. ميجان; Edrîsî الميزان; Mokaddasi المنجار. *b*) A. et B. والخصب et sic videtur legisse auctor vers. Pers. quae habet ونعمت. *c*) B. وكر ينحترقها, C. ودحروب, D. وبجيرفت. *d*) E. رودى بزرك. *e*) A. et B. h. l. بدورود, C. بديوزوذى. *f*) C. et D. add. ورباط. *g*) C. et D. رساتيقها. *h*) C. et D. زروعهم. Deinde E. ارزن, quod secundum Vullers quoque speciem milii designat quae دخن appellatur. *i*) Abulfeda, p. ٣٣٧, مجمع التجار الواصلين من. *k*) A. et B. om. E. وترنج. *l*) E. add. بسيار. *m*) C. et D. add. مشهورة. *n*) A. et B. sine punctis. Secutus sum D. *o*) B. النواريين. *p*) A. et B. باهل, E. اهل سنت. *q*) A. et B. om. القلعة.

الجامع للخوارج بيت مالهم للصدقات وشُرَاتهم قليلون الّا انّ لهم يسارًا وبمُ اكبر من جيرفت[a] ✿ وامّا الشِيرَجَان فمياههم من القنىّ فى المدينة ومياه رساتيقها من الآبار وهى اكبر مدينة بكرمان وابنيتها آزاج[b] لقلّة الخشب بها والغالب على اهل الشيرجان مذهب اهل الحديث والغالب على اهل جيرفت الراى والغالب على اهل الروذبار[c] وقوهستان ابى غانم والبلوص والمنوجان التشيُّع ✿ ومن حدّ مغون[d] وولاشجرد الى ناحية هرموز يزرع النيل والكمُّون ويحمل منها الى الآفاق ويتّخذ بها الفانيد[e] وقصب السكّر والغالب على طعامهم الذرة وبها نخيل كثير حتّى ربّما بلغ بها وبسائر الجروم من جيرفت التمر مائة منًّا[f] بدرهم، ولهم سنّة حسنة لا يرفعون من تمورهم[g] ما اسقطه[h] الريح * فياخذه غير اربابه[i] وربّما كثرت الرياح فيصير الى الضعفآء[k] من التمور فى التقاطهم ايّاها اكثر ممّا يصير للارباب[l] وليس عليهم فيها الّا العشور للسلطان مثل ما بالبصرة ✿ وامّا ناحية رويست فانّه بلد قشف والغالب على اهلها اللصوصيّة ✿ وشهرو[m] قرية على البحر بها صيّادون وانّما هى منزل لمن اراد ان ياخذ من فارس الى هرموز وليس بها منبر ✿ ولسان اهل كرمان الفارسيّة الّا انّ القفص لهم مع لسان الفارسيّة لسان القفصيّة وكذلك البلوص والبارز لهم مع لسان الفارسيّة لسان آخر ✿ ويرتفع من بمّ ثياب قطن تحمل

---

a) C. addit: ونهرج (ونهرج .l) مدينة صغيرة وعامة حشيشها النرجس والسوسن وحطبهم كله من الآس. In exemplari Abulfeda videtur exstitisse locus de urbe بيمند (vid. p. ٣٣٥). b) Abulfeda, p. ٣٣٧, اقبآء .l اقبَّآء. c) A. et B. ut supra الرودال. Male C. الروذبار وهو قوهستان. d) A. معور, B. دعون, E. معان. Deinde A. et B. ولاساحرد, D. ولاشاجرد, E. ولاشكرد. e) C. add. السجزى i. e. السجرى. f) C. et D. منّ. g) A. et B. جرومهم. h) C., D. et Jacut, II, p. ١٧٤, 6 اسقطَتْه. i) Jacut بل هو للصعاليك. k) Jacut الفقراء, C. الضعفى. l) C. et Jacut الى الارباب. m) C. وسهروا, D. وسورو; vid. supra ad p. ١٦٣ c.

الى الآفاق ومن ناحية *زَرَنْد[a] ترفع بطائن[b] معروفة تحمل الى فارس والعراق ۞ والخواش[c] انّما هى اخواش مثل النواحى وهم بادية اصحاب ابل ومراعٍ ولهم اخصاص[d] ينزلون فيها ولهم نخيل كثيرة وامّا الاخواش فانّه يرفع منها الفانيذ الّذى يحمل الى سجستان[e] ۞ وامّا نقودهم فانّ الغالب عليها الدرهم[f] ولا يستعملون الفلوس ولا شيئًا من النقرة[g] والدنانير فيما بينهم كالعرض[h] لا يتبايعون بها ۞

وامّا المسافات بين مدن كرمان فانّ من الشيرجان الى رستاق الرستاق من حدّ فارس من الشيرجان الى كاهون مرحلتين ومن كاهون الى حسناباذ نحو من فرسخين ومن حسناباذ الى رستاق الرستاق نحو من مرحلة، ومن الشيرجان الى الروذان ممّا يلى فارس منها الى بيمند[i] ۴ فراسخ ومن بيمند الى كردكان[k] فرسخين ومن كردكان الى أُناس مرحلة كبيرة ومن أُناس الى الروذان من حدّ فارس مرحلة خفيفة، ومن الشيرجان الى رباط السرمقان[l] وحدّ فارس مرحلتين كبيرتين[m] وليس فيما بينهما منبر *وبشت خم[n] بين الشيرجان وبين رباط السرمقان، ومن الشيرجان الى بمّ اوّل مرحلة منها الشامات[o] وتعرف بكوهستان ومن الشامات الى بهار[p] مرحلة خفيفة ومن بهار الى خنّاب مرحلة خفيفة ومن خنّاب الى غُبَيرا مرحلة خفيفة ومن غبيرا

---

a) A. زنرد, B. زيبود. Deinde C., D. et Abulf. habent ترتفع. b) C. مطاسر sic. E. كرباس وآستر. c) A. et B. والحوش et deinde احواش et الاحواس; C. والحواس et حواس; E. حواس. Vid. supra ad p. ١٦٣ d. d) A. et B. احواص. e) E. habet: اما حواس قومى صحرانشين باشند واشتر بسيار دارند ودرخت خرما دارند وفانيد نيز (sic) ومدار خورش ايشان ازين دو خيز بود واكثر معيوب (sic) باشند بدين سبب واز آنجا فانيد بسيستان برند. f) C. et D. الدراهم. g) C. النقود. h) C. كالعوض. i) B. et C. ممند, E. همد. Recte D. et Mokaddasí. k) Mokaddasí كردن. l) E. مشرقان. m) E. سه مرحله (sed Ous. habet ut recepi). n) A. وبشت حم, B. ودست خم. o) E. سيماب. p) A. et B. h. l. دهاره.

الى كوغون فرسخ ومن كوغون[a] الى رائين مرحلة ومن رائين الى سروستان مرحلة خفيفة ومن سروستان الى دارجين مرحلة ومن دارجين الى بَمّ مرحلة ۞ ومن الشيرجان الى جيرفت ان شئتَ سرتَ على طريق بَمّ الى سروستان ثمّ تعطف عن يمينك الى هرمز قرية الجوز مرحلة ومنها الى جيرفت مرحلة، وان شئتَ خرجتَ من الشيرجان الى ناجت مرحلتين ومن ناجت الى خير مرحلة ومن خير الى جبل الفضّة مرحلة ومن جبل الفضّة الى دربای مرحلة ومن دربای الى جيرفت مرحلة ۞ ومن الشيرجان الى خبيص منها الى فرزين مرحلتين ومن فرزين الى ماهان مرحلة ومن ماهان الى خبيص ٣ مراحل، ومن الشيرجان الى زرند منها الى بردشير مرحلتين ومن بردشير الى جنزروذ مرحلة كبيرة ومن جنزروذ الى زرند مرحلة كبيرة ومن زرند الى حدّ المفازة مرحلة كبيرة ۞ ومن بَمّ الى المفازة منها الى نرماشهر مرحلة ومن نرماشهر الى الفهرج على طرف[b] المفازة مرحلة، ومن بَمّ الى جيرفت منها الى دارجين مرحلة ومن دارجين الى هرمز مرحلة ومن هرمز الى جيرفت مرحلة ۞ ومن جيرفت الى فارس منها الى * قناة الشاه مرحلة ومن قناة الشاه الى[c] مغون مرحلة ومن مغون الى ولاشكرد[d] مرحلة ومن ولاشكرد الى السورقان[e] مرحلة ومن السورقان الى المرزقان مرحلة[f] ومن مرزقان الى جيروقان فرسخ ومن جيروقان الى كشيستان مرحلة خفيفة ومن كشيستان الى روبين مرحلة خفيفة ومن روبين الى فارس[g] مرحلة خفيفة ۞ ومن جيرفت الى هرموز تسير الى ولاشكرد ثمّ تعدل منها الى يسارك الى كومين مرحلة كبيرة ومنها الى نهر زنكان[h] مرحلة ومن نهر زنكان الى المنوجان مرحلة ومن

---

a) C. h. l. عركون; E. كرعان. Pro رائين, quod sequitur, Mokaddasí habet اناس. b) C. et D. طريق ut Jacut, IV, p. ٧٧٥, 6, sed Mokaddasí ut recepi. c) A. et B. haec om., neque in C. et E. leguntur, sed hi habent منها الى مغون مرحلتان. d) A. et B. اولاشكرد. Deinde E. دو مرحله. e) A. et B. السرمقان. f) A. om. g) C. et E. تارم. h) Jacut, IV, p. ٣٣. نهر راغان; vid. supra p. ١٦٣ a.

المنوجان الى هرموز مرحلتان، والطريق من هرموز الى فارس من هرموز الى شهرو على شطّ البحر مرحلة ومن شهرو الى رويست ٣ مراحل ومن رويست الى تارم ٣ مراحل فهذه جوامع مسافات كرمان ۞

بلاد السند

وامّا بلاد السند وما يضافها ممّا قد جمعناه فى صورة واحدة فهى بلاد السند وشىء من بلاد الهند ومُكْران وطُوران والبُدْهَة [a] وشرقىّ ذلك كلّه بحر فارس وغربيّه كرمان ومفازة سجستان واعمال سجستان وشماليّه بلاد هند وجنوبيّه مفازة [b] بين مكران والقفص ومن ورآئها بحر فارس وانّما صار بحر فارس يحيط بشرقىّ [c] هذه البلاد والجنوبىّ من ورآء هذه المفازة من اجل انّ البحر يمتدّ من صَيْمُور [d] على الشرقىّ الى *نحو تيز [e] مكران ثمّ ينعطف على هذه المفازة الى ان يتقوّس على بلاد كرمان وفارس [f] ۞

والّذى يقع من المدن فى هذه البلاد فبناحية مكران التِّيز وكيز وقَنَزْبُور [g]

---

a) E. semper ندهه. b) C. et D. add. لما. c) A. الشرقى. d) B. صممور, C. صممون, E. صمون. e) A. et B. بحر دمن. In C. et D. deëst نحو, ut quoque apud Mokaddasí, p. 232. f) B., C. et Mokaddasí يعطف. g) Hic sequi deberet mappa Indiae. h) A. h. l. نمردور, infra A. et B. العمردون et نمردون; C. نمردون; D. قنزبور et semel القنخجْبُور; E. فيربون, عمردون et نمردون. Vid. porro ad Beládsorí, p. ۴٣٩ *d* et Jakubí, p. ٩٥ *d*. Dimaschkí, ed. Mehren, p. ١٧٥ seq. habet قرنبوس s. قرنبوز. Mokaddasí ex hoc nomine et praecedente كيز unum nomen fecit كبرتون, sed in itinerario habet قمروز, alio loco العمردون. Jacut habet قيربون (IV, p. ٢١٢ et ٣٣٣, 1, II, p. ٥٩٩, 10). *Djihân-Numa*, p. ٢٣٥ پنجپور.

وذَرَك [a] وراسك [b] وهى مدينة الخروج [c] وبه وبند [d] وقصرقند [e] واصففة وفهلفهرة [f]
ومَشْكى وقنبلى وأرمائيل، وامّا طوران فانّ مدنها محالى [g] وكيزكانان [h] وسورة [i]
وقُصْدار، وامّا البدهة فانّ مدينتها قنْدَابيل [k]، وامّا مدن السند فانّها المنصورة

---

a) Andreson, qui hoc caput cum versione edidit in *Bibliotheca Indica*, 1852, p. 49—74, pronunciat *Duzuk* et sic exstat in mappa Epit. Paris., sed omnes Codices habent ut recepi. *Djihân-Numa* l. l. solus habet ديزك. *b*) A. et B. hic et deinde راسل. *c*) A. et B. الخروج, D. h. l. الخوارج; cf. Jacut sub راسك. *d*) Sic semper A., B. et Mokaddasí. C. et E. دبد, D. نبيد et بيد (ut semel E.). Vullers habet sub بُنْد 21) nomen regionis. *e*) E. قصر دبد et قصر فند, Ous. et Edrîsî, I, p. 166 قصر بند. Mappa Epit. Paris. قصرقند et sic legere vult Gildemeister, qui in opusculo *Scriptorum Arabum de rebus Indicis loci* Ibn Haucalis caput cum versione edidit, p. 176. Andreson, p. 58, dicit nomen genuinum esse كسرى كند, et *Djihân-Numa*, p. ۲۳۹, 3 habet قصر كند. Locus sequens in eodem opere l. l. 7 appellatur اسپكه. *f*) A. et B. bis قلفهرة, sed in itinerario فهلفهرا. C. فهلفهوه et in mappa فهلهوية; D. et E. ut recepi et sic Jacut. Mokaddasí scribit فهل فهرة. Andreson legit قلفهرة et nomen genuinum esse dicit كلپورة. *g*) Secutus sum A., B., Ous. p. 147 et Mokaddasí. C. et E. h. l. om., sed C. in mappa محالى habet cum var. lect. بحال. In mappa Epit. Paris. المحجال legitur. D. محجال. Adreson dat var. lect. محالرى et proponit legere بنجارة. *h*) A. كنيكانان, B. كنينكانان, infra A. et B. كيرادابان et كمركامان. D. h. l. كبزدكانان, infra كبزكانان et كيزكانيان. E. bis habet كيزكانان, semel كبزكانان ut apud Mokaddasí exstat. Jacut habet كبزكابان, sed videtur ibi legendum esse كيزكابان. *Djihân-Numa*, p. ۲۳۴ كبر كاكيان. Cf. Reinaud, *Mémoire sur l'Inde*, p. 240. *i*) D. habet شورة. *k*) A. et B. قندابيا quod fortasse corruptum est ex كندابيل.

واسمها بالسنديّة برهمناباذ [a] والدّيبل والبيرون وقالّرى [b] وأنّرى وبلرى والمسواهى [c] والبهرج وبانية [d] ومنحاترى [e] وسدوسان والزور [f] وامّا مدن الهند فهى قامهل [g] وكنباية [h] وسوبارة [i] وسندان وصيمور والملتان وجندراور [k] وبسمد [l] فهذه من مدن

---

a) A. et B. دامبوان, D. ذ.امـيرامـان. Vid. Reinaud, p. 60 seq., Beládsorí, p. ٤٣٩, ٤٤١. Jacut, IV, p. ٦٦٣, 13 بِرْهَمَنابَاذ. *b*) A. et B. والدى, C. والوى, E. قالوى, D. الرى, Mokaddasí قارون. Jacut habet قلرى, ut infra in itinerario scribitur. De hoc nomine et seq. vid. Abulfeda, p. ٣٤٧. Pro انرى, *Djihán-Numa*, p. ٢٣٣ آنرى habet. *c*) A., B. et C. h. l. والمسراهى, Mokaddasí, p. 228 المشراهى, sed p. 28 ut recepi. Nomen sequens in A. et B. semper scribitur المهرج. E. بهرج, D. الفهرج. Mokaddasí, p. 28 البهراج, p. 228 ut recepi. *d*) A., B. et D. semper sine punctis (D. semel بادىه). E. infra باينة, Mokaddasí دادىه et دابىه. Edrîsî, I, p. 163 ut recepi. *e*) A., B. et D. semper sine punctis; C. h. l. et Edrîsî منحابرى, E. h. l. منحادرى; Mokaddasí منحادرى et منجابرى. Gildemeister, p. 170, laudat locum ex M'Murdo, *Account on Sind*, p. 234, ubi urbs Mánhatára s. Mankátara appellatur. *f*) A. et B. saepe الدور et الزور, E. h. l. روز, D. الزور et الروز, C. الرون, Abulfeda, p. ٣٤٧ ازور, *Djihán-Numa*, p. ٢٣٣ آزور. Vid. Mokaddasí, Jacut in v., Beládsorí et Mas'udí, I, p. 378. Sub الزّور a Jacut memoratur templum Indicum in regione الداوَر. *g*) Codd. interdum فامهل, Edrîsî semper فامهل. *Djihán-Numa*, p. ٢٣٤ فامهل s. تامهل. *h*) A. et B. hic et infra كنبانة, C. كنبانة, E. كينانة. In *Djihán-Numa* l. l. كسانة. *i*) *Djihán-Numa*, p. ٢٣٢ چوپاره. *k*) A. et B. h. l. sine punctis. C. حندرور et in mappa حندراوار. E. h. l. جندور, infra جندوار; D. الجندرور. Mappa Epit. Paris. et Edrîsî الجندور; Mokaddasí جندرون. Jacut, IV, p. ٦٩٠, 1 جندراون. D. semel habet جندرم quod recepit Reinaud, p. 239. Potius huc referendum Birunii جهراور, *ibid.* p. 65 et 117. *Djihán-Numa*, p. ٢٣٣ habet چنداور. *l*) A. et B. h. l. دسمب, infra دسمد et سمد;

هذه البلاد الّتى عرفناها ۞ ومن كنباية الى صيمور من بلد بَلْهَرا بعض ملوك الهند وهى بلاد كفر الّا انّ هذه المدن فيها المسلمون ولا يلى عليهم من قبل بلهرا الّا مسلم وبها مساجد يجمّع فيها الجمعات ومدينة بلهرا الّتى يقيم فيها مانكير[a] وله مملكة عريضة ۞

والمنصورة مدينة مقدارها فى الطول والعرض نحو من ميل فى ميل ويحيط بها خليج من نهر مهران وهى فى شبيه بالجزيرة واهلها مسلمون وملكهم من قريش يقال انّه من ولد هبّار بن الاسود تغلّب عليها هو واجداده الّا انّ الخطبة بها للخليفة وهى مدينة حارّة بها نخيل وليس لهم عنب ولا تُفّاح ولا كمثرى ولا جوز ولهم قصب سكّر وبارضيهم ثمرة على قدر التفّاح تسمّى الليمونة[b] حامض شديد الحموضة ولهم فاكهة تشبه[c] الخوخ يسمّونها الانبج[d] تقارب طعم الخوخ واسعارهم رخيصة وفيها خصب، ونقودهم القاهريّات[e] كلّ درهم نحو خمسة دراهم ولهم درهم يقال له الطاطرىّ[f] فى الدرهم وزن درهم وثُلْثَيْن ويتعاملون بالدنانير ايضًا وزيُّهم زىّ اهل العراق الّا انّ زىّ ملوكهم يقارب زىّ ملوك الهند من الشعور والقراطق ۞ وامّا المُلْتان[g] فهى مدينة نحو نصف المنصورة وتسمّى فرج بيت الذهب وبها صنم تعظّمه الهند وتحجّ اليه من اقاصى بلدانها وتتقرّب الى هذا الصنم فى كلّ سنة بمال

C. h. l. دسمىد, infra سمند et sic in mappa; E. habet بسميد ut in mappa Epit. Par. legitur. D. habet ut recepi (et sic Mokaddasí), addens وتكتب بالباء والفاء. Cf. Beládsorí, p. ۴۳۹. *Djihán-Numa* l. l. بسمند. *a*) Cf. Mas'udí, I, p. 374. Pro بلهرا in E. substitutum est nomen, quod nunc corrupte legitur دىادىهرح, دمهرح et بهرج, sed quod fortasse مَهراج esse debuit. *b*) Jacut, IV, p. ۹۹۴, 4 البَهْلوية et deinde melius شديدة الحموضة. *c*) A. سدح, B. سيح. E. مانند. *d*) A. et B. h. l. الانيج. *e*) Sic habent E., Jacut l. l. 6 et Mokaddasí, p. 230. A. et B. العادهردات, D. القندهاريات. Cf. Reinaud, *Mémoire*, p. 203, 219 seq., 236. *f*) E. rursus قاهرى. Mokaddasí الطاطرا. *g*) A. saepe الملّتان. E. مولتان. Jacut المُولْتان.

عظيم لينفق على بيت الصنم *والعاكفين عليه[a] منهم وسُمّيت الملتان بهذا الصنم وبيت هذا الصنم قصر[b] مبنيٌّ فى اعمر موضع بسوق الملتان بين سوق العاجيّين وصفّ الصفّارين وفى[c] وسط هذا القصر قبّة[d] والصنم فيها وحوالى القبّة بيوت يسكنها خدم هذا الصنم ومن يعكف[e] عليه وليس بالملتان من الهند والسند الّذين يعبدون الاوثان غير هاؤلآء الّذين هم فى هذا القصر مع الصنم وهذا الصنم صورة على خلقة الانسان متربّع[f] على كرسىّ من جصّ وآجرّ والصنم قد *أُلبس جميعُ جسده جلدًا[g] يشبه السختيان احمر[h] حتّى لا يبيّن[i] من جُثّته شىء الّا عيناه فمنهم من يزعم انّ بدنه خشب ومنهم من يزعم انّه من غير الخشب الّا انّه لا يترك بدنه ينكشف وعيناه جوهرتان[k] وعلى راسه اكليل ذهب[l] متربّع على ذلك الكرسىّ قد مدّ[m] ذراعيه على ركبتيه وقد قبض كلّ *يد له[n] كما تحسب اربعة وعامّة ما يحمل الى هذا الصنم من المال فانّما ياخذه امير الملتان وينفق على السدنة منه[o] فاذا قصدهم الهند *للحرب وانتزاع هذا الصنم منهم[p] اخرجوا الصنم فاظهروا

---

a) C. والمتعكفين به, D. والمعتكفين فيه, E. كى آنجا معتكف باشد, Jacut IV, p. ٦٨٩ 7, والمعتكفين عليه. *b*) E. كوشك. *c*) A. et B. فى sine copula. *d*) E. گنبدى عظيم. Moḳaddasí, p. 231, add. حسنة. *e*) C. et Jacut يعتكف, E. ورهبانان. *f*) Abulfeda, p. ٣٥١ مربّع. Apud Jacut praecedit جالس. *g*) A. et B. لبس — جلدٌ; Moḳaddasí وقد البسوه جلدًا. Pro جسده Jacut بدنه. *h*) C., Jacut et Kazwíní, II, p. ٨١, الاحمر, sed lect. A., B., D. et Moḳaddasii confirmatur ab E. qui habet جامهٔ سرخ — مانند سختيان. Pro السختيان Mokadd. habet السنجاب. *i*) C. يتبيين, D. يبيين. *k*) A. جوهريان. *l*) Jacut insert وهو et mox habet السرير pro الكرسى. *m*) C. جعل. Pro ذراعيه Mokadd. باعيه habet. *n*) C. et D. يديه. Mokadd. وقبض اصابع يديه كانه وجعل كلتَىْ يَدَيْه كما يعقد فى الحساب اربعة قد لَفّ. Jacut يحسب ويرفع الباقى: البِنْصِرَ والوُسْطَى وبَسَطَ الخِنْصِرَ والسَّبّابة. *o*) Jacut l. l. 19 addit: لنفسه. *p*) Jacut بحرب او انتزاع البلد.

كسره واحراقه فيرجعون[a] ولولا ذلك لخرّبوا الملتان، وعلى الملتان حصون
منيعة وهى خصبة الّا انّ المنصورة اخصب واعمر منها والملتان انّما سُمّى
فرج بيت الذهب لانّها لمّا فُتحت فى اوّل الاسلام كان * فى المسلمين[b]
ضيق وقحط فوجدوا فيها ذهبًا كثيرًا فاتّسعوا به، وخارج الملتان على مقدار
نصف فرسخ ابنية كثيرة تسمّى جندراور وهى معسكر للامير لا يدخل الامير
منها الى الملتان الّا فى الجمعة فيركب الفيل ويدخل الى صلاة الجمعة
واميرهم قرشىّ من ولد سامة بن لُؤىّ قد تغلّب عليها ولا يطيع صاحب
المنصورة[c] الّا انّه يخطب للخليفة ۞ وامّا بسمد فهى مدينة صغيرة وهى
والملتان وجندراور عن[d] شرقىّ نهر مهران وبين كلّ واحدة منها وبين النهر
نحو فرسخ[e] ومآؤهم من الآبار وبسمد خصبة ۞ ومدينة الرُّور تقارب الملتان فى
الكبر عليها سوران وهى على شطّ نهر مهران[f] وهى من حدّ المنصورة ۞
والدَّيْبُل هى غربىّ[g] مهران على البحر وهى متجر كبير وفرضة لهذه البلاد
وغيرها وزروعهم مباخس وليس لهم كثير شجر ولا نخيل وهو بلد قشف
وانّما مقامهم للتجارة ۞ والبِيرُون مدينة بين الديبل والمنصورة على نحو من
نصف الطريق وهى[h] الى المنصورة اقرب، ومَنْحَاتَرى على غربىّ مهران وبها،
يعبر من جآء[i] من الديبل الى المنصورة وهى بحذائها ۞ والمسواهى والبهرج
وسَدُوسان هذه كلّها غربىّ مهران ۞ وامّا * أَنّرِى وقالّرِى[k] فهما شرقىّ مهران
على طريق المنصورة الى الملتان وهما بعيدتان من شطّ مهران ۞ وامّا بُلرى
فهى على شطّ مهران عن غربيّه بقرب الخليج الّذى ينفجر[l] من مهران على
ظهر المنصورة ۞ وامّا بانية فهى مدينة صغيرة ومنها[m] عمر بن عبد العزيز

---

a) Jacut addit عنهم. *b*) Jacut male بالمولتان. *c*) Male C. add. احدًا. Jacut addit ولا غيره. *d*) A. et B. add. عين (?يمين). E. بر جانب شرقى. *e*) C. فرسخين. *f*) Jacut, II, p. ٨٣٣, 19 male addit على البحر, et verba sequentia de urbe Daibol huc transtulit. *g*) E. شرقى et sic Abulfeda, p. ٣٤٩, et D. *h*) Abulfeda l. l. وربما هى. *i*) A. ومنها. *k*) A. et B. ابرى ولرى, E. ادرى والوى. *l*) Abulfeda, p. ٣٤٧, ann. 3. ينفتح. *m*) A. et B. وفيها. Sed E. كه — بنا كرده است.

الهبّاريُّ القرشىُّ جدُّ هاؤلآء المتغلّبين على المنصورة ۞ وقامُهُل مدينة من اوّل حدّ الهند الى صيمور فمن صيمور الى قامهل من بلد الهند ومن قامهل الى مكران والبدهة[a] وما والى[b] ذلك الى حدّ الملتان هى كلّها من بلد السند ۞ والكفّار فى حدود بلد السند انّما هم البدهة وقوم يعرفون بالمَيْد[c] واما البدهة فهى مفترشة ما بين حدود طوران ومكران والملتان ومدن المنصورة وهم فى غربىّ مهران وهم اهل ابل وهذا الفالج الّذى يحمل الى الآفاق بخراسان وفارس وسائر البلاد *الّتى يكون بها البخاتى[d] انّما يحمل منهم[e] ومدينة بدهة الّتى يتّجرون اليها قَنْدَابِيل وهم مثل البادية لهم اخصاص وآجام، والميد فهم[f] على شطوط مهران من حدّ الملتان الى البحر ولهم فى البرّيّة الّتى بين مهران *وبين قامُهُل مراعٍ[g] ومواطن كثيرة ولهم عدد كثير ۞ وبقامهل وسندان[h] وصيمور وكنباية مسجد جامع وفيها احكام المسلمين ظاهرة وهى مدن خصبة واسعة وبها النارجيل والموز وانبج والغالب على زروعهم الارزُّ وبها عسل كثير وليس بها نخيل ۞ والراهوق[i] وكلوان رستاقان متجاوران وهما بين كيز وارمائيل فامّا كلوان فهى من مكران واما الراهوق فهى من حدّ المنصورة وهى مباخس قليلة الثمر قشفة الّا انّ لهم مواش كثيرة ۞ والطُوران *قصبتها القُصْدار[k] وهى مدينة لها[l] رستاق ومدن والغالب

---

a) A. et B. والمدبهه, E. تا بدهه, D. فللبدهة. b) C. et D. وراء. c) Sive بالمند. Codices optionem dant inter has duas lectiones; cf. Reinaud, ***Mémoire***, p. 43, Beládsorí l. l. in indice, Mas'udí, I, p. 378. d) Pro his Jacut, IV, p. ٧٧٢ ult. seq. (sub الندهة): ذو السَّنَامَيْن يجعل فَحْلاً للنوق العربية فيكون عنها البَخَاتى. e) A. et B. منه. Jacut من بلادهم فقط. f) A. et B. بهم, Jacut l. l. ٧٧٣, 3 والمند وهم طايفة كالزُّطّ. g) Jacut وبرّ قامهل ناحية بالسند مزارع. h) A. et B. وسمداف, E. وسدوسان. i) A. et B. الداهوق. Jacut et Edrîsî (I, p. 165) راهون. Mokaddasí الزاهوق. Secutus sum D., E. et *Djihán-Numa*, p. ٢٣٤. k) A. et B. قصبة القصدان, sed. vid. e. g. Abulfeda, p. ٣٤٩. l) A. et B. بها.

عليها رجل يعرف *بمُغير بن احمد[a] يخطب للخليفة فقط ومقامه بمدينة تعرف بكيزكانان وهى ناحية خصبة واسعة الاسعار وبها اعناب ورُمّان وفواكه الصرود وليس بها نخيل ✿ وبين بانية وقامهل مفاوز ومن قامهل الى كنباية ايضًا مفاوز ثمّ يكوّن حينئذ من كنباية الى صيمور قرى متّصلة وعمارة للهند[b] ✿ وزىّ[c] المسلمين والكفّار بها واحد فى اللباس وارسال الشعر ولباسهم الازر والميازر لشدّة الحرّ ببلدانهم وكذلك زىّ اهل الملتان لباسهم الازر والميازر ولسان اهل المنصورة والملتان ونواحيها العربيّة والسنديّة ولسان اهل مكران الفارسيّة والمكريّة ولباس القراطق فيهم ظاهر الّا التجار فانّ لباسهم *القمص والأَرْدية[d] وسائر زىّ اهل فارس والعراق ✿ ومُكْران ناحية واسعة عريضة الغالب عليها المفاوز والقحط والضيق[e] والمتغلّب عليها رجل يعرف بعيسى بن معدان[f] ويسمّى بلسانهم مهراج[g] ومقامه بمدينة كيز[h] وهى مدينة نحو النصف من الملتان وبها نخيل كثيرة وفرضة[i] مكران وتلك النواحى تيز وتعرف بتيز مكران واكبر مدينة بمكران *القَنْزَبُور وبه وبند[k] وقصرقند ودَرَك وفهلفهرة كلّها مدن صغار وهى كلّها جروم ولهم رستاق يسمّى الخروج ومدينتها راسك ورستاق يسمّى جَدْران[l] وبها فانيذ كثير ونخيل وقصب سكّر وعامّة الفانيذ الّذى يحمل الى الآفاق منها الّا شيئًا[m] يحمل من ناحية ماسكان[n] وبقصدار ايضًا فانيذ، وماسكان[o] هذه رستاق الشراة ✿ ويتّصل بنواحى كرمان ناحية تسمّى

---

a) D. eum احمد بن معين, E. (sic. Ous. معمر) احمد بن دغمر, Jacut, IV, p. ١٠٥, 7 معمّر بن احمد appellat. b) Secundum D.; A. et B. الهند. E. هندوان دارند. c) A. et B. وراى. d) E. درّاعه ودستار. e) Jacut, IV, p. ٤١٩, 5 والضرّ. f) A. بمعدان. g) A. et B. مهيا, E. مهرا ut Jacut l. l. h) A. et B. كمهره, E. كيز, Jacut كيزة. i) Jacut وهى فرضة. k) A. et B. القيزبون وبها بيّد, Jacut العمزنوب ونبه ودند, E. عمزبون دبه بند et deinde وقصر قند. l) D. حردان, E. جربان, Mokaddasí, p. 231, خرزان, Jacut جربان. m) A. et B. شى; Jacut addit يسير. n) Cf. Jacut sub ماسكان et مكران. In edit. Dimaschkii, p. ١٧٩, vs. 1. male هاسكان receptum est. o) A. et B. ومسكن, D. ويسكن الرستاق هذا. Edrîsî, I, p. 166, et E. confundunt ماسكان et مشكى.

مَشْكَى[a] وهى مدينة قد تغلّب عليها رجل يعرف بمطهر[b] بن رجآء وهو لا يخطب الّا للخليفة ولا يطيع احدًا من الملوك المصاقبين له وحدود عمله نحو ثلاث مراحل وبها نخل قليل وشىء من فواكه الصرود على * انّها من الجروم ۞ وارمائيل وقنبلى مدينتان كبيرتان[c] وبينهما مقدار منزلتين وبين ارمائيل والبحر مقدار نصف فرسخ وهما بين ديبل ومكران ۞ وقَنْدَابيل مدينة كبيرة ليس بها نخيل وهى فى برّيّة وهى ممتار البدهة[d] ۞ وبين كيزكانان وقندابيل رستاق يعرف بايل[e] وفيه مسلمون وكفّار من البدهة واكثر زروعهم الباخوس[f] ولهم كروم ومواشٍ وهى ناحية خصبة وايل هو اسم رجل تغلّب على هذه الكورة فنُسبت اليه ۞

واما المسافات بها فمن تيز الى كيز نحو ٥ مراحل ومن كيز الى قنّزبور مرحلتان ومن اراد من قنّزبور الى تيز مكران فطريقه على كيز ومن قنّزبور الى دَرَك ٣ مراحل ومن درك الى راسك ٣ مراحل ومن راسك الى فهلفهرة ٣ * مراحل ومن فهلفهرة[g] الى اصفقة مرحلتان خفيفتان ومن اصفقة الى بند مرحلة ومن بند الى به مرحلة ومن به الى قصرقند[h] مرحلة، ومن كيز الى ارمائيل ٦ مراحل ومن ارمائيل الى قنبلى مرحلتان ومن قنبلى الى الديبل ٤ مراحل، ومن المنصورة الى الديبل ٦ مراحل ومن المنصورة الى الملتان ١٢ مرحلة ومن المنصورة الى طوران نحو ١٥ مرحلة ومن قصدار[i] الى الملتان

---

a) A. et B. h. l. مسكا, Jacut مسكى. Mokaddasí مشكه. *b*) E. طفر (Ous. ظفر), Jacut, IV, p. ٥٣٢, 2 بمظفر (pro رحالا lege رجاء لا). *c*) A. et B. pro his inde ab انها habent: انها خير من وارمابيل. Pro انها من الجروم Jacut habet انهار تجرى. *d*) Additur in A. ومن كزدراباد, in B. ومن كندراباد, quod ortum videtur e repetitione verborum seqq., ubi A. et B. habent كيزادابان, *e*) A. بمابل, B. بمابمل, D. باتل; E. نابل, Edrîsî ايل habet. Fortasse cum hoc nomine componendum est nomen urbis سابيل in *Djihân-Numa*, p. ٢٣٤, 5. Nomen principis in A. scribitur ادل, in B. et C. ايل ut apud Edrîsî; D. habet أُنل, E. نابل. *f*) C. مباخس. *g*) A., B. et E. om. *h*) A. et B. قصر قبل. *i*) A., B., C. et E. hic et deinde قصدان. Mokaddasi in suo Istakhrii Codice

نحو ٢٠ مرحلة وقصدار[a] مدينة طوران ومن المنصورة الى اوّل حدّ البدهة ٥ مراحل ومن كيز مسكن. عيسى بن معدان الى البدهة نحو ١٠ مراحل ومن البدهة الى التيز نحو ١٥ مرحلة، وطول عمل مكران من تيز الى قصدار نحو ١٢ مرحلة ۞ ومن الملتان الى اوّل *حدود الاسنان[b] المعروف ببالس نحو ١٠ مراحل، وتحتاج الى عبور مهران اذا اردتّ بلاد البدهة من المنصورة الى مدينة تسمّى سَدُوسان على شطّ مهران، ومن قندابيل الى مَسْتَنْج *مدينة بالس[c] ٤ مراحل ومن قصدار الى قندابيل نحو ٥ فراسخ ومن قندابيل الى المنصورة نحو ٨ مراحل[d] ومن قندابيل الى الملتان مقدار[e] نحو ١٠ مراحل وبين المنصورة وبين قامهل ٨ مراحل ومن قامهل الى كنباية ٤ مراحل وكنباية على نحو فرسخ[f] من البحر ومن كنباية الى سوبارة نحو ٤ مراحل وسوبارة من البحر على نصف فرسخ وبين سوبارة وسندان نحو ٥ مراحل وهى ايضًا على نصف فرسخ من البحر وبين صيمور وبين سندان نحو ٥ مراحل وبين صيمور وسرنديب نحو ١٥ مرحلة ۞ وبين الملتان وبسمد نحو مرحلتين ومن بسمد الى الرور ٣ مراحل ومن الرور الى أَنْرِى[g] ٤ مراحل ومن انّرى الى قلّرى[h] مرحلتان[i] ومن قلّرى الى المنصورة مرحلة ومن الديبل الى بيرون[k] ٤ مراحل ومن بيرون الى منحاترى مرحلتان ومن قلّرى الى بُلّرى[l] نحو

---

legit قصران nec vidit eundem locum hoc nomine significari qui nomine قزدار quo ipse utitur. *a*) A. et B. add. نحو quod fortasse inserendum est post البدهة. *b*) A. et B. حدو السنان, C. حدود اللسان, E. حد سمار (Ous. تتار). *c*) E. (Ous. مسنج در ناحيت بير (بين. Lectio بالس suspecta est, nam ut infra videbimus in capite de Sedjestán, inter provinciae Bális urbes hujus nominis urbs non exstat. Jacut s. v. مستنج habet السرار. *d*) A. et B. فراسخ. *e*) C. مغازة, D. مغاوز, E. om. *f*) C. فرسخين et sic Jacut, IV, p. ٣٣٣, 20 qui pro كنباية legit كيفانه. *g*) A. ابرى, B. et D. ايرى, C. et E. sine punctis. Mokaddasí انرى habet. *h*) A., B. et C. فلدى, D. والرى, E. قلوى, Mokadd. فلوى. *i*) E. م ٤. *k*) Jacut بيرون memorat sub forma نيروز. *l*) Sic recte Abulfeda, p. ٣٤٧. A. et B. مكرى, C. et E. بلدان, D. ايرى.

٤ فراسخ وبانية[a] هى بين المنصورة وبين قامهل على مرحلة من المنصورة ✿ وامّا انهارها فانّ لهم نهرًا يعرف بمهران وبلغنى انّ مَخرجه من ظهر جبل يخرج منه بعض انهار جيحون فيظهر مهران بناحية الملتان فيجرى على حدّ * بسمد والرور[b] ثمّ على المنصورة حتّى يقع فى البحر شرقىّ الديبل وهو نهر كبير عذب جدًّا ويقال انّ فيه تماسيح مثل ما فى النيل وانّه مثل النيل فى الكبر وجريه مثل جريه يرتفع على وجه الارض ثمّ ينصب فيزرع عليها مثل ما ذكرناه فى ارض مصر ✿ والسنْدَرُوذ من الملتان على نحو من ثلاث مراحل وهو نهر كبير عذب بلغنى انّه يَفْرَغ[c] الى مهران ✿ وامّا مكران فانّ الغالب عليها البوادى والمباخس وهى قليلة الانهار جدًّا ولهم ما بين المنصورة ومكران[d] مياه من مهران كالبطائح عليها طائفة من السند يعرفون بالزُّطّ فمن قارب منهم هذا المآء فهم فى اخصاص وطعامهم السمك وطير المآء فى جملة ما يتغذّون به ومَنْ بَعُدَ منهم فى البرارىّ فهم * مثل الاكراد[e] ✿

قد انتهينا فى حدّ المشرق الى آخر حدود الاسلام ونرجع الى حدّ الروم غربًا فنَصِفُ اقاليمها الى آخر الاسلام فى حدّ المشرق فالّذى نبتدأ به ارمينية والران واذربيجان وقد جعلناها اقليمًا واحدًا ✿

ارمينية والران واذربيجان

فامّا ارمينية والران واذربيجان فانّا[f] جمعناها فى صورة واحدة وجعلناها اقليمًا واحدًا والّذى يحيط بها ممّا يلى المشرق الجبال[g] والديلم[h] وغربىّ

---

a) A. et B. h. l. سامه, C. دادند, D. et E. sine punctis. b) Kazwíní ut videtur e duobus unum locum fecit, scribens على حد سمندور (II, p. ٩٣ et ٨٣), sed Jacut, IV, p. ٩٩٨, 4 habet سَمَنْدُور والرور. c) A. et B. يفرّغ; cf. Jacut, III, p. ١٩٧, 16. d) A., B., C. et E. om. e) E. مثل عربان. f) A. فانّها. g) E. كوهستان. h) In C. additur والبحر. Deinde pro وغربى omnes

بحر الخزر والّذى يحيط بها ممّا يلى المغرب حدود الارمن واللّان[a] وشى؟
من *حدّ الجزيرة[b] والّذى يحيط بها ممّا يلى الشمال اللّان وجبال القبْق[c]
والّذى يحيط بها ممّا يلى الجنوب حدود العراق وشى؟ من حدود
الجزيرة[d] ۞

فامّا أذْرَبِيْجَان فانّ اكبر مدينة بها أرْدَبيل وبها المعسكر ودار الامارة وهى
مدينة تكون ثُلُثَى فرسخ فى مثلها وعليها سور *فيه ثلاثة ابواب[e] وبنآؤها
الغالب عليه الطين وهى مدينة خصبة واسعارها رخيصة وبها رساتيق وكور[f]
*وبها جبل نحو فرسخين[g] يسمّى سَبَلّان[h] عظيم مرتفع[i] لا يفارقه الثلج شتآءً
ولا صيفًا ولا يكون به[k] عمارة ۞ وتلى اردبيل فى الكبر المَرَاغة وكانت فى
قديم الايّام المعسكر ودار الامارة والمراغة نزهة جدًّا خصبة كثيرة البساتين
والرساتيق والزروع وكان عليها سور خرّبه ابن ابى الساج ۞ ثمّ تلى المراغة
فى الكبر أُرْمِيَة[l] وهى مدينة نزهة كثيرة الخير رخيصة الاسعار على شطّ
بحيرة الشُّرَاة[m] ۞ وامّا *المَيَانِج والخُونَج[n] وأُجَن وداخَرّقان[o] وخُوَىّ

Codd. habent وشرقى excepto E. *a*) C. والران. *b*) C. حدود الخزر. *c*) E. قينتاق (Ous. p. 156 قبشاق i. e. قپچاق), Abulfeda, p. ٣٨٩ القينق. *d*) Hic sequi deberet mappa Armeniae. *e*) C. وعليه اربعة ابواب et E. وچهار دروازه دارد. Fortasse differentia explicanda est ex eo quod Ibn Haucal docet, murum nempe anno 331 esse dirutum. Quum hic auctor scriberet, murus nondum restitutus fuit. Mokaddasí MS. p. 181 dicit urbem esse مصلّبة الى اربعة دروب والجامع وسط الصليب. *f*) Secundum C. addendum est كثيرة; D. add. جليلة. *g*) Abulfeda, p. ٣٩١، وعلى فرسخين منها جبل et sic habet E. Lectionem vero A., B. et C. confirmat D. *h*) A., B., C. et D. سيلان; E. recte سبلان. Cf. Jacut in v. et Abulfeda l. l. *i*) C. add. جدا. *k*) C. بهذا الثلج fortasse respiciens id quod alii e. g. Jacut de contrario affirmaverunt. *l*) E. ارمية. *m*) A. et B. السراة, C. عين السراة. *Lacus schismaticorum* appellatur ab incolis insulae parvae كبوذان in hoc lacu sitae; cf. Jacut, I, p. ٥١٣, 19. *n*) E. ميانه وخونه. Nomen sequens in A. et B. sine punctis est, apud ceteros desideratur. Cf. Jacut sub أُجَان. *Djihán-Numa*, p. ٣٨٣, 3 اوجان. *o*) A. et B. واحرامان, E.

وسَلَمَاس *a* ومَرَنْد وتِبْريز * وبَرْزَنْد ووَرْثَان *b* ومُوقَان وجابَرْوَان *c* وأُشْنُه فانّها مدن صغار متقاربة فى الكبر، * وامّا جابروان *d* وتبريز وأُشْنُه الأَذَرِيّة *e* فانّ هذه الثلاثة المدن وما تحتفّ به * تعرف بالرُّدَيْنىّ *f* وامّا بَرْذَعَة *g* فانّها مدينة كبيرة جدًّا تكون اكبر *h* من فرسخ فى فرسخ وهى نزهة خصبة كثيرة الزرع والثمار جدًّا وليس فيما بين العراق وخراسان بعد الرىّ واصبهان مدينة اكبر ولا اخصب ولا احسن موضعًا ومرافق من برذعة ومنها على اقلّ من فرسخ موضع يسمّى الأَنْدَراب *i* ما بين كُرْنة *k* ولَصُوب ويقطان *l* اكثر من مسيرة يوم فى يوم مشتبكة البساتين والباغات كلّها فواكه وفيها البندق الجيّد اجود من بندق سمرقند وبها شاهبَلُّوط اجود من شاهبلّوط الشام ولهم فاكهة تسمّى الزوقال *m* فى تقدير الغُبَيْرآء وله نوى حلو الطعم اذا ادرك وفيه مرارة *n* قبل ان يدرك،

وبروانان. Infra A. et B. خَرّقان et داخرقان, C. et E. دير خرقان et sic *Djihán-Numa*, p. ٣٨٩, paen. D. habet semper داخرقان (Epit. Paris. دَهْخَرْقان); cf. Quatremère, *Hist. des Mongols*, p. 317. Jacut, *Moschtarik*, p. ١٥٤ ده خَرّقان, *Modjam* دِهْنَخيرْجان et خَرّقان (vid. in vv.). *a*) E. add. ونشوى i. e. والنَّشَوِى. *b*) A. et B. وبروند ودربان. *c*) A. et B. وحابروار, E. وخابران, D. الجابران. *d*) A. et B. وحابروار, D. داخرقان. *e*) A. الادرنه, B. الارونه. *Djihán-Numa*, p. ٣٨٥ locum appellat اشنويه. Apud Ibno 'l-Kaisaráni, ed. de Jong, p. ١. pronunciatur أُشْنَه. *f*) A. يعرف ببنى, B. يعرف للردينى, D. يعرف للرودسى. الردينى. *g*) E. بردع. *h*) B. اكثر. *i*) B. et D. الاندرات. *k*) A., B. et D. كوده, E. كرده (Ous. كريه). Jacut, I, p. ٥٥٨, 22 كُرِّه. Cf. *Merácid*, II, p. ٤٩٣ 8), nisi in loco ibi laudato potius legendum sit كرنا. Deinde E. habet لصوت. *l*) A. et B. sine punctis, E. يقطان (Ous. بقطان). Jacut نَفْطان. Praetuli lectionem cum ي quoniam hanc quoque Ibn Haucal ante oculos habuisse videtur, ut efficio e voce اقطاره. *m*) Sic A., B., D. et Mokaddasi MS. p. 183; C. الزوقال, Kazwíní, II, p. ٣٤٤ الدرقال, Jacut, I, p. ٥٥٩, 1 الدَّوْقال, Edrîsî, II, p. 321 رقبان. *n*) B. مزازة quod fortasse praeferendum.

واما الشاهبلوط فانّه على تقدير نصف جوزة سوداء يقارب طعمه طعم البندق الرطب [a] ، وببرذعة تين يحمل من اَصُوب يفضّل على جنسه ويرتفع [b] من الابريسم شىء كثير *يربّى على [c] توت مُباح لا مالك له ويجهّز منه الى فارس وخوزستان *شىء كثير [d] ، وعلى *ثلث فرسخ [e] من برذعة نهر الكُرّ وبنهر الكرّ السرماهى [f] الّذى يحمل الى الآفاق مالحًا ويرتفع من نهر الكرّ سمك يسمّى الزراقن [g] والعشوبة [h] سمكان يفضّلان على اجناس السمك بتلك النواحى ، وعلى باب برذعة يسمّى *باب الاكراد سوق [i] يسمّى الكُرْكىّ [k] مقدار فرسخ فى فرسخ يجتمع فيه الناس كلّ يوم احد وينتابه الناس [l] من كلّ مكان [m] حتّى من العراق [n] وهو اكبر من سوق كولسره [o] وقد غلب على هذا اليوم لدوامه اسم [p] الكُرْكىّ حتّى انّ كثيرًا منهم اذا عدّ ايّام الجمعة قال السبت والكركىّ

---

a) B. الرطب sine copula. *b*) D. add. بها, Jacut منها. *c*) Jacut مستاحدث من. *d*) Jacut et D. جهازًا واسعًا. Pro فارس E. habet عراق. *e*) C. et D. ثلاثة فراسخ et sic Jacut. Abulfeda, p. ٦. ثلث فراسخ *sic*. Kazwíní, II, p. ٣٤٤ على ثلثة فراسخ, I, p. ١٨٤ primum ويجرى على باب برذعة, deinde وبقربها من برذعة موضع الشورماهيج. Lectio quam recepi, licet errorem contineat, revera tamen Istakhríi esse videtur. Mokaddasí MS. p. 180 habet ونهر الكر منها على فرسخين ut Jacut, IV, p. ٢٥٩, 12. *f*) Sic A., B. et D. Magis accurate Jacut الشورماهى. Kazwíní l. l. الشورماهيج et deinde pro مالحًا habet مملّحًا. *g*) A. الورادن, B. الدرادن, C. et Kazwíní الدراقن, D. الزراقى, E. راقن, Abulfeda, p. ٦٠, الرازقى, Jacut l. l. الدَّوَاقِن. *h*) A. والعسوده, B. والعسوده, C. والعسب, D. والعسوبه, E. وعشر, Kazwíní والعشب et sic Jacut. *i*) A. et B. باب يسمّى بالاكراد وسوق. *k*) Vocales in A. et B. adduntur. Jacut الكُرّكى. *l*) A. الكماس *sic*. *m*) Jacut et D. add. واوب. *n*) E. add. وخراسان. *o*) A. لوكسره, B. كولسره, E. نُوره سره, D. كـور سره, Jacut كُور سره. In itinerario A. et semel B. كواسره, B., C. et E. كولسره, D. كورسره et كوسره, Edrîsî, II, p. 170 كواسر, Mokaddasí p. 184 كويسره. *Djihán-Numa*, p. ٣٨٩, 3 a f. لوسره. *p*) A. et B. بيوم.

والاثنين والثلثآء حتّى يعدّ ايّام الجمعة، وبيت مالهم فى مسجد الجامع على رسم الشام فانّ بيوت اموال الشام فى مساجدها وهو بيت مال مرصّص السطح وعليه باب حديد وهو على تسعة اساطين ودار الامارة بجنب[a] مسجد الجامع فى المدينة والاسواق فى ربضها ۞ وامّا باب الابواب[b] فانّها مدينة على البحر وفى وسطها مرسى للسفن وبين هذا المرسى وبين البحر قد بُنى على حافتى البحر سدّان[c] حتّى ضاق مَدْخَلُ السفن وجُعل المدخل مُلْتَوِيًا وعلى هذا الفم سلسلة ممدودة[d] لا يخرج المركب ولا يدخل الّا بامرٍ[e] وهذان السدّان من صخر ورصاص، وباب الابواب على بحر طبرستان *هى مدينة تكون اكبر من اردبيل[f] ولهم زروع كثيرة وثمار قليلة الّا ما يحمل[g] اليهم من النواحى وهى مدينة عليها سور من حجارة وآجُرّ وطين وهى فرضة بحر الخزر من السرير[h] وسائر بلدان الكفر وهى ايضًا فرضة جرجان وطبرستان والديلم ويرتفع منها ثياب كتّان وليس بالران وارمينية واذربيجان ثياب كتّان الّا هناك وبها زعفران ويقع اليها رقيق من سائر دور الكفر[i]۞

a) A. sine punctis, B. بحيث. b) E. دربند. Versio Persica hujus loci datur a Melgunoff, *das Ufer des Kaspischen Meeres*, p. 295. c) A., B. et C. سدين et sic Jacut I, p. ۴۳٧, 16. d) A. et B. ممدودة. Deinde C. et D. ولا, Jacut فلا. e) C. et Jacut باذن. f) Jacut addit نحو ميليين فى ميلين. Abulfeda, p. ۳۹. plane contrarium ex Ibn Haucale dat: وتكون فى القدر اصغر من اردبيل. E. pro اردبيل habet دبيل. g) C. يجلب. h) Abulfeda البحر من الخزر والسرير. i) C., E. et Jacut multo plura dant. Textum dabo imprimis secutus Jacut: وعلى المدينة سور من الحجارة ممتدّ من الجبل طولانىّ[1] *غير ذى عرض[2] لا مَسْلَكَ على جبلها الى بلاد[3] المسلمين لدروس الطرق وصعوبة المسالك من بلاد الكفر الى بلاد المسلمين ومع طول السور قد[4]

[1]) Male in ed. Jacut طولا فى. C. ممتدة et طولانية et deinde ذات pro ذى.
[2]) E. وباروى ديكر از كِل. [3]) C. ناحية من. [4]) Jacut فقد.

وتَفْلِيس مدينة دون باب الابواب فى الكبر وعليها سوران[a] من طين ولها

a) C. سور.

مدّ قطعة من السور[1] فى البحر *شبه انف طولانىّ[2] ليمنعَ من تقارب السفن من السور وهى[3] محكمة البناء موثّقة[4] الاساس من بناء انوشروان[5]، وهى احد الثغور الجليلة *العظيمة لانّها[6] كثيرة الاعداء الّذين قد[7] حفّوا بها من امم شتّى والسنة مختلفة وعدد كثير، والى جنبها جبل عظيم *يعرف بالذِئب[8] يجمع *فى راسه[9] فى كلّ عام[10] حطب كثير ليُشعلوا فيه النار ان احتاجوا اليه فينذروا[11] اهل اذربيجان واَرّان وارمينية *بالعدوّ ان دَهَمَهم[12]، وربّما اصاب ماء البحر سور[13] هذه المدينة، وقيل انّ *فى اعلى جبلها[14] الممتدّ *المتّصل بباب[15] الابواب نيّفا وسبعين *امّة لكلّ امّة لغة لا يعرفها مجاوِرُهم[16] وكانت الاكاسرة *كثيرة الاهتمام بهذا الثغر لا يفترون عن النظر فى مصالحه[17] لعظم خطره *وشدّة خوفه[18] واقيمت لهذا المكان حفظة من ناقلة *البلدان واهل الثقة عندهم لحفظه[19] واطلق لهم عمارة ما قدروا عليه *بلا كُلْفة للسلطان ولا موامرة فيه ولا مراجعة[20] حرصًا على *عمارة هذا الثغر[21] وصيانته من اصناف الترك[22] والكفر والاعداء فممّن رتّبوا هناك *من

1) E. از ديوار سنگى. 2) C. طولانية et deinde لتمنع. E. habet مقدار شش برج. 3) C. على انها. 4) C. وثيقة. 5) Secundum E. addendum العادل. 6) C. om. E. وايں شهرى بزرگ ومعروفست. 7) Jacut om. 8) C. om.; E. كوه اديب بان. 9) C. عليه. 10) C. سنة. 11) Jacut ينذرون. 12) C. من عدوهم. Quae sequuntur usque ad المدينة in Jacut desunt. 13) C. صور. E. آب اين دريا بسر باروى [?]. 14) C. على هذا الجبل, E. در نواحى آن كوه et Melgunoff addit: كه بشهر دربند پيوسته است. 15) C. الذى عليه باب. 16) C. لغة لا يعرف الجار لغة جاره لاختلاف الالسن. 17) C. يهمّها امر هذا الثغر. 18) C. وجلالة قدره. E. واز دشمنان بسيار رعايت كرده اند [?]. Deinde C. واقامت. 19) C. بلدانهم وثقاتهم. 20) C. om. 21) Jacut om. 22) C. om. و الترك.

ثلاثة ابواب وهى خصبة جدًّا كثيرة[a] الفواكه والزروع[b] وهى ثغر وبها حمّامات

a) B. لكثرة. b) In E. uno loco additur وكشاورزى بسيار دارد (cf. Ous.), altero وعسل از روى آب كردند (*sic*) بآنجا آرند.

الحفظة[1] امّة يقال لها طَبَرْسَرَان وامّة الى جنبهم تعرف بفيلَان وامّة[2] يعرفون باللّكز[3] كثير عددهم *عظيمة شوكتهم[1] واللّيزان[4] وشَرْوان وغيرهم *وجُعل لكلّ صنف من هاؤلآء مركز يحفظه[5] وهم اولو عدد وشدّة رجّالة وفرسان[6] وباب الابواب فرضة لذلك البحر يجتمع اليه الخزر والسرير *وشَنْدَان[7] وخَيْزان وكُرْج ورُقْلان[8] وزريكران وغميك[9] هذه من جهة شماليّها ويجتمع اليه ايضًا من[10] جرجان وطبرستان *والدَّيْلَم والجبل[11] *وقد يقع بها شغل ثياب كتّان وليس باران وارمينية واذربيجان كتّان الّا بها وبرساتيقها[12] وبها زعفران *ويقع بها من الرقيق من كلّ نوع[13] وبجنبها ممّا يلى بلاد[14] الاسلام *على ساحل البحر[15] رستاق *يقال له مَسْقَط[16] ويليه بلد الملكز وهم امم[17] كثيرة ذوو خلق واجسام وضياع عامرة وكورة ماهولة فيها احرار يعرفون[18] بالخماشرة

1) C. om. 2) C. وقوم. 3) E. الكزان, Melgunoff لكزان. 4) E. البيزان, Melgun. ليزان. 5) C. ووكل كل صنف من هاولاء بحفظ مركز من تلك المراكز et paucis omissis haec verba corrupta sequuntur: وهذه المدينة اعنى باب الابواب ولشكر ايس. 6) E. ووردكران وعمك وكلها فرضة جرجان — والحمل وموقان طوايف پياده باشند وسوار كمتر باشند (اندك باشد). 7) Jacut وسِنْدان *sic*. Cf. Beládsorí, p. ٢٠٨, vs. 4 a f. et ann. *d*, ubi quoque شندان restituendum; vid. Jacut, II, p. ٣٢٨ in v. 8) *Sic*. Deinde male وزرنكران. 9) Male عميك; vid. Beládsorí, p. ٢٠٦ et Masudí, II, p. 40. 10) Haec inde a وسندان in E. desunt. 11) E. وكرج وقيتاق. Melgun. وكرج وقباق. Jacut male الجبل. 12) C. om. 13) C. وتحتها ودجها i. e. ويرتفع منها من انواع الرقيق انواع شتى. Deinde habet ودجها i. e. وتحتها. 14) C. om. 15) Jacut om. 16) C. يعرف بالمسقط. 17) C. امة. 18) C. ويعرفون.

مثل حمّامات طبريّة[a] مآؤها سخن من غير نار، وليس بالران[b] مدينة اكبر من برذعة والباب وتفليس فامّا بَيْلَقان ووَرْثان وبَرْديج[c] وبَرْزَنْج والشَّمَاخِيَة[d] وشَرْوان والأَبْخازُ[e] والشابَرَان وقَبَلَة وشَكِّى وجَنْزَة[f] وشَمْكُور وخُنَان[g] فانّها صغار متقاربة

a) A. et B. كمده. b) C. et E. ut solent باران. c) B. بىلمان وروىان وىروىح. In E. uno loco tantum est برزنج, altero tantum بىردنج, in A. et D. deëst برزنج, quod in B. scribitur وىرىدح. In C. tantum habetur ىررىح. d) E. ut Jacut شماخى. e) B. والاىحان, D. والابجان, E. والاىحان et لايجان (Ous. والانجان). St. Martin, *Mémoires sur l'Arménie*, I, p. 222 seq. proponit legere apud Ibn Haucal الابخان et شاه ابخان minus feliciter, ut mihi quidem videtur. f) E. كنجه. g) E. omisso خنان h. l. addit: (Ous. وشروشند) وشتروشه,

وفوقهم الملوك ودونهم[1] المشاق *ثمّ الاكرة والمهان[2] وبينهم وبين باب الابواب بلد طبرسران[3] وهم بهذه الصفة من *البأس والشدّة والعمارة الكثيرة[4] الّا ان اللكز اكثر عددًا واوسع بلدًا وفوق ذلك فيلان وليس بكورة كبيرة، وعلى ساحل هذا البحر دون المسقط مدينة الشابِرَان صغيرة حصينة[5] كثيرة الرساتيق[6] وفوقها رستاق جشمدان[7] وورآء ذلك ضياع الجبل[8] وشَرْوان الى حدّ باكُوه[9] ودرنيق[10] واللكز ومجمع النهرين ثمّ اللِّيرَان خلف ذلك وبها قلعة كبيرة حصينة *يذكر انّ فى هذه القلعة عيون خرّارة وهى منيعة جدًّا[11].
Memoratu autem dignum est, hunc excursum apud E. exstare juxta versionem textus A. et B. quae sic incipit: ... ديوارى از سنگ دارد ويكى از گِل فرضه گاه وسرير وگرگان وطبرستان وديلمان باشد وبهمه اران واذربايگان جامه كتان از دربند برند وبرده انجا افتد وزعفران آنجا خيزد، تفليس الخ.

[1]) C. ثم دونهم. [2]) Jacut om. [3]) Jacut add. شاه. Deinde C. هم. [4]) C. الناس والكثرة. [5]) Secundum E. خصيبة. [6]) Locus apud Jacut h. l. explicit. Deinde in C. deëst copula. [7]) C. حمشدان, E. جسمدان. Cf. Beládsorí, p. ٢٠٦, vs. 2. [8]) C. sine punctis. [9]) Secundum E.; C. باكه. [10]) Sic E. (Ous. دريتق), C. sine punctis. [11]) In E. haec habentur: وچهار پايان آن ولايت درآن نواحى چرا كنند وهيچ بنگاه داشتن حاجت نيفتند.

فى الكبر خصبة واسعة المرافق ۞ وامّا دَبيل فانّها مدينة اكبر من اردبيل وهى قصبة ارمينية وبها دار الامارة كما انّ دار الامارة بالران برذعة ودار الامارة باذربيجان اردبيل وعليها سور[a] والنصارى بها كثير ومسجد الجامع لجنب[b] البيعة[c] ويرتفع بها ثياب الصوف من بسط ووسائد ومقاعد وتكك وغير ذلك من اصناف الارمينىّ[d] ولهم صبغ يسمّى القرمز[e] به يصبغ الصوف وبلغنى انّه دودة تنسج على نفسها مثل دودة القزّ[f] وبلغنى انّه يرتفع بها بزيون[g] كثير، وهى قصبة ارمينية وكان بها سَنْبَاط بن أَشُوط ولم تزل فى ايدى الكبرآء من النصارى وهم الغالب على اهل ارمينية وهى مملكة الارمن متاخمين للروم *فحدّ لهم[h] الى برذعة وحدّ لهم الى الجزيرة[i] وحدّ لهم الى اذربيجان[k] والثغر الّذى يلى الروم من ارمينية قاليقَلا واليها يغزو اهل اذربيجان *والجبال والرىّ وما والاها[l] ولهم مدخل الى الروم يعرف بطَرَابَزُنْدَة[m] يجتمع فيه التجّار فيدخلون[n] بلد الروم للتجارة فما وقع من دبابيج وبزيون وثياب الروم الى تلك النواحى فمن طرابزندة، وامّا نَشَوى وبَرْكَرى وخِلاط ومَنَازْكِرْد[o] وبَدْليس وقاليقَلا وأَرْزَن ومَيّافَارِقين وسِرَاج[p] فهى بلدان صغار متقاربة فى المقدار خصبة كلّها عامرة[q] كثيرة الخير[r]، وميّافارقين يعدّها قوم من الجزيرة الّا انّها دون دجلة وخلفها حدّ الجزيرة فيما صوّرنا ما بين دجلة والفرات فلذلك جعلناها بارمينية ۞

---

duo nomina ut videtur, quorum prius شَتَر est (cf. Jacut, III, p. ٢٥٩). *a*) E. add. گوينند بس فراخ وعجايب. *b*) A. بجنب, D. الى جنب. *c*) E. addit: دربن شهر ارمنيان يا جهودان منار دزديده اند. *d*) A. الارمين, B. الارمن. *e*) C. et D. add. احمر. *f*) C. et D. add. اذا نسجت على نفسها القزّ. *g*) Ex D. et E.; A. et B. om. *h*) A. et B. وحدهم. *i*) C. الخزر; cf. supra p. ١٨١ *b*. *k*) C. add.: وجبل اذربيجان ممتد حتى يجوز الرى وما والاها deinde omissis والجبال — والاها. *l*) C. وهى ارزن الروم pro his. *m*) A. et B. h. l. بطربزنده. *n*) C. et D. add. الى. *o*) C. ملازكرد, D. منازجرد. *p*) E. et D. وسروج, C. om.; A. et B. وسرج. Secutus sum Edrîsî, II, p. 326. Non videtur differre a سراج الطير. *q*) A. et B. عامر. *r*) A. et B. الخزر.

واما الانهار بهذه البلاد التى تجرى فيها السفن فنهر الكُرّ ونهر الرّسّ فامّا سَبيذرُود [a] التى بين اردبيل وزَنْجان فنهر يصغر عن جرى [b] السفن فيه، والكرّ نهر عذب مرئ خفيف يخرج من ناحية الجبل على حدود جنزة وشمكور الى قرب تفليس ثمّ يقع فى بلدان الكفر [c] واما نهر الرّسّ فانّه نهر عذب طيّب يخرج من ارمينية حتّى ينتهى الى باب ورثان ثمّ ينتهى الى خلف موقان [d] وخلف مخرج نهر الكرّ فيقع فى البحر [e] ۞ واما بحارها فانّ باذربيجان بُحَيرة تعرف ببحيرة أُرْمِيَة [f] مالح الماء وفيها سمك وفيها دابّة تسمّى كلب الماء [g] وهى كبيرة وحواليها كلّها عمارة وقرى ورساتيق وبين هذه البحيرة

---

a) C. et E. اسفيذرود. *b*) Jacut, I, p. ٢٣٩ جريان. *c*) Hanc descriptionem ridiculam esse observavit Abulfeda, p. ٦٠, dicens فعلى هذا يكون جريانه من الجنوب الى الشمال, atque Ibn Haucal eam correxit. In C. haec leguntur: يجرى ساكنًا من ناحية اللان من جبال بين ...... [ثمّ يمرّ على [addatur مدينة تفليس ثمّ على خُنان قلعة تعرف بقلعة التراب تلّ عظيم عليه قلعة ثمّ الى [l. على] شَكِى ومن جانبه جنزة وشمكور (vid. Jacut, II, p. ٤٧٤) فيجرى على باب برذعة الى برزنج. E. facit cum A. et B. et sic verbotenus transcripsit Mokaddasí. *d*) A. et B. مدوان, Abulfeda, p. ٦٠ موغان. *e*) C. haec habet: ونهر الرسّ دون نهر الكرّ فى الكبر والعذوبة ومخرجه من وراء ارمينية حتّى ينتهى الى باب ورثان ثمّ ينتهى الى موقان حتّى يصاقب رستاقًا من شروان يعرف بدربمق (بدرنيق) وهى ارض زكيّة فيجتمع مع الكرّ فيقع فى البحر الطبرىّ، ونهر السّمُور هو نهر يجرى ببلد اللكز يعرف بالسمور مخرجه من جبال تصاقب شطى (l. شكّى) ربّما زادت [مياهه .add] وربّما نقصت الّا انّه نهر واسع كثير الخلجان، وليس لسائر الانهار من الغائلة والكبر ما يذكر فى عداد هذه. Fluminis السمور mentio fit apud Beládsorí, p. ٢٠٦ et ٢٠٨. Vid. *Djihán-Numa*, p. ٣٩٧. *f*) E. ارميّة. *g*) Sic A., B. et C.; E. e contrario ودرو هيچ جانور نباشد گويند كه سگ آبى باشد, et D. وليس فيها سمك ولا دابّة. Cum his faciunt Jacut, I, p. ٥١٣ et Kazwíní, II, p. ١٩٤. Cf. *Djihán-Numa*, p. ٣٨٧ in fine. Pro فيها A. et B. bis habent فيه.

*وبين مراغة[a] ثلاثة فراسخ *وبينها وبين ارمية[b] فرسخان[c] وبين داخَرّقان[d] وشطّ البحيرة نحو اربعة فراسخ[e] وطولها نحو اربعة[f] ايّام سيرَ الدوابّ وامّا للريح فانّه ربّما يُسَار فى ليلة، وبحيرة بارمينية *تعرف ببحيرة أَرْجِيش[g] يرتفع منه سمك الطّرّيخ يحمل الى الآفاق، ولهم بحر طبرستان وعليه من المدن باب الابواب وباكُوه[h] وبباكوه النفط، فامّا دجلة فانّ شيًا يسيرًا ينتهى منها الى ارمينية وقد صوّرنا دجلة فى صورة جزيرة[i] والعراق ۞ ويرتفع من نواحى بَرْذَعَة بغال[k] تجلب الى الآفاق ويرتفع منها *هذه القُوّة الّتى تجلب[l] الى بلاد الهند وسائر المواضع ۞

وحدُّ الرّان من باب الابواب الى تفليس الى قرب نهر الرسّ مكان يعرف بحجيران[m]، واذربيجان حدُّها الجبل حتّى ينتهى الى ظهر الطّرْم[n] الى حدّ زَنْجان الى ظهر الدّينَوَر ثمّ يدور الى ظهر حُلْوان وشَهْرَزُور حتّى ينتهى الى قرب دجلة ثمّ يطوف على حدود ارمينية، وقد بيّنّا حدّ ارمينية قبل هذا[o] ۞

---

a) A. et B. ومراغة; vid. quoque Abulfeda, p. ۴۳. *b*) E. اورمى. *c*) C. add. وفى وسطها قرى كثيرة. *d*) C. ut semper دير خرقان. *e*) In B. haec desunt inde a وبينها. *f*) E. پنج. Pro سير ut A., C. et Mokaddasí habent, B. et D. بسير. *g*) E. نزديك ارزيز وخلاط. *h*) A. et B. h. l. ودابحره. Post باب الابواب in C. haec leguntur: وعليه معدن النفط الاخضر والنفط الابيض بباكوه ومدينة موقان وبينهما فوهة البحر منها يصطاد السوماهى (cf. Masudí, II, p. 25) (السرماهى .l) الّذى يحمل الى الآفاق وعلى جنباتها موقان وهى قرى كثيرة كانت لقوم من المجوس والى جنبها الحبل (الجيل .l) ثمّ قطعة من الديلم بحدّ سالوس ثمّ طبرستان الى آبسكون ثمّ تدور حول البحر مفاوز الى جبل سياكوه الى ان ينتهى الى الخزر ثم الى باب الابواب. *i*) A. et B. دجله. *k*) C. ثعالب. *l*) C. هذه القرمز الذى يصبغ به الثياب يحمل. *m*) A., B. et C. sine punctis. Vid. Abulfeda, p. ۳۸۷. Editor ibi proposuit legere نخجوان et sic revera exstat in E. *n*) C. الظرم, E. تارم. *o*) C. addit: والظرم فيه منبت الزيتون والرمّان والتين وهى حرمه (جرومية ?) وفيها معادن الحديد

وبهذه المدن من السعر الرخيص ما يبلغ فى بعض المواضع الشاة بدرهمين وربّما بلغ العسل فى بعض اقاليمها *المنويْن والثلاثة بدرهم[a] وبها من الخصب ما ان ذُكر لمن لم يشاهده انكره لعظمه[b]، وبها ملوك فى[c] الاطراف اماكنهم مثل الممالك *لهم مملكة واسعة واموال[d] منهم ملك شَرْوان يعرف بشروان‌شاه وملك الابخاز[e] يعرف بالابخازشاه[f] ✿ والغالب على اذربيجان وارمينية والرّان الجبال[g] وبدَبيل جبل عظيم يسمّى الحارِثُ لا يرتقى الى اعلاه من *ارتفاعه وصعوبة[h] مسلكه والثلوج عليه دائمة ودونه جبل صغير[i] يسمّى الحُوَيْرث وتخرج من الحارث مياههم ومحتطبهم ومتصيّدهم فيه ويقال انّه لا يعرف جبل اعلى منه بهذه المدن ✿ ومنّ اردبيل الف درهم واربعون درهمًا مثل منا شيراز الّا *انّ بشيراز يسمّى المنا وباردبيل يسمّى الرطل[k] ✿ ولسان اذربيجان

والزاج والذهب والفضّة والنحاس وغيرها وبها قلاع حصبه (l. حصينة) منيعة وملكها يلقّب بسلّار وهو [من] آل لنجر يقال انّهم من العرب وقعوا على قديم الدهر الى هناك فتغلّبوا عليها وبينهم وبين آل خُنان الملك مصاهرة والطرم هى قرى مشتبكة عامرة ويجرى نهر اسفيذروذ فى وسط مملكتها حتّى يدخل البحر. Ad ultima verba cf. Jacut, I, p. ٢٣٩, 11, Masudí, II, p. 76.

a) C. المنّ بدرهم. *b*) B. تعظمه. *c*) C. et D. om. *d*) C. الواسعة ولهم اموال ومواش كثيرة. *e*) A. et C. اللابحان, B. الدايحان. Deinde A. والبرانشاه (l. وليرانشاه), B. ثلاثحاشاه, C. بلادبحان شاه. *f*) C. addit: بلادبحادساه وطبرسرانشاه وثيلانشاه وحندقانشاه (*sic*) واعرراورنشاه (*sic*) وهو صاحب السرير والغالب على هاولآء الملوك حسن العشرة لمن يختلط بهم من الغرباء. Pro حندقانشاه fortasse legendum est خيزانشاه coll. Beládsori, p. ٢٠٤, et ann. *a*. Locus Masudii ibi laudatus est in ed. Paris. II, p. 39, 40 (ubi male جيدان). Princeps qui portat titulum صاحب السرير exstat apud Beládsori, p. ١٩٩ sub forma وهرارزانشاه. Tempore Masudii (II, p. 41) Filánscháh erat hujus tituli possessor. *g*) C. والجبال شدة البرد et E. وابن حدود همه سردسيرست. Deinde A., B. et E. male وباردبيل (C. لدبيل). *h*) C. صعوبته وتعذر. *i*) C. اصغر منه. *k*) C. انه بشيراز يسمى رطلا. Deinde in C., ceteris ad المسافات omissis, haec

وارمينية والرّان الفارسيّة والعربيّة غير انّ اهل دبيل[a] وحواليها يتكلّمون بالارمينيّة ونواحى برذعة لسانهم الرانيّة، ولهم جبال يسمّونها القَبْق وتحيط بها السنة مختلفة كثيرة للكفّار ٭ ونقود اذربيجان والرّان وارمينية الذهب والفضّة جميعًا ٭

المسافات بهذه النواحى[b] الطريق من بَرْذَعَة الى اردبيل من برذعة الى يُونَان[c] ٧ فراسخ ومن يونان الى بَيْلَقان ٧ فراسخ ومن بيلقان الى وَرْثان ٧ فراسخ ومن ورثان الى بَلْخَاب[d] ٧ فراسخ ومن بلخاب الى بَرْزَنْد ٧ فراسخ ومن برزند الى أَرْدَبيل ١٥ فرسخًا ٭ الطريق من برذعة الى باب الابواب من برذعة الى بَرْزَنْج[e] ١٨ فرسخًا ومن برزنج *الى مَعْبَر[f] الكُرّ الى الشَّمَاخِيَة[g] ١٤ فرسخًا ومن الشماخية الى شروان[h] ٣ ايّام ومن شروان الى الابخاز[i] يومان ومن الابخاز الى جسر سُمُور[k] ١٢ فرسخًا ومن جسر سمور الى باب الابواب

---

sequuntur: وورآء برذعة وشمكور صنف من الارمن يقال لهم السياوردية اهل العبث والفساد والتلصّص. De hoc populo vid. Masudí, II, p. 75 et Beládsorí, p. ٢٠٣, ubi edidi السَّاوَرْدِيَّة. *a*) E. اردبيل. *b*) Titulus in A. et B. deëst. *c*) A. et E. نوبان, B. دومان, C. توبان et تومان, D. يومان, Mokaddasí MS. p. 183 تويار et توبار. Apud Beládsorí p. ٢٠٩, vs. 7 a f. scribatur يونان pro (*sic*) دونان. Abulfeda, p. ٣٩. male ورثان ut *Djihán-Numa*, p. ٣٩٩, 18. *d*) A et D. بلحاب, B. بلحاب, E. بلحان, C. تهلاب, Edrîsî, II, p. 323 ناجاب. Mokaddasí بلخان. Jacut sub بَلْخَابُ tantum habet موضع. *e*) A. بردىح, E. بردنج, B. بررىح, Mokaddasí بزددج, *Djihán-Numa* l. l. 19 برزند. *f*) Jacut, III, p. ٣١٧, 20 ثم يعبر; C. ومنها يعبر. *g*) Jacut add. وليس فيها منبر. *h*) C. semel et Jacut legunt شابران (C. semel in textu et in mappa شابروان) et addunt: وهى مدينة صغيرة فيها منبر. Pro ايام, E. ut *Djihán-Numa* habet فرسنك. Mokaddasí habet ut recepi. *i*) A. اللادىحان, B. اللايجان, E. لايجان ut *Djihán-Numa*. C. om.; Mokaddasí اللادجان. *k*) E. et *Djihan-Numa* (پول) جسر ميمون. C. السمور cum articulo. Mokaddasí شمور.

*٢٠ فرسخًا ۞ الطريق من برذعة الى تفليس من برذعة الى جَنْزَة [a] مدينة ٩ فراسخ ومن جنزة الى شَمْكُور [b] ١٠ [c] فراسخ ومن شمكور الى خُنَان مدينة ٢١ [d] فرسخًا ومن خنان الى قلعة ابن كندمان [e] ١٠ فراسخ ومن القلعة الى تفليس ١٢ فرسخًا ۞ الطريق من برذعة الى دبيل من برذعة الى قلقاطوس [f] ٩ فراسخ ومن قلقاطوس الى متريس [g] ١٣ فرسخًا ومن متريس الى دَوْميس [h] ١٢ فرسخًا ومن دوميس الى كيل كوى [i] ١٦ فرسخًا ومن كيل كوى الى سِيسَجَان [k] ١٦ فرسخًا ومن سيسجان الى دَبِيل ١٦ فرسخًا، والطريق من

---

a) C. دون عشرة فراسخ et addit: وعلى جبال باب الابواب قصور بناها الاكاسرة فيها قوم رُتّبوا لحفظ تلك المسالك التى كانت الخزر تسلكها الى بلد الاسلام وهى اربعة عشر قصرًا يسكنها قوم من الموصل وديار ربيعة والشام معروفة بهم واللغة باقية فى اعقابهم لا يد لاحد عليهم وهم مطلّون على الباب وبين اللكز وشروان حدّ وبين شروان والليزان (l. الليران) حدّ متاخم وبين الليران والموقانية حدّ وكذلك بلاد العيسية وهى كورة ليست بكثيرة القرى فيها قلعة حصينة ظهرها ممّا يلى بلاد اللكز الى جبالها وهم يحامون عليها لميل صاحب العيسية اليهم وحسن جواره ايّاهم ثمّ بلاد شكّى ثمّ العبرية ثم السارية ثمّ تفليس. b) E. كنجه. c) E. et *Djihán-Numa* چهار; cf. Abulfeda, p. ٣٩٠, vs. 2. Mokaddasí habet مرحلة. d) E. يازده et sic *Djihán-Numa*. e) B. كمدبان, E. كندمان, C. كىلوس. Cf. *Journal Asiat.* 1849, I, p. 504. *Djihán-Numa* l. l. 22 قلعهٔ ابر كندمان. f) A., B. et C. sine punctis. E. فلعاطرين (Ous. قلقاطرين). Mokaddasí فلقاطوس. g) A., B. et C. h. l. sine punctis. Deinde C. ut D. ميريس. E. مىرس et ميرس; Mokaddasí ميردش; *Djihán-Numa* مريس. Vid. Jacut in v., unde efficeres paullo supra pro تسعة legendum esse سبعة. h) *Djihán-Numa* دوميش; Mokaddasí دمىس. i) A. sine punctis; B. semel كيل كوره; D. كنكلوين; E. كليكون; Mokaddasí كيلواى; *Djihán-Numa* كلكون. Jacut habet كلكوى. Cf. *Journ. Asiat.* 1849, II, p. 468 et 495 ubi كيلكون legitur. (k E. سيسبان.

برذعة الى دبيل فى الارمن وهذه القرى كلّها مملكة سَنْباط بن أَشُوط ۞ الطريق من اردبيل الى زنجان من اردبيل الى قنطرة سَبِيذْرُوذ مرحلتان ومن القنطرة الى السّرّاة[a] يوم ومن السراة الى نوى[b] يوم ومن نوى الى زَنْجَان يوم ۞ ومن اردبيل الى المراغة من اردبيل الى المَيَانج ٢٠ فرسخا ومن الميانج الى خُونَج مدينة ٧ فراسخ ومن خونج الى كولسره رستاق سوق عظيم لا منبر فيه ٣[c] فراسخ ومن كولسره الى المَرَاغَة ١٠ فراسخ ۞ الطريق من اردبيل الى آمد من اردبيل الى المراغة ٤٠ فرسخا ومن المراغة الى داخَرّقان[d] منبر مرحلتان *ومنها الى أُرْمِيَة مدينة مرحلتان ومن ارمية الى سَلَمَاس مرحلتان[e] ومن سلماس الى خُوَىّ ٧ فراسخ ومن خوىّ الى بَرْكَرِى ٣٠ فرسخا ومن بركرى الى أَرْجِيش يوم واحد[f] ومن ارجيش الى خِلَاط *٣ ايّام[g] ومن خلاط الى بَدْلِيس يوم[h] ومن بدليس الى مَيَّافَارِقِين *٣ ايّام[i] ومن ميّافارقين الى آمد يومان[k] ۞ الطريق من المراغة الى دبيل[l] من مراغة الى ارمية ٣٠ فرسخا ومن ارمية الى سلماس ١٤ فرسخا ومن سلماس الى خوىّ ٧ فراسخ ومن خوى الى نَشَوَى ٣ ايّام ومن نشوى الى دبيل ٤ مراحل ۞ ومن المراغة الى الدينور ٦٠ فرسخا[m] ليس فيها منابر ۞

---

a) Jacut سراو. Hujus loco C., E. et *Djihán-Numa*, p. ٣٨٩ ult. ponunt stationem خونج (خونه) distantia bidui. b) C. توى, D. sine punctis, Edrîsî, II, p. 171 بوى, Mokaddasí ثوى. E. توث سواران et *Djihán-Numa* l. l. 20 ديه بوب سوارى. c) C. et E. عشرة et sic *Djih.-Numa* l. l. 3 a f. Mokaddasí habet مرحلة. d) A. et B. خَرَّقان, C. et E. ut semper دير خرقان. e) C. et E. pro his inde a ومنها habent: ومنها الى تبريز مرحلتان ومنها الى مرند مرحلتان ومن مرند الى سلماس مرحلة (مرحلتان .E) Mokaddasí utramque viam memorat. f) D., E. et Abulfeda, p. ٣٩٩ يومان et sic Mokaddasí. g) A. et B. يوم. h) D., E., Abulfeda et Mokaddasí ثلاثة ايام. i) D. et Abulfeda اربعة ايام et sic Mokaddasí in suo Codice habuit. C. (coll. D.) ومن بدليس الى ارزن يومان ومن ارزن الى ميافارقين اربعة ايام. k) E. et Abulfeda يوم et sic videtur legisse Mokaddasí. l) A., B. et Abulfeda اردبيل. E. دينور. m) In C. additur: نسخة اخرى من مراغة الى الدينور ٣٠ فرسخا et sic habet *Djih.-Numa*, p. ٣٨٩, 4 a f.

الجبال[a]

واما الجبال فانها تشتمل على ماه * الكوفة والبصرة[b] وما يتصل بهما مما ادخلناه فى اضعافها فحدها الشرقى مفازة خراسان وفارس واصبهان وشرقى خوزستان وحدها الغربى اذربيجان وحدها الشمالى حدود الديلم وقزوين والرى وانما افردنا الرى وقزوين وأبهر وزنجان عن الجبال وضممناها الى الديلم لانها محتفة[c] بجبالها على النقويس وحدها الجنوبى العراق وخوزستان[d] ۞

والجبال تشتمل على مدن مشهورة ومعظمها همذان والدينور واصبهان وقم[e] ولها مدن اصغر من هذه مثل قاشان ونهاوند[f] واللور والكرج[g] والبرج[h] واشباهها وسنذكر ما تقع الحاجة الى معرفته ۞ فاما المسافات بها فالطريق من همذان الى حلوان فمن همذان[i] الى أسداباذ مدينة ٧[k] فراسخ ومن اسداباذ الى * قصر اللصوص[l] ٧ فراسخ فيه منبر بناه مونس ومن قصر اللصوص الى * ماذران[m] ۴ فراسخ ومن ماذران الى قنطرة ابى النعمان ٥ فراسخ ومن[n] قنطرة ابى النعمان الى[o] ابى ايوب ۴ فراسخ ومنها الى بيستون[p] فرسخان

---

a) E. ذكر ديار كوهستان. *b*) A. البصرة والكوفة. *c*) A. مختصة, B. محمعه, D. مختفية, C. محملطه (l. مختلطة); Abulfeda, p. ۴٠٨ تحف; E. مى كردد. *d*) D. وبعض خوزستان et E. وبهرى از خوزستان. Deinde A. et B. addunt: وصورة الجبال فى الورقة التى بعدها, quod in capitibus praecedentibus semper omiserunt. *e*) C. add. h. l. ونهاوند, quod infra omittit. *f*) Pro hoc C. واردستان. *g*) E. كره. *h*) C. et E. addunt وجرباذقان. *i*) A. et B. male حلوان. *k*) D. ١٥, E., Abulfeda, p. ۴١٧ et *Djihán-Numa*, p. ٣٠٥ ult. ٩. Mokaddasí مرحلة. *l*) E. قصر دزدان. *m*) D. habet ٧. *n*) In E., Mokaddasí et *Djih.-Numa* haec desiderantur. Non ausus sum delere ابى in nomine قنطرة ابى النعمان, licet tantummodo sic exstat apud A. et B. Sed de origine nominis non constat; vid. Jacut, IV, p. ١٨٠ et ١٩١. *o*) Secundum E. et D. addendum est قرية. *p*) D. ut vulgo بهستون. D. et Mokaddasí addunt جبل.

والقرية بها تسمّى ساسانيان[a] ومن بيستون الى قِرْمَاسين[b] ٨ فراسخ ومن قرماسين الى الزُّبَيْدىّ[c] منزل ٨ فراسخ ومن الزبيدىّ الى مَرْج القَلْعة ٩ فراسخ ومن المرج الى حُلْوَان ١٠ فراسخ ۞ الطريق من همذان الى الدينور تجىء الى ماذَران ومن ماذران ٤ فراسخ الى صُحْنَة[d] ومن صُحْنة الى الدِّيْنَوَر ٤ فراسخ ۞ الطريق من همذان الى الرىّ من همذان الى ساوة ٣٠ فرسخا مدينة ومن ساوة الى الرىّ ٣٠ فرسخا ۞ الطريق من همذان الى اذربيجان من همذان الى نارستان[e] ١٠ فراسخ ومن نارستان الى أُوذ[f] ٨ فراسخ ومن أُوذ الى قَزْوِين يومان، وليس بين قزوين وهمذان مدينة، ومن قزوين الى أَبْهَر[g] ١٢ فرسخا ومن ابهر الى زَنْجان[h] ٢٠ فرسخا وهذا الطريق اذا كان الخوف فاذا امنوا فانّهم ياخذون من همذان الى زنجان على سُهْرَوَرْد ٣٠[i] فرسخا ۞ الطريق من همذان الى اصبهان من همذان الى رامن[k] ٧ فراسخ ومن رامن الى بَرُوجِرْد[l] ١٤ فرسخا ومن بروجرد الى الكَرَج ١٠ فراسخ ومن الكرج الى البُرْج ١٢ فرسخا ومن البرج الى خُونَجان[m] منزل ١٠ فراسخ ومن

---

a) A. et B. سايسادان, D. ساسيان. Secutus sum Jacut, I, p. ٧٩٩, 18; cf. Add. ad *Merácid*, IV, p. 411. b) D. قرميسين, E. كرمانشاهان, *Djihán-Numa* كرمانشاه. c) D., E. et Mokaddasí ut vulgo الزبيدية. d) Vocales in A. et B. Jacut in v. habet سَحْنة, sed cf. IV, p. ٣٨١, 5. In mappis recentioribus *Sihna* exstat. e) A. نارستان, E. et *Djih.-Numa*, p. ٣٠٩, 2 بارسبين. f) A., B., D. et E. أُوذ; vid. Jacut in v. g) E. اوهر. h) E., Jacut, I, p. ١٠٥, 14, Abulfeda, p. ٤١٩ et *Djih.-Numa* l. l. ١٥. i) Secundum E. (سى واند) legendum est نيف وثلاثون. k) A. et B. رامر (B. semel رادس). Vid. Jacut in v. In mappa C. memoratur var. lect. راميس. l) E. وروكرد (*Djih.-Numa* ورد كرد) et deinde ١١ pro ١٤, ut quoque Jacut, II, p. ٧٣٧, 22. *Djihán-Numa* habet ١٠. m) B. حومىحان (خونيجان ut Edrîsî, II, p. 166 ?); C. in mappa حومىحان cum var. lect. نودمدجان. E. حومىحان (Ous. خومنجان, *Djih.-Numa* حومنجان); D. جوسجان. Mappa Mokaddasíi حودىحان. Vid. Jacut in v. et inde corrige IV, p. ٢٥١, 10 ubi male نوبنجان.

خونجان الى اصبهان ٣٠ فرسخًا لا مدينة فيها ۞ الطريق من همذان الى خُوزِسْتان من همذان الى رُوذَرَاوَر ٧ فراسخ ومن رُوذراور الى نهاوَنْد [a] ٧ فراسخ ومن نهاوند الى لَاشْتَر ١٠ فراسخ ومن لاشتر الى شابُرْخاسْت [b] ١٢ فرسخًا ومن شابرخاست الى اللُّور ٣٠ فرسخًا لا مدينة فيها ولا قرية ومن اللُّور الى قنطرة أَنْدَامِش [c] مدينة فرسخان ومن قنطرة اندامش الى جُنْدَيْسَابُور فرسخان ۞ المسافات ما بين مدن الجبال من همذان الى ساوة ٣٠ فرسخًا ومن ساوة الى قُمّ ١٢ فرسخًا ومن قمّ الى قاشان ١٢ فرسخًا ومن الرّىّ الى قزوين [d] ٣٠ فرسخًا ومن همذان الى الدينور نيف وعشرون فرسخًا ومن الدينور الى شَهْرَزُور ٤ مراحل ومن حلوان الى شهرزور ٤ مراحل ومن الدينور الى الصَّيْمَرَة ٥ مراحل ومن الدينور الى السِّيروان [e] ٤ مراحل ومن السيروان الى الصيمرة مسيرة يوم ومن [f] اللُّور الى الكرج ٦ مراحل ومن اصبهان الى قاشان ٣ مراحل ومن قُمّ الى قاشان مرحلتان ۞

المدن بالجبال هَمَذان رُوذَرَاوَر رامن [g] بُرُوجِرْد فراوندة [h] وزاذقان [i] شابُرْخَاسْت لاشتر نهاوَنْد قصر اللصوص أَسَدَاباذ الدِّيـنَـوَر قِرْماسِين الكَرْج طَزَر [k] حورمة

---

a) E. et *Djih.-Numa* ٩; sed vid. quoque Jacut, II, p. ٨٣٢, 7. *b*) Vulgo شابر خواست (سابور خواست) ut habent E. et C. in mappa. *c*) A. et B. اندامس. *d*) E. et *Djih.-Numa* ٢٧. *e*) A. h. l. et in mappa الشيروان. Deinde E. چهار فرسنک. *f*) A. et B. من. *g*) A. et B. om. In C. praecedit رومل اوز *sic*. *h*) D. ورارونده, ubi و copula est, sed lectio A. et B. quoque exstat in mappa C. et in textu. In E. legitur وفراونده et apud Mokaddasí MS. p. 193 وفرايده (non فرايده ut habet Sprenger, p. 55) sive fortasse وفراويده, l. وفراونده. De راوند cogitari nequit. *i*) A. et B. ورادقان; C. وارقان et in mappa راوقان; D. et E. om.; Mokaddasí l. l. in itinerario habet دارقان. Jacut tantum nomen زاذقان memorat, sed auctor *Merácid* dicit كانها من بلاد همذان. *k*) B. طرز, C. طرر. Deinde C. حرومه et in mappa حومه. E. حومه. D. hoc et sequens nomen jungens habet وطرز حونه (Edrîsî, II, p. 165 طور حوسه). Secutus sum A. et B.

شهرورد زنجان ابهر[a] سمنان قمّ قاشان روذة[b] بوسنة الكرج البرج[c] سراى[d]
ودوان اصبهان المدينة[e] واليهوديّة خان لنجان بارة[f] الصيمرة سيروان دور
بنى[g] الراسبىّ الطّالقان[h] ✿ وامّا صفات المدن وغير ذلك بها امّا همذان
فمدينة كبيرة مقدارها فرسخ فى فرسخ ولها مدينة وربض ولمدينتها اربعة
ابواب حديد وبناؤهم من طين ولهم مياه وبساتين وزروع كثيرة خصبة[i] ✿
وامّا الدّينور فانّها مثل ثلثى همذان وهى مدينة كثيرة الثمار والزروع خصبة
واهلها احسن طباعًا[k] من اهل همذان ولها مياه ومستشرف نزه ✿ واصبهان[l]
هى مدينتان احداهما اليهوديّة[m] والاخرى المدينة[n] وبينهما مقدار ميلين[o]
وفى كلّ واحدة منهما مسجد جامع واليهوديّة اكبرهما وهى وحدها اكبر من

---

a) C. add. ساوه. *b*) E. زرده. Deinde A. دوسمه, B. دوسمد, C. دوستة sed in mappa sine punctis. Mokaddasí MS. p. 189 دوستة. Secutus sum D. et E. *c*) E. h. l. بره (ut semper كره pro كرج). Deinde in E. additur كربايكان, Ous. كربادكان i. e. جرباذقان; vid. Jacut, II, p. ۴۹ et *Djih.-Numa*, p. ۲۹۹, 2 ubi formae دربايكان et كل بادكان memorantur. *d*) E. addit متصل *sic*. De nomine sequenti incertus sum. Cum A. et B. facit Jacut, qui habet وزوان (vid. IV, p. ۹۲۹ in v.); sed e lectione C. مردان, E. مودان (Ous. بردان) suspicor legendum esse مزدقان (*Djihân-Numa*, p. ۲۹۸, 6 مزوقان). *e*) A. et B. والمدينة, C. et E. om. *f*) A., B. et C. نارة, sed C. in mappa تارة; E. پارة, D. بارمه. C. addit مدينة محدثة. In A. subscribitur مدينة, quod in D. sequitur, tamquam si بارة esset nomen urbis ditionis aç-Çaimara. *g*) C. et D. om. *h*) C. et E. addunt قصر البرازنين et deinde C. شهرزور, E. زنجان. *i*) C. addit: وبها مواش وفواكه ولها اسعار رخيصة. *k*) C. et D. طبعا et sic Jacut, II, p. ۷۱۴, 7. *l*) E. semper سپاهان. *m*) E. semper جهودستان. *n*) E. شهر سباهان, D. شهرستانة. Nomen proprium hujus est جى. *o*) C. ميل et sic Jacut, III, p. ۳۴۳, 1 et Dimaschkí, p. ۱۸۳. Sed alio loco Jacut habet نحو الميل او اكثر (IV, p. ۴۵۲, 5) et tertio نحو ميلين (II, p. ۱۸۱, 10).

همذان والمدينة اقلّ من نصف اليهوديّة فى الكبر[a] وبناؤهما من طين وهما اخصب مدن الجبال واوسعها عرصة واكثرها اهلًا ومالًا، وهى فرضة لفارس والجبال وخراسان وخوزستان وليس بالجبال كلّها اكثر حمّالًا للمحمولات منها ويرتفع منها من العتّابىّ والوشى[b] وسائر الثياب الحرير[c] والقطن ما يجهّز الى العراق وفارس[d] وخراسان وغير ذلك من الامصار[e] وبها زعفران وفواكه تجلب الى العراق وغيرها[f] وليس من العراق الى خراسان بعد الرىّ مدينة اكبر من اصبهان واكثر[g] خيرًا منها ۞ والكَرَج[h] مدينة متفرّقة ليس لها اجتماع المدن وتعرف بكَرَج ابى دُلَف كانت مسكنًا له ولاولاده الى ان زالت ايّامهم * والبناء بها بناء الملوك قصور وابنية واسعة متفرّقة[i] وهى مدينة بها زروع ومواش[k] فامّا البساتين والمتنزّهات فليست[l] بها وامّا فواكههم فمن بروجرد وغيرها وبناؤهم من طين وهى مدينة طويلة نحو فرسخ ولها سوقان سوق على باب الجامع وسوق آخر بينهما * صحن كبير[m]، وبُروجِرْد[n] مدينة اتّخذ فيها المنبر حَمُولة وزير ابى دُلَف وهى مدينة خصبة كثيرة الخير تحمل فواكهها الى الكَرَج وغيرها وطولها اكثر من عرضها وطولها نحو نصف فرسخ وبها زعفران[o] ۞ ونهاوَنْد على جبل وهى مدينة بناؤها من طين ولها[p] انهار وبساتين وفواكه كثيرة تحمل الى العراق لجودتها وكثرتها وبها جامعان احدهما عتيق والآخر محدث ويرتفع بها زعفران ۞ ورُوذراوَر[q] اسم رستاق[r] والمنبر منها فى الكَرَج يعرف بكَرَج رُوذراوَر وهى مدينة صغيرة بناؤها من طين وهى خصبة لها مياه

---

a) C. addit: والتجار والاسواق باليهودية. *b*) C. addit: والحلل والسقلاطون المرتفع et habet deinde وغير ذلك من سائر. *c*) C. et D. الابريسم. *d*) C. et D. addunt: وسائر الجبال. *e*) C. امصار المسلمين. *f*) C. et D. وسائر النواحى. *g*) C. ولا اكثر. *h*) E. semper كره. *i*) E. pro his: وهنوز اثار كوشكهاى ايشان مانده است. *k*) C. addit كثيرة. *l*) A. et B. فليس. Correxi e Jacut, IV, p. ٢٥١, 13. *m*) C. صحراء واسعة, D. صحراء كبيرة, Jacut l. l. صحراء. E. مسافتى دورست. *n*) E. semper وروكرد. *o*) C. et D. addunt كثير. *p*) A. لها. *q*) E. semper روذآور, sed Ous. رودراور. *r*) C. الرستاق.

وانهار[a] وزروع ويرتفع منها[b] من الزعفران ما لا يرتفع لغيرها[c] من مدن الجبال فيجهّز الى العراق وسائر النواحى لكثرته وجودته ۞ وامّا حُلْوان فانّها مدينة فى سفح الجبل المطلّ على العراق وقد صوّرناها فى صورة العراق وهى مدينة بناؤها من طين وفيها ايضًا بناء حجارة وهى مدينة نحو نصف الدينور والثلج منها على مرحلة وهى مع ذلك حارّة فيها نخيل وتين كثير[d] ورمّان ۞ وامّا الصَّيْمَرَة والسِّيرَوان فهما صغيرتان غير انّ بناءهما الغالب عليه الحجارة والجصّ يجتمع[e] فيهما التمر[f] والجوز وما يكون فى بلاد الصرود والجروم وفيهما[g] مياه واشجار وزروع وهما نزهتان يجرى الماء فى خلال *الدور والمحالّ[h] ۞ وامّا شَهْرَزُور فانّها مدينة صغيرة قد غلب عليها الاكراد *على قربها[i] من العراق ولا يكون بها امير ولا عامل وهى فى يد الاكراد وكذلك سُهْرَوَرْد[k] الغالب عليها الاكراد وهى مدينة صغيرة[l] ۞ وامّا قَزْوِين فانّها مدينة عليها حصن ولها مدينة[m] داخلة والجامع فى المدينة[n] وهى ثغر[o] الديلم وبينها وبين مستقرّ ملك الديلم مرحلتان[p] اثنا عشر فرسخًا[q] والطّالقان[r] اقرب

---

a) C., D., Abulfeda, p. ٤١. et Jacut, II, p. ٨٣٢, 6 ثمار (Jacut quoque مروج pro مياه). *b*) C. et Jacut بها. *c*) C. بغيرها, D. من غيرها. *d*) C. add. موصوف ut quoque D. et Abulf. p. ٣٠٧. *e*) B. يجمع. *f*) Jacut, III, p. ٤٤٣, 15 الليمون. *g*) A., B. et C. وفيها. Post مياه C. et Jacut addunt كثيرة. *h*) C., D. et Jacut دورهم ومحالهم (ومنازلهم) et fere sic quoque Abulfeda, p. ٤١٣. *i*) C. وهى قريبة. *k*) C. et D. شهرورد. *l*) C. addit: وذكر لى ثقة ان بشهرورد كرم (sic) يحمل (sic) العنب سنة والسنة الثانية يحمل (sic) الودع اصله. Ad quae verba conf. Kazwíní, II, p. ٢٦٦, ubi de monte prope شهرزور idem narratur. *m*) C. add. كبيرة. E. habet قزوين شهرى بزرگست شارستان وحصار دارد. Deinde A., B. et C. داخل. *n*) C. add. الداخلة. *o*) A. تغزو. *p*) C. et D. om. *q*) C. addit: ولها مدينتان اخريان صغيرتان ومن عجائبها ان لها ضيعتين على اربعة فراسخ منها يعرف (تعرفان .l) بالرست وتاشفين وبينهما جادّة الطريق ينطبع الحديد بالرست ولا ينطبع بتاشفين وينصبغ الصبغ بتاشفين ولا ينصبغ بالرست واذا حوّل احدهما الى موضع صاحبه من الضيعتين لم ينجع فعلهما. *r*) C. add. وهى فوقها.

الى الديلم منها[a]، وليس *لقزوين مآءٌ جارٍ[b] الّا مقدار ما يشرب ويجرى هذا المآء فى مسجد الجامع فى قناة وهو مآءٌ وبىٌّ غير انّ لهم اشجارًا وكرومًا[c] وزروعًا كلّها عذى *تزكو حتّى يحمل الى الآفاق[d] ✿ وامّا قُمّ فانّها مدينة[e] عليها سور وهى خصبة[f] ومآؤهم من[g] آبار ومآؤهم للبساتين على سوانٍ[h] وبها فواكه واشجار فستق وبندق وليس بتلك النواحى بندق الّا بمدينة لاشْتَر فانّ بها بندقًا[i]، وليس بجميع الجبال نخيل الّا بالصيمرة والسيروان وشابُرْخاسْت وهى نخيل قليلة واهل قمّ كلّهم شيعة *والغالب عليهم العرب[k] ✿ وقاشان مدينة صغيرة بنآؤها وبنآء قمّ الغالب عليه الطين[l]، وامّا سائر ما ذكرنا من مدن الجبال سوى الرىّ فانّها صغار متقاربة ✿

---

a) C. add. وهى صغيرة. *b*) C. بقزوين كثير ماء. *c*) C. add. حول البلد. *d*) C. pro his تزكوا جدًّا; E. بسيار خيزد (Ous. موز ومويز) واز آنجا بادام. Deinde C. addit: ويتخذ بها اكسية مرعزى وجورب ما علمت انه يتخذ فى غيرها مثله et E.: وگليمهاى بُزوشم يعنى مَرْعِز نيكو بافند. *e*) C. add. كبيرة ليس et ليس etiam in suo Codice habuit Jacut (IV, p. ١٧٥, 9); cf. quoque Ous.; cum A. et B. faciunt D., E. et Abulfeda, p. ۴٢١. *f*) A. et Abulfeda حصينة. *g*) A. et B. om. من. Post ابار C. addit: وخبرتُ ان ماء ابارهم مالح فاذا حفروها صيّروها واسعة مربّعة ثم رفعت (l. رفع) من قعرها بالاحجار بنآء حتى يبلغ ذروة البئر وجاء (l. فاذا جاء) الشتاء واوان المطر اجروا ماء واديهم الى هذه الابار فلا يزال طول الشتاء وايام المطر يدخل الابار من ذلك الماء العذب فاذا انقطع عنهم الماء العذب والمطر استقوا من تلك الابار ماء عذبًا الى ان ينفد ما وقع فى البئر ويعود الى ملوحته. Jacut, IV, p. ١٧٥, 10—13 fere eadem habet. Pro مربعة hic legit مرتفعة. Cf. Kazwíní, II, p. ٢٩٧. *h*) Abulfeda سواقى. *i*) Secundum D. addendum est كثيرًا. *k*) E. pro his: واصل ايشان گويند ديشمر (پيشتر?) از عربست. C. pro العرب habet الاشعريون. Recte; cf. Jacut l. l. 22 seqq. *l*) C. addit: وبقاشان عقارب كبار سود منكرة; cf. Jacut, IV, p. ١٥, 18 et Kazwíní, II, p. ٢٩٠, et vid. Ous. p. 171.

وليس بجميع الجبال بحر صغير ولا كبير ولا بها نهر تجرى فيه السفن والغالب عليها كلّها الجبال الّا ما بين همذان الى الرىّ والى قمّ فانّ الجبال هناك[a] قليلة ۞ وامّا الّذى يحيط بالجبال من حدّ شهرزور ممتدًّا على حلوان والصيمرة والسيروان واللور الى اصبهان[b] وحدّ فارس راجعًا على[c] قاشان الى همذان حتّى ينتهى الى قزوين وسُهْرورد على حدود اذربيجان الى ان يعود الى شهرزور فانّها كلّها جبال لا يكاد يوجد فيها فضآءٌ كبير لا يُرى منه جبل[d]، فامّا الرىّ فانّا ضممناها الى الديلم وان كانت قائمة بنفسها لانّ اتّصالها بها اتّصال واحد وليس بينهما حاجز يستحقّ *به الانفراد عنها[e] فمرّةً من الجبال ومرّة من عمل[f] خراسان، والرىّ مدينة ليس بعد بغداد فى المشرق اعمر منها *الّا انّ[g] نيسابور اكبر عرصة منها فامّا اشتباك البناء والبساتين[h] والخصب والعمارة فهى اعمر وهى مدينة مقدارها فرسخ ونصف فى مثله الغالب على بنآئها الطين[i] ۞ ومن الجبال المذكورة بهذه الكورة[k] جبل دُنْباوَنْد[l] جبل مرتفع *يُرَى فيما بلغنى من خمسين فرسخًا[m] لارتفاعه وما بلغنى انّ احدًا ارتقاه ويتحدّث فى خرافات الفرس انّ الضحّاك حىٌّ فى هذا الجبل وانّ السحرة من جميع اقطار الارض تأوى اليه[n] ۞ وجبل

Deinde in C. haec sequuntur: واردستان مدينة عليها سور وهى كل محلة (*sic*) حصن حصين وبها اثار قديمة من اثار المجوس مثل انوشروان وكسرى ولها قناة عظيمة عجيبة جدًّا واهلها اصحاب الحديث والادب والكتبة. Jacut, I, p. ١٩٨ nomine Istakhrii multo plura dat. *a*) C., D. et Abulfeda, p. ۴١١ بها. *b*) E. addit وشابر خواست. *c*) A. et B. الى. *d*) In C. additur: فهذا ماه البصرة والكوفة. *e*) C. افراده, D. افرادا به. *f*) C. سهل. *g*) C. et Jacut, II, p. ٨٩۴, 18 وان كانت. *h*) C. et D. واليسار et sic tum E. et Jacut, tum infra A. et B. in capite sequenti, ubi eadem fere verba recurrunt. *i*) C. et Jacut addunt والخشب. *k*) C. كور. *l*) E. دماوند. *m*) C. جدًّا يرى من خمسة وسبعين فرسخًا وقيل انه رُمِىَ من شيراز فارس. *n*) C. addit اعنى الجبل et habet تجتمع pro تاوى.

بيستون[a] جبل ممتنع لا يرتقى الى ذروته وطريق الحاجّ تحته سوآء ووجهه من اعلاه الى اسفله املس حتّى كانّه منحوت ومقدار قامات كثيرة من الارض قد نُحت وجهه ومُلّس ويزعم[b] الناس انّ بعض الاكاسرة اراد ان يتّخذ جوفَ[c] هذا الجبل موضع سوق ليدلّ به على قوّته[d] وسلطانه، وعلى ظهر هذا الجبل بقرب الطريق مكان يشبه الغار فيه عين مآء تجرى وهناك صورة دابّة كاحسن[e] ما يكون من الصور زعموا انّه صورة دابّة كسرى المسمّاة[f] شبْديز وعليه كسرى وصورة شيرين ۞ وليس بهذه النواحى جبل عظيم مذكور الّا ما ذكرنا غير انّ جبل سَبَلان اعظم من دنباوند والحارث[g] بدبيل اعظم منهما، وامّا جبال الخُرّمية فانّها جبال ممتنعة وفيها الخُرّميّة وكان منها[h] بابك وفى قراهم مساجد وهم * يقرءون القرآن[i] غير انّه يقال انّهم لا يدينون فى الباطن بشىء[k] الّا الاباحة ۞

وامّا نقودهم فالذهب والفضّة جميعًا والغالب على نقودهم الذهب، وامّا اوزانهم فانّ منا همذان والماهات[l] اربع مائة درهم ۞ ولا اعلم بجميع الجبال معادن ذهب ولا فضّة[m] غير انّ بقرب اصبهان معدنًا للكحل[n] والغالب على اهل الجبال كلّها اقتنآء الاغنام وعلى اطعمتهم الالبان وما يكون منه حتّى انّ جبنهم يحمل الى الآفاق ۞

---

a) C. et D. بهستون. Deinde, omisso جبل, C. habet عالٍ ut Kazwíní, I, p. ١٥٤. Jacut, I, p. ٧٦٩, 20 عالٍ مرتفع. *b*) D. et Jacut addunt بعض. *c*) C. et Jacut حول. In E. legitur: كه اين كوهرا كوشكى سازد. *d*) C. et Jacut عزته. *e*) A. باحسن. *f*) A. المسمى. *g*) E. addit: كه بپارسى آنرا ماست كوه گويند. *h*) C. فيها. *i*) A. et B. ىعرون الران. *k*) C. et D. بدين. *l*) A. et B. والماهان. *m*) Hinc in A. et B. lacuna exstat usque ad وسنذكر in capite sequenti, quam supplevi ex D. et E.; C. pro his inde a ولا اعلم habet: وبالجبل معدن ذهب وبشروان جبل طين يغسل بالماء ترابه فيوجد فيه ما بين الدينار الى المذرّة، نسخة اخرى وبالجبل معدن ذهب بسراوان وكذلك بناحية شهرورد جبل طين يغسل ترابه بالماء فيوجد فيه ما بين الدينار الى المذرّة. *n*) C. addit: وبدنباوند معدن الكحل والزاج وبقرب البخ.

الديلم

وأمّا الدَّيْلَمُ وما يتّصل بها فمن ناحية الجنوب قزوين والطَّرْم[a] وشىء من اذربيجان وبعض الرىّ ويتّصل بها من جهة المشرق بقيّة الرىّ وطبرستان ويتّصل بها من جهة الشمال بحر الخزر ومن جهة المغرب شىء من اذربيجان وبلدان الرّان وقد ضممنا الى ذلك ما يتّصل بها من جبال الروينج[b] وقاذُوسيّان[c] وجبال قارن وجرجان وامّا بحر الخزر فقد افردنا له صورة *وقد صوّرنا الديلم وما يتّصل بها[d] ۞

وسنذكر من مدنها وما يقع فى اضعافها مثل ما ذكرنا من غيرها، امّا الديلم فانّها سهل وجبل امّا السهل فهم الجيل وهم مفترشون على شطّ البحر[e] تحت جبال الديلم وامّا الجبل[f] فللديلم المحض وهى جبال منيعة والمكان الّذى يقيم به الملك يسمّى رُوذبار[g] وبه يقيم آل جستان[h] ورياسة

---

a) C. الظرم, E. طارم. *b*) ? A. et B. الرودمح (semel اررودمح), C. الرودمح, اوردج, الدرومج et ووردج, D. الروبيبج, الرودح, زرويح et رردمح. E. روبست et semel روزمج; Edrîsî, II, p. 179 رونج; Dimaschkí, p. ٢٢٩ الدرنونج cum var. l. الدرونج. *c*) A. et B. primum قادراسان, deinde نادوسمان; C. فادوسان, فادوسمان et دادوسمان; D. دادوسمان et باذوسيان; E. نادوسنان; Jacut, I, p. ٢١٩ قاذُوسيان. Cf. Melgunof, *Das südliche Ufer des Kaspischen Meeres*, p. 50 „vielleicht die alten Cadusii قادوسيان," coll. Dorn, *Gesch. Tabarist. nach Chondemir*, p. 71. Apud Dimaschki l. l. corrupte بادمسان legitur; cf. ann. editoris. Apud Ibno 'l-Athír, VII, p. ٨٩, vs. 8 a f. قاوشان editum est. Fieri autem potest duo nomina esse confusa nempe nomen gentis in monte degentis قادوسيان et nomen principis پادوسبان s. بادوسبان e. g. Melgunof, p. 56, Beládsori, p. ٣١٣ (الفادوسبان s. الفادوسفان). *d*) Secundum C.; E. habet واذربايگان واران در جمله ارمينيه نگاشته ايم بر حد طبرستان وديلمان واين صورت طبرستان وديلمان است. Hic desinit lacuna in A. et B. *e*) C., D. et E. بحر الخزر. *f*) C. et D. الجبال. Deinde A. et C. فالديلم, B. بالديلم. *g*) In A. et B. lacuna. Dant nomen C., E., Jacut, II, p. ٨٣١, 16, Abulfeda, p. ۴٢٩ et *Djihán-Numa*, p. ٣۴۴. *h*) A. et B. حستان, D. حسان. Secutus sum C. et E., coll.

الديلم فيهم وزعم بعض الناس انّ الديلم طائفة من بنى ضبّة ومواضعهم كثيرة الاشجار والغياض واكثر ذلك للجبل فى الوجه الّذى يقابل البحر *وطبرستان وقراهم مفترشة [a] *وهم اهل زرع وسوائم ولبس عندهم من الدوابّ ما يستقلّون بها [b] ولسانهم مفرد غير العربيّة والفارسيّة وفى بعض الجبل [c] فيما بلغنى طائفة منهم يخالفون بلسانهم لسان الجيل والديلم والغالب على خلقة الديلم النحافة [d] وخفّة الشعر والعجلة وقلّة المبالاة وقد كان الديلم [e] دار كفر يسبى من رقيقهم الى ايّام الحسن بن زيد فتوسّطهم العلويّة واسلم بعضهم وفيهم [f] الى يومنا هذا [g] كفّار بالجبال المتّصلة بها ۞ والروينج وجبال [h] قادوسيان وقارن هى جبال ممتنعة لكلّ جبل منها رئيس والغالب عليها الاشجار العالية والغياض والمياه وهى خصبة جدّا فامّا جبال قارن فانّها قرى لا مدينة بها الّا سهمار [i] على مرحلة من سارية ومستقرّ آل قارن بموضع يسمّى فِرِيم [k] وهو موضع حصنهم وذخائرهم ومكان ملكهم *ويتوارث

Abulfeda, p. ۴۲۹ seq. et Ibno 'l-Athír, VII, p. ۱۱۹, ۱۸۳ et ۳۹۱. a) E. pro his: وطبرستان زمينى هامونست. b) C. pro his: وطعامهم الارز والسمك ومواشيهم قليلة. c) A. الجيل. E. haec habet: وبهرى هستند كه ديگر ديلمان زبان ايشان ندانند. d) C. add. والسمرة. Deinde A. et B. والشعر omisso خفة. e) A. et B. للديلم, E. مردمان طبرستان وديلمان. f) In E. hoc introducitur verbis وگويند كه. g) C. add. قوم يقال لهم الشرز. Vid. Jacut, III, p. ۲۷۵, 8 et 9 et p. ۵۰۴, 4. h) A. et B. omittunt copulam, C. om. جبال. i) Alii شهمار e. g. Ous. p. 175 (Male شمهار apud Jacut, III, p. ۳۳۴). C. الا شهما وفريم على مرحلة من سارية ومستقر ال قارن بمدينة فريم. In E. ab initio quaedam desunt, excidit nempe quoque nomen سهمار (cf. Ous.). Supersunt: واز قرتم (sic) تا سارى يك مرحله دارند وقرارگاه مازن (sic) كه پادشاه ايشان بود اين جايگاهست. Cum his denique cf. Jacut, III, p. ۸۹, ubi الا شهْمار وفِرِيم ut quoque p. ۳۳۴. k) A. et B. بريين, D. بردين; Jacut فى مدينة فريم.

صاحب الجبل المملكة بها منذ [a] ايّام الاكاسرة ✿ وجبال قادوسيان جبال مملكة ورئيسهم يسكن قرية تسمّى أُرَم [b] وليس بجبال قادوسيان منبر وبينها وبين سارية مرحلة ✿ وامّا جبال الرويَنج فانّها كانت ممالك الّا انّ فى هذا العصر قد فنيت ملوكهم وهم من [c] الرىّ وطبرستان فما كان فى وجه الرىّ فمن حدود الرىّ وما كان من وجه طبرستان فمن طبرستان ✿ والمدخل الى الديلم من طبرستان على سالُوس [d] وهى على البحر ولها منعة اذا استوثق منها بالشحنة صَعُبَ المسلك على اهل الديلم الى طبرستان، وبينُ هذه الجبال من حدّ الديلم الى أَسْتَرَاباذ [e] الى البحر اكثره [f] يوم وربّما ضاق حتّى يضرب الماء الجبل فاذا قرب [g] الديلم الى الجبل [h] اتّسع حتّى صار بينه وبين البحر مسيرة يومين واكثر ✿

وامّا نواحى قزوين فانّ الّذى يتّصل بها من المدن [i] أَبْهَر وزَنْجان والطالَقان [k] ويتّصل بالرىّ الخُوار وشَلَنْبَة [l] ووِيمَة وتقع فى قومس سِمْنان والدامَغان وبِسْطام وتقع فى طبرستان آمُل وناتِل وسالُوس وكَلّار والرُّويان وميلة [m] وبرجى [n]

---

a) C. et Jacut pro his يتوارثونه من. D. habet اصحاب pro صاحب. b) C. addit ومنصورة. c) C. et D. بين. d) C. et D. شالوس. e) E. استاراباذ. f) C. et D. اكثر من. g) C. جُزت. h) A., B., C. et D. الجَبَل. i) C. ونواحى قزوين والرى (والذى l.) يتصل بها من المدن; D. ومن الرى الى قزوين; E. ناحيت قزوين بناحيت رى پيوسته است. k) C. addit: وقصر برادين (براذين), E. وقصر البرادين; cf. supra p. ١٩٨ ann. h. l) A. et B. sine punctis. C. وشانيه; D. وسليبه et infra سلينه (ut male apud Jacut, III, p. ٥٠٤, 10). Jacut praefert شلمبة. Deinde A. et B. ودممه. m) Codd. h. l. sine punctis, sed in itinerario C., D. et E. مبيلة ut quoque habet Edrîsî (II, p. 169 et 179) et exstat in mappa Epit. Paris. In Codice Mokaddasii p. 169 scribitur مِيْلة (cf. Sprenger, p. 52). n) ? A. et B. برحى (B. semel برخى); C. نوحى, نرجى et in mappa نرحى; D. درحى et دربكى; E. نوحى et توجى (Ous. ترجى); Edrîsî ترجى et بربكى. Mokaddasi MS. p. 27 بوبرجا, p. 169 et in mappa

وعين الهَمّ ومامَطِير وسارِيَةa وطَمِيسة وتقع فى عمل جرجان جُرْجان وأَسْتَرَاباذ وآبَسْكُون ودِهِسْتان وامّا جبال الروبنج وقادوسيان وقارن فلستُ اعرف بها منبرًا غير سهمارb وهى فى جبال قارن ۞ واعظم هذه المدن الرَّىّ وهى مدينة اذا جاوزتَ العراق الى المشرق فليس مدينة اعمر ولا اكبر ولا ايسر اهلًا منها الى آخر الاسلام الّا نيسابور فانّها فى العرصة اوسع فامّا اشتباك الابنية والعمارة واليسار فانّ الرىّ تفضلها وطولها فرسخ ونصف فى مثله وبناؤها طينc وقد يستعمل فيهاd الجِصّ والآجُرّ، ولها ابواب مشهورة منها باب طاقe يخرج منه الى الجبال والعراق وباب بليسانf يخرج منه الى قزوينg وباب كُوهَكينh يخرج منه * الى طبرستان وباب هشام يخرج منه الى قومس وخراسان وباب سين يخرج منه الى قُمّi، ومن اسواقهاk المشهورة رُوذة وبليسان ودهك نوl

---

درجى, p. 178 برجى ut recepit Sprenger, p. 53. — Deinde vocalem ad عين الهم addidi secundum Abulf. p. ۴۳۷. *a*) E. وممطير وسارى. Deinde C. et E. addunt: ومهروان ولمراسك. De مهروان vid. Jacut in v. et III, p. ٥٠۴, 10. Nomen posterius eodem modo scribitur ab C. et E. in itinerario et in mappa (Ous. المراسك), sed ceteri ibi aliam stationem memorant. Pro وطميسه E. habet وتميشه i. e. نميشه, vid. Mokaddasí l. l. et Jacut, IV, p. ٨١٥, 3. C. habet طميس, sed in mappa طميشه ut Abulfeda, p. ۴۳۷. *b*) C. et E. سمنان et in E. additur وقرذم i. e. وفريم. *c*) C. من طين وخشب; cf. supra p. ٢٠٣ ann. *i*. *d*) A. بها. *e*) C. et E. باطان, i. e. ut videtur, باطاق; D. ماطا; *Djihán-Numa*, p. ٢٩٣, 5 ساطر. *f*) A. دلمسان, B. نلسان, C. فليسان, D. بلنسان, E. تليسان (Ous. قليسان), *Djihán-Numa* l. l. 4 مرسيان. *g*) C. addit والديلم واذربيجان. *h*) Sic A. et B.; C., D. et E. كوهك ut *Djihán-Numa*. *i*) E. omnia haec om. exceptis verbis الى قم. C. eorum loco habet: الى قم واصبهان وشيراز وباب دولاب يخرج منه (Cod. male semper منها) الى طبرستان وجرجان وباب خراسان يخرج منه (deëst) الى خراسان. *k*) C. محالها; E. بازارها ومحلهاى معروف. *l*) A. et B. ودهل بر, C. ودهك نوم, D. ودهك. Recte scribitur in E. et *Djihán-Numa* l. l.

ونَصْرَاباذ وسَارْبَانان [a] وباب الجبل وباب هشام *وباب سين [b] واعمرها الرُّوذة فانّ بها معظم التجارات والخانات وهو شارع عريض مشتبك الخانات والابنية ولها مدينة [c] عليها حصن [d] وفيها مسجد الجامع [e] واكثر المدينة خراب والعمارة فى الربض ومياههم من الآبار ولها ايضا قُنِىٌّ ولهم فى المدينة نهران [f] للشرب احدهما يسمّى سورقنى [g] يجرى على الروذة والآخر الجيلانىّ [h] *يجرى على ساربانان ومنهما [i] شربهم [k] *ولهم قنىّ كثيرة ما يفضل عن مشربهم [l] ويتفرّع الى ضياعهم، ونقودهم الدراهم والدنانير وزىّ اهلها زىّ [m] العراق ويرجعون الى مروّة ولهم دهآء وتجارب [n] *وبها قبر محمّد بن الحسن الفقيه والكسَائىّ المقرئ والفَزَارىّ المنجّم [o] ۞ وامّا الخُوَار فانّها مدينة صغيرة نحو ربع ميل وهى عامرة وبها ناس يرجعون الى شرف ولهم مآء جارٍ يخرج من

---

a) A. et B. وسرودداروار, D. وسروس وان, C. et E. وساردادان. Infra A. et B. ساردادان, D. يساربابان, E. ساربانان (ut Ous. et *Djih.-Numa* l. l.). Jacut, III, p. ٩٧ habet سَرْبَان. Fortasse lectio A., B. et D. h. l. est شتر بزار „camelorum statio." b) E. در آهنين. C. وباب الصين وباب آهنين وباب الطبريين وباب بعضها جوف المدينة وراس الروذة. Apud Ous. additur در عتاب. c) C. addit: وگرد بر گرد شارستان. d) C. addit: ويسمى القلعة. e) E. addit: بعض ديوارست. f) C. قناتان. g) Jacut سُورِين. D. سوريني. In C. et E. nomen desideratur. h) E. كيلانى. i) A. et B. ومنها. k) C. pro his habet: يجرى وسط المدينة ومنه شربهم وقناة نصراباذ وقناة شهى تجرى وسطها وقناة عبد الوهّاب تجرى على روذة وقناة قرشى تجرى على فليسان. l) Haec ex C. restitui, coll. E. وكاريزهاست كى بران كشاورزى كنند et Abulfeda, p. ۴۲۱ وبها قنى ايضا. Deinde A. وتنفرّغ, B. وتفرّع, C. وينعرج. m) C. addit اهل. n) C. وتجارات. o) E. وكور ومقبرة محمد بن الحسن الفقيه على بن حمزه الكسائى المقرى وهشام بن عبيد الله وابراهيم الخواص وفزارى المنجّم بشهر رى نهاده اند. Deinde addit: واز متاعها كرباسهاى نرم ويرتفع منها قطن كثير الى العراق; Abulfeda l. l. خيزد كى بافاق از آن ببرند.

ناحية[a] دُنْبَاوَنْد ولهم ضياع ورساتيق[b] ، وامّا وِيمَة وشَلَنْبَة فهما من ناحية دنباوند وهما مدينتان صغيرتان اصغر من الخوار واكبرهما ويمة ولهما زروع ومياه وبساتين ولهم اعناب كثيرة وجوز وهى[c] اشدّ تلك النواحى بردًا ۞ وللرىّ سوى هذه المدن قرى تزيد فى الكبر على هذه المدن كثيرًا مثل سُدّ[d] ووَرَامين وأُرْنَبُويَة[e] وورزنين[f] ودِزاه[g] وقوسين وغير ذلك من القرى الّتى بلغنى انّ فيها ما يزيد اهله على عشرة آلاف رجل[h] ، ومن رساتيقها المشهورة قَصْرَانْ[i] الداخل والخارج وبهْزَان[k] والسِّنّ وبشاويه[l] ودُنْبَا * ورستاق قوسين

---

a) C. addit وادى. Deinde E. ut solet دماوند. b) C. ضياع كثيرة خصبة.
c) Sic quoque legit Abulfeda, p. ۴۳۰, pronomen referens ad ويمة, quod E. facit perspicue, dicens: ودرين حدود هيچ جاى سردسيرتر از ويمه نيست. C. habet وهما; D. وخوار quod ortum esse potest e praecedenti وجوز quod in D. deëst.
d) In E. nomen desideratur. C. haec sex nomina omisit. Jacut articulum addit, vid. in v. et II, p. ۸۹۴, 21. Probe distinguatur ab السِّرّ, de quo loco vid. Jacut in v. et III, p. ۹۷, 9. e) A. et B. sine punctis. f) A. et B. ودرسن, D. وسردن. E. وورزمن (Ous. ودرسين). Vid. Jacut in v. g) A., B. et D. ودرك; E. ودرا. Vid. Jacut in v. Infra in capite مفازة خراسان A., B. et E. habent درة, C. درا. Forma درك pro درة (cf. Jacut, II, p. ۵۷۳, 17) servari potest. — Pagus sequens a Jacut non memoratur, nisi forte eundem locum designet nomine قوهذ (II, p. ۸۹۴ et in v.). h) C. addit et eadem habet Jacut, III, p. ۵۱۳: وبالسدّ منها على فرسخين يقال ان مفاتيح بساتينها المعروفة اثنا عشر الف مفتاح وكان يذبح فى هذه (Jacut بهذه) القرية فى كلّ يوم مائة وعشرون شاة واثنا عشر ثورًا (Jacut واثنتا عشرة بقرة وثور). i) E. قصرانين. k) A., B., D. et E. وبهمان; Mokaddasí MS. p. 186 دهنان (male ibi addito nomine سن scribitur سردهنان, unde Sprenger, p. 54, vs. 4 edidit سرتهنان). Secutus sum Jacut, II, p. ۸۹۴, ult., p. ۸۹۹, 3 et I, p. ۷۹۹. Fortasse autem edendum est تهران pro طهران, vid. Jacut III, p. ۵۹۴, 15 et *Djihán-Numa*, p. ۳۹۳. l) A. et B.

وغير ذلك[a]، ويرتفع من الرىّ ممّا يجلب الى غيرها القطن الّذى يحمل الى بغداد واذربيجان ومن الثياب المنيّرة والابراد والاكسية ❊

وليس لجميع هذه النواحى نهر يجرى فيه السفن، وامّا الجبال فانّ من حدّ عمل الرىّ دنباوند وهو جبل رايتُهُ انا من وسط رُوذَة بالرى[b] وبلغنى انه يُرَى من قرب ساوة[c] وهو جبل وسط جبال يعلو فوقها كالقبّة ويحيط بالموضع الّذى يعلو على الجبال نحو اربعة فراسخ ولم يصحّ عندى انّ احدًا ارتقى[d] اعلاه وفى حماقات الاوّلين[e] انّ الضحّاك المَلك مُقيّد بها وانّ السحرة يجتمعون * اليه فى اعلاه[f] ويرتفع من اعلاه دخان دائم الدهر كلّه وحوالى هذه القلّة قرى منها قرية دبيران[g] ودرمية[h] وبرا[i] وغيرها من القرى وكان علىٌّ بن شَرْوِين[k] الّذى أُسر على وادى جيحون من قرية درمية الّا انّ القلّة الّتى ترتفع عن هذه البقعة جبل اقرع ليس عليه كثير شجر ولا نبات * ولا يعلم بسائر الجبال ونواحى الديلم[l] جبل اعظم منه[m] ❊

وامّا قُومِس فانّ اكبر مدينة بها الدّامَغان وهى مدينة اكبر من خُوَار

ودساده, D. ونساوده, E. وبسا. Secutus sum Jacut, II, p. ٨٩٤ ult. Deinde E. ودماوندكها et Jacut ودنباوند l. l. et p. ٨٩٩, 20. *a*) E. horum verborum loco habet ومركوى, i. e. idem pagus quem Jacut l. l. vocat مَرْجَبَى. Cf. *Merácid*, III, p. ٧٩, ubi vs. 5 l. مركويه pro مركونة. Jacut, IV, p. ٤٩٠, 4 مركبويه. *b*) C. الرى. E. nomine روذة partim ortum videtur quod habet E. وهمواره از سر دماوند روذى مى خيزد. *c*) B. سماره, C. اصبهان. *d*) D. ارتقاه; C. et D. addunt الى. *e*) C. الفرس. *f*) A. et B. اعلاها. *g*) Puncta ex E. *h*) Secundum D. et E. (in E. *teschdíd* inscribitur litterae ي); C. درمنه; A. et B. h. l. sine punctis, deinde دومنة. *i*) E. om.; A. et B. sine punctis, D. ونوا. Secutus sum C. Cf. de his pagis Edrîsî, II, p. 181. *k*) A. et B. سروين. Videtur intelligi filius ejus qui memoratur ab Ibno 'l-Athír, VIII, p. ١٥٥, 5. *l*) C. et D. addunt: وما يتصل بها. *m*) E. pro his: وكوه دماوند بر همه كوههاى طبرستان مشرف بود واز همه جاى آنرا بتوان ديد.

الرىّ * وسِمْنان اصغر منها وبِسْطام اصغر من سمنان *a* والدامغان قليلة الماء وهى متوسّطة العمارة وبسطام اكبر منها عمارة *b* واكثر فواكه ويحمل الى العراق من بسطام فواكه كثيرة ويرتفع من قُومس اكسية معروفة تحمل الى الامصار ✿ وامّا قَزْوين فانّها مدينة عليها حصن وداخلها *c* مدينة صغيرة *d* عليها حصن *e* ومسجد الجامع فى * المدينة الداخلة *f* وهى مدينة مآؤها من السماء والآبار * وليس بها نهر الّا قناة صغيرة للشرب لا يفضل لزروعهم *g* وهى خصبة مع قلّة مياهها وهى ثغر الديلم *h* * وبها فواكه واعناب كثيرة وزبيب *i* يحمل الى الآفاق وتكون نحو ميل فى ميل *k* ✿ وأَبْهَر وزَنْجان صغيرتان خصيبتان *l* كثيرتا المياه والاشجار والزروع وزنجان اكبر من ابهر غير انّ اهل زنجان الغالب عليهم الغفلة ✿

وامّا طبرستان فانّ اكبر مدنها آمُل وهى مستقرّ الولاة *m* فى هذا العصر وكانوا فى قديم الايّام يسكنون سارية وطبرستان بلد كثير المياه *n* والثمار والاشجار الجبليّة والسهليّة والغالب عليها الغياض والغالب على ابنيتها الخشب والقصب * وهى كثيرة الامطار شتاء وصيفًا *o* وسطوحهم مُسَنَّمَة لذلك *p*،

---

a) E. وسمنان كم از دامغانست وبسطام كم از دامغان. Abulfeda, p. ٤٣٧ منها quoque ad Damaghán refert. *Djihán-Numa*, p. ٣٣٩, habet ut recepi. *b*) Conjectura addidi عمارة ut tollatur contradictio cum iis quae praecedunt. *c*) A. et B. وداخله. *d*) C. كبيرة. *e*) C. addit ولها مدينتان صغيرتان. Vid. supra p. ٢٠٠ ann. *q*. *f*) C. الكبرى. *g*) C. وقناة غير كثيرة الماء وليس بها نهر. *h*) C. et E. addunt: وبها اختلاف كثير وحروب بين القنىّ الا هذه دائم. *i*) E. وميوه بسيار دارند انگور وبادام ومويز چندان بود كه اهلها دائمة. *k*) C. addit: وبر زمين وباغ كه در قصبه E. ,وليس عليهم خراج فى القصبة دارند خراج نبود. *l*) B. et C. خصبتان. Deinde A., B. et C. كثيرة. *m*) C. الولاية. *n*) C. addit والغياض. *o*) E. پيوسته باران باشد (ridicule Ous. „it adjoins Aran"). *p*) C. لكثرة الامطار. Deinde C. et E. واما سارية (سارى) فهى اكبر. Sed cf. quoque Abulfeda, p. ٤٣٥.

وآمُل اكبر من قزوين مشتبكة العمارة لا يعلم بقدرها[a] اعمر منها فى هذه النواحى، ويرتفع من[b] طبرستان من الابريسم ما يُميرُ[c] الآفاق *وليس فى الاسلام مدينة اكثر منها ابريسم[d] وبها خشب كثير *من اصلب[e] الخشب ينحت[f] منها آنية واطباق تنقل الى الآفاق، والغالب على اهلها وفور الشعر واقتران الحواجب وسرعة الكلام والعجلة *وخبزهم اكثره من الارزّ واكثر طعامهم السمك[g] وكذلك الديلم[h] والجيل، *ويعمل بطبرستان ثياب كثيرة من الحرير تنقل الى الآفاق وكذلك من الصوف والفُرُش والاكسية وغير ذلك[i]، وليس بجميع طبرستان نهر تجرى فيه سفينة الّا انّ البحر منها قريب على اقلّ من يوم غير انّ بجميع طبرستان الماء والغياض غالب بها الّا فى الاماكن الجبليّة وامّا بطن طبرستان فالغالب عليها *الندى والنزوز[k]☆

وامّا جُرْجان *فاكبر مدنها جُرْجان وهى اكبر من آمُل وبناؤها من طين وهى ايبس تربة من آمُل[l] واقلّ مطرًا واندآء[m] من طبرستان واهلها احسن وقارًا ومروّة ويسارًا فى كبرآئهم[n] *وهى قطعتان احداهما المدينة والاخرى

---

a) C. et D. على قدرها. b) C., D. et E. بجميع. c) Abulfeda, p. ۴۳۳ ult. يعمّ. d) E. خاصه بآمل. e) C. اصلب من سائر. f) A. دمحب, C. يعمل. Pro آنية واطباق quod sequitur, E. habet چوبينه, cui voci male substitutum est موينه in excerptis quae e libro Nacíro'd-dín Tusii dedit Melgunof l. l. p. 296. g) E. وبيشتر طعام ايشان نان برنج وماهى بود وسير بسيار خورند et sic D. et Mokaddasí MS. p. 169. C. habet: والغالب على طعامهم الارز والسمك والباقلى وقصب السكر. h) C. طرف من الديلم. i) C. et partim D.: ويرتفع من طبرستان اصناف ثياب الابريسم والاكسية الصوف والمدارع والميازر والمناديل والغرش الطبرى واثواب الحرير والسموك والاترج والنارنج مع الات من الخشب. k) A. et B. الندا والنزور; C. addit وعلى جبالها البيس; cf. Abulfeda, p. ۴۳۳. l) C. et Jacut, II, p. ۴۹ pro his: فانها اكبر مدينة بنواحيها et deinde وهى اقل. m) C. et Jacut ونديى. n) A. كبريابهم. C. et Jacut pro فى male habent من.

بكْرَاباذ[a] بينهما نهر يجرى كبير يحتمل ان تجرى فيه السفن، ويرتفع فيها[b] من الابريسم[c] شىء كثير *وابريسم طبرستان يحمل بزر دوده من جرجان ولا يرتفع من بزر طبرستان ابريسم[d]، ولهم مياه كثيرة وضياع عريضة وليس *فى المشرق[e] بعد ان تجاوز العراق مدينة *بقدر جرجان اخصب ولا اجمع منها[f] وذلك انّ بها الثلج والنخيل وبها فواكه الصرود والجروم من التين والزيتون وسائر الفواكه[g] واهلها اصحاب مروّة يتبارون[h] فى المروّات *وياخذون انفسهم بالتانّى[i] والاخلاق المحمودة وقد خرج منهم رجال كثيرون موصوفون بالسرو[k] منهم العَمْركى صاحب المامون[l]، ونقودهم ونقود طبرستان الدنانير والدراهم واوزانهم المنا ستّمائة درهم وكذلك بالرىّ وطبرستان وقومس مناها ثلاثمائة درهم ❁ وأَسْتَرَاباذ *يرتفع بها ابريسم كثير[m] ولهم فرضة على البحر يركبون منها الى[n] الخزر والى باب الابواب والجيل والديلم وغير ذلك[o] وليس فى

---

a) E. وبيرون شهر محلتى هست كى آنرا بكراباذ خوانند. b) C. بها, Jacut منها. c) C., E. et Jacut addunt: وثيابه (s. وثياب الابريسم). Deinde C. ما يحمل الى جميع الافاق, Jacut اصناف شتى. d) E. وتخم ابريشم كركانى را در طبرستان ابريشم نخيزد; Jacut وابريسم جرجان بزر دودة (l. دوده) يحمل الى طبرستان ولا يرتفع من طبرستان بزر ابريسم (l. من بزر طبرستان ابريسم). e) C. et Jacut بالمشرق. f) C., D. et Jacut: اجمع ولا اظهر على مقدارها من جرجان (D. خصبًا) حسنًا. g) E. supra omissis بها الثلج والنخيل h. l. habet: وبنابستان برف نزديك بود. h) A. et B. يبارون. i) A. et B. om. Deinde C. et D. للاخلاق. (Jacut habet بالتانى). k) B. باليسر, Jacut بالستر والسخاء. Deinde hic البرمكى pro العمركى. Melgunof l. l. عمركى. l) C. addit وغيره. m) C. مدينة طيبة الهواء جبلية سهلية بحرية وبها قلاع كثيرة منيعة وهى عامرة خصبة ويرتفع بها ثياب ابريسم تحمل الى الافاق من المقانع والوقايات وغيرها. n) E. pro his: بدريا نزديكست واز آنجا بدريا سوى بابسكون روند. o) C. addit تعرف بابسكون et sic videtur legisse Abulfeda, p. ٤٣٩.

هذه الناحية[a] الّتى ذكرتُها غرضة اجلّ من آبَسْكُون ولهم ثغر يعرف برباط دِهِسْتان وبها منبر[b] وهو ثغر للغُزّيّة الاتراك[c] ويتّصل حدّ جرجان بالمفازة الّتى تلى خوارزم ومنها[d] تجيئهم الاتراك ❊

ذكر مسافات هذه الديار[e] الطريق من الرىّ الى حدّ اذربيجان من الرىّ الى قزوين ٤ مراحل ومن قزوين الى ابهر مرحلتان خفيفتان ومن ابهر الى زنجان يومان *اوفر ممّا بين قزوين[f] وابهر' ومن اراد *الطريق القصد[g] لم يدخل قزوين وجعل الطريق على قرية تسمّى بزداباذ[h] من رستاق دَسْتَبَى[i] ❊ والطريق من الرىّ الى الجبال من الرىّ الى قُسْطَانَة[k] مرحلة ومن قسطانة الى مشكويه[l] مرحلة ومن مشكويه الى ساوة مرحلة ٩ فراسخ وساوة ربّما كانت من عمل الجبال وربّما كانت من عمل الرىّ ❊ الطريق من الرىّ الى طبرستان *من الرىّ الى برزيان[m] مرحلة خفيفة ومن برزيان الى نامهند[n] مرحلة كبيرة[o] *ومن نامهند الى اسك[p] مرحلة ومن اسك الى بلور[q]

---

a) C. et D. النواحى. b) C. addit: ولها ضياع كثيرة. c) C. addit وهى حصينة جدا, Vers. Pers. apud Melgunof l. l. جاى پر نعمت i. e. خصيبة. d) C. فمنها. e) Addidi titulum ex D. et E. f) A. et B. فيما بين قُرَى. g) C. الاقتصاد, D. الطريق بالقصد. h) A. دوناباذ, B. بوزاباد, D. بودابار, E. زيراباذ (Ous. بدراباذ). Recte C., vid. Jacut in v. In C. ante قرية legitur قرب الرى, quae verba post قرية suo loco erunt. i) A. et B. ستة, D. وسمه; C. et E. om., Ous. contra: „one of the villages of the waste or desert" (?) k) E. كشتانه, Mokaddasí MS. p. 193 كسكانة. Vid. Jacut, IV, p. ٩٤, 20. l) A. et B. sine punctis, C. مسكويه. In textu *Merâcid* minus recte مشكونة receptum est. In edit. Jacut, IV, p. ٥٤٣ مُشْكُوِيَه scribitur. m) A. et B. دوربان, D. بربريان et بربريان. Secutus sum lectionem Edrîsîi, II, p. 169 coll. Mokaddasí MS. p. 178 دوزبان (cf. Sprenger, p. 52). n) Secutus sum Mokaddasí l.l.; A. et B. دامهند, C. دامهز, D. ماهيد, E. نامهين, Edrîsî تانهند. o) C. et E. pro his من الرى الى نامهند ٩ فراسخ, omissa statione برزيان. p) A. et B. اسل (A. semel اسنل ut Mokaddasí), D. اسك, Edrîsî اشك; cf. Melgunof, p. 24 et Sprenger, p. 52. q) A. دكور, B. بكور, Edrîsî بلّون, Mokadd. دلور. Ceteri ut recepi.

مرحلة[a] ومن بلور * الى كنازل[b] مرحلة ٦ فراسخ ومن كنازل الى قلعة اللَّارِز[c] مرحلة ٥ فراسخ ومنها الى فرست[d] مرحلة ٦ فراسخ ومنها[e] الى آمُل مرحلة[f] ✿ الطريق من الرىّ الى خراسان على قومس من الرىّ الى افرنديين[g] مرحلة ومن افرنديين الى كهدة[h] مرحلة ومن كهدة الى خُوّار[i] مرحلة ومن خوار الى * قرية الملح[k] مرحلة ومن قرية الملح الى راس الكلب مرحلة ومن راس الكلب الى سِمْنان مرحلة ومن سمنان الى علياباذ مرحلة ومن علياباذ الى جرمجوى[l]

---

a) C. et E. pro his: (يكمرحلة) ومن نامهند الى بلور ٥ فراسخ. *b*) Sic E. (Ous. كلازل). Fortasse legendum est كنازك et componendum cum nomine pagi inter Sária et Asterabádh جُنَّازه (Jacut, II, p. ١٢٤, 6). Deinde verba مرحلة ٦ فراسخ ومن كنازل ex solo Ous. addidi. C. ومنها (بلور) الى عند قلعة اللارز. *c*) E. لازر. Vid. Jacut, IV, p. ٣٤. (لارجان) et ٣٤١ (لارز) et Abulfeda, p. ٤٣٤. Secundum eos hoc castellum jacet medium inter Raij et Amol, ab unaquaque 15 (18) Paras. distans, quod non prorsus quadrat in nostrum itinerarium. Scribitur quoque الأرْز, vid. Jacut, I, p. ٣٠٤ et III, p. ٥٠٤, 5. — Deinde E. om. ٥ فراسخ. *d*) C. فرسب, Ous. قهرست. Deinde C. ٥ فراسخ. E. distantiam om. *e*) In A., B., D., Mokaddasí et Edrîsî haec omnia inde a كنازل الى desunt. Cf. Sprenger l. l. *f*) C. ٩ فراسخ. *g*) A., B. et C. sine punctis (B. semel افريدون ut in *Merácid* editum est). Jacut locum memorat. E. امرىدىن. Edrîsî, II, p. 175 فراندين et hic inter hanc stationem et Raij aliam stationem collocat, dicens: a Raij ad معقل اباذ 6 Par., hinc ad Afrandín 8 Par. — Kodáma legens افريدين habet: من الرى الى فصلاباذ ٤ فراسخ ومن فصلاباذ الى افريدين ٦ فراسخ. Ibn Khordádbeh, p. 46 legit مصقل اباذ (Sprenger p. 12 مفصلاباذ) et فرخدين. *h*) C. et E. كهنده. Kodáma et Ibn Khord. (كاسب) كاست. *i*) C. addit مدينة. Kodáma حراسب. *k*) E. ده نمق. Kodáma et Edrîsî قصر الملح. *l*) A. et B. (جومجوى) خومجوى D. حرمحون, E. جرم جرى, Mokaddasi p. 177 حرمجوى. Secutus sum C. (جرم جوى). Edrîsî (II, p. 176), Ibn Khordádbeh

مرحلة ومن جرمجوى الى الدامغان مرحلة ومن الدامغان الى الحدّادة مرحلة[a] ومنها الى بَذَش[b] مرحلة ومنها الى الموزجان[c] مرحلة كبيرة ومنها الى هَفْذَرَ[d] مرحلة ومنها الى أَسَدَاباذ مرحلة وهى من عمل نيسابور[e] ✿ الطريق من طبرستان الى جرجان من آمل الى ميلة فرسخان ومن ميلة الى برجى *٣ فراسخ وهما جميعًا مرحلة[f] ومنها الى سارية مرحلة ومنها الى مارست[g] مرحلة ومنها الى اباذان[h] مرحلة ومنها الى طَميسَة[i] مرحلة ومنها الى

---

et Kodáma eandem, ut videtur, stationem appellant احرين s. جزين ,اجوين. Kodáma inter راس الكلب et سمنان memorat stationem سيرج (Sprenger, p. 13 قرية دابة) et inter جرم جوى (احرين) et الدامغان (قومس) stationem سيرج.
a) Sive 7 Paras., vid. Kodáma et Jacut, II, p. ٢١٧, 6. Inter hanc et sequentem stationem Kodáma collocat كورسان. b) A. بدمن, B. برمن, C. et Ous. بدليس, E. بدلس, D. بدس, Mokaddasí بدش, Ibn Khordádbeh حدس, Kodáma كرمن et دوس (cf. Sprenger, p. 13). Vid. Jacut in v. et Edrîsî. c) A. الموفرجان, B. الموفوحان; C., D. et E., ut Edrîsî مورجان; Mokaddasí مرحان. Kodáma et Ibn Khordádbeh hanc stationem omittunt, sed deinde aliam memorant, quam Kod. منجد, Ibn Khord. ميمل, Edrîsî مبرار appellant; cf. Sprenger l. l. d) Secutus sum B., C., D. et Mokaddasí l. l. (vocales ex D.). A. هقدر, E. هفت در, Kodáma قهندر, Ibn Khord. هم كند, Edrîsî هشكيده, Mokaddasí p. 166 هقدر. e) B. نيشاپور. Deinde in C. et E. additur: ودخل اسداباذ وقف على المارّة فمن اقام بها ثلاثة ايام أُجرى عليه ومن اعتلّ بها أُنفق عليه حتّى يبرأ. f) C. تمام المرحلة. E. minus correcte مرحله. g) A. et B. مارست, D. بادست, Edrîsî, II, p. 177 نارست, Mokaddasi p. 178 انارست (Sprenger, p. 53 legit امارست). E. habet نامه (Ous. بامية) i. e. نامية ut videtur legendum esse apud Beládsorí, p. ٣٣٤ ult., nam Jacut, III, p. ٥٠٤, 10 et ٥٠٥, 6 sic habet (cf. نامشة, IV, p. ٧٣٣) et Mokaddasí p. 27 مامية, p. 167 نامية. — In C. مهروان exstat, cf. supra ad p. ٢٠٧ a. h) A. et B. sine punctis, Edrîsî باباذان. C. et E. لمراسك, cf. supra p. ٢٠٧ a. i) C. et E. طميشه.

أَسْتَرَاباذ مرحلة ومنها الى رباط حفص مرحلة ومنه الى جرجان مرحلة، *ومن اراد ان يخرج من آمل الى مامطير مرحلة ومنها الى سارية مرحلة ولا يجعل طريقه على برجى فهو اقصد وانّما ذكرنا الاطول لانّ فيه منبرين a ✿ الطريق من آمل الى الديلم من آمل الى ناتل b مرحلة ومن ناتل الى سالوس مرحلة خفيفة ومن سالوس الى كلار مرحلة ومنها الى الديلم مرحلة، ومن آمل الى البحر الى عين الهُمّ مرحلة خفيفة ✿ الطريق من جرجان الى خراسان من جرجان الى دينارزارى c مرحلة ومن دينارزارى الى املوتلو d مرحلة ومن املوتلو الى اجغ e مرحلة ومن اجغ الى سيبداست f مرحلة ومن سيبداست الى أَسْفَرَائين g مرحلة ✿ الطريق من جرجان الى قومس من جرجان الى جُهَينة h مرحلة ثمّ منها الى بسطام مرحلة ✿

بحر الخزر

وامّا بحر الخزر فانّ شرقيّه بعض الديلم وطبرستان وجرجان وبعض المفازة الّتى بين جرجان وخوارزم وغربيّه أرّان i وحدود السرير وبلاد الخزر وبعض مفازة

---

a) C. haec om.; E. وهر كه خواهد از استراباد بيرون آيد تا رباط وذاره مرحله تا جرجان مرحله واگر از آمل تا مامطير مرحله تا سارى مرحله. Edrîsî, II, p. 178 aliud itinerarium dat inde a مامطير ad استراباذ. b) E. بابل. c) E. دنمار زارى, C. دنمار حالى, Mokaddasí ديناراري. d) A. et B. املق يلوا, E. امرويلو (Ous. امرويلو), D. املوادلوا, C. املوتلوا, Edrîsî, II, p. 181 املويلو, Mokaddasí املوتا (cf. Sprenger, p. 53). e) A., B. et D. اجع, Edrîsî اخغ, Mokaddasí اجّر. Secutus sum E. Apud C. haec et sequens statio desiderantur. f) A. سيبداسب, B. سيداسب, D. سيداست, E. سنداست (Ous. سنداىست). Edrîsî سنداسب, Mokaddasí سنبداست. g) C. سبرادىن, E. سرادىن. h) A., B. et C. sine punctis. Vid. Jacut in v. i) E. addit وموقان. C. habet اللآن ارّان من جبال القبق الى حدود السرير. Jacut, I, p. ٥٠, idem, sed substituto اللآن pro اران.

الغُزِّيَة وشماليُّه مفازة الغُزِّيَة[a] بناحية سِيَاكُوه وجنوبيُّه الجبيل والديلم[b] ✿
وهذا البحر ليس له اتّصال بشىء من البحار على وجه الارض فلو انّ رجلًا طاف بهذا البحر لرجع الى مكانه الّذى ابتدأ منه لا يمنعه مانع الّا *نهر عذب يقع[c] فيه وهو بحر مالح ولا مدّ له ولا جزر وهو بحر مُظْلم قعره طين بخلافِ بحر القلزم[d] وسائر بحر فارس فانّ فى بعض المواضع من بحر فارس ربّما يرى عمقه لصفآء ما تحته من الحجارة البيض، ولا يرتفع من هذا البحر شىء من الجواهر من لؤلؤ او مرجان او غيره ممّا يرتفع من البحار ولا ينتفع بشىء ممّا يخرج منه سوى السموك ويركب فيه التجّار من اراضى المسلمين الى ارض الخزر وما بين الران والجيل وطبرستان وجرجان، وليس فى هذا البحر جزيرة مسكونة فيها عمارة كما ذكرنا فى بحرى فارس والروم الّا انّ فيها جزائر فيها غياض ومياه واشجار وليس بها انيس ومنها جزيرة سِيَاكُوه وهى جزيرة كبيرة بها عيون واشجار وغياض[e] وبها دوابّ وحش[f]، ومنها جزيرة *بحذآء الكُرّ[g] وهى كبيرة بها غياض واشجار ومياه ويرتفع منها الفُوّة ويُخْرَج اليها من نواحى بَرْذَعَة فيحملون[h] منها الفُوّة ويحملون فى السفن اليها[i] دوابّ من نواحى برذعة وسائر المواضع فتُسْرَح فيها حتّى تسمن[k] ✿

---

a) Jacut addit: وهم صنف من الترك. b) C. et Jacut وبعض الديلم; D. addit: وما دانى ذلك quod quoque Ous. p. 184 habet. Ad seqq. collatum quoque est excerptum ex opere Nacíro'd-dín Tusí, apud Melgunof, p. 291 seq., jam saepius laudato, quod opus nuda est versio nostri. — Hic sequi deberet mappa maris Caspii. c) Jacut ان يكون نهر يصبّ. Cf. supra p. ٨, 8. d) Vers. Pers. (vid. Melgunof, p. 291 et Ous.; in E. est lacuna) addit وعمان ut quoque *Djih.-Numa*, p. ٣٩٩, 4. e) C. ورياض, E. زراعات واشجار. Jacut addit (III, p. ٢٠٨) ومياه عذبة. f) E. ودد ودام ماوا گرفته اند, Melgunof ودد ودام انجا. g) E. بحد لكزان. h) B. ينتحملون. i) A. et B. منها. k) C. et Jacut addunt: وجزيرة تعرف بجزيرة الروسية (وجزائر صغار).

وليس من آبَسْكُون الى الخزر *عن اليمين[a] على شطّ البحر قرية ولا مدينة سوى موضع من آبسكون على نحو خمسين فرسخًا يسمّى دِهِسْتان[b] وهى دخلة فى البحر[c] تستتر[d] فيها السفن فى هيجان البحر ويقصد هذا الموضع خلق كثير من النواحى ويقيمون[e] بها للصيد وبها مياه، ولا اعلم غير هذا المكان مكانًا يقيم به احد الّا ان يكون سياه كوه فانّه تقيم به طائفة من الاتراك وهم قريبو العهد بالمقام به لاختلاف وقع بين الغُزّيّة وبينهم فانقطعوا عنهم واتّخذوه مأوًى ومرعًى ولهم عيون ومراعٍ۞ هذا ما عن يمين هذا البحر من آبسكون ومن آبسكون *عن يسارها[f] الى الخزر فى عمارة متّصلة الّا شيًّا[g] بين باب الابواب والخزر وذلك انك اذا اخذتَ من آبسكون مضيتَ على حدود جرجان وطبرستان والديلم والجيل *ثمّ تدخل فى حدود الران واذا[h] جزتَ مُوقان[i] الى باب الابواب على يومين فهو بلد شَرْوان شاه[k] ثمّ تتجاوزه الى سَمَنْدَر[l] اربعة ايّام ومن سمندر الى اِثل[m] سبعة ايّام مفاوز۞ ولهذا البحر بناحية سياكوه زنقة يخاف على السفن *الداخلة بها الريح[n] ان تنكسر واذا انكسرت السفن هناك لم يتهيّأ جمع شىء منها من الاتراك فانّهم يستولون على ذلك۞

---

a) Jacut للآخذ على يمنى يديه. *b*) A. et B. دهسانمادسر et infra in itinerario دهستابناسر. C. ibi دسان دماسر et D. دهستابيناس. E. hic et infra habet nudum nomen دهستان, nec scio quid illud additamentum significet. Vestigium ejus superesse videtur in illo وبناء apud Jacut; vid. ann. seq. Observandum autem est Dihistán proprie esse nomen regionis, non urbis الرباط. *c*) C. وهى داخلة البحر, Jacut وبناء داخل البحر. *d*) Jacut تستّر, C. يستر, D. فيرسون. *e*) C. et Jacut فيقيمون. Deinde D. et Jacut به. *f*) A. et B. om. *g*) A. et B. شى ut saepius. *h*) Ex E.; A., B. et D. sine copula. *i*) A. et B. من باب. *k*) A. سمبذان شاه, B. سيندان شاه. Pro his C. et Jacut: وموقان وشروان والمسقط والباب (باب الابواب). *l*) A. et B. h. l. سمندان, mox A. سمندربا. Deinde A. et B. سبعة. *m*) A. et B. h. l. آمل. *n*) Videtur addendum اليها. Jacut اذا اخذتها الريح اليها; D. اذا اخذتها ريح هناك; E. کشتی را بدان جانب اندازد (Melgunof اگر باد کشتی بآنجا برد).

واَمّا الخَزَرُ فانّه اسم *الاقليم وقصبته[a] تسمّى اِتِل واتل اسم النهر الّذى يجرى اليه[b] من الرّوس وبُلْغار واتل قطعتان قطعة على غربىّ هذا النهر المسمّى اتل وهى اكبرهما وقطعة على شرقيّه والمَلِك يسكن فى[c] الغربىّ منهما ويسمّى الملك بلسانهم بك[d] ويسمّى ايضا باك وهذه القطعة مقدارها فى الطول نحو فرسخ ويحيط بها سور الّا انّه مفترش البناء وابنيتهم خركاهات لبود الّا شيئًا يسيرًا بُنى من طين ولهم[e] اسواق وحمّامات[f] وفيها خلق[g] من المسلمين يقال انّهم يزيدون على عشرة آلاف مسلم[h] ولهم نحو ثلاثين مسجدًا وقصر الملك بعيد من شطّ النهر وقصره من آجُرّ وليس لاحد بناء من آجُرّ غيره ولا يسوّغ[i] الملك لاحد ان يبنى بالآجُرّ ولهذا السور ابواب اربعة *منها الى ما[k] يلى النهر[l] *ومنها الى ما[m] يلى الصحراء[l] على ظهر هذه المدينة، وملكهم يهودىّ يقال[n] انّ له من الحاشية نحو اربعة آلاف رجل *والخزر مسلمون ونصارى ويهود[o] وفيهم عبدة اوثان[p] واقلّ الفِرَق اليهود واكثرهم المسلمون والنصارى الّا انّ الملك وخاصّته يهود والغالب على اخلاقهم اخلاق اهل الاوثان يسجد بعضهم لبعض عند التعظيم واحكام *خصّوا بها[q] على رسوم قديمة مخالفة[r] لدين المسلمين واليهود والنصارى، *وللملك من

---

a) C., Jacut (II, p. ٤٣٦) et E. اقليم من قصبة. Lectio اتل propria est A. et B.; ceteri اتل habent. *b*) C. et D. اليها, Jacut الى الخزر, E. بدريا. *c*) C. et Jacut om. *d*) C. et Kazwini, II, p. ٣٩٣ بلك, Jacut يلك, versio Pers. apud Melgunof ملك. *e*) E. همين, Melgunof چون. *f*) E. addit ومساجد. *g*) Jacut addit كثير. *h*) E. addit: باقى همه جهود. *i*) Jacut يمكن. *k*) E. et Jacut احدها, Mokaddasí MS. p. 172 احدهن. *l*) A. et B. البحر. *m*) E. et Jacut واخرها, Mokaddasí والاخر. *n*) C., D., E. et Jacut copulam addunt. *o*) A. et B. يقال انهم مسلمون ونصارى; E. از مسلمان وباقى از ترسا وجهود; Melgunof از مسلمان وخزر وترسا وبت پرست باشند. Apud D. et Jacut deëst ويهود, quod habent C. et Epit. Paris. *p*) C., D. et Jacut الاوثان. *q*) A. خُصّونها, B. حصونها, C. مصرها, D. يمضونها, Jacut مصرهم. *r*) A. et B. مخالف, D. تخالف.

الجيش[a] اثنا عشر الف رجل واذا مات منهم رجل أُقيم آخر مكانه[b] وليست لهم جراية[c] دارّة الّا *شىء نزر[d] يسير يصل اليهم فى المدّة الطويلة اذا كان لهم حرب او حزبهم امر يجتمعون[e] له، وابواب مال هذا الملك من الارصاد وعشور التجارات على رسوم لهم من كلّ طريق وبحر ونهر ولهم وظائف على اهل المحالّ والنواحى من كلّ صنف ممّا يحتاج اليه من طعام وشراب وغير ذلك، وللملك سبعة[f] من الحُكّام من اليهود والنصارى والمسلمين واهل الاوثان اذا عرض للناس حكومة قضى فيها هاؤلآء ولا يصل اهل الحوائج الى الملك نفسه وانّما يصل[g] الى هاؤلآء الحُكّام *وبين هاؤلآء الحكّام يومَ القضاء وبين الملك[h] سفير يراسلونه فيما يجرى* من *الامر وينتهون[i] اليه فيَرِد[k] عليهم امره ويمضونه، وليس لهذه المدينة قرى الّا انّ مزارعهم مفترشة يخرجون فى الصيف *فى الزروع نحو[l] عشرين فرسخًا *ليزرعوا ويجمعوا[m] بعضه على[n] النهر وبعضه على[n] الصحارى *فينقلون غلّاتهم بالعجل وفى النهر[o] والغالب على قوتهم الارزّ والسمك[p] وهذا الّذى يحمل منهم من العسل والشمع انّما يحمل اليهم من ناحية الروس وبلغار وكذلك هذه الجلود الخزّ الّتى تحمل الى الآفاق لا تكون الّا فى تلك الانهار الّتى بناحية بلغار والروس وكُويَابَة[q] ولا تكون فى شىء من الاقاليم فيما علمتُهُ، والنصف الشرقىّ من[r] الخَزَر فيه

a) Jacut وجريدة جيش الملك. b) A., B. et C. om.; Jacut مقامه, addens فلا ينقص هذه العُدّة ابدًا. c) B. male خزانة. Deinde Jacut دايرة. d) A. بر, B. يدّ, C. قليلة نزرة, D. شىء يسير; vid. Jacut. e) B. et Jacut يجمعون. f) C., E. et Jacut تسعة; cf. Mas'udí, II, p. 11. g) C. يصلون. Jacut deinde اليه. h) C. فاذا جلسوا للقضاء كان بين الملك وبينهم. i) Jacut الامور ينهون. k) Jacut ويردّ. l) Jacut الى المزارع نحوًا من. m) Jacut فيزرعون ويجمعونه اذا ادرك. n) D. et Jacut الى. o) Jacut فيحملونه على العجل والنهر. p) E. add. وعسل. q) A. sine punctis, B. hic et infra كردادة; C. كوثابة et sic Jacut, IV, p. ٣١٨; Abulfeda, p. ٢٢٣ كوتابا. E. كونادة. Frähn, *Ibn Foszlan*, p. 143 seqq. probavit intelligi urbem *Kiew*. r) Jacut addit مدينة. Vers. Pers. apud Melgunof اتل pro الخزر habet.

معظم التجّار والمسلمين والمتاجر والغربىّ[a] خالصةٌ للملك[b] وجنده والخزر[c] الخُلّص، ولسان الخَزَر غير لسان الترك والفارسيّة ولا يشاركه لسان فريق من الامم ۞ وامّا نهر اتل فانّه فيما بلغنى يخرج من قرب خَرْخِيز فيجرى فيما بين الكَيْماكيّة والغُزّيّة وهو الحدّ بين الكيماكيّة والغُزّيّة ثمّ يذهب غربًا[d] على ظهر بُلْغار ويعود راجعًا الى ما يلى المشرق حتّى يجوز على الرُّوس ثمّ يمرّ على بُلْغار ثمّ على بُرْطاس حتّى يقع فى بحر الخَزَر ويقال انّه ينشعّب من هذا النهر[e] نيف وسبعون نهرًا ويبقى عمود النهر يجرى على الخَزَر حتّى يقع فى البحر ويقال انّ هذه المياه اذا كانت مجموعة[f] فى نهر واحد اعلاه[g] يزيد على جَيْحون وبلغ[h] من كثرة هذه المياه وغزارتها[i] انّها تنتهى الى البحر فتجرى فى البحر داخلًا مسيرة يومين وتغلب على ماء البحر حتّى يجمد فى الشتاء لعذوبته وحلاوته ويبين[k] فى البحر لونه من لون ماء البحر ۞ وللخزر مدينة تسمّى سَمَنْدَر فيما بينها وبين باب[l] الابواب لها[m] بساتين كثيرة ويقال انّها تشتمل على نحو من اربعة آلاف *كرم الى حدّ السرير[n] والغالب على ثمارها الاعناب وفيها خلق من المسلمين ولهم بها مساجد وابنيتهم من خشب قد نسجت[o] وسطوحهم مسنّمة وملكهم من

---

a) C. et E. والقرى (ودهها). b) C. et D. addunt وحاشيته. c) A. h. l. الخَزّر. Deinde C. الخاص. d) A. غربيا, Jacut, I, p. ١١٣, 19 مغربًا. e) A. البحر. f) C. مجتمعة, D. et Jacut اجتمعت. g) D. باعلاه, Jacut فى اعلاه, C. كان اعلاه. h) C. et D. ويبلغ. i) Jacut addit وجدّة جريها. k) A. sine punctis; C. et D. يتبين. l) C. اتل وباب male; cf. Jacut, III, p. ١٤٣, 13. Mokaddasí vero, p. 172 in f. habet وسمندر بلد كبير عند البحيرة بين نهر الخزر وباب الابواب. m) B. et D. بها, Jacut ذات, Mokaddasí لهم. n) Jacut بستان كرم وهى ملاصقة لحد ملك السرير. E. post كثيرة habet: چنانک از دربند تا (حد) سرير باغستان بود (باغ وبوستان اين شهر است) گويند چهل (چهار) هزار باغ معمور دارد. Uncinis inclusa secundum Melgunof sunt. — D. quoque habet اربعين الف. o) Jacut شخت. Mokaddasí quoque بنيانهم خشب منسوجة بالقصبان.

اليهود قرابة ملك الخزر وبينهم وبين حدّ السرير فرسخان وبينهم وبين صاحب[a] السرير هدنة ❁

* والسّرير هم[b] نصارى ويقال انّ هذا السرير هو[c] لبعض ملوك الفرس[d] من ذهب فلمّا زال ملكهم * حمل الى السرير وحمله[e] بعض ملوك الفرس بلغنى انّه من اولاد بَهْرَام جُوبِين[f] والملك الى يومنا هذا فيهم ويقال انّ هذا السرير عُمِل لبعض الاكاسرة فى سنين كثيرة وبين السرير وبين المسلمين هدنة[g] ، ولا اعلم فى عمل الخزر مجمع[h] ناس سوى سَمَنْدر ❁ وبُرْطاس هم أُمّة[i] متاخمون للخزر ليس[k] بينهم وبين الخزر امّة اخرى وهم قوم مفترشون على وادى اتِل وبرطاس اسم الناحية[l] وكذلك الروس والخزر والسرير اسم للمملكة لا للمدينة ولا للناس ❁

والخَزَر لا يشبهون الاتراك وهم سود الشعر[m] وهم صنفان صنف يسمّون قراخزر[n] وهم سمر يضربون لشدّة السمرة الى السواد كانّهم صنف من الهند وصنف بيض ظاهرو الحسن والجمال والّذى يقع من رقيق الخزر هم[o] اهل الاوثان الّذين يستجيزون بيع اولادهم واسترقاق بعضهم بعضًا[p] فامّا اليهود منهم والنصارى * فانّها تدين[q] بتحريم استرقاق بعضهم بعضًا مثل المسلمين ، وبلد الخزر لا يرتفع شىء منه يحمل الى الآفاق غير الغِرَى وامّا الزيبق[r] والعسل

---

a) A. et B. om. Pro هدنة, C. مهادنة. b) C., E. et Jacut (III, p. ٨٨) واهل السرير. c) C., D., E. et Jacut كان. d) C. addit وانه سرير, D. وهو, Jacut وهو سرير. e) Jacut حمل السرير. f) Jacut جور. g) C. مهادنة. Deinde hic et Jacut addunt: وان كان كل واحد (منهما) يحذر (حذرًا من) صاحبه. h) C. مجتمع. i) A. et B. امته. k) C. et Jacut (I, p. ٥٩٧) وليس. l) Jacut للناحية ut infra A. et B. m) C., D. et Jacut (II, p. ٤٣٨) الشعور. n) A. قرا جور, C. قرى خزر. o) C. فهم من, Jacut وهم, D. om. p) C. et Jacut لبعض. q) B. فانهما بدين, D. فانهم يتدينون, Jacut فانهم يدينون. r) D. الرقيق, Jacut الدقيق. E. enumerat موئينه وپشم وبرده وعسل وشمع, quod confirmat lectionem D.

والشمع والخزّ والاوبار فمجلوب اليها، ولباس الخزر وما حواليها القراطق والاقبية وليس يكون عندهم شىء من الملبوس وانّما يحمل اليهم من نواحى جرجان وطبرستان وارمينية واذربيجان والروم، وامّا سياستهم وامر المملكة بهم[a] فانّ عظيمهم يسمّى خاقان خَزَر وهو اجلّ من ملك الخزر الّا انّ ملك الخزر هو الّذى يقيمه واذا ارادوا ان يقيموا هذا الخاقان جآءوا به فيخنقونه[b] بحريرة حتّى اذا قارب ان ينقطع نفسه قالوا له كم تشتهى[c] مدّة الملك فيقول كذا وكذا سنة فان مات دونها والّا قُتل اذا بلغ تلك السنة ولا تصلح الخاقانيّة عندهم الّا *فى اهل[d] بيت معروفين وليس له من الامر والنهى شىء الّا انّه يعظّم ويسجد له اذا دُخل اليه ولا يصل اليه احد الّا نفر يسير مثل الملك ومَنْ فى طبقته ولا يدخل عليه الملك الّا لحادثة فاذا دخل عليه[e] تمرّغ فى التراب وسجد وقام من بُعْد حتّى ياذن له بالتقرّب واذا *حزبهم حَزْب[f] عظيم أُخرج فيه خاقان[g] فلا يراه احد من الاتراك ومَنْ يصاقبهم من اصناف الكفر الّا انصرف ولم يقاتله تعظيمًا[h] له واذا مات ودُفن لم يمرّ بقبره احد الّا ترجّل وسجد ولا يركب *ما لم يَغِبْ عن[i] قبره، ويبلغ[k] من طاعتهم لملكهم انّ احدهم ربّما يجب عليه القتل ويكون من كبرآئهم فلا يحبّ الملك ان يقتله ظاهرًا فيامره ان يقتل نفسه فينصرف الى منزله ويقتل نفسه، والخاقانيّة فى قوم معروفين ليس لهم مملكة ويسار[l] فاذا انتهت الرياسة الى احدهم عقدوا له ولم ينظروا الى ما عليه حاله ولقد اخبرنى من أثِق به انّه رأى فى بعض اسواقهم شابًّا يبيع الخبز[m] كانوا يقولون انّ خاقانهم اذا مات فليس احد احقّ منه بالخاقانيّة الّا انّه كان مسلمًا ولا تعقد الخاقانيّة[n] الّا لمن يدين باليهوديّة والسرير والقبّة الذهب الّتى لهم لا

---

a) A. لهم, D. فيهم. b) B. et C. فخنقوه. c) B. تستمد. Deinde C. هذا. d) C. لاهل. e) C. addit الملك. f) C. دهمهم خطب; cf. Dimaschkí, p. ٢٦٣. g) C. الخاقان. h) C. addit واجلالا. i) C. حتى يغيب عنه ut Dimaschkí l. l. k) C. وبلغ. l) C. ولا يد, D. ولا يسار. m) In A. et B. deëst. E. چيزى. n) C. بالخاقانية.

تضرب الّا للخاقان ومضاربه اذا برّزوا[a] فوق مضارب الملك ومسكنه فى *البلد ارفع من منزل[b] مسكن الملك ۞

وبُرْطاس اسم للناحية وهم اصحاب بيوت خشب وهم مفترشون ۞ وبَسْجِرْت[c] هم صنفان صنف فى آخر الغُزِّيَّة على ظهر بلغار ويقال انّ مبلغهم نحو الفى رجل ممتنعون فى مشاجر[d] لا يقدر عليهم[e] وهم فى طاعة بلغار وبسجرت *آخر هم[f] متاخمون لبجناك وهم وبجناك اتراك، وهم متاخمون للروم، ولسان بلغار مثل لسان الخزر ولبرطاس لسان آخر وكذلك لسان الروس غير لسان الخزر وبرطاس۞ وبُلْغار اسم المدينة وهم مسلمون وفيها[g] مسجد جامع وبقرب[h] مدينة اخرى تسمّى[i] سُوار فيها ايضًا مسجد جامع واخبرنى[k] من كان يخطب بها انّ مقدار عدد الناس بهاتين المدينتين نحو عشرة آلاف رجل ولهم ابنية خشب يأوونها[l] فى الشتآء وفى الصيف يفترشون فى الخركاهات واخبرنى الخاطب بها انّ الليل عندهم[m] لا يتهيّأ ان يسير فيه الانسان اكثر من فرسخ[n] فى الصيف وفى الشتآء يقصر النهار ويطول الليل حتّى يكون نهار الشتآء مثل ليالى الصيف ۞

والرُّوس هم ثلاثة اصناف فصنف هم اقرب[o] الى بلغار وملكهم يقيم بمدينة

---

a) Vocalis ex A. *b*) C. البلدان فوق; D. البلدان ارفع من; E. وكوشك وآن ملك از خاقان از جميع ابنيه بلندتر بود et additur apud Melgunof: از ملك خاقان كمتر باشد. *c*) A. et B. ولسجرت (B. semel ولشجرت), C. وبسجرد. E. nomen omisit. Jacut, I, p. ٧٢٣, 5 بَشْجِرْد, sed in voce باشْغِرْد et باشجرد (I, p. ٤٩٨ seqq.). *d*) A. دساحر, B. مساحر. *e*) C. على موضعهم. *f*) C. اخرون. *g*) D. et Jacut, qui haec male sub برطاس collocavit I, p. ٥٩٧, ولهم. *h*) Jacut وبالقرب منها, D. وبقربهم. *i*) In A. et B. desideratur. Deinde B. et D. سواد, Jacut سوارا, E. سو. Sed Melgunof et Mokaddasí ut recepi. Vid. Frähn, *Drei Münzen der Wolga-Bulgharen*, p. 35 (Mém. de l'Acad. Imp. d. Sc. de St. Pétersbourg, VIme Série, t. I). *k*) A. واخبر لى. *l*) Jacut يأوون اليها. *m*) C. يقصر عندهم حتى. *n*) C. نصف فرسخ, D. فرسخين. *o*) Jacut, IV, p. ٣١٨, 20 قريب et mox ut C. مقيم pro يقيم.

تسمّى كُوِيَابَة وهى اكبر من بلغار وصنف ابعد[a] منهم يسمّون الصّلاوية[b] وصنف يسمّون الأرْثانيّة[c] وملكهم مقيم بأرْثا[d] *والناس يبلغون فى التجارة الى كويابة فامّا ارثا فانّه لا يذكر انّ احدًا دخلها من الغرباء[e] لانّهم يقتلون كلّ من *وطئ ارضهم[f] من الغرباء وانّما ينحدرون فى الماء يتجرون[g] فلا يخبرون[h] بشىء من امورهم ومتاجرهم ولا يتركون احدًا يصحبهم ولا يدخل بلادهم. ويحمل من ارثا[i] *السمّور الاسود والرصاص[k]* والروس قوم يحرقون انفسهم[l] اذا ماتوا وتحرق مع مياسيرهم الجوارى بطيبة من انفسهنّ *وبعضهم يحلق اللحى وبعضهم يفتله مثل الذوائب[m] ولباسهم القراطق القصار ولباس الخزر وبلغار وبجناك القراطق التامّة *وهاؤلآء الروس يتجرون الى الخزر ويتجرون الى الروم وبلغار الاعظم وهم[n] متاخمون للروم فى شماليّها وهم عدد كثير يبلغ من قوّتهم انّهم ضربوا خراجًا على ما يلى بلادهم من الروم وبلغار الداخل هم نصارى ✡

المسافات بين بلاد بحر الخزر ونواحيها من آبَسْكُون[o] الى بلاد الخزر عن اليمين نحو ٣٠٠ فرسخ ومن آبسكون عن يسار السائر الى الخزر نحو ٣٠٠ فرسخ ومن آبسكون الى دِهِسْتان[p] ٦ مراحل ويقطع هذا البحر اذا

a) D. et Jacut اعلى. b) Secundum C., D., Melgunof et Jacut. E. جلابه; A. et B. العلاربه. Cf. Frähn, *Ibn Foszlan*, p. 170, 259 seq. c) A. et B. sine punctis; C. الاوثانيه, D. الارثابيه, E. اوثانى, Jacut الاربانية. Cf. Frähn l. l. p. 162 seqq. d) A. بادارقا, B. بابارقا, C. et Jacut باربا. Mox A. ادارقا, B. ادارقا, C. اربا. e) A. et B. الغزّ. E. pro his: بازركانان تا آنجا نرسند. f) E. بارثا رود. g) B. فيتجرون, Jacut للتجارة. وزيادة تا بلغار روند. h) C. et Jacut addunt احدًا. i) A. ارفا, B. ارقا. k) E. سمور سياه نيكو. (وعادت دارند كه) ريش يكديكررا وارزيز وقلعى. l) C. اجسامهم. m) E. بتراشند. n) C. pro his: ما بين الخزر وبلغار الاعظم (l. وارثا) واربا. Pro وهم coll. D. et E. mallem هم. o) Melgunof ابشكون. p) Vid. supra p. ٢١٩ ann. b. Deinde C., D. et Vers. Pers. apud Melgunof addunt نحو. Supra distantia 50 Par. data est; cf. Jaubert ad Edrîsî, II, p. 337 coll. 334.

طابت الريح عرضًا من طبرستان الى باب الابواب *فى اسبوع[a] وامّا من آبسكون الى بلاد الخزر فانّه زائد على العرض لانّه مزوّى، ومن اتِل الى سمندر ٨ ايّام ومن سمندر الى باب الابواب ۴ ايّام وبين مملكة السرير وباب الابواب ٣[b] ايّام ومن اتل الى اوّل حدّ برطاس مسيرة ٢٠ يومًا ومن اوّل برطاس الى آخره نحو ١٥ يومًا ومن برطاس الى بجناك نحو ١٠ مراحل ومن اتل الى بجناك مسيرة شهر ومن اتل الى بلغار على طريق المفازة نحو شهر وفى الماء نحو شهرين فى الصعود وفى الحدور نحو ٢٠ يومًا ومن بلغار الى اوّل حدّ الروم نحو ١٠ مراحل ومن بلغار الى كويابة نحو ٢٠ مرحلة ومن بجناك الى بسجرت[c] الداخل ١٠ ايّام ومن بسجرت الداخل الى بلغار ٢٥ مرحلة ۞

مفازة خراسان

وامّا مفازة خراسان وفارس فانّ الّذى يحيط بها من شرقيّها حدود مكران وشىء من حدود سجستان وغربيّها حدود قومس والرىّ وقمّ وقاشان وشماليّها حدود خراسان وشىء من سجستان وجنوبيّها حدود كرمان وفارس وشىء من حدود اصبهان[d] ۞

هذه المفازة من اقلّ مفاوز الاسلام سكّانًا وقرى ومدنًا على قدرها لانّ مفاوز البادية فيها مراعٍ واحياء العرب ومدن وقرى لا يكاد يخلو بنجد وتهامة وسائر الحجاز مكان الّا وهو فى حيّز قبيلة يتردّدون فيها على المراعى وكذلك عامّة اليمن الّا شيئًا بين عمان واليمامة ممّا يلى البحر الى حدود اليمن فانّ ذلك الموضع خالٍ عن ديار العرب وكذلك المفاوز الّتى فى اضعاف كرمان ومكران والسند عامّتها مسكونة بالاخبية والاخصاص وغيرها ومفاوز البربر ايضًا مسكونة باحياء البربر فى مراعيها وليس يستدرك من مفازة فارس وخراسان الّا عَلَمُ الطريق وما يعرض فى اضعاف طرقها[e] من المنازل[f]

a) Ex D. et Edrîsî. A., B., C. et E. om. b) C. اربعة. c) E. cum ش (Melgunof بسخرت). d) Sequitur in A. et B. وهذه صورة مفازة ما بين فارس وخراسان. e) A. et B. sine punctis, C. طرقها. f) C. et D. addunt والبقاع.

ان ليس فيما عدا طرقها[a] كثير عمارة ولا سكّان[b] ، وهذه مفازة من اكثر المفاوز لصوصًا وفُسّادًا وذلك انّها ليست فى حيّز اقليم بعينه فيرعاها اهل ذلك الاقليم بالحفظ لانّه يحيط بهذه المفازة ايدٍ كثيرة من سلاطين شتّى فبعض هذه المفازة من عمل خراسان وقومس وبعضها من عمل سجستان وبعضها من عمل كرمان وفارس واصبهان[c] وقمّ وقاشان والرىّ فاذا افسد القطّاع فى عمل دخلوا عملًا آخر ومع ذلك فهى مفازة يصعب سلوكها بالخيل وانّما يقطع بالابل فامّا * الدوابُّ للاحمال[d] فلا تسلك الّا على طرق[e] معروفة ومياه معلومة ان تجاوزوها[f] فى اعراض هذه المفازة[g] هلكوا ، وللّصوص فى هذه المفازة مأوى يعتصمون به ويأوون اليه ويخفون فيه المال[h] والذخائر يعرف بجبل * كركس كوه وكركس[i] اسم المفازة الّتى تتاخم الرىّ وقمّ الى مسيرة ايّام من شرقىّ هذه المفازة فامّا جبل كركس كوه فهو[k] جبل ليس بالكبير وانّما هو جبل منقطع[l] عن الجبال تحيط به المفازة[m] وبلغنى انّ دور اسفله نحو فرسخين فقط وبهذا الجبل ماء يسمّى آب بنده[n] ووسط هذا الجبل مثل الساحة وفى شعاب هذا الجبل مياه قليلة وهو جبل وعر المسلك الى ذراه[o] فيه معاطف ومسالك وحشة لا يكاد يظهر على من توارى فيه[p] وآب بنده اذا صرت عند هذا الماء كنت كانّك فى حظيرة والجبل محيط بك ، وسياه كوه جبل يمتدّ ويتّصل بجبال الجيل[q] ۞ وفى هذه المفازة قرى ولا اعلم بها مدينة سوى سَنِيج[r] فانّها مدينة من عمل[s] كرمان فى المفازة على طريق

---

a) C. طرفها. *b*) C. ولا كثير سكان بها. *c*) C. وبعضها من عمل اصبهان. *d*) C. دواب الاحمال. *e*) C. طريقة. *f*) C. et D. تجاوزها. *g*) C. البرية. In E. legitur: وبى دليل نتوان دران رفت. *h*) C. et D. الاموال. *i*) A. كركسى كوه وكركسى. *k*) C. et D. فانه. *l*) B. ينقطع. *m*) C. et D. المفاوز. *n*) A. et B. h. l. sine punctis; C. h. l. نبده, infra بيده ut E. et Jacut, IV, p. ٣٩٣, 18; D. h. l. بنده, deinde بيده. *o*) C. دواره. Deinde A. et B. فيها. *p*) A. et B. om. *q*) A., B. et C. الجبل; cf. quoque Jacut, III, p. ٢٠٨, 19. *r*) B. h. l. اسنح, infra سنح; E. h. l. سپنح, infra سنج, C. سننح; Mokaddasí MS. p. 233 سفند, p. 235 اسفند; *Djihán-Numa*, p. ٣٢٧, 5 سپنج, p. ٣٢٨, 6 a f. سبيح. Jacut habet سنيح, sed *Lobbo 'l-lobáb* praescribit سنيج (cf. ann. ad *Merácid*, II, p. ٩١). Vid. supra p. ١٩٣ ann. *c*. *s*) E. et Jacut اعمال.

سجستان وتحيط بها من جميع نواحيها هذه المفازة، وفى المفازة على طريق اصبهان الى نيسابور[a] موضع يعرف بالجَرْمَق[b] وهى ثلاث قرى ۞ وتحيط بهذه المفازة من مشاهير المدن ممّا يلى فارس نائين[c] ويَزْد وعُقْدة وأَرْدِسْتان من اصبهان ومن حدّ كرمان خَبيص وزاور[d] ونَرْماشير ومن حدّ الجبال قمّ وقاشان ودِزَه[e] وكذلك الرىّ والخُوار[f] جميعًا سوادهما يتاخم المفازة وسِمْنان والدّامَغان *من قُومِس[g] ومن خراسان مدن قوهستان[h] وهى الطَّبَسَيْن والطَّبَس[i] وقاين فانّ حدّ سوادها[k] ينتهى الى المفازة ۞

والطرق[l] المعروفة فى هذه المفازة طريق من اصبهان الى الرىّ وهو اقربها وطريق من كرمان الى سجستان وطريق من فارس وكرمان الى خراسان فمنها طريق يزد فى حدّ فارس وطريق شُور[m] وطريق زاور وطريق خبيص من حدود كرمان الى خراسان وطريق يسمّى الطريق الجديد من كرمان الى خراسان فهذه هى الطرق المعروفة لا اعلم بها طريقًا مسلوكًا غير الّذى ذكرناه وهناك طريق قلّ ما تسلك من اصبهان يخرج على[n] قومس لا تسلك الّا عند ضرورة والمسلك فيها على السمت، وسأصف مسافات[o] هذه الطرق وما فيها ان شآء الله ۞ الطريق من الرىّ الى اصبهان من الرىّ الى دِزَه

---

a) B. ut solet نيشاپور, E. نشابور. *b*) A. et B. h. l. بالجردق; vid. Jacut, II, p. ٩۴. *c*) A. باس, E. مايين. In mappa C. نسايين خ مايين. *d*) A. h. l. زور; B., D. et Mokaddasí رور; C. et E. زوزن. Infra E. روان, Epit. Paris. زوز. Edrîsî, I, p. 435 دوار. Mappa C. روزن خ روذان, Mokaddasí p. 235 الرور. *Djihán-Numa*, p. ٣٢٧, 3 a f. راون. Infra A., B. et C. habent ut recepi. — Deinde D. et E. وبرماشير, B. sine punctis. *e*) B. et E. دره, C. درا. *Djihán-Numa*, p. ٣٢٧, 11 دره, Mokaddasí حره. Supra p. ٢٠٩ edidi دزاه; cf. ann. *g*. *f*) E. خار. *g*) A. et B. وقومس. *h*) E. كوهستان. *i*) Secutus sum C.; A. والحمسين, D. et E. om.; Mokaddasí enumerat طبس كرى قاين. *k*) C., D. et E. سوادهما. *l*) A., C. et D. والطريق. *m*) A., B. et C. h. l. سور. *n*) D. الى. *o*) A. et B. مفازات.

مدينة فيها منبر مرحلة * ومن الرىّ اليها عمارة الّا مقدار فرسخين فى وسط الطريق[a] ومن دزه الى دَيْر الجصّ[b] مرحلة وبين دزه ودير الجصّ مفازة محاذية كركس كوه وسياه كوه ودير الجصّ رباط من جصّ وآجُرّ يسكنه بذرقة السلطان وهو[c] منزل للمارّة وليس به[d] زرع ولا شجر وفيه بئر ماء[e] ملح لا يشرب ومآؤهم من المطر فى حوضين خارجين من هذا الدير والمفازة تحيط به من كِلَى الجانبين، ومن دير الجصّ الى كاج[f] ايضًا مفازة[g] وكاج كانت قرية فخربت ولا سكّان بها وانّما هى منزل ومآؤها من الامطار ايضًا فى حياض والآبار بها مالحة، ومن كاج الى قُمّ منزل[h] والطريق فى مفازة الى ان تنتهى على فرسخَيْن من المدينة ثمّ تنتهى الى قرية ثمّ الى المدينة ايضًا مفازة، ومن قُمّ الى *قرية المجوس[i] طريق عامر مرحلة وفى هذه القرية مجوس ومن هذه الى قاشان مرحلتان فى عمارة على جنب من المفازة، ومن قاشان الى حصن يسمّى بدرة[k] مرحلتان والطريق بعضه مفازة تحيط بها العمارة وبدرة حصن لهم زرع وفيها نحو خمسين مسكنًا، ومن بدرة الى رباط ابى[l] علىّ ابن رستم مرحلة كبيرة مفازة تتّصل بمفازة كركس كوه ويسكن هذا الرباط رجّالة على النوب[m] وهو منزل للمارّة وله ماء جارٍ من قرية بالقرب منه[n] الى حوض فى الرباط، ومن هذا الرباط الى دانجى[o] مرحلة ودانجى

a) D. plane contrarium habet: وليس من الرى اليها عمارة غير مقدار فرسخين فى وسط الطريق. Sed cf. *Djihán-Numa* l.l. (ex vers. Norberg, I, p. 426 „cujus duae parasangae desertum"). b) E. دير كَجين; *Djihán-Numa* بركجين. c) A. om. هو. d) A. et B. بها. e) Supplevi ex E. f) A. h. l. كاح, B. كاخ ut Mokaddasí, p. 234. g) Epit. Par., *Djihán-Numa* et Edrîsî, I, p. 441 addunt مرحلة. h) D. مرحلة. i) E. ده كبران (*Djih.-Numa* male قريةٔ كران). k) A., B. et Mokaddasí بدره (B. semel بردة); E. رباط بدره, *Djihán-Numa* بدر. Sed D. ىدره, Edrîsî دُرّه, Epit. Par. ذره. l) E. et *Djih.-Numa* om.; Mokaddasí om. ابى على. m) E. بنوبت راه نگاه دارند. n) A., B. et D. منها. o) A. sine punctis (semel دادخى), D. دالجى, Mokaddasí دآنجى.

قرية كبيرة عامرة ومن دانجى الى اصبهان مرحلة خفيفة والطريق من دانجى الى المدينة عامرة ۞ والطريق من الرىّ الى اصبهان بين سياه كوه وكركس كوه وكركس كوه عن يسار السائر وسياه كوه عن يمينه وسياه كوه ايضًا مأوى للّصوص ليس بها عمارة ومن كركس كوه الى دير الجصّ ۴ فراسخ ومن دير الجصّ الى سياه كوه ٥ فراسخ وبين سياه كوه وكركس كوه ٩[a] فراسخ على دير الجصّ ومن كركس كوه الى دِزَه[b] ٧ فراسخ ۞

واما طريق نائين[c] فانّ من نائين الى مزرعة فى المفازة مرحلة وهذه مزرعة ربّما كان بها نفسان او ثلاثة وتسمّى بُونة[d] وفيها عين ماء يزرع عليه، ومن بونة الى جَرْمَق[e] ۴ مراحل وفى الطريق فى كلّ فرسخين او ثلاثة جنبذة[f] وبركة ماء وجرمق هذه تسمّى سِيدَه[g] وتفسيرها ثلاث قرى اسم احداها بيادق[h] والاخرى جرمق والثالثة[i] ارابة[k] وهى تعدّ من خراسان وفيها[l] نخيل وزروع وعيون ماء ومواش كثيرة وفى القرى الثلاث[m] نحو الف رجل وكلّها فى رأى العين قريبة بعضها من بعض، ومن جرمق الى نوخانى[n] ۴ مراحل فى كلّ ثلاثة او اربعة فراسخ جنبذة[f] وبركة ماء ومن نوخانى الى رباط خَوْران[o] مرحلة ومن الرباط الى قرية تسمّى انشكهان[p] مرحلة خفيفة ومن انشكهان الى الطَّبَسَيْن[q] مرحلة، ومن اراد من نوخانى الى دسكروان مرحلة ومن دسكروان الى بنّ[r] مرحلة كبيرة ومن بنّ الى تُرْشِيز[s] مرحلتان ومن ترشيز

a) B. سبعة. b) D. دره, significans nempe locum qui supra بدرة vocatur. c) A. et B. sine punctis; E. مادين, *Djihân-Numa* مايين. d) A. et B. h. l. بونه, deinde sine punctis. e) A. et B. خرمن. E. خرمه i. e. جرمه ut habet *Djihan-Numa*, p. ٣٢٧, 8 et 21. f) A. حنىده, E. كنبدى. g) Pro سه ده A. et B. سيرده. h) A. et B. دمادق. Vid. Jacut, II, p. ٦۴, 16 ubi بيانق. i) A. et B. والثالث. k) A. et B. ارانه. l) D. et Jacut وبها. m) D. et Jacut الثلاث قرى. n) A. et B. نوحاى, D. نوجاى. Secutus sum E. (*Djihân-Numa* نوحانى). A. bis habet برحاى. o) Sic omnes, excepto D. qui habet خوزان. p) E. مسكهان. q) D. طبس. r) A. نىن, D. بىر. Vid. quoque Edrîsî l. l. s) A. نرسنر, B. دزسمر, D. برسبين; E. بردسبير ut quoque *Djihân-*

الى نيسابور ٥ مراحل۞ وطريق من يَزْد وشُور ونائين تجتمع بكُرِى[a] وهى قرية فيها نحو الف رجل ولها رستاق كبير وبين طَبَس[b] وكرى ٣ فراسخ۞

وامّا طريق شُور فانّ شور اسم مآء مالح فى المفازة وليس باسم قرية ولا مدينة ووراس مفازة شُور قرية تسمّى ببرة[c] وهى قرية صغيرة بها دون عشرة انفس من حدود كرمان، ومنها الى عين مآء تسمّى مغول[d] مرحلة وليس بها بنآء، ومنها الى غَمْر سُرْخ[e] وهو غمر كثير[f] فى وهدة طين احمر وجبله[g] احمر مرحلة، ومنها الى منزل يسمّى جاه بر وهو بئر وقباب[h] مرحلة ليس بها احد، ومنه الى حوض هزار حوض يجتمع[i] فيه مآء المطر مرحلة، ومن حوض هزار الى شُور وهو عين مآء مالح الّا انّه شروب وعليها قباب[k] وليس بها احد، ومن شور الى منزل يسمّى مغول ايضًا عين مآء وقباب مرحلة، ومن مغول الى كُرِى مرحلة كبيرة وعلى ٤ فراسخ من كرى بركة يجتمع[i] فيها مآء السيل، وفى مفازة شور *وبين مآء شور[l] وبين برّ عن يمين الذاهب من خراسان الى كرمان على نحو فرسخين[m] حجارة فى صُوَر الفواكه من[n] اللوز والتفّاح ونحوه وفيها صور تقارب الناس والاشجار وغير ذلك۞

Numa. Vulgo ترشيش s. طرثيث (طريثيث) scribitur. Vid. infra ad caput de Khorasán. In mappa Kiepert Terschiz appellatur. *a*) Sic A., B., D. et E.; Epit. Paris., Edrîsî et Jacut كرين. *b*) B. طبسين. *c*) Secundum D. et Mokaddasí, p. 235. A. et B. sine punctis, E. دره. *d*) Lectio incerta est. Secutus sum B. et D. (A. et Edrîsî, I, p. 438 habent معول hic et infra). Sed E., Mokaddasí et *Djihán-Numa* legunt معزل. *e*) A. et B. عمر وسرح, D. غمر سرج, E. عمر در سرح, Ous. عمرو برسرح, Edrîsî عمو سوح, *Djihán-Numa* منزل عمر دربن منزل مغاكى عظيم است از كل .E. *f*) A. et B. وهو ركس. بن سرح سرخ. *g*) A. sine punctis. *h*) A. et B. وقنات. *i*) B. يجمع. *k*) A. hic et mox deinde قبات et altera manus in B. قنات. *l*) B. om. Fortasse pro وبين legendum est ما بين. Utrum برّ sit eadem statio quae supra جاه بر vocatur, nescio. *m*) E. addit آنجا گويند هست درختستانى; vid. Ous. p. 197 ann. et cf. *Djihán-Numa*, p. ٣٢٧, 5 a f. *n*) A. et B. ومن.

واما طريق زاور فانّ زاور قرية عامرة عليها حصار ولها مآء جارٍ وهى من حدود كرمان فمنها الى مكان يسمّى دركوجرى[a] وفيه مآء عين ضعيف وليس هنالك بنآء مرحلة، ومنه الى شور دوازده[b] مرحلة *وهناك رباط قد خرب[c] وهو شعث فيه نخيل وليس به احد وهو مكان مخوف وقلّ ما يخلو من اللصوص، ومنه الى دير بردان[d] وهناك آبار وهو صحرآء لا بنآء فيه مرحلة[e] ومنه الى منزل فيه حوض يجتمع فيه مآء السيول مرحلة وليس هناك بنآء، ومن هذا الحوض الى نابند[f] وهو رباط فيه مقدار عشرين مسكنًا وفيه مآء عليه رحًى صغيرة[g] ولهم زرع على مآء عين ولهم نخيل وقبل نابند بفرسخين عين مآء وعنده اصيلات[h] نخيل وقباب[i] ليس بها احد وعن يمين نابند على هذا النهر نخيل كثيرة ومراعٍ ليس فيها احد وهو مأوى للّصوص[k] غير انّ اهل نابند يتعاهدون هذه النخيل ويجنونها، وتسير من نابند مرحلتين الى مكان يسمّى بئر شك[l] وما بين كلّ فرسخين او ثلاثة قباب[m] وحياض ليس

---

a) A. sine punctis, D. درکرحری, E. دکوحوی (Ous. دکوخوی), *Djihán-Numa* بد جوی, Mokaddasí درکود. Fortasse legendum est درکوجوی. b) A. شور سور دوازده; B. شوردراورده, D. سوره وارکه (Epit. Par. ستورد وازکه), E. درانرده, *Djihán-Numa* سود دواره, Mokaddasí سور داور. Cf. Ous. p. 198. c) E. ثنا رباط ويران مرحلة (*Djihán-Numa* رباط ويرانه). d) A. et B. در دردان, Mokaddasi دْير دان, E. ديىر برقان ut *Djihán-Numa;* D. ديىر دار, Epit. Par. در بردان. Edrîsî دردان. e) A., B. et D. om. f) A. h. l. فاسد, deinde ناسد; B. h. l. فاند, deinde دانند; D. يابيد (semel بابيد), Epit. Paris. نابند; Mokaddasí بمادند et دمادند; in E. nomen desideratur. Idem, ut videtur, locus in mappis hodiernis appellatur *Nehbendán*. Ejusdem nominis locus jacet in Perside, vid. Abulfeda, p. ٣٣٢ et Jacut, II, p. ۴٨٩, 1. g) A. et B. صغير. h) A. اصلاف s. احلاف, B. اصملات. i) B. وقنات. k) B. اللصوص. l) A. et B. sine punctis. E. habet پرشك (Ous. بيرشك), *Djihán-Numa* ترشك, Epit. Paris. تُرْشَك, Mokaddasí دوشك, D. semel بيرشك. m) B. قنات.

بها احد وبئر شك بئر طيّبة الماء، ومن بئر شك الى خُور[a] مرحلة ليس بها شيء ومن خور الى خَوْسْت[b] مرحلتان ومن خوست[c] الى كرى نحو ٣ مراحل ٭

وامّا طريق خَبِيص فانّ خبيص مدينة على شفير المفازة من جروم كرمان وهى مدينة صغيرة مآؤها جارٍ وبها نخيل كثيرة وهى خصبة رخيصة الاسعار فمنها الى مكان يسمّى الدّرْوازق[d] مرحلة فيه ابنية مدّ البصر منهدّمة[e] وبها تلال تدلّ على انّها كانت ابنية فهدمت وليس بها نهر ولا بئر ولا عين، ومنها الى مكان يسمّى شُورروذ مرحلة وهو وادٍ تجرى فيه سيول الامطار ولا يجرى الّا من مطر ولكنّه يجرى على ارض سبخة فيجرى السيل مالحًا وهذه المفازة مالحة[f] التربة، ومنه الى مكان يسمّى بارسك[g] وهو جبل صغير مرحلة، ومنه الى مكان يسمّى نيمة[h] مرحلة وفيه بئر حفرتها ابنة عمرو بن اللّيث فبلغ الماء نحو عشرين باعًا فخرج[i] مآؤها اسود وبلغنى انّه سقى منها كلب فمات، ومن هذا المكان الى مكان يسمّى الحَوْض وهو حوض يجتمع فيه ماء المطر، ومنه الى راس الماء مرحلتان وفيه عين ماء يجتمع فى حوض ويسقى زراعاتها[k] وهو رباط يكون فيه الواحد والاثنان، ومن راس الماء الى كُوكُور[l] قرية عامرة وهى من حدّ قوهستان مرحلة[m] ومن كوكور الى

---

a) E. جور, *Djihán-Numa* چور; Mokaddasí خر. De oppido hoc et sequenti quae in Kuhistán jacent, vid. infra in capite de Khorasán. b) E. بست ut *Djih.-Num.*, D. درست, Epit. Par. حَرست, Edrîsî حوسب. Ejusdem nominis locus jacet in Tokharistán. c) D. خور, E. جور. d) A. et B. الدروارق, E. دوراق, *Djihán-Numa* دروارق. Alius locus ejusdem nominis memoratur a Jacut. Mokaddasí hunc locum appellat قرية سلم, nescio utrum recte. e) B. مْنْهَدمة. f) A. et B. مالح. g) A. h. l. دارستل, B. دارسل, D. بارسل, E. ut *Djih.Num.* ارسل. Infra A. دارسك, B. et D. ut recepi. h) A. ديمه, D. بيمه (Epit. Par. ديبه), Mokaddasí دمه. Secutus sum B. Apud E. deëst. i) A. وبحوح, B. يخرج. k) A. زرعا بها. l) A. et B. كوكون, Mokaddasí كوبور. Vocales ex Epit. Paris. m) Addidi ex E.

خَوْسْت مرحلتان، وفى مفازة خبيص على فرسخين من راس الماء ممّا[a] يلى خراسان حجارة صغار سود نحو اربعة فراسخ، ومن بارسك الى قبر الحاجّى[b] حصى صغار بعضه فى لون الكافور بياضًا وبعضه اخضر فى لون الزجاج، وليس فى هذه المفازة اذا جزتَ فرسخين من راس الماء الى جبل نحمرده[c] نحو مرحلة نبات[d] ❁

الطريق من يزد الى خراسان من يزد الى آبخيزه[e] مرحلة وآبخيزه ليس بها قرية ولا ساكن وبها عين ماء وحوض يجتمع فيه ماء المطر وليس من يزد الى آبخيزه عمارة، ومن آبخيزه الى خزانة[f] مرحلة ليس بينهما عمارة وخزانة قرية فيها نحو مائتى رجل وبها زرع وضرع وبساتين، ومن خزانة الى تلّ سياه سبيذ[g] مرحلة وليس بها عمارة وهو خان ليس به احد وبه حوضان تحوى مياه الامطار، ومن تلّ سياه سبيذ الى ساغند[h] مرحلة وليس بينهما عمارة وساغند قرية فيها نحو اربع مائة رجل وبها عين ماء وهى عامرة الّا انّ خزانة اعمر منها، ومن ساغند الى *پُشْت باذام[i] مرحلة كبيرة وليس بينهما عمارة وبها خان ومنزل ومآؤها من الآبار، ومن پشت باذام الى رباط محمّد مرحلة خفيفة وليس بينهما عمارة وهو رباط فيه نحو ثلاثين رجلًا ولهم زرع وعيون ماء، ومن رباط محمّد الى الريك[k] مرحلة وهو منزل فيه حوض ماء وخان ليس فيه سكّان والرمل مقداره فرسخان، ومن الرمل الى

a) A. مما. b) Secundum A., B. et Mokaddasí p. 234. D. et E. habent قبر الخارجى (كور خارجى). Pro بارسك Mokaddasí habet باسك. c) Sic A. et B.; D. نحو به. d) In A. et B. lacuna (in B. superest ده). Apud D. نبات ante نحو ponitur. e) Vid. supra p. ١٣. et ann. b. f) A. et B. حرانه (A. semel حرابه, B. semel خرابه); Edrîsî خرانه, E. خوانه (*Djih.-Numa* حوان), Mokaddasí حرانه. Secutus sum D. g) B. سيذ, E. وسفيد et sic *Djih.-Numa.* h) A. sine punctis, B. ساعند, E. سباعيد ut *Djih.-Numa.* Secutus sum D. et Mokaddasí. i) Locus in mappis notatus. Pro پشت A. et B. شت. k) A. الرجل, B. الرديل, E. ريك: *Djih.-Numa* ريك بيابان.

مهلب a مرحلة وهو خان وعين ماء وعنده جبل وليس بينهما عمارة، *ومنه
الى رباط حوران مرحلة b ورباط حوران رباط من جصّ وحجارة يكون فيه
ثلاثة نفر او اربعة يحفظونه وبها عين ماء وليس بها زرع، ومن رباط حوران c
الى زادآخرة d مرحلة وزادآخرة بئر ماء جارٍ e ليس فيه ساكن وليس بينهما
عمارة، ومن زادآخرة الى بستادران f مرحلة وهى قرية فيها نحو ثلاث مائة
رجل وفيها ماء جارٍ من قناة *وزرع وضرع g وكروم وليس بينهما عمارة، ومن
بستادران الى بنّ h مرحلة خفيفة وليس بينهما عمارة وبنّ قرية عامرة فيها نحو
خمس مائة i رجل وفيها ماء جارٍ وزرع وضرع وخصب k*، ومن بنّ الى رادويه l
مرحلة وليس بها عمارة ورادويه منزل فيه بئر ماء وخان ليس فيه ساكن، ومن
رادويه الى ريكن m مرحلة وليس بينهما عمارة وريكن رباط فيه زرع وماء جارٍ
وفيه ثلاثة او اربعة نفر، ومن ريكن الى استلشت n مرحلة وليس بينهما عمارة

---

a) A. sine punctis, B. مهلت, D. et Edrîsî المهلب, Mokaddasí المهلبى. Apud E. deëst. *b*) A. et B. haec om. Pro حوران D. habet خوران, E., Mokaddasí et *Djihán-Numa* كوران. Recepi lectionem A. et B., putans fortasse hanc stationem non differre ab ea quae p. ٢٣١ (cf. ann. *o*) memoratur. *c*) Apud E. et *Djihán-Numa* intermedia statio exstat رباط كره. *d*) A. زاذاخره, B. زاداحره et زاداخره, D. زاد اجره. Jacut, I, p. ٨٩, 22 ازاد خرّه. Secutus sum E. et *Djih.-Numa*, ubi زادآخرت. *e*) Pro جارٍ D. habet وخان, E. كاروانسراى. *f*) A. et B. دسادران, D. دسمادران, E. et *Djihán-Numa* بستاداران. Vid. Edrîsî et Jacut l.l. qui habet بُسْتَاذَرَان. *g*) Jacut زروع وبساتين. *h*) A. sine punctis. E. habet tantum دهى ديگر „pagus alius." (Hinc in *Djihán-Numa* ده ديرل). Sed Jacut habet ابرقوه. *i*) Jacut سبعمائة. *k*) Jacut وهى خصبة جدّا. *l*) A. et B. sine punctis, D. رادونه. Secutus sum E. et Edrîsî (*Djih.-Numa* دادويه). Jacut habet زادويه (I, p. ٨٧, 2). *m*) A. et B. رنكن, D. ريكر. E. et *Djih.-Numa* رباط زنكى, Jacut زيكن. Secutus sum Edrîsî. *n*) A. اسلست, B. اسلمت, D. اسقيست; Jacut اسْتَلْشْت; *Djih.-Numa* آستلشت. Secutus sum E.

واستلشت منزل فيه حوض ماء للمطر وخان وليس فيه سكّان، ومن استلشت الى ترشيز[a] مرحلة وهى حومة بُشْت[b] نيسابور مدينة وفى هذا الطريق على كلّ فرسخين او ثلاثة خان وحوض ماء ۞

وطرق هذه المفازة على الترتيب من اصبهان الى الرىّ طريق ثمّ يليه طريق[c] نائين الى خراسان ويليه طريق يزد الى خراسان ويلى ذلك طريق شور ثمّ طريق زاور[d] ثمّ طريق خبيص ثمّ الطريق الجديد ثمّ طريق سجستان الى كرمان، وامّا الطريق الجديد فان تاخذ من نَرْماشير[e] الى دارُستان مرحلة وهى قرية فيها نخيل وليس وراءها عمارة والى راس الماء[f] مرحلة ومن راس الماء وهو عين ماء يجتمع فى بركة ثمّ[g] يقطع عرض المفازة من راس الماء الى قرية سلم[h] ۴ مراحل مفازة كلّها ويقال انّ قرية سلم من كرمان ومن قرية سلم الى هراة *١٠ ايّام[i]، وان شئتَ اخذتَ من نَرْماشير الى سَنيج ٥ مراحل ومن سنيج الى قرية سلم نحو ٥ ايّام فى *عيون ماء[k] قليلة ۞ وامّا طريق سجستان فانّ المدخل اليها من نرماشير الى سنيج ٥ ايّام فى حدّ كرمان ومن سنيج الى سجستان ٧ مراحل وقد بيّنّاها فى صفة سجستان وكرمان ۞

---

a) A. et B. h. l. دوسنر (ut cogitare quis posset de نوشير s. نوشهر, vid. *Merácid*, III, p. ٢٣٨), D. درسىر, E. et *Djihán-Numa* برسير. Jacut l.l. habet ترشيش. Vid. supra p. ٢٣١ *s* et infra in capite de Khorásán. *b*) A. sine punctis, B. نسب, D. بسب, E. بست. Vid. Jacut, I, p. ٩٢٨, 18 seqq. — Deinde B. ut solet نيشاپور, E. نشابور. *c*) A. من. Deinde A. et B. داى, E. مادس. *d*) A. et B. رادن, D. رود, E. om. *e*) A. نرماشير, B. بَرْماشير, D. et E. برماسير. *f*) E. سرآب. *g*) A. et B. om. *h*) Mokaddasí de hoc pago prorsus eadem dicit quae supra p. ٢٣۴ (cf. ann. *d*) de الدروازق leguntur. *i*) E. دو روز. *k*) A. ماء عيون.

سجستان [a]

وامّا سجستان وما يتّصل بها ممّا قد جُمع اليها فى الصورة فانّ الّذى يحيط بها ممّا يلى المشرق *مفازة [b] بين مكران وارض السند [c] وشىء من عمل الملتان وممّا يلى المغرب خراسان وشىء من عمل الهند وممّا يلى الشمال ارض الهند وممّا يلى الجنوب المفازة الّتى بين سجستان وفارس وكرمان وفيما يلى خراسان والغُور والهند تقويس [d] ۞

وامّا مدنها وما يقع فى اضعافها ممّا يحتاج الى معرفته فلها من المدن زَرَنْج وكِشّ [e] ونه والطاق والقَرْنين [f] وخُوَاش وقَرَه [g] وجَزَه وبُسْت [h] ورُوذان وسَرْوَان

---

a) E. سبستان. Editum est hoc caput ex B. cum versione a W. Anderson in *Journal of the Asiatic Society of Bengal*, 1852, p. 365—383. b) C. كرمان ومفازة. c) Abulfeda, p. ۳۴۰: وهى مفازة هى بين سجستان وبين مكران وهى مفازة بين كرمان وارض D.; المفازة التى تفصل ما بين مكران والسند ايضا السند وبين سجستان. d) In A. et B. vocabulum corruptum est. Hic sequitur وهذه صورة سجستان. e) Beládsorí habet كشّ et sic Mokaddasí. D. hoc et nomen sequens in unum conjungens habet كسرويه. Eodem modo E. habet كروانه, sed prior pars non ex كس corruptum est, sed hoc modo. Mokaddasí urbes hujus regionis enumerans habet p. 27 كراود كرونه et p. 145 كرواد كروانه. Scribendum est, ni fallor, كرواذكن ونه, nempe كرواذكن est alia forma nominis كرواتكن s. جرواتكن (vid. Jacut, II, p. ٩۴, 17). P. 149 primum describitur كرواذكر *sic*, deinde نه. f) B. hic et infra الغزنين. g) B. وقره. Deinde B. وخره, A. et D. sine punctis. Anderson, p. 374 habet *Churruh*, addens „was seen and determined by Colonel Sanders." Vid. autem Jacut sub جزه et كزه. h) A. et B. ودشت et deinde وزردان.

وصالِقان [a] وبَغْنِين [b] ودَرْغَش [c] وتَلّ [d] وبِشْلَنْك [e] وبَنْجَوای [f] وكُهك وغَزْنَة والقصر وسيوى [g] واسفنجاى [h] وجامان [i] ☆ ومدينتها العظمى تسمّى زَرَنْج ولها مدينة وربض وعلى المدينة حصن وخندق وعلى الربض ايضًا سور [k] والمآء الّذى فى

---

a) D. والزالقان (cf. Jacut sub زالق); E. طالعان. Quoque scribitur جالقان. *b*) A. et B. semper بعس; C. بعىں et بغنين, in mappa ىمعىس; D. h. l. بغبيين, deinde بغنين; E. h. l. indistincte بعمر, infra تغبى, in itinerario secundum Ous. نعس. (Cf. Sprenger, p. 47). Edrîsî, I, p. 442, 449 et 450 habet باغنين, sed p. 457 بغنين. Cf. Anderson, p. 373 *Bugnee.* *c*) A. et B. درعس, D. درعش; E. درغش et sic *Djihán-Numa*, p. ٢٥٠, 1, Abulfeda, p. ٣٤٤ seq. et Edrîsî. C. habet h. l. درغور, ut Jacut, II, p. ٥٤١, 4 et in v., sed infra rectius درغوز. Anderson l. l. p. 374: „Durghosh — a large place some thirty or forty miles East of the river." *d*) A. et B. sine punctis; cf. Jacut l.l.; D. semper درنك, Edrîsî, I, p. 442, 450 et 457 درتل, sed p. 444 بك. Legendum igitur apud D. videtur درتل. Anderson audacter substituit *Gurmabuk* legens بك pro تل. *e*) A. وبسلمل, B. وسلمل (semel A. بنشك, B. بشتك); C. فسلك et تسلك, in mappa بمسلك cum var. l. فلك; D. بسند, deinde بيسند; E. سلتك; Edrisî بشنك. Cf. Anderson l. l. p. 373 et Vullers in v. *f*) A. et B. h. l. وسحواى. Anderson p. 373: „a celebrated village or rather town ten miles West of Kundahar." C. infra in marg. nomen explicat per پنج واى (quinque putei). *g*) A. et B. h. l. وسوهى; E. سوى; C. سنوى et سردن, in mappa دسوى. *h*) A. et B. h. l. اسعحاى, deinde سعحان et سفتجان; C. سعحاى et اسعحاق (et sic in mappa); D. اسعحاى et اسفيخاى; E. اسعحاى et سفنجاى; *Djihán-Numa*, p. ٢٥٠, 18 سفنكاى; Edrîsî اسفنجاى; Mokaddasí MS. p. 145 سعحاوى. Anderson proponit: „*Espangulee* — a well known village." *i*) Secundum E. et mappam C.; D. ماهكان, ceteri omnes h. l. حامان. Vid. infra ad itinerarium. *k*) C. حصن.

الخندق ينبع من مكانه ويقع فيه[a] ايضًا فضل من المياه ولها خمسة ابواب احدها الباب[b] الجديد والآخر الباب العتيق وكلاهما يخرج منهما الى فارس وبينهما قريب والباب الثالث باب كُرْكُويَه[c] يخرج منه الى خراسان والباب الرابع باب نيشَك[d] يخرج منه الى بُسْت[e] والباب الخامس يعرف بباب الطعام يخرج منه الى الرساتيق واعمر هذه الابواب باب الطعام وهذه الابواب كلّها حديد[f]، وللربض ثلاثة عشر بابًا فمنها باب مينا[g] ياخذ الى فارس ثمّ يليه باب جرجان ثمّ يليه باب شيرك[h] ثمّ يليه باب شتاراق[i] ثمّ يليه باب شُعَيب[k] ثمّ يليه باب نوخيك[l] ثمّ يليه باب الكان[m] ثمّ يليه باب نيشك ثمّ يليه باب كركويه ثمّ يليه باب استريس[n] ثمّ يليه باب غنجرة[o] ثمّ يليه باب بارستان[p]

---

a) A. منه. b) C. باب hic et mox deinde. E. legens الحديد habet آهنين et sic *Djihán-Numa*, p. ٢٤٩, 5. c) A. et B. sine punctis; C., E. et *Djihán-Numa* كركونه. Vid. Jacut in v. d) A. et B. sine punctis. Vid. Jacut in v. et sub ارك. Anderson, p. 373 scribens *Meeshuk:* „This is the gate leading to the districts watered by the river of this name. Edward Conolly calls the river Chabulk; but adds, that it rises at a spring called Meeshuk." e) A. بشت, B. شت. f) E. همه درهارا از آهن ومس وضبوط كرده اند. g) A. h. l. سمنا, B. شتا, *Djihán-Numa* ميتا. h) A. et B. سمرل, D. سيرك, Ous. p. 203 اشيرك, *Djihán-Numa* اشرك. Anderson *Sarol.* Apud C. et E. nomina harum portarum desiderantur. Fortasse legendum est سيزك = سيزج (Jacut, III, p. ٢١٥, 19). i) A. et B. شتاراو, D. ساراق; Ous. شان, *Djihán-Numa* درسارا. k) A. et B. سعنب. l) A. دوحيك, B. sine punctis, D. نوخيرك, Ous. خويك et sic *Djih.-Numa;* Anderson *Lookheek.* m) Sic A. et D.; B. الكاز, Ous. الكار. n) A. اسردس, B. اسرلس, D. ut recepi; Ous. ابريس, *Djih.-Numa* اسپر. Anderson *Esrees*, quae lectio fortasse praeferenda est. o) A., B. et D. عنجره, Ous. عنجوه, *Djih.-Numa* غنجوه. p) A. بارسمان, B. بارسان, D. فارستان, Ous. et *Djih.-Numa* رستان. Hanc fortasse portam vult Jacut sub نشنان (IV, p. ٧٨٣).

ثمّ يليه باب روذكران[a] وابنيتها كلّها طين آزاج معقودة لانّ الخشب بها يتسوّس *ولا يثبت[b] ومسجد الجامع فى المدينة دون الربض اذا دخلتَ من باب فارس ودار الامارة فى الربض بين باب الطعام وبين باب فارس خارج المدينة والحبس فى المدينة عند مسجد الجامع وهناك ايضًا دار امارة على ظهر مسجد الجامع وعند الحبس ولكنّها نُقلت الى الربض، وهناك بين باب الطعام وبين باب فارس قصر ليعقوب بن اللّيث وقصر لعمرو بن الليث ودار الامارة فى دار يعقوب بن الليث وداخل المدينة بين باب كركويه وباب نيشك بنيّة[c] عظيمة تسمّى أَرْك[d] كانت خزانة بناها عمرو بن الليث، واسواق المدينة الداخلة حوالى مسجد الجامع وهى اسواق على غاية العمارة واسواق الربض اسواق عامرة ايضًا[e] منها سوق يسمّى سوق عمرو بناه عمرو بن الليث وَقَفه على مسجد الجامع والبيمارستان والمسجد الحرام وغلّة[f] هذه السوق فى كلّ يوم نحو الف درهم، وفى المدينة الداخلة انهار منها نهر يدخل من الباب[g] العتيق والثانى من الباب الجديد والثالث يدخل من باب الطعام ومقدار هذه الانهار[h] اذا جمعت ما يدير الرحى، وعند مسجد الجامع حوضان عظيمان يدخلهما الماء الجارى[i] ويخرج ينصرف[k] فى بيوت الناس وسراديبهم ومعظم دور المدينة والربض فيها ماء جارٍ وبساتين وفى ربضها انهار تاخذ منها هذه الانهار الّتى تدخل المدينة والسوق ممتدّة من باب فارس من المدينة الى باب مينا[l] متّصل نحو نصف فرسخ ۞

وارضها[m] سبخة ورمال وهى حارّة بها نخيل ولا يقع بها ثلوج[n] وهى[o] ارض

---

a) Secutus sum D.; A. et B. رودكران, Ous. زنكيان, *Djih.-Num.* زنكبار.
b) A. et B. om. Vid. quoque Abulfeda, p. ٣٤٣ ubi يسوس ut apud D. c) D. ابنية et sic Jacut, I, p. ٢١٠, 10. d) A. et B. ارل; vid. Jacut l. l. Cf. quoque Ibno 'l-Athír, VIII, p. ٥٢, 3 a f. e) Ex D. addidi ايضا. f) Abulfeda, p. ٣٤٣ اجرة, E. حاصل. g) C. hic et mox deinde بباب. h) C. addit الثلاثة. i) B. الحار. k) C. et D. ويتفرق. l) C. addit للربض. m) Jacut, III, p. ٤١ ارض سجستان. n) C. et Jacut الثلج. o) A. addit فى. Videtur igitur pronomen in ارضها referre ad زرنج.

سهلة لا يرى فيها[a] جبل واقرب جبالها بناحية فَرَه وتشتدّ رياحهم وتدوم حتّى انّهم قد نصبوا عليها *طواحين يديرها الهوآء[b] وينتقل رمالهم من مكان الى مكان فلولا[c] انّهم يحتالون فيها لطمّت[d] على القرى والمدن بها، وبلغنى انّهم اذا احبّوا نقل الرمل من مكان الى مكان من غير ان يقع على الاراضى[e] الّتى الى جانب الرمل جمعوا حول الرمل مثل الحائط من خشب[f] وشوك وغيرهما بقدر ما يعلو على ذلك الرمل وفتحوا[g] فى اسفله بابًا فتدخله الريح وتُطَيِّر *اعلى الرمل[h] مثل الزَّوْبَعة فيرتفع *حتّى يقع[i] على مدّ البصر حيث لا يضرّهم ۞ ويقال انّ المدينة القديمة فى ايّام العجم[k] كانت فيما بين كرمان وسجستان اذا جزتَ دارك[l] بحذآء راسك[m] عن يسار الذاهب من سجستان الى كرمان على ثلاث مراحل[n] وابنيتها وبعض بيوتها قائمة الى هذه الغاية واسم هذه المدينة رامَ شَهْرِسْتَان ويقال انّ نهر سجستان كان يجرى عليها فانقطع *فانقلع عليها[o] بثق كان سُكر من هِنْدَمَنْد[p] وانخفض المآء عنه[q] ومال فتعطّلت وتحوّل الناس عنها وبنوا زَرَنْج ۞

وامّا انهارها فانّ اعظم نهر لها هندمند ويخرج من ظهر بلد الغُور حتّى يخرج على حدّ رُخّج وبلدى الدّاوَر ثمّ يجرى على بُسْت حتّى ينتهى

---

a) A. et B. منها; vid. quoque Abulfeda, p. ٣٤٠. b) C. ارحية تديرها الرياح; Jacut ارحاء يسيرونها بها; D. ارحية تطحن بالريح; Abulfeda ارحية تدور بها. Deinde B., C., Abulfeda et Jacut وتنقل s. وينقل. c) C. et Jacut ولولا. d) C. et Jacut لطمست. Deinde C. et D. om. على. e) C., D. et Jacut الارض. f) C. et Jacut حطب. A. et B. inverso ordine شوك وخشب. g) Secundum D. et Jacut; A. et C. ودفتحوا, B. ويفتحون. h) C. الرمل اعلاه; D. الرمل على اعلاه; Jacut الرمال الى اعلاه. i) Jacut فيقع, C. ويقع. k) C. et Jacut (II, p. ٧٣٧, 9) add. الاول. l) A. et B. ارل, C. om. m) A. et B. راسكر. n) Jacut addit من زرنج. o) C., D. et Jacut om. Deinde C. et D. ببثق. p) C. الهيد ميد. E. semper هيرمند. Mokaddasí هيرميد. q) Jacut عنها.

الى سجستان ثمّ يقع فى بحيرة زرّه[a] وزره هذه بحيرة يتّسع الماء[b] فيها وينقص على قدر زيادة الماء ونقصانه وطولها نحو ثلاثين فرسخا من ناحية كُرين[c] على طريق قُهِسْتان[d] الى قنطرة كرمان[e] على طريق فارس وعرضها مقدار مرحلة وهى عذبة الماء ويرتفع منها سمك كثيرة[f] وقصب وحواليها كلّها قرى الّا الوجه الّذى يلى المفازة[g]، ونهر هندمند هو نهر واحد[h] من بُسْت الى ان ينتهى الى مرحلة من سجستان ويتشعّب منه[i] مقاسم الماء فاوّل نهر ينبثق[k] منه نهر الطعام فياخذ[l] على الرساتيق حتّى ينتهى الى حدّ نيشَك ثمّ ياخذ منه نهر باشترود[m] فيسقى رساتيق كثيرة ثمّ ياخذ منه نهر يسمّى سَنارُود[n] فيجرى على فرسخ من سجستان وهو النهر الّذى يجرى فيه السفن من بست[o] الى سجستان اذا امتدّ الماء ولا يجرى اليهم السفن الّا فى زيادة الماء[p] وانهار مدينة سجستان كلّها من سناروذ، ثمّ ينحدر فياخذ منه نهر شعبة[q] فيسقى مقدار ثلاثين قرية ثمّ ياخذ منه نهر يسمّى مِيلَى[r]

---

a) C. addit الفاضل, Jacut, IV, p. ٩٩٣, 3 منه الفاضل. b) A. et B. المياه; vid. quoque Jacut, I, p. ٥١٤ et Abulfeda, p. ٤٤. c) A. et B. كردن, C. كودن, D. كوثى, E. كوده, sed Ous. كويد. Supra p. ٢٣٢ كُرِى (cf. ann. a). d) Jacut قوهستان, C. et E. كوهستان. e) C. كردهان, Jacut l. l. p. ٥١٥ كَريهان; vid. infra in itinerario. f) C., Abulf. et Jacut كثير. g) Jacut addit: فليس فيه شىء et E. كه عمارتى نيست. h) A. et B. اخذ. i) A. et B. منها. Jacut, IV, p. ٩٩٣, 3 تشعب منه. Hic ut supra سجستان est pro زرنج quod ei substituit Mokaddasí MS. p. 159. k) C., D. et Jacut ينشق. l) A. et B. ماخذ. m) A. et B. باشترود, Mokaddasí داشترود; E. بيسرود (Ous. سبيرود), D. باسنرود; Abulfeda, p. ٥٩ باسيرود; C. راسين sic. *Djihán-Numa*, p. ٢٥١, 7 باسروز. Cf. Beládsorí, p. ٣١٣ ناشروذ et hinc Jacut in v. n) A. et B. سارود, C. سيارود, E. سيّاره رود. *Djihán-Numa* ستاروز. Vid. Jacut, III, p. ١٥٤. o) B. نشت. p) A. et B. om. q) A. et B. sine punctis. r) Sic D.; C. et E. ميلا; *Djihán-Numa* ملا; A. et B. ممان.

فيسقى رساتيق كثيرة ثمّ ياخذ منه زالق[a] فيسقى رساتيق كثيرة وما يبقى من هذا النهر يجرى فى نهر يسمّى كزك[b] وقد سُكّر هناك سكر يمنع الماء ان يجرى الى بحيرة زَرَه حتّى يجىء المدّ فاذا جآءت ايّام المدّ زال السكر ووقع فضل هذا النهر الى بحيرة زره، وعلى نهر[c] هندمند على باب بُسْت جسر من السفن كما يكون على انهار العراق ۞ ويقع فى بحيرة زره الفاضل[d] من وادى فَرَه وغيره من تلك النواحى، ومن انهار سجستان نهر فره يخرج من قرب[e] الغُور حتّى يسقى تلك النواحى ويقع فضلته[f] فى بحيرة زره۞ ونهر نِيشَك يخرج من قرب الغُور فيسقى تلك النواحى وقلّ ما يفضل منه لبحيرة زره۞ وسجستان خصبة كثيرة الطعام والتمور[g] والاعناب واهلها[h] ظاهرو اليسار ويرتفع من مفازة سجستان فيما بينها وبين مكران غلّة عظيمة من الحلتيت[i] حتّى انّه قد غلب على طعامهم[k] ويجعلونه فى عامّة اطعمتهم۞

وبالس[l] اسم الناحية ومدينتها سِيوى[m] غير انّ الوالى مقيم بالقصر، واسفنجاى اكبر من القصر۞ ورُخّج[n] اسم الاقليم ومدينتها بَنْجَوَاى ولها من المدن كُهَك، ورخّج اقليم بين بلدى الدّاوَر وبين بالس وعامّتها صواف يرتفع لبيت المال منها مال عظيم ويتّسع اهل تلك النواحى بغلّاتها وهى على غاية الخصب والسعة۞ وبلاد الداور اقليم خصب وهو ثغر للغُور وبغنين

---

a) Conjectura scripsi. A. et B. رالو, ceteri om. *b*) Sic edidi secundum Jacut, IV, p. ۲۷۲, 1 et ۹۹۳, 6. A. et B. دسول, C. ببرك quod recepit Juynboll in *Merácid*, III, p. ۳۲۴, 3 a f. ubi Codd. كرك. Beládsorí l. l. نسوق, Mokaddasi دسود, Jacut, III, p. ۱۵۴ ult. كركر. *c*) A. et B. سكر. Vid. Abulfeda l. l. *d*) E. (Ous. p. 207) رود عامل. Pro من وادى C. habet ووادى. *e*) C. نحو. *f*) C. فاضله, D. فضله. *g*) A. sine punctis, B. والثمور, C. et Abulfeda, p. ۳۴۱ والتمر. Cf. Jacut, III, p. ۴۲, 6. *h*) A. et B. اهلها. *i*) A. الحليتت. *k*) C. addit وعلوفتهم. *l*) D. وباليش, Edrísi, I, p. 444 باليس. *Djihán-Numa*, p. ۲۵۱, 9 a f. بالينه. Cf. supra p. ۱۷۹ *c*. Hodie *Siwistán* appellatur. *m*) A. et B. سرى, C. سنوى. *n*) A. et B. وزرنج.

وخَلَّج وبِشْلَنْك وخاش[a] وليس عليها سور ولها[b] قلعة، وبلد الداور اسم الاقليم ومدينتها تلّ ولها من المدن دَرْغش وهما[c] على مجرى هندمند على الشطّ غير انّ بغنين وخلج وكابل والغور وهذه النواحى بعض هاؤلآء قد اسلموا وبعضهم مسالمون وهى من الصرود، والخلج صنف من الاتراك وقعوا فى قديم الأيّام الى الارض الّتى بين الهند ونواحى سجستان فى ظهر الغور وهم اصحاب نعم على خلق الاتراك وزيّهم ولسانهم[d] ☆ وامّا بُسْت فانّها مدينة ليس فى اعمال سجستان بعد زرنج اكبر منها الّا انّها وبيّة وزيّهم زىّ العراق[e] يرجعون الى مروّة ويسار وبها متاجر الى بلد الهند والسند وبها نخيل واعناب وهى خصبة جدًّا ☆ وامّا القَرْنين[f] فانّها مدينة صغيرة لها قرى ورساتيق وهى على مرحلة من سجستان عن[g] يسار الذاهب الى بست على فرسخين من سروزن[h] منها الصّفّارون الّذين تغلّبوا على فارس وكرمان وخراسان وسجستان وكانوا اربعة اخوة يعقوب وعمرو وطاهر وعلىّ بنو اللَّيْث فامّا طاهر فانّه قُتل بباب بُسْت وامّا يعقوب فانّه مات بجُنْدَيْسَابُور[i] بعد رجوعه من بغداد وقبره هناك وامّا عمرو بن الليث فانّه قُتل ببغداد وقبره هناك[k] وامّا علىّ

---

a) A., B., C. et E. وحادمن (e qua lectione ortum videtur nomen مابين apud *Djihân-Numa*, p. ٢٥٠, 18). Secutus sum Edrîsî, I, p. 450 (ubi حاش) et p. 457 coll. D. qui habet خواش. Vix autem opus est dicere hoc nomine hic non intelligi urbem خواش de qua mox fiet mentio, sed urbs provinciae Kabul, caput regionis خواست. Vid. Jacut, IV, p. ٣٢٠ ult., sed praesertim *Djih.-Numa*, p. ٢٣٧, ٢٣٨. b) B. وبها, c) A. et B. وهنا. d) Jacut, IV, p. ٣٢٠, 18 ولباسهم. e) C. كالعراق, D. اهل العراق. Deinde C. et D. ويرجعون. f) A., B. et E. الغزنين. Vid. Jacut, IV, p. ٧٣, 20 seqq. g) C. et D. على. Pro seq. على Jacut habet عن. h) A. et B. h. l. سرودن; C. et Jacut سرور, ut quoque infra in itinerario habent C. et E. et ut recepit Sprenger p. 46. Anderson proponit legere Tschagnasur. Mokaddasí, p. 166 habet quoque ut recepi. i) E. بحدود نشابور. Jacut l. l. p. ٧٤, 3 addit بعد ان ملك اكثر بلاد العاجم. k) Jacut واما عمرو

ابن الليث فكان استامن الى رافع بجُرْجان ومات بدِهِسْتان وقبره هناك، ويعقوب كان اكبرهم وكان[a] غلامًا لبعض الصفّارين[b] وامّا عمرو فانّه كان مكاريًا وبلغنى[c] انّه كان فى بعض[d] ايّامه بنّاءا وكان علىّ بن الليث اصغرهم سنًّا، وكان السبب فى خروجهم وارتفاع امرهم انّ خالًا لهم يسمّى كثير بن رقّاق[e] كان قد تجمّع[f] اليه جمع فى وجوه الخوارج فحوصر فى قلعة * تسمّى قفيل[g] وتخلّص هاؤلاء ووقعوا[h] الى ارض بُسْت وكان بتلك الناحية رجل عنده جمع كثير يظهرون الحسبة * فى الغزو[i] وقتال الخوارج يسمّى درهم بن نصر[k] فصار هاؤلاء الاخوة فى جملة اصحابه فقصدوا سجستان والوالى بها * ابراهيم ابن الحسين من قبل الطاهريّة[l] وكان فى ضعف فنزل على باب المدينة وكان درهم بن نصر هذا يظهر انّه من المُطّوعَة[m] وانّه قاصد[n] لقتال الشراة محتسبًا فاستمال[o] العامّة حتّى مالوا اليه ودخل المدينة وخرج منها واليها الى بعض النواحى[p] فتمكّنوا من البلد وقاتلوا الشراة وكان للشراة رئيس

قُبض عليه فى حرب وحُمل الى بغداد وطِيفَ به على فالج ومات. Deinde habet وقيل ان كان. a) C. وقيل ان كان. b) Jacut addit: يخدمه فى عمل الصفر. c) A. et B. بلغه. d) C. قديم. e) Sic C. et E.; A. et B. رباد, D. نقاد, Jacut كُثير بن رِفاق. f) A. et B. يجمع. Pro فى Jacut habet من, D. فيه. g) Sic A., B. et E.; C. تسمى ملاذه فقتل; D. وبلغ السلطان خبره فانفذ من حاصره فى قلعة; Jacut كانت له وقتل تسمى ملاذه وضيّق عليه حتى قُبض عليه وقُتل. h) Jacut وفرّوا et post يظهرون الزهد addit: وقد صار لهم ذكر وصيت ببست. i) C. والغزاة; Jacut والقتال على الحسبة فى الغزو للخوارج. k) E. نصر, et النصر habent Ibno 'l-Athír, VII, p. ٤٣ et Ibn Khallicán, n. 838. Secundum hos confunduntur hic درهم بن الحسين et صالح بن النضر. Jacut l. l. 11 et deinde دُرَيم pro درهم. l) A. et B. tantum الطاهرى. m) C. الطوعة. n) C. et D. قصد; deinde C. قتال. o) Jacut واظهروا من الزهد والتقشف ما استمال اليهم; cf. Ibno 'l-Athír, VII, p. ١٢٤, 12. p) E. ابراهيم از درى ديگر بدر رفت. Jacut حتى صاروا فى دريم بن نصر واصحابه من البلد.

يعرف بعمّار[a] بن ياسر فانتدب لقتاله يعقوب فقاتله[b] وقتل عمّارًا وكان لا يحزبهم[c] امر شديد الّا انتدب له يعقوب فكان يرتفع ذلك *الامر له على[d] ما يحبّه فاستمال اصحاب درهم بن نصر حتّى قلّدوه الرياسة وصار الامر له وكان درهم بن نصر بعد ذلك من[e] جملة اصحابه وما زال محسنًا الى درهم ابن نصر حتّى استاذنه فى الحجّ[f] واقام ببغداد مدّة ثمّ رجع رسولًا من *امير المؤمنين[g] اليهم فقتله واستفحل امرهم بعد ذلك حتّى استولوا على سجستان وما يتّصل بها من اطراف السند والهند ومهّدوا تلك الثغور واسلم على يدى يعقوب خلق منهم ثمّ استولى بعد ذلك على كرمان وفارس وخوزستان وبعض العراق وعلى خراسان ۞

وامّا الطاق فانّها مدينة على مرحلة *من زرنج[h] تكون على ظهر الجائى من سجستان الى خراسان وهى مدينة صغيرة ولها رستاق وبها اعناب كثيرة يتّسع بها اهل سجستان ۞ وخُوَاش هى من قرنين على مرحلة عن يسار الذاهب الى بُسْت[i] وبينها وبين الطريق[k] نحو نصف فرسخ وهى اكبر من قرنين وبها نخيل واشجار وبها وبالقرنين مياه جارية وقنىّ ۞ وامّا قَرِه فانّها مدينة اكبر من هذه المدن ولها رستاق يشتمل على نحو[l] من ستّين قرية وبها نخيل وفواكه وزروع وعليها نهر قره[m] وابنيتها طين وهى فى ارض سهلة ۞

---

a) D. et E. بعمارة. *b*) A. et B. فقتله. Jacut فظهر منه فى ذلك نجدة فجعلوا بعد ذلك لا يَعْزُوهم وعزم وحزم حتى قتل عمّارًا واباد ذكره. *c*) Jacut وصار — من اتباعه. *d*) B. الا منزله. *e*) C. et D. فى; Jacut فاذن له فحجّ وعاد فاقام. *f*) Jacut له فحجّ وعاد فاقام. *g*) C. الخليفة, Jacut السلطان. Deinde pro اليهم Jacut: الى يعقوب فنقم عليه. *h*) Ex D. addidi. E. نزديك بست. *i*) Apud Jacut, II, p. ٤٨٩, 11 male تستر. *k*) Abulfeda, p. ٣٤٣ الطاق et hinc fortasse *Djih.-Numa*, p. ٢٤٩, 3 a f. حصن الطاق خواشه نصف مرحلة قريب يرده. Quod apud Edrîsî, I, p. 445 legitur قارعة videtur esse pro قارعة الطريق. *l*) Jacut, III, p. ٨٨٨, 11 اكثر. *m*) Jacut l. l. ولها نهر كبير عليه قنطرة (ut *Djih.-Numa*, p. ٢٥٠, 12 et 13), sed quae deinde addit: وهى على يمين القاصد

وجِزّه[a] متّصلة بعمل فره عن يمين الذاهب من سجستان الى خراسان على نحو من مرحلة وهى مدينة صغيرة نحو القرنين ولها قرى ورستاق وهى خصبة ومآؤها من القنىّ وابنيتهم من طين ✿ وسَرْوان مدينة صغيرة[b] نحو القرنين الّا انّها اعمر من القرنين وبها فواكه كثيرة ونخيل واعناب وهى من بُسْت على نحو مرحلتين * احدى المنزلتين[c] تسمّى فيروزقند[d] * والاخرى هى سروان[e] على طريق بلدى الداور، وصالَقان[f] من بست على مرحلة وبها فواكه ونخيل وزروع واكثر اهلها حاكة ومآؤهم من انهار وبنآؤهم من طين وهى نحو القرنين فى الكبر ✿ ورُوذان[g] هى اصغر من القرنين وهى بقرب فيروزقند عن يمين الذاهب الى رُخّج واكثر غلّاتها الملح ولهم مع ذلك زروع وفواكه ومياه جارية ✿

واما المسافات بها فانّ الطريق من سجستان الى هراة اوّل مرحلة تسمّى كَرْكُوَيه[h] على ٣ فراسخ ومنها الى بَشْتَر[i] ٤ فراسخ ويعبر على قنطرة يجرى فيها ما فضل من مياه هندمند ومن بشتر الى جُوَين[k] مرحلة ومن جوين الى بست مرحلة[l] ومن بست الى كنجر[m] مرحلة ومن كنجر الى

من سجستان الى خراسان, aeque ut apud C. e descriptione urbis Giza huc migrasse videntur. *a*) A. et B. وحره et deinde متصل. Scribitur quoque كنزه. *b*) C. addit خصبة. *c*) C. et Jacut (III, p. ٨٤ paen.) احد المنزلين et deinde والاخر. *d*) C. نمرور قمدر; Jacut فيروز مند ut habet E.; *Djih.-Numa* l. l. 19 فروهند. *e*) A. et B. والاخر هرشروان l. ut C. recte habet والاخر هو سروان et cf. ann. *c*. Jacut om. هو. *f*) B. وصالفان, C. ومالقان, E. وطالقان; D. والزالقان. *g*) A. et B. وررفان. *h*) A., B. et C. sine punctis; E. كركونه. *i*) Sic scribo secutus Anderson, p. 374, qui ab hoc loco canalem باشتروذ nomen accepisse opinatur. A., B., C. et E. سدر, سبر, دسمر et دسبر, D. تستر, Mokaddasi MS. p. 166 دشمن, *Djih.-Numa*, p. ٢٥١, 16 بستر, Edrîsî, I, p. 447 دستر habent. *k*) Vocales in A.; C. et E. حودر, *Djih.-Numa* جزير; Mokaddasi حودر. *l*) A. et B. om. Deinde B. ومنه. Quod attinet ad lectionem بست (A. et C. دست) sana esse nequit, ut mappa inspecta facile patebit. Quae vero genuina sit lectio, nescio. D. habet السب; Anderson legit بَسَنْت. Edrîsî, I, p. 447 انسنت. In E. et *Djih.-Numa* haec et duae stationes seqq. desiderantur. Mokaddasí habet بست. *m*) A. كعاجر, B. كفاجر, C. كماحر, D. كنكره, Mokaddasí كياجن.

سرشك[a] مرحلة ومن سرشك الى قنطرة وادى فَرّه مرحلة ومن قنطرة الوادى الى فره مرحلة ومن فره الى دَرّه مرحلة ومن دره الى كوستان[b] وهى آخر عمل سجستان ومن كوستان الى خاستان[c] وهى من الأَسْفُزار[d] مرحلة ومن خاستان الى قناة سَرِى[e] مرحلة ومن قناة سرى الى الجبل الاسود[f] مرحلة ومن الجبل الاسود الى جامان[g] مرحلة ومن جامان الى هراة مرحلة ۞ وامّا الطريق من سجستان الى بُسْت فاوّل[h] مرحلة الى زانبوق[i] ومن زانبوق الى سروزن قرية عامرة سلطانيّة مرحلة[k] ومن سروزن الى حرورى[l] قرية عامرة سلطانيّة وبينهما نهر نيشَك وعليها قنطرة معقودة من آجرّ مرحلة[m] ومن حرورى الى دَهَك[n] والمنزل رباط من حدّ دهك ومن هذا الرباط المفازة فمنزل منها رباط يسمّى آب شُور[o] ومن آب شور الى رباط كرودين[p] ومن رباط[q] كرودين

---

a) A. et B. سرسد, D. سرسك, Edrîsî شرشك. C. et Mokaddasí ut recepi. b) A. et B. كوسمار, كوسار et كوكسكار; C. كوىسمان, D. كونسان, E. et *Djih.-Numa* كوسان; Mokaddasí ut recepi. Edrîsî habet كوشكان cui lectioni favet كوكسكار in A. et B. (alius locus ejusdem nominis apud Jacut, II, p. ١٥٣, 14). c) A. حادستان et, ut B., حاستان; C. حاس; D. خاديان et خاسان; E. حادمان; *Djih.-Numa* خاسان; Mokaddasí خاشان; Edrîsî جاشان ut Anderson. d) Cf. Jacut sub دره. e) Pro قناة, E. كاريز. f) E. سيباه كوه. g) A. et B. حاقان, حاىان et حامان; C. حارمان, D. حاربان, Edrîsî جدمان. Mokaddasí habet ut recepi. Vid. supra p. ٢٣٩ ann. *i*. h) A. et B. اول. i) Secutus sum Mokaddasí. A. راسوق, B. et C. رادموق, D. رانيوق; E. et *Djih.-Numa* رسوق; Edrîsî زينون. k) A. et B. om. l) Sic habent A., B., D., Ous. et *Djih.-Numa*. C. حذورى, E. حرودى. Mokaddasí habet حروزن. Quae sit vera lectio nescio. m) A., B., D. et Mokaddasí om. n) A. et B. دهل, C. رهك, Mokaddasí دهبك. o) A., B. et C. ارسور (C. semel السور). E. سود, sed Ous. ازسور; *Djih.-Numa* ارسوره. In D. superest tantum اب. Vid. Edrîsî l. l., Sprenger, p. 46 et Anderson. p) A. et B. كوروىمو, C. كردوين, D. كروزين, Edrîsî كرورين, Mokaddasí كرودن, *Djih.-Numa* كرود. Secutus sum E. q) A. et B. om.

*الى رباط قهستان a ومن رباط قهستان الى رباط عبد الله ومن رباط عبد الله الى بست، ومن رباط دهك الى فرسخ من بست كلّها مغازة ۞ وامّا الطريق من بُسْت الى غَزْنَة فانّ من بست الى رباط فيروزقند b منزل ومنه الى رباط ميغون c منزل ومنه الى رباط كبير d منزل ثمّ الى مدينة الرُّخّج المسمّاة بَنْجَوَاى e منزل ومنها الى تكين اباذ f منزل ثمّ الى خرسانة g منزل ثمّ الى رباط سراب h منزل ثمّ الى الاوقل i وهو رباط منزل ثمّ الى رباط جنكل اباذ منزل ثمّ الى قرية غوم k منزل *ثمّ الى قرية خاست منزل ثمّ الى قرية جومة منزل l ثمّ الى خابسار m منزل وهو اوّل حدّ غزنة ثمّ الى قرية

---

a) Sic A. et B.; C. هسمان, D. هفستان, E. et *Djih.-Numa* هستان, Edrîsî هفشيان, Mokaddasí هفشان. Secundum Anderson et Sprenger legendum est هفت چاه. *b*) E. فيروزمند ut *Djih.-Numa;* Mokaddasí فيروز فند. Vid. supra p. ٢٤٨ et cf. ann. *d.* *c*) A. منعون s. ميعون s. ut B. ميغون; C. معون et سغور; D. مغور; E., Edrîsî et *Djih.-Numa* معون; Mokaddasí ميعوف. *d*) A. om.; B. et Mokaddasí كممر; C. كدر, Edrîsî كيبر; E. et *Djih-Numa* كنر. Secutus sum D., coll. Sprenger, p. 47 ult. *e*) A. et B. بمحواى, Mokaddasí دمحوان. *f*) Sic D. et A. et B. (qui habent دكمن ابساذ); C. بكير ابساذ; E. تسكيين اباذ; *Djih.-Numa* كمين اباذ; Edrîsî مكين اباذ; Mokaddasí نكراباذ. *g*) A. et B. خومسناه; C. et E. خراسانه (C. semel حرساه); Mokaddasí حرساد; *Djih.-Numa* خراسان. In D. et Edrîsî lacuna est. *h*) Sic omnes, exceptis A. et B. qui habent سرام. Secundum Anderson legendum est سراسپ. *i*) A. et B. الاوند; D. et Edrîsî الاوق; E. ادفى; *Djih.-Numa* اوفى; Mokaddasí الاوز. Cf. porro Sprenger. Secutus sum Anderson. *k*) A. et B. عرمر, C. خومه, D. عزم, E. et *Djih.-Numa* عوم, Mokaddasí عرم, Edrisi عزير. *l*) A., B. et D. haec om. Pro خاست C. habet خاسب, Edrisi جابست, Mokaddasí حاسب, *Djih.-Numa* خواست. Pro جومة, quae statio apud Mokaddasí desideratur, C. habet خومه, Edrisi حومه, *Djih.-Numa* حوم. *m*) A. حاسمان, B. et C.

خشباجى[a] منزل ثمّ الى رباط هزار[b] منزل وهى قرية عامرة ثمّ الى غزنة منزل ۞ ومن سجستان الى بالس[c] طريق على المفازة تاخذ من مدينة الزّخّج المسمّاة بَنْجَوَاى الى رباط الحَجَرِيّة[d] منزل ثمّ الى[e] رباط جنكى[f] منزل ثمّ الى رباط برو[g] منزل ثمّ الى رباط اسفنجاى[h] منزل ۞ وامّا الطريق من سجستان الى كرمان وفارس فانّ اوّل *حدّ ينزل[i] من سجستان خاوران[k] والثانى رباط يسمّى دارك ومن دارك الى برين[l] منزل ومنه الى كاونيشك[m] منزل وهما رباطان ثمّ الى رباط الناسى[n] ثمّ الى رباط القاضى[o] منزل ثمّ الى رباط كرامخان[p] منزل ثمّ الى سَنيج[q] منزل وسنيج مدينة من كرمان، وحدّ

حابسان, D. خايسار, E. خابسار, Edrîsi حابسان; *Djih.-Numa* خوابسار, Mokaddasí چاشان. *a*) A. حساحى, B. حساحى (et sic C. sed alieno loco), D. خشباذى, E. حنساجى, Edrîsi l. l. جسراجى (cf. p. 459). *b*) A. et B. هدوه, D. هُدُو, Edrîsi هروى. Secutus sum E. Hic in E. exstat واز آنجا بغزنى شانزده مرحله i. e. totius itineris distantia. *c*) D. باليش. *d*) A. et B. sine punctis, Edrîsi الحجر. E. et *Djih.-Numa* سنكين. *e*) A. et B. om. *f*) A. et B. حملى, D. et Edrisi كنكر; Mokaddasí حسكى, C. حنكى, E. et *Djih.-Numa* ut recepi. *g*) E. بم, *Djih.-Numa* يم, C. et Mokaddasí tantum رباط. *h*) A. et B. اسحاب, E. بيسجان, *Djih.-Numa* سنجان. In E. additur totam viam esse 14 dierum (جمله چهارده مرحله است). *i*) C., D. et E. منزل. *k*) A. حاودث, B. حارث, C. حراوران, D. حاوزن, Edrîsi, I, p. 431 جارون. Secutus sum E. et *Djih.-Numa*. *l*) A., B. et E. sine punctis, C. بردر, D. بردن, Edrîsi بزردين. Secutus sum Ous. p. 211. *m*) A. et B. h. l. كاربسك, infra A. كادنيسك, B. كاربنسك; C. et E. sine punctis; D. كاوبيسك; Ous. كاو پلنك. Cf. Sprenger, p. 47. *n*) A. et B. الناس, E. ماسى. *o*) C. القاهر. *p*) A. et B. كوراعان, C. كرانجان, D. كراغان, E. كرامحان, *Djih.-Numa* كرامجان. Cf. Sprenger l. l. *q*) A., B. et C. sine punctis; D. مستنج. In E. additur viam hucusque esse 8 dierum (جمله هشت مرحله).

سجستان اذا جزت كاونيشك بينها وبين كُنْدُر رباط بناه عمرو[a] وهذا المكان يعرف بقنطرة كرمان وليس هناك قنطرة ولكن تسمّى كذلك ۞ وسائر المسافات بسجستان من سجستان الى جِزّه[b] ٣ مراحل وهى بين فَرَه والقَرْنين وبينها وبين فره ايضًا مرحلتان وبين نه وفره نحو مرحلة راجحة وهى بحذآئها ممّا يلى المفازة، وبين كِشّ[c] وبين سجستان ٣٠ فرسخا فيما[d] يلى حدّ كرمان والطاق على طريق كشّ[e] على ٥ فراسخ، وخُوَاش *على نحو[f] فرسخ من طريق بست وبينه وبين القرنين منزل، ومن بست الى سَرْوان[g] مرحلتان على طريق بلد الداور ثمّ تعبر هندمند على مرحلة من سروان فتدخل تلّ[h] وتمضى مرحلة الى دَرْغَش[i] على شطّ هندمند كلاهما من جهة واحدة ومن تلّ الى بَغْنِين يوم فى *قبليّ تلّ[k] وبشْلَنْك فى جنوبيّ بغنين، وبَنْجَوَاى[l] على ظهر غزنة وبينها وبين كُهَك مقدار فرسخ عن غربيّ بنجواى ومن بنجواى الى اسفنجاى ٣ مراحل والقصر بحذآئها وبينهما فرسخ ومن اسفنجاى الى سيوى مرحلتان ۞

---

a) D. et E. addunt بن الليث. *b*) A. et B. حُرّه, D. دره. Conjectura addidi وهى coll. Edrisí, I, p. 446. Aliter corrigit Sprenger, p. 46, legens وبين جره pro بين فره et addens مرحلتان post والقرنين propter sequens ايضا. Fortasse post والقرنين inserendum est: وبينها وبين بست نحو مرحلتين. E. habet (sed alieno loco) از سيستان تا حره ٣م ميان فره وقرنين وميان قرنين وتاخذ من زرنج : Mokaddasí. وفره ٣م وفره برابر قرنين است نزديك بيابان الى فره ٣ مراحل ثم الى فره او الى القرنين مرحلتين مرحلتين. *c*) C. كمس. *d*) B. مما. *e*) E. كوير; Mokaddasí inter urbis Sidjistáni memorat كوير (p. 27) كوير (p. 145) كودن (p. 148). Vera lectio est كوين = جوين sub qua forma memoratur supra p. ٢٤٨ *k* (cf. Jacut, II, p. ١٩٤, ult.). Mokaddasí h. l. ومن زرنج الى الطاق مرحلة ومن زرنج الى كش ٣٠ فرسخا. *f*) A. et B. عككن. Pro فرسخ C. habet فرسخين. *g*) A. et D. h. l. شروان. *h*) B. بل, D. درنک (l. درتل). *i*) A. دوعس, B. روعس, C. درغوز. *k*) C., D. et Edrisí l. l. قبائل, D. et Edr. omissa deinde copula. Cf. Sprenger, p. 47. B. habet فملى بلى, A. sine punctis. *l*) A. et B. ومبكواى. Deinde C. على ظاهر.

ذكر خراسان[a]

وامّا خُرَاسَانُ فانّها تشتمل على كور وهو اسم الاقليم والّذى يحيط بها من شرقيّها نواحى سجستان وبلد الهند لانّا ضممنا الى سجستان ما يتّصل بها من ظهر الغُور كلّه الى الهند وجعلنا ديار خَلَّج[b] فى حدود كابُل ووَخّان[c] فى ظهر الخُتّل[d] كلّه وغير ذلك من نواحى * بلد الهند[e] وغربيّها مفازة الغُزّيّة[f] ونواحى جرجان وشماليُّها ما ورآء النهر وشىء[g] من بلد الترك يسير[g] على ظهر الخُتّل وجنوبيُّها مفازة فارس وقُومِس، وضممنا قومس الى نواحى جبال[g] الديلم مع جرجان وطبرستان والرّىّ وقزوين[h] وما يتّصل بها وجعلنا ذلك كلّه اقليمًا واحدًا[g] وضممنا الخُتّل الى ما ورآء النهر لانّها بين نهر وَخْشاب[i] وجرياب وضممنا خوارزم الى ما ورآء النهر لانّ مدينتها ما[k] ورآء النهر وهى اقرب الى بخارا منها الى مدن خراسان، وبخراسان فيما يلى المشرق زَنْقَة فيما بين مفازة فارس وبين هراة والغور الى غَزْنَة[l] ولها زنقة فى المغرب من حدّ قومس الى ان يتّصل بنواحى فَرَاوَة[m] فيقصر هاتان الزنقتان عن تربيع سائر خراسان وفيها من حدّ جرجان وبحر الخزر الى خوارزم تقويس على العمارة[n] ۞

وامّا كور خراسان الّتى تجمع على الاعمال وتفرق[o] فانّ اعظمها نَيْسَابُور[p]

a) Editum est hoc quoque caput ab Anderson in Journal of the As. Soc. of Bengal. 1853, p. 152—193. b) Vocales in A. et B.; Jacut habet خَلْج in v., الخلْج, IV, p. ٣٢٠, 16. c) A. ووحان, B. وفرحان, C. ورحان. Secutus sum D. d) A. sine punctis, C. الحتل; B. et D. الجبل. e) A. et B. om. f) A., B. et E. غزنة. g) A. et B. om. h) C., D. et E. om. i) A. وخشنار, B. وخشتار. Deinde A. et B. وحوبار. k) C. et D. om., D. addens العظمى. l) C. الغزّيه. m) B. فواوه, C. نواره, E. قواره. Deinde A. ويعصى, B. وسعص (وينقص?), D. et Abulfeda, p. ٤٤١ فيصير. E. habet جدا كرده اند تا تربيع خراسان درست آيد. n) Hic in A. et B. sequitur صورة خراسان. o) A. ويعرف; secundum E, legendum esset وتعرف. p) B. ut solet نيشاپور, E. نشابور.

ومرو وهراة وبلخ وبخراسان كور دونها فى الكبر[a] فمنها قوهستان وطوس ونسا وأبيورد وسرخس وأسفزار[b] وبوشنج وباذغيس وكنج رستاق ومروروذ وجوزجان[c] وغرج الشار والباميان وطخارستان[d] وزمّ وآمل ٭ وامّا خوارزم فانّها نذكرها فيما ورآء النهر لانّ مدينتها ورآء النهر وهى الى مدن ما ورآء النهر على السمت اقرب منها الى مدن خراسان، ولنيسابور كور لم نفردها لانّها مجموعة اليها * فى الاعمال[e] سنذكرها فى صفة نيسابور، وافردنا طخارستان عن بلخ وان كانت مجموعة اليها لانّها مفردة فى الذكر والدواوين فيقال بلخ وطخارستان وليس فى تفريقنا هذه الكور وجمعها درك اكبر من استيعابها[f] وتاليفها فى الصور ومعرفة مكان كلّ شىء منها فى صورة خراسان ٭

فامّا نيسابور فهى أبرشهر[g] وهى مدينة فى ارض سهلة ابنيتها طين وهى مفترشة البنآء ومقدار عرضتها[h] نحو فرسخ فى فرسخ ولها مدينة[i] وقهندز وربض وقهندزها ومدينتها عامرتان ومسجد جامعها فى الربض بمكان يعرف بالمعسكر[k] ودار الامارة بمكان يعرف بميدان الحسينيّين[l] والحبس عند دار الامارة وبين الحبس ودار الامارة وبين المسجد الجامع نحو فرسخ ودار الامارة من بنآء عمرو بن الليث وللقهندز بابان وللمدينة اربعة ابواب احدها يعرف * بباب راس القنطرة[m] والثانى بباب سكّة[n] معقل والثالث بباب[o] القهندز والرابع * بباب قنطرة تكين[p] وقهندزها خارج عن مدينتها ويحيط[q] بالمدينة

---

a) C. et D. الصغر. b) C. واسفران, E. واسفراين; A., B. et D. واسفرار. c) E. وكوزكانان. Nomen sequens, quod etiam غرجستان (ut in E.) et غرشستان scribitur, apud A. et B. deëst, apud C. et D. corruptum est. d) C. وطبرستان, E. وتخارستان. e) A. et B. والاعمال. f) A. اسعادها, B. اسعادها, D. استيفايها. g) C. ايران شهر ut quoque Jacut, II, p. ۴۰۹, 21 et Mokaddasí MS. p. 152. h) A. et B. عرضها. i) C. et E. مدينتان. k) E. لشكرگاه. l) A. et B. h. l. الحسنى; C., D. et E. الحسين; vid. infra. m) E. در پول. n) A. et B. om. Inserui ex D. et E. (ubi كوى). o) A. et B. باب. p) Secundum E. ubi دروازهٔ پول تكين. A., B. et D. habent (ميكين) قنطرة در منكين, in quibus در est vocab. Persicum pro باب. q) C. ودحتف, D. ويحف.

والقهندز جميعًا الربض، وللربض ابواب[a] فامّا الباب الّذى يخرج منه الى العراق وجرجان فانّه يعرف بباب القِبَاب[b] والباب الّذى يخرج منه الى بلخ وما ورآء النهر فانّه[c] يعرف بباب جَنْك[d] والباب الّذى يخرج منه الى فارس وقوهستان فانّه يعرف بباب * احوص اباذ[e] والباب الّذى يخرج منه الى طوس ونسا عدّة ابواب فمنها باب سُوختنه وباب يعرف بسر شيرين وغيرهما، وامّا اسواقها فانّها خارجة[f] من المدينة والقهندز فى الربض واعظم اسواقها سوقان احداهما تعرف بالمربّعة الكبيرة والاخرى[g] بالمربّعة الصغيرة واذا اخذتَ من المربّعة الكبيرة نحو المشرق فالسوق[h] يمتدّ الى ان تجاوز مسجد الجامع واذا اخذتَ من المربّعة نحو المغرب فالسوق[h] يمتدّ الى ان تجاوز المربّعة الصغيرة واذا اخذتَ من المربّعة نحو الجنوب فالسوق ممتدّة الى قرب مقابر الحُسَينيّين[i] ويمتدّ السوق من المربّعة فى شماليّها حتّى ينتهى الى راس القنطرة والمربّعة الصغيرة بقرب ميدان الحسينيّين جنب دار الامارة، واكثر[k] مياهها قُنِىٌّ تخرج تحت مساكنهم وتظهر خارج البلد فى ضياعهم ولها[l] قُنِىٌّ تظهر فى البلد وتجرى فى دورهم وبساتينهم داخل البلد وخارجًا عنه ولهم نهر كبير يعرف بوادى سغاور[m] يسقى منه بعض البلد ورساتيق كثيرة وعلى هذا الوادى قوّام[n] وليس لهم فى البلد نهر اعظم منه، وليس بخراسان مدينة اصحّ[o] هوآء ولا اكبر من نيسابور ويرتفع منها من اصناف ثياب القطن والابريسم[p] ما ينقل[q] الى سائر بلدان الاسلام وبعض بلاد الشرك لكثرتها

---

a) C. add. كثيرة; Mokaddasí p. 153 ودروب المصر تجاوز الخمسين. *b*) E. باب عتاب (Ous. عشاب); Mokaddasí درب القبات. Vid. Jacut, IV, p. ٢٥, 8 seqq. *c*) A. et B. om. *d*) Ex E.; A. حمل, B. حبل, D. جيَكْ, Mokaddasí الجيق. Deinde B. والثالث. *e*) D. حوص اباذ. *f*) A. et B. خارج. *g*) A. et B. والاخر. *h*) B. بالسوق. *i*) Sic A.; B. الحسسس; D. et E. الحسين (كورستان حسين). *k*) A. et B. واكبر. *l*) B. وبها. *m*) B. سغادر, C. سعاوه, D. بشغاوذ, E. سقا, Epit. Par. سغَارِن. Est fluvius qui a pago بُشْتَنِقَان venit. *n*) A. et B. تُوّام. *o*) A. et B. om. *p*) B. والابريشم. *q*) A. ىىنقل.

وجودتها ۞ ولنيسابور حدود واسعة ورساتيق عامرة ولها [a] مدن منها البُوزْجَان [b] ومالِنْ [c] المعروف بكواخرز [d] وجايمند [e] وسلومك [f] وسَنْكان [g] وزُوزَن وكُنْدُر وتُرْشِيز [h]

---

a) B. وبها. b) A. et B. النوريحان; E. بوزكان. c) E. مآلن et sic altera manus in B.; Jacut مالين; *Djih.-Numa* مالين et پالان (p. ٣١٩, 19). d) A. دكراحرر, B. دكراحرد, D. بكواجرد. Infra in itinerario A. et B. لواخرر, D. كواجون. Edrîsî, I, p. 452 كواخور; Mokaddasí p. 166 كواخون (quod Sprenger male legit كوافون). Anderson legit كسراجرد. Jacut, IV, p. ٣٩٨, 10 habet كياخون. Idem locus intelligi videtur apud Jakubí, p. ٥٥, 10 ubi, pro وما بحذو editoris, Cod. habet ومالحز. Sprenger, p. 51 dat باخَرْز, nimis audacter. Attamen inter duo nomina relatio arcta esse videtur. Idem de hoc nomine valere videtur quod de بالا = كوالا (Jacut, I, p. ٤٧٧, 20). e) A., B. et D. وحادمند; infra in itinerar. A. et B. حوادمن, D. حايمن, Mokaddasí جاتمن. Edrîsî, I, p. 452 حايمن, II, p. 182 جايمند. Vix facere audeo cum E. qui (ut *Djih.-Numa*, p. ٣٣٠, 5) in itinerario habet nomen urbis notae جام s. زام. f) A. et B. h. l. وسلومل, D. وساومل (Edrîsî ساومك). In itinerario A. et B. سكومل, D. سلومل. E. h. l. سلونك (Ous. سلوتك), in itiner. سلومد. In mappa C. fortasse سورمك. g) A. et B. h. l. وسكان, infra in itinerario سكان. D. h. l. بشكان, infra سيكان ut Edrîsî. E. h. l. وسركان (Ous. تركان), infra سكان Mokaddasí سكان. (Cf. Sprenger, p. 50 seq. et vid. Jacut sub سنجان (*Djih.-Numa*, p. ٣١٩, 12 quoque سنجان habet). h) A. sine punctis, B. وبرسير, D. وبرنيس, Edrîsî برسين s. برشين s. برشيق. In itinerario A., B. et E. habent برشير, Edrîsî برشين, solus D. ترشير. Sprenger, p. 50 recte corrigit ترشيز = تُرْشِيش quam formam Jacut memorat pro طُرَيْثِيث. Idem jam suspicatus est Jaubert, I, p. 454 ann., II, p. 184. Cf. Quatremère, *Hist. des Mongols*, p. 177 et vid. supra p. ٢٣١ ann. s et ٢٣٧ ann. a. Mokaddasí in textu p. 27 et 154 طرثيث sed in mappa برشير.

وخان روان [a] وأزاذوار وخسروكرد وبهمناباذ ومزنيان [b] وسابزوار [c] وديوارة [d] ومهرجان وأسفرائين وخوجان [e] ورزيلة [f] ✿ وان جمعنا طوس الى نيسابور فمن مدنها الزاذكان [g] والطابران وبزدغور [h] والنوقان التى بها قبر علىّ بن موسى الرضى

---

a) A. et B. وحان رااں, D. وحان راوں, E. وحان وداوں. Infra in itin. A. et B. خان روان ut Edrisi, II, p. 184 et Sprenger, p. 49; D. خان زوان. In mappa C. حان روان; E. خاوران ut *Djih.-Numa*, p. ٣٣., 7. Cogitari autem nequit de خاوران apud Kazwíní, II, p. ٢٤١. *b*) Secundum D.; A. et B. h. l. ومرسان, in itiner. A. مرىدان, B. موردان; E. h. l. وخرينان, in itiner. مرىسان (Ous. مرسان); D. in itiner. مرنبان; Epit. Paris. مرنيان. In mappa C. مويدان. Sprenger, p. 49 موريان. Cf. Anderson, p. 172. Idem esse nequit locus ac is qui memoratur supra sub nomine المورجان, vid. p. ٢١٩ et ann. *c* et cf. Edrisi, II, p. 176 seq. — Mokaddasí p. 166 in itinerario habet مرىنان, in mappa مزقيان. *c*) A. et B. وساذوار, E. وساروان ut C. in mappa (cum var. l. سفروار). Vid. Jacut, I, p. ٨٠٤, 13 et II, p. ٤٤١, 18. *Djih.-Numa*, p. ٣٢٢ سبزوار (ut fortasse restituendum est apud Jacut, I, p. ٨٠٤, 14 pro سَبْزور). *d*) A. وديوازه, B. وديواذه, E. ورمواذه. In itinerario A. ربواذه, B. ربواذه. D. ut recepi. Jacut ديوَرَه. In mappa C. est ريوند (cf. Sprenger, p. 49) cum var. lect. ربواره. In mappa Mokaddasii زيواذه. Anderson putat eam non differre a „the Rewat of the maps" (Jacut ريوَنْد). *e*) Sic recte D.; A. et B. وحىرحان, mappa C. حرحان, E. وكركان; mappa Mokaddasii جرجان, sed in textu p. 154 recte خوجان. Jacut, II, p. ٤٨٧, 21, dicit urbem quoque appellari خوشان. Videtur legendum خَبُوشان, vid. II, p. ٤.., 4. seq. et *Djih.-Numa*, p. ٣٢٣. In mappis hodiernis *Kabuschan* exstat. *f*) A. et B. وورزىله, E. وريدين (Ous. زيدين). Ceteri omnes omittunt, nec certi quidquam statuere possum, licet suspicor nihil esse nomen quam lectionem corruptam regionis ريوند. *g*) A. et B. الرانكان, D. et E. الرايكان; Epit. Par. زانكان. Deinde A. et E. والطايران. *h*) A. et B. وبروعور

عمّ وقبر هارون الرشيد ومنها يرتفع البرام *وقبر الرضى من المدينة على نحو ربع فرسخ بقرية يقال لها سَنَابَاذa، وفى جبال نيسابور وطوس يكون الفيروزجb، *وكانت دار الامارة بخراسان بمرو وبلخ الى ايّام الطاهريّة فنقلوها الى نيسابور فعمرت وكبرت وكثر مالها من توطّنهم ايّاهاc*

وامّا مَرْوُ فانّها تعرف بمرو الشّاهجان وهى قديمة البنآء يقال انّ قهندزها من بنآء طَهمورَثd وانّ المدينة القديمةe من بنآء ذى القَرْنَيْن وهى فى ارض مستوية بعيدة عن الجبال لا يرىf منها جبل وليس فى شىء من حدودها جبل وارضها سبخة كثيرة الرمال وابنيتهاg طين وفيها ثلاثة مساجد للجماعاتh امّا اوّل مسجد اقيمت فيه الجمعة فمسجد بُنى منi داخل

(A. fortasse بردعور); E. برعود; in D. fortasse pars nominis superest وغول; C. in mappa بردعمد; Mokaddasí in mappa بروغرك in textu p. 27 بروعنه, p. 155 بروغمه. Edrisí, II, p. 183 بردغور. Edidi بزدغور, putans intelligi locum quem Jacut, I, p. ٩٠۴, 17 بِزْدِيغَرَة appellat, quique in *Lobbo'l-lobáb* بِزْدِيغَر vocatur. *a*) Secundum Abulfeda, p. ۴٥١ وعلى رُبع فرسخ من طوس قبر على بن موسى الرضى واما قبر الرشيد ففى قرية تسمى سناباذ. Utrumque sepulcrum in eo pago collocatur a Jacut, III, p. ١٥٣ et Kazwíní, II, p. ٣٦٢, quocum consentit *Djih.-Numa*, p. ٣١٨ et ٣١٩. E. habet: ونوقان كه مشهد مقدس امام على بن موسى الرضا عليه التحيه والثناست وكور هارون انجاست بچهار فرسنكى. Abulfeda, p. ۴٥٣ et D. addunt: وعلى قبر على بن موسى حصن وفيه قوم معتكفون. *b*) E. addit وديگر كانهاست. *c*) Haec in A. et B. desunt. Exstant apud C., D., E. et Abulfeda, p. ۴٥٣. Deinde E. addit: وبزرگان ودبيران معروف از نشابور خواسته اند چون حصر نتوان كرد ذكر نشد, quae in suo Codice quoque videtur legisse D. *d*) A., C. et Abulfeda, p. ۴٥٧ طهمورث. *e*) A. et B. قديمة. *f*) D. addit بالقرب, quod quoque abest apud Abulfedam. *g*) C. et D. add. من. *h*) B. الجماعات, C. للجمعة, D. للجمعات. *i*) Ex solo A.

المدينة فى اوّل الاسلام فلمّا كثر الاسلام بُنى المسجد المعروف[a] بالمسجد العتيق على باب المدينة ويصلّى فيه اهل الحديث وتركت[b] الجماعات فى المسجد الاوّل ويعرف بمسجد بنى ماهان ثمّ بُنى بعد ذلك المسجد الّذى على ماجان ويذكر انّ ذلك المسجد والسوق[c] ودار الامارة من بناء ابى مُسْلِم، ودار الامارة على ظهر هذا المسجد وفى هذه الدار قبّة بناها ابو مسلم كان يجلس فيها والى هذه الغاية يجلس فى هذه القبّة امرآء مرو وهى قبّة من آجرّ وسعة هذه القبّة *خمسة وخمسون ذراعًا[d]* وتنصّ لها من داخل نصبة السطح[e] وللقبّة اربعة ابواب كلّ باب الى ايوان *سمك كلّ ايوان ....[f]* وبين يدى كلّ ايوان صحن مربّع، والقهندز *فى الكبر مثل مدينة[g]* الّا انّه خراب وهو *مرتفع وعلى[h] ارتفاعه قد سيقت اليه قناة ماء جارٍ[i] الى يومنا هذا وربّما زرع عليه مباطخ[k] ومباقل وغير ذلك، وامّا اسواقها فانّها فى القديم كانت على باب المدينة جنب المسجد العتيق فانتقلت فى ايّام ابى مسلم الى ماجان واسواقها من انظف اسواق الامصار ومصلّى العيد فى محلّة راس الميدان فى مربّعة ابى الجهم ويطيف بهذا المصلّى من جميع وجهاته البنيان والعمارات وهو بين نهر هُرمزفرّه وماجان۞ وارباع البلد *معروفة الحدود ولارباعه[l] انهار معروفة فمنها نهر هرمزفرّه وهو نهر عليه ابنية كثيرة من البلد وهو ممّا يلى سَرَخْس فى اوّل ما يدخل الداخل من سرخس وهى ابنية كثيرة كان *الحسين بن طاهر[m] بنا فيها تلك الابنية واراد ان ينقل اليها السوق ودار الامارة ومن هذا النهر شرب محلّة راس

---

a) E. خوانند — كه امروز آنرا. *b*) B. ونزلت. *c*) E. وبازار ماجان. *d*) Ex C.; in textu A. et B. لم يذكر المساحة. Cf. *Djih.-Numa*, p. ٣١٦ infra. *e*) In his A. ودمض et دصمه, B. وننص et لصمه. Ceteri om. *f*) Ex A. et B.; ceteri om. (In A. lacuna non indicatur). *g*) C. بقدر المدينة عظمًا. *h*) C. مربع فى. *i*) C. et D. تجرى فيه. *k*) B. مباطيخ l. مباطيخ. *l*) Addidi ex D. et partim ex E. *m*) Videtur legendum esse طاهر بن الحسين; E. بر كنار اين رود Pro فيها ibi حسن طاهر.

الشّابَاى الّذى فيه دور الشيخ الجليل ابى الفضل محمّد بن عبيد الله *a*، ومنها نهر يعرف بالماجان وعليه دار الامارة والمسجد الجامع المُحْدَث والحبس وعلى هذا النهر دار آل ابى النجم مولى آل ابى مُعَيْط وهى الدار الّتى فيها القبّة الّتى صبغ *b* فيها سواد دعوة بنى العبّاس والقبّة باقية الى اليوم، ومنها نهر يعرف بالرِّزيق ومجراه على باب المدينة ومن هذا النهر يشرب اهل المدينة *بسياق من *c* هذا النهر الى حياض فيها وعلى هذا النهر المسجد العتيق ومن اسفل هذا النهر *قصور آل خالد بن احمد بن حمّاد *d* الّذى كان على امارة بخارا، ومنها نهر *يعرف باسعدى الخراسانىّ *e* وعليه شرب محلّة باب سَنْجان *f* وبنى ماهان وغيرها وعلى هذا النهر كانت دور مرزبان مرو، فهذه انهار مرو الّتى عليها محالُّ البلد وابنيتها وعلى هذه الابنية سور يحيط بها وبهذه الاربعة الانهار ويحيط بهذه المدينة ورساتيقها سور آخر يشتمل على جميع رساتيقها يعرف بالراى وتُرَى آثارُ هذا السور الى هذه الغاية، وللمدينة الداخلة اربعة ابواب فمنها باب يُعْرَف *بباب المدينة *g* ممّا يلى مسجد الجامع وباب *يسمّى باب *h* سَنْجان وباب يسمّى

---

a) Notus vezirus Samanidarum, vulgo البَلْعَمِىُّ dictus; vid. Ibno'l-Athír, VIII, p. ٩٩, ١٩٩, ٢٠٧ (ubi male البلغمى) et ٢٨٣ (ubi male عبد الله البلغمى) et Jacut, I, p. ٧٣٣, 19 seqq. Obiit anno 329. *b*) A. et B. صنع. *c*) A. لسما ومن, B. سان ومن. Cf. Abulfeda, p. ٤٥٧. *d*) Avum hujus عمرو appellat Abu'l-Mahásin, II, p. ٣٩, خالد Ibn Khallicán, n. 580, p. ١٠١. In E. legitur: سراهاى كه بباسعيد خراسان باز خوانند آل خالد وكوشكها احمد بن حمادست. *e*) E. Apud Mokaddasí MS. p. 159 in f. nomen canalis scribitur اسعدى. *f*) A. et B. سحان, E. indist. شهان. Mokaddasí pro باب سنجان habet بارسحان. Vid. Jacut in v. Deinde pro وبنى ماهان A. et B. وبرماهان, Mokaddasí ومرماهان, E. وحدود ماهان. Cf. *Lobbo'l-lobáb* ميرَمَاهان قرية بمرو. *g*) Supplevi ex Mokaddasí MS. p. 151 et Ous. p. 216 ubi بباب شارستان. *h*) A. et B. corrupte حمى باب (وباب). Deinde A. et B. سنجار, E. شيخان (ut *Djih.-Numa*,

باب بالين[a] وباب در مشكان[b] ومن هذا الباب يخرج الى ما ورآء النهر وعلى هذا الباب مسكن[c] المأمون ومضربه أيّام مقامه بمرو الى ان انتهت الخلافة اليه ❁ ولمرو نهر عظيم يتشعّب هذه الانهار كلّها وانهار الرساتيق منه ومبتدآه من[d] ورآء الباميان ويعرف هذا النهر بمرْغاب[e] اى *مآء مَرْو[f] ومن الناس من يزعم انّ النهر منسوب الى مكان يخرج منه المآء يسمّى مرغاب ومنهم[g] من يقول[h] تفسير مرغ اجمة ومجرى هذا النهر على مَرْوَرُوذ وعليه ضياعهم واوّل حدّ هذا النهر من عمل مرو كوكين[i] بين خوزان والقَرِينَيْن[k] *فخوزان من مرو[l] الروذ والقرينين من مرو ومقاسم هذا المآء من زرّق[m] قرية بها مقسم مآء مرو وقد جُعل لكلّ محلّة وسكّة[n] من هذا النهر نهر صغير عليه الواح خشب فيها ثُقَب يتساوى بها الناس فى تناول حصصهم من المآء فان زاد اخذ[o] كلٌّ شرب نصيبه من الزيادة وكذلك اذا نقص ويتولّى هذا المآء امير على

---

p. ٣١٧, 7 a f.; Ous. (شهجان), D. سيحان, Mokaddasí سنحان. *a*) D. المن, E. بر (ut *Djih.-Numa*, l. l.), Mokaddasí مالـن. *b*) A. et B. مسكان; D. مشكان, Mokaddasí مينكان. Secutus sum E. et *Djih.-Numa*. Deinde E. واين دروازهٔ خراسانست. *c*) D. عسكر, E. سراى, Mokaddasí قصر. Deinde pro مضربه E. habet لشكرگاه. *d*) A. et B. om. Vid. quoque Abulfeda, p. ٤٤٦. *e*) C. et D. addunt مرواب (وتفسيره) واصله. Pro بمرغاب A. بمرزاب. *f*) A. تامر, B. دامر. *g*) A. et B. ومنه. *h*) B. add. بزعم. *i*) D. كوكر. Deinde pro خوزان, A., B. et E. حوران habent. *k*) A. et B. والفرسى, D. والقرنين, E. وفربعمن. *l*) B. om. Deinde in A. et B. deëst الروذ. *m*) A. hic et paullo infra ررق, B. رزق, E. زريق; D. دور et الورق. Infra in itinerario A. et B. زرُق, D. درق (et دزق). Nomina زرّق (زرّق) et رَزيق (زريق) confunduntur a Jacut, nam II, p. ٧٧٧ narrat regem Jezdedjird occisum fuisse Razíko, p. ٩٢٥ Zarko. Est autem رزيق nomen fluvii (cf. Jacut, IV, p. ٥٠٨, 19 seq.), cui adjacet pagus Zark. Sed hinc quoque fluvius نهر الزرق appellatur, ut e. g. apud Mokaddasí p. 159 in fine. *n*) B. وصَنكة. *o*) A. et B. sine punctis.

حدّة وهو اجلّ من والى المعونة بلغنى انّه يرتزق على هذا المآء زيادة على عشرة آلاف رجل لكلّ واحد منهم على هذا المآء عمل ❋ وكانت مرو معسكر الاسلام فى اوّل الاسلام وفيها[a] استقامت مملكة فارس للمسلمين لانّ يزدجرد ملك الفرس قُتل بها فى طاحونة زَرْق ومنها ظهرت دعوة[b] بنى العبّاس وفى دار آل ابى النجم المُعَيْطىّ[c] صُبغ اوّل[d] سواد لبس[e] المسوّدة وفيها جآءت المامونَ[f] الخلافة وظهر على * اخيه محمّد بن زُبَيدة[g] ومنها عامّة قوّاد الخلافة وكُتّابها[h] بالعراق ووُلاة[i] خراسان * ومنها ائمّة من الفقهآء واهل الادب معروفون[k] ولولا انّا بنينا[l] كتابنا على التجوُّز[m] وأنّ الّذى تركنا شرحه هو معروف فى الاخبار والكتب المؤلّفة لشرحنا من طبقات الناس وسائر ما اجملنا[n] ذكره ٬ وفى ايّام العجم كانوا مقدّمين من بين نواحى أَبْرَشَهْر[o] فى الطبع والتادّب حتّى كان طبيبهم المعروف[p] ببرزويه مقدّمًا على سائر اطبّآء العجم ومُلْهيهم المعروف بالبارْبَد مقدّم على سائر مَنْ صاغ[q] الالحان وتعاطى الملاهى[r] ٬ ثم هى من اطيب بلاد خراسان اطعمةً امّا خبزهم فليس بخراسان[s] انظف خبزًا والذّ طعمًا منه[t] حتّى انّ اليابس من فواكهها من الزبيب وغير ذلك يفضّل على سائر الاماكن وانّما يذكر[u] من قَرَاة الكثرة[v] وانّه يكثر فى الآفاق فامّا الطعم والجودة فانّ المروزىّ يفضله ٬ ومن صحّة فواكههم انّ البطّيخ يقدّد ويُحْمَل الى العراق[w] ولم اعلم هذا يمكن ببلد غيره[x] ٬ وبلدهم من النظافة وحسن الترصيف وتقسيم الابنية والمحالّ فى

---

a) C. وبها. b) C., D., E. et Abulfeda, p. ۴۴٦ دولة. c) C. add. مولى لهم. d) A. et B. om. e) Abulfeda لبسته et sic D.; deinde C. للمسودة. f) C. لعبد الله المامون. g) C. الامين. Deinde Abulfeda وخرج منها. h) C. وكماتها. i) A. ولولاة. k) C. ومنها جماعة من ائمّة الفقه والادب معروفين. l) B. بينا. m) A. التجور. n) A. et B. احملنا. o) C. ايرانشهر. p) A. et B. om. q) B. ضلع. r) E. addit: وهمكنان هنوز بوى اقتدا كنند. s) C. مدينة. t) C. منها. u) A. sine punctis. v) B. اكثره. w) C., D. et E. الافاق. x) E. addit كه تباه شود.

خلال الانهار والغروس وتمييز اهل كلّ سوق من غيره بحيث يفضل سائر مدن خراسان فى حسنه، وفى مفازتهم يكون الأُشْتُرغاز[a] الّذى يحمل الى سائر الدنيا ويرتفع من مرو الابريسم[b] والقزّ الكثير وبلغنى[c] انّ اصل الابريسم بجرجان وطبرستان انّما نُقل فى القديم من مرو وربّما حمل من بزر دود القزّ منها الى طبرستان ومنها يرتفع القطن الّذى ينسب اليه القطن الليّن والثياب الّتى تجهّز الى الآفاق ۞ ولها[d] منابر قديمة وحديثة فبمرو منبران وبكُشْمَيْهَن[e] منبر وبهُرْمُزفَرّه[f] منبر وبِسِنْج[g] منبر وبجِيرَنْج[h] منبر وبالدَّنْدَانقَان[i] منبر وبالقَرِينَيْن[k] منبر وبَباشَان[l] منبر وبخَرَق منبر وبالسُّوسَقَان منبر فهذه منابر مرو الّتى اعرفها ۞

امّا هَرَاة فانّها اسم المدينة ولها اعمال ومن مدنها مالِن[m] وخَيْسَار[n] واستربيان[o]

---

a) A. et B. الاسترعار, C. الاشرغار, D. sine punctis; E. اشترقاز. b) B. et E. cum ش. c) E. ومشهوراست. d) B. et C. وبها. e) A. وبكشيمهن, B. وبلشمهن. Apud Ibno'l-Athír, VIII, p. ٣٤٧ كشماهن legitur. f) A. et B. ونهر مرفوه. g) A. وسمح, B. وسخ, E. سمح. h) A. et B. وبحرنح, E. وبحرنح. Vid. Jacut in v. Apud Mokaddasí p. 27 كسرنك, p. 146 كيوبك, p. 151 كبربك scribitur i. e. كيرنك. i) E. sine punctis; A. وبالزندادعان. B. hoc et duo seq. nomina om. k) A. وبالعرسن, E. عرنعس. l) A. وبباسنان; D. وبياسان, E. باسان. Mokaddasí p. 27 et 146 باسيان, p. 151 باسان. Vid. Jacut in v. et sub فاشان. m) A. et B. مالق, E. مالف. n) A. وحساب, B. وحسان ut E. Nomen hoc (de quo vid. Jacut in v.) fere ubique corruptum legitur. Infra Codd. habent حسان s. حسان; Mokaddasí habet semel حسان, bis حتسان, sed in itinerario p. 165 حتبار. o) Sic h. l. B., A. واستربدان. Deinde A. et B. sine punctis. In itinerario D. et Edrisi (I, p. 460) اسبارابان s. اسنارابان, C. سمريان. Mokaddasí استربيان, اسفريان et اسفريان.

وأَوْفَه[a] ومارأباذ وباشان[b] وكَرُوخ وخشت[c]، وبأَسْفِزار[d] ادرسكر[e] وكواران وكوشك[f] وكواشان[g] واسفزار اسم للكورة[h] لا اسم مدينة ومدنها هذه الاربعة الّتى ذكرناها ۞ وامّا هراة فانّها مدينة عليها سور[i] وثيق وحواليها مآءٌ وداخلها مدينة عامرة ولها ربض وفى مدينتها قلعة[k] ومسجد الجامع * فى مدينتها[l] ودار الامارة خارج الحصن بمكان يعرف بخراسان اباذ منقطع عن المدينة بينه[m] وبين المدينة اقلّ من ثُلْث فرسخ على طريق[n] بوشنج على غربىّ هراة وبنآؤها من طين وهى مقدار نصف فرسخ فى نحوه ولمدينتها الداخلة اربعة ابواب الباب الّذى يخرج منه الى بلخ[o] ممّا يلى الشمال يسمّى باب سَرَاى والباب الثانى الّذى يخرج منه الى نيسابور غربىٌّ يسمّى باب زياد[p] والباب الثالث الّذى يخرج منه الى سجستان جنوبىٌّ[q] يسمّى باب فيروزاباذ

---

a) E. اوبه ut Jacut. *b*) A. وداسمان. Vid. Jacut in v. et sub فاشان *c*) A. et B. sine punctis; E. حسن. In itinerario A. حسب, B. حسب, D. خشت, C. et Mokaddasí حسن. Idem locus videtur appellari جشن in *Djih.-Numa*, p. ٣١٢, 20. Sed Anderson et Sprenger ediderunt ut D. habet. *d*) A. et B. وداسترار. E. ماسوان s. ماستوان (Ous. ماشران). *e*) A. et B. h. l. ادرسكر ut Mokaddasí p. 27 et 145. Infra A. اردسكر, B. ادرسلر; D. ادرسكن; Dimaschkí p. ٢٢٥ اردسكن; Edrisi, I, p. 463 ادراشكن; E. et mappa C. ادرسكر. Ous. p. 217 ادرشكر (ut recepit Sprenger, p. 38). Deinde B. ودواران. *f*) A. وكوسند, B. وكوسند; infra A. et B. كوسك; E. كوسف; D. كوسد; Mokaddasí كوسد et كوشد. Edrîsî كوشد. Cf. Anderson, p. 174. *g*) Secutus sum D. et Mokaddasí. A., B. et E. h. l. وخراسان. Infra A. et B. كواسان; Edrîsî كراسان, Ous. اشران. Deinde A. واسمرار, B. واسراد i. e. واسبزار. *h*) B. الكورة. *i*) C. et D. حصار et sic E. *k*) C., D. et E. قهندز. *l*) A. et B. om. *m*) C. et D. بينها. *n*) C. ظهر. Deinde A. et B. سمح. *o*) A. et B. h. l. male نيسابور et mox inverso ordine بلخ pro نيسابور. *p*) Ex D. et Mokaddasí p. 149; A. et B. iterum سراى. *q*) Ex D.; A. et B. om.

والباب الّذى يخرج منه الى الغُور شرقىّ يسمّى باب خُشْك[a] وابوابها من خشب غير باب سراى[b] فانّه حديد وعلى كلّ باب سوق * يشتمل بما[c] يليه من المحالّ وفى داخل المدينة والربض مياه جارية وللحصن اربعة ابواب بحذآء كلّ باب من ابواب المدينة باب لهذا الحصن ويسمّى باسم ذلك الباب وخارج الحصن جدار يطوف بالحصن كلّه اطول من قامة وبينهما مقدار ثلاثين خطوة والمسجد الجامع من المدينة فى وسطها وحواليها اسواق والسجن على ظهر قبلة مسجد الجامع وليس بخراسان وما ورآء النهر وسجستان والجبال مسجد اعمر بالناس على دوام الأيّام من مسجد هراة ثمّ مسجد بلخ ثمّ مسجد سجستان فانّ بهذه المساجد حلق الفقهآء والناس يتزاحمون على رسم الشام والثغور وسائر المساجد بهذه الاماكن انّما ينتابها الناس فى الجمعات، وهراة[d] مطرح الحمولات من فارس الى خراسان وهى فرضة لخراسان وسجستان وفارس، والجَبَلُ من هراة على نصف فرسخ على طريق بلخ ومحتطبهم من مفازة بينها وبين اسفزار وليس بهذا الجبل محتطب ولا مرعى وانّما يرتفقون منه بالحجارة للارحية والفرش وغير ذلك وعلى راس هذا الجبل بيت نار يسمّى سِرِشك وهو معمور وبينها وبين المدينة كنيسة للنصارى[e] وليس بينها وبين المدينة مياه ولا بساتين الّا نهر المدينة على باب المدينة يعبر بالقنطرة ثمّ لا يكون بعده مآء ولا خضرة، وعلى سائر الابواب مياه وبساتين اعمرها باب فيروز اباذ ومخرج مآئهم من قرب رباط كروان فاذا خرج عن الغور الى هراة ينشعب[f] منه انهار فمنها نهر يسمّى برخُوى[g] يسقى رستاق سنداسنك[h] ونهر يسمّى بارست[i] يسقى رستاق

a) A. et B. sine punctis; Mokaddasí حمسك; D. كشك; E. درخشك; vid. Jacut in v. — *Djih.-Numa*, p. ٣١٠ درب خوش. b) E. بلخ. c) B. يشمل مما.
d) E. h. l. هرى quae lectio quoque exstat in mappa C. e) B. النصارى.
f) A. sine punctis. g) A. et B. برخُوى, E. يرخوى, D. وحوى, Edrîsî, I, p. 462 وحرى; Mokaddasí MS. p. 159 دوحون. Deinde A. et B. male يسمى.
h) A. et B. سنداسنك, Ous. سپيداسنك; Mokaddasí شبيدان; Edrîsî سنداسنة. Secutus sum D. i) A. et B. sine punctis; E. بار; D. باريست; Mokadd. بارشت.

كواشان a وسياوشان b ومالن وتيزان c وروامز d ونهر اذربيجان e يسقى رستاق
سوسان f ونهر يسمّى سكوكان g يسقى رستاق سله h * ونهر يسمّى كَرَاغ يسقى
رستاق كوكان i ونهر يسمّى غوسمان k يسقى رستاق كرك l ونهر يسمّى كنك m
يسقى رستاق غُوبان n وكربكرد o * ونهر يسمّى فَغَر يسقى رستاق بغاوردان
وفيرد p ونهر يسمّى آنجير q يسقى *مدينة هراة r ‹ والبساتين متّصلة على
طريق سجستان مقدار مرحلة ۞ واكبر مدينة بهراة بعد هراة كَرُوخ وأَوْفَه
ويرتفع من كروخ الكشمش الّذى يجلب الى الآفاق والزبيب الطائفىّ الّذى
يحمل الى الآفاق معظمه يرتفع من مالن وكروخ مدينة صغيرة واهلها شراة

---

a) A., B. et Mokaddasí كواسان, E. كراسان. *b*) Secundum E.; A. et B. وساوسان; Mokaddasí وسياوسان. Vide *Djih.-Numa*, p. ٣١١, 19. *c*) A. وممران, B. ومموان; Mokaddasí وبمران; D. et E. om. Anderson, p. 183, 19 edidit ميزان, sed in vers. p. 158 habet Teezan. Vid. Jacut in v., *Djih.-Numa* l. l. 22. *d*) Sic A.; B. ودوابر (Anderson وروابر). Ceteri om. *e*) A. اذربسجان, B. اذربيجان, E. ارديحان (Ous. اردنجان), Mokaddasí اذربنجان. *f*) D. سوسار, E. ميرستان, Ous. سبرشيان, Mokaddasí مموسان. *g*) E. بسكوكان, Ous. نسكوكان. *h*) Sic A. et B; E. زرّين, Ous. ورين. *i*) Edrîsî كوركان (D. et E. om.). Mokaddasí ut recepi. *k*) B. غودسمان, Edrîsî غوسيجان, Mokaddasí عوسجان. Fortasse conferendum est Jacuti غوسنان. *l*) Edrisî et Mokaddasí كوك. *m*) A. et B. كفل, D. كند, E. كبك. Secutus sum Edrisî et Mokaddasí. *n*) A. et B. عُوبان, D. عوبان, E. عريان, Edrîsî غزنان, Mokaddasí عربان. *o*) Secundum B. et Mokaddasí. A. sine punctis, D. وكودكرد, E. وكردكر (Ous. كربكر), Edrîsî كرنكرد. *p*) Pro his D., E., Edrîsî et Mokaddasí habent: E. ونهر يعرف بسنقر (سعركى (Mokaddasi سوخنه) يسقى رستاق سرخس (Edrisî شيغر, Mokaddasí سعف فى حد بوشنج. *Djih.-Numa* l. l. pagum سقر memorat. *q*) B. النجيز, E. حمركى (Ous. خهركى), Edrîsî جير, Mokaddasí الحمن s. الحمف. *Djih.Numa* انجيل. *r*) Mokaddasi القصبة.

ومسجد الجامع بمحلّة منها تسمّى سبيدان[a] وبناؤها طين وهى فى شعب بين جبال وحدّها مقدار *عشرين فرسخا[b] كلّها مشتبكة البساتين والمياه والاشجار والقرى العامرة، وأوْفة اهل جماعة وهى نحو[c] كروخ ولها بساتين ومياه وبناؤها من طين ۞ ومالن[d] اصغر من كروخ وهى مشتبكة البساتين والمياه والكروم عامرة جدًّا ۞ وخَيْسار[e] قليلة الاشجار وهى اصغر من مالن واهلها اهل جماعة ۞ واستربيان اهلها خوارج وهى اصغر من مالن ولها مياه وبساتينهم قليلة والغالب عليهم الزرع دون الكروم وهى فى الجبال ۞ وماراباذ كثيرة البساتين والمياه وهى مدينة اصغر من مالن ويرتفع منها ارزّ كثير يجلب الى النواحى ۞ وباشان[f] مدينة اصغر من مالن ولهم زرع وهى قليلة البساتين على كثرة مياهها ۞ وبأسْفِزار اربع من المدن واكبرها كواشان وهى مدينة اصغر من كروخ ولها ماء وبساتين كثيرة وكواران وكوشك وادرسكر هى متقاربة فى الكبر ولها مياه وبساتين واسفزار مقدارها ثلاث مراحل فى مرحلة وهى كلّها عامرة ليس فى ظهرانيّها مفازة، وباسفزار شعب يسمّى كاشكان وفيها قرى عامرة كلّهم شراة فامّا مدن اسفزار فانّ اهلها اهل جماعة ۞

وامّا بُوشَنْج فانّ بها من المدن خَرْكَرْد[g] وفَرْكَرْد[h] وكوسوى[i] وكره واكبرها[k]

---

a) Sic A.; B. يسمى بسيدان; D. يعرف بسندار; Mokaddasí MS. p. 149 تسبدان. b) A. et B. فرسخ; vid. quoque Abulfeda, p. ٤٥٩, Jacut IV, p. ٢٧٠, 4 et Mokaddasí l. l. سعة رستاقها عشرون فرسخا. c) B. بحر, D. بحذا. E. چند. d) A. h. l. ومادى, B. ومابين; E. مالان, Mokaddasí مالين. e) Vid. supra p. ٢٦٣, ann. n. f) A. وداسمار, B. وناسار, Edrîsî, I, p. 462 ناشان. Vid. supra ad p. ٢٦٤, ann. b. g) Jacut et Abulfeda خرجرد. E. خروكرد, Ous. خسروكرد, quod recepit Anderson. *Djih.-Numa*, p. ٣١٣, 5 a f. خردكرد. h) A. وقركود h. l., B. وفزكرد; E. نوكرد; Abulfeda, p. ٤٥٣ مزكرد. Mokaddasi p. 145 et 150 فركرد, sed p. 27 فلاجرد et sub hac forma memoratur a Jacut. i) Sic Abulfeda l. l., Anderson p. 174 et Sprenger p. 38, secundum Mokaddasí. A. h. l. كودمر, B, كوسر. Deinde omnes كوسرى. *Djih.-Numa* l. l. كوسويه. k) A. et B. واكثرها.

بوشنج وهى مدينة نحو النصف من هراة وهى وهراة فى مستوٍ ومن بوشنج الى الجبل نحو فرسخين وهو هذا الجبل الّذى من هراة اليه نصف فرسخ وبناؤهم[a] من جنس بناء هراة ولهم مياه واشجار كثيرة وبها من اشجار العرعر ما ليس بجميع خراسان فى بلد ويحمل هذا الخشب الى سائر النواحى ومآؤهم من نهر هراة وهو النهر الّذى يخرج الى سرخس غير انّه ينقطع المآء دون سرخس ويستعمل الّا فى بعض السنة ولبوشنج سور وخندق وثلاثة ابواب باب يسمّى باب علىّ الى نيسابور وباب هراة الى هراة وباب قوهستان الى قوهستان ۞ واكبر المدن بها بعد بوشنج كوسوى وهى مدينة لها مآء وبساتين قليلة وهى نحو الثلث من بوشنج وبناؤهم من طين ۞ وخَرْكَرْد لها مآء وبساتين كثيرة وهى اصغر من كوسوى ۞ وفَرْكرد اصغر من خركرد *ولها مآء جارٍ[b] وهم اصحاب سوائم وليس لهم بساتين كثيرة ولهم مآء جارٍ قليل ۞ وكره لها بساتين ومياه كثيرة وهى نحو من فركرد فى الكبر ۞

وامّا بَاذَغيس فانّ بها من مدنها جَبَلُ الفضّة[c] وكُوفَا[d] وكوغناباذ[e] وبُشْت وجاذَوى[f] وكابرون[g] وكالوون[h] ودِهِسْتان والسلطان يكون مقامه بكوغناباذ واعمرها واكبرها دهستان وتكون نحو النصف من بوشنج وبناؤها من طين

---

a) A. et B. وبناوه. E. hic semper هرى pro هراة habet. b) Videntur omittenda. c) E. كوه سيم. d) A., B. et D. h. l. كوه, mox كو; B. et D. semel كر; E. om. Sed Mokaddasí recte habet كوفا; vid. Jacut, IV, p. ٣٢٠, paen. e) A. sine punctis. E. كوى عماباد. Mokaddasí كون عناداد ,كوغزاباذ ,كوغاناباذ et كوغيباداد. Deinde A., B., D. et E. وبست. Vid. Jacut, I, p. ٦١٣٠, 2. f) A. h. l. وحاذو, B. وجارو, D. وجادو, E. وحارور. Infra A. et B. حادوى, D. وجادوا, E. وحارو. Mokaddasí p. 27 حادواذ, p. 145 جاداوا, p. 150 corrupte حاكورا. g) A. وىحابرون h. l., B. وىحايرون, infra A. et B. كافورن. Mokaddasí bis كادوون, semel كازون. Apud D. كابوون, كايوون et كادوون exstat, sed incertum est uter locus, hic an sequens, intelligatur. Secutus sum E.; Edrisi, I, p. 464 seq. habet كانوور. h) Jacut habet كالوان. E. كالون. A. et B. infra كالزون. Mokaddasí bis ut recepi, semel كاليون.

ولهم اسراب كثيرة فى الارض وهى على جبل ولهم ماء جارٍ قليل وليست لهم بساتين ولا كروم وانّما هى مباخس وكذلك كوفا وجبل الفضّة وكوفا اكبر من جبل الفضّة وجبل الفضّة على جبل كان فيه معدن الفضّة وتعطّل لفناء[a] الحَطَب وامّا كوفا فانّها فى صحراء ۞ وبكوغناباذ[b] وبشت وجاذوى بساتين ومياه ولهم مباخس كثيرة، وكالوون وكابرون ليس لهم بساتين ولا مياه جارية وانّما مياههم من الامطار والآبار وهم اصحاب زروع مباخس واصحاب اغنام، وجبل الفضّة على طريق سرخس من هراة، وباذغيس[c] اهل جماعة الّا خُجُسْتان قرية[d] احمد بن عبد الله[e] فانّ اهلها شُراة ۞

وامّا كَنْج رُسْتَاق فانّ مدينتها بَبَن[f] ولها كيف وبَغْشُور والسلطان منها بببن وهى اكبر هذه المدن وببن اكبر من بوشنج وبغشور نحو بوشنج[g] فى الكبر وكيف نحو نصف بغشور وببن وكيف لهما مياه كثيرة جارية وبساتين وكروم وبناؤهما[h] من طين، وامّا بغشور فانّها فى مفازة وهى عذى وزروعهم كلّها مباخس ومآؤهم من الآبار *وهم اصحاب[i] زروع وهى مدينة[k] صحيحة التربة والهواء، وهذه المدن كلّها على طريق مَرْوَرُوذ ۞

ومَرْوَرُوذ بها من المدن قصر أَحْنَف[l] ودِزَه[m] ومَرْوَرُوذ واكبرها مروروذ وهى اصغر[n] من بوشنج ولها نهر كبير وهذا النهر الجارى[o] الى مرو ولهم عليه

---

a) A. لقياء, B. لقناء. b) A. et B. وبكرعناباذ. c) Jacut, II, p. ٤٠٤ واهل باذغيس. d) A. et B. وقرية. e) E. addit خارجى. Cf. Ibno'l-Athír, VII, p. ٢٠٤ seqq. f) A. sine punctis, B. et D. بين. Locus appellatur quoque بَوَن, بَوْن et بَبْنَة (sic restituatur apud Beládsorí, p. ٤٠٥, 7). Nomen urbis sequentis in editione Jacuti, IV, p. ٣٠٨, 13 et ٣٣٣, 17 pronunciatur كَيْف. g) A. et B. فرشنج (Anderson infelicissime edidit فرسخ). h) A., B. et D. وبناوها. Mokaddasí p. 150 بناوهم. i) Ex D.; A. et B. وهى اخصب. k) A. et B. مدن; Abulfeda, p. ٤٥٧ وهى من المدن الصحيحة. l) E. كوشك احنف. m) A. et B. ودره. Jacut درق. Deinde in A. et B. copula deëst ante مروروذ. n) C. et D. habent اكبر et sic Abulfeda, p. ٤٥٧. o) In B. additur من. In Epit. Paris. من الباميان legitur. Cf. supra p. ٣٦١, 4.

بساتين وكروم كثيرة وهى طيّبة التربة والهوآء ۞ وقصر احنف على مرحلة منها على طريق بلخ ودزه على طريق أَنْبار[a] على اربعة فراسخ وقصر احنف لها مآء جارٍ ولها[b] بساتين وكروم وفواكه حسنة، ودزه يشقّ نهر مروروذ وسطها وهى نصفان وبينهما قنطرة ولها بساتين وكروم وفواكه حسنة، ومن مروروذ الى النهر غلوة ۞ والطّالَقان مدينة نحو من مروروذ فى الكبر ولها مياه جارية وبساتين قليلة وبنآؤها وبناء مروروذ من طين وهى اصحّ هوآء من مروروذ، ومن مروروذ الى الجبل ثلاثة فراسخ ممّا يلى المغرب *ومن جانب[c] الجبل منه على فرسخين ممّا يلى المشرق، والطالقان فى الجبل ولها رساتيق فى الجبل ۞ والفارِيَاب[d] مدينة اصغر من الطالقان الّا انّها اكثر بساتين ومياهًا[e] من الطالقان وبنآؤها من طين ۞

والجُوزَجان اسم للناحية ومدينتها اليَهُودِيّة وشَبُورْقان وأَنْخُذ[f] رستاق ومدينتها أُشْتُرْج[g] وكُنْددرم[h] وأَنْبَار وسَانُ[i] واكبرها انبار وبها مقام السلطان وهى مدينة على الجبل وهى اكبر من مروروذ ولها مياه وكروم وبساتين كثيرة وبنآؤها طين ۞ وسان مدينة صغيرة لها مياه وبساتين والغالب على ثمارها الجوز وهى فى الجبل ۞ واليهوديّة[k] اكبر من سان ولها مياه وبساتين وهى

---

a) A. انسار, B. اسار, D. الشار. Cf. Jacut sub الانبار et انبير. b) B. وبها. c) Sic A., B. et D., ceteri om. Fortasse legendum وقصر احنف. d) A. والقاربان, B. والقاريان. e) Abulfeda, p. ٤٦١ ومياهها. f) A. plerumque انحد, semel ut B. انجد et sic erat in Codice quo usus est Jacut (III, p. ٢٥٧, 1). Regio a Jacut appellatur اندخود et memoratur sub forma نخذ (D. habet النخذ). g) A. السلج, B. اسلج. Infra A. اسمرج, B. سيرج, C. السلج; ceteri om. Lectio mea mera est conjectura, sed cf. Jacut in v. h) Sic semper A.; B. كندررم; E. كندرم, ut Jakubí, p. ٦٦, et sic D. sed una littera ante م expuncta. Edrîsî, I, p. 469 كيدرم. i) A. et B. h. l. et deinde سسان. Lectio constat collatis inter se Jacut, III, p. ٢٣, 15 et *Djih.-Numa*, p. ٣١٦, 22. Jakubí l. l. habet اسان. k) E. جهودان. Postea haec urbs ميمنة dicta est.

فى الجبل ۞ وكندرم فى الجبل وهى مدينة كثيرة الكروم والجوز ولها مياه كثيرة ۞ وشُبُورْقان لها ماۤء جارٍ والغالب عليهم الزروع وبساتينهم قليلة وهى اكبر من كندرم *ومن شان[a] وهى نحو من اليهوديّة فى الكبر ۞ وأُشْتُرْج مدينة أَنْخُذ[b] مدينة صغيرة فى مفازة لها سبع قرى وبيوت للاكراد[c] اصحاب اغنام وابل منها *شعر مَرْوَ[d]، ويرتفع من ناحية الجوزجان الجلود[e] الّتى تحمل الى سائر خراسان وهى عامّة الخصب ۞ فمن شبورقان[f] الى انبار مرحلة فى ناحية الجنوب ومن شبورقان الى اليهوديّة يحتاج ان يرجع الى *فارياب مرحلتين[g] ثمّ منها الى اليهوديّة مرحلة ومن شبورقان الى انخذ مرحلتان فى الشمال ومن شبورقان الى كندرم اربع مراحل ثلاث مراحل الى اليهوديّة[h] ومرحلة اليها ۞

وغَرْج الشار[i] لها مدينتان احداهما تسمّى نشين[k] والاخرى سُورمين وهما

---

a) A. et B. ومرسان tamquam nomen oppidi habent et Ibn Haucal hinc revera oppidum finxit, quod in Codd. مرقان, مرتان, مرنان (ut apud Edrisi, I, p. 470) s. برنيان (ut in Epit. Par.) appellatur, et ponitur inter اليهودية et الفارياب. Me recte Ibn Haucalo hoc oppidum reddidisse, itaque pronomen وهى pertinere ad شبورقان, patet quoque e C. qui habet (l. وشبورقان) وسان صغيرة وسعور بقدر اليهودية. *b*) A. et B. وانجد. *c*) C. et D. الاكراد. *d*) Conjectura scripsi. A. et B, اسعر مدد. *e*) D. addit المدبوغة; E. سختيان كوركانى. *f*) A. h. l. سبوركان. *g*) A. et B. ماريار من حمس. Cf. Jacut, III, p. ٢٥٩ ult. Hic addit فى الشمال. *h*) Ex D.; A. الهرز, B. النهر. *i*) A. et B. السار, C. السان. E. غرجستان. Vid. praeter Jacut, III, p. ٧٨٥, qui transscripsit locos Istakhríi et Mokaddasíi, Ibno 'l-Athír, IX, p. ١٠۴. *k*) A. دسمر, B. سسر et سير, C. دسير, D. افشين, E. بشين, دشن et semel دشنر; Edrîsî, I, p. 470 بشين (ut legit Sprenger, p. 38). Mokaddasí MS. p. 27 ادسن, p. 150 انسن, p. 165 in itin. ادشين. Jacut l. l. 12, 16 et 20 بشير, sed Codd. *Merácid*, II, p. ٣٠٧ دسين; Kazwíní, II, p. ٢٨٥ نشين ut habet Ous. — Pro سورمين D. et Mokaddasí habent شورمين.

متقاربتان فى الكبر وليس بهما مقام للسلطان[a] والشار الّذى ينسب اليه المملكة مقيم بقرية فى الجبل تسمّى بَلْكيان[b] وهاتان المدينتان لهما مياه وبساتين ويرتفع من نشين ارزّ كثير يحمل الى البلدان ويرتفع من سورمين زبيب كثير يسافر بها وبين نشين وبين دِرّة مروروذ مرحلة *فى المطلع وهى[c] من نهر مروروذ على غلوة عن شرقيّه ومن نشين الى سورمين مرحلة ممّا يلى الجنوب وهى فى الجبل ۞

وامّا الغُور فانّها دار كفر *وانّما ذكرناه فى الاسلام لانّ به مسلمين[d] وهى جبال عامرة ذات عيون وبساتين وانهار *وهى خصيبة منيعة[e] وفى اوائلهم ممّا يلى المشرق قوم يظهرون الاسلام وليسوا بالمسلمين ويحتفّ بالغور عمل هراة الى فَرَّه[f] ومن فره الى بلدى داور ومن بلدى داور الى رباط كروان من عمل ابن فريغون[g] ومن رباط كروان الى غرج[h] الشار ومنها الى هراة فهذا الّذى يطوف بالغور *كلّها مسلمون[i] وانّما ذكرناها لانّها فى وسط الاسلام ۞

وامّا سَرْخَس فانّها مدينة بين نيسابور ومرو وهى فى ارض سهلة وليس لها ماءٌ جارٍ الّا *نهر يجرى[k] فى بعض السنة ولا يدوم مآؤه وهو فضلُ مياه هراة وزروعهم مباخس وهى مدينة على *نحو النصف[l] من مرو وهى عامرة صحيحة

---

a) C. السلطان. b) A. نكماكان, B. بكنكان, C. دلمكان, D. بلكتان, Mokaddasí بليكان et دلمكان. Vid. Jacut in v. c) Ex D.; A. et B. والمطلع وهو. E. وآب مرو از نشين بدرّه مرو الرود آيد et deinde ودشمر بالاى درّه است. d) C. الا انها بين المسلمين وبها مسلمون. e) C. supra عيون omisso, habet: كثيرة الاشجار والعيون وهى منيعة جدّا; Abulfeda, p. ٤٦٤ حصينة pro خصيبة. f) A. et B. فرّه, D. فراوه. g) A. قريغور, B. قريغون, D. افريغون, E. افريقون. Secutus sum C. et Ibno 'l-Athír, IX, p. ١٠٣ ult. Cf. supra p. ١٤٨, 3. h) A. et B. حسان. Ceteri h. l. غرجستان. i) C. كله اسلام et addit: ولسان الغور غير لسان اهل خراسان وجبالهم خصيبة كثيرة الزروع quae in A. et B. infra leguntur. k) A. et B. om.; cf. quoque Jacut, III, p. ٧٣, 4; Abulfeda, p. ٤٥٥. l) C. نصف فرسخ. E. واين شهر در حساب كه چند همهٔ مرو الرود بود.

التربة والغالب على نواحيها المراعى وهى قليلة القرى ومعظم املاكهم الجمال وهى مطرح لحمولات ما يحيط بها من مدن خراسان ومآوهم آبار[a] وارحيتهم على الدوابّ وابنيتها طين ❀ وامّا نَسَا فانّه اسم المدينة وهى خصبة كثيرة المياه والبساتين وهى فى الكبر نحو سرخس ولهم مياه جارية فى دورهم وسككهم نزهة جدًّا ولها[b] رساتيق واسعة خصبة وهى فى اضعاف الجبال ❀ وفَرَاوَة ثغر فى وجه * البرّيّة على[c] الغُزّيّة وهى منقطعة عن القرى وفيها منبر يقيم بها المرابطون وهم[d] عدد يسير الّا انّهم يرجعون الى عدّة وافرة ينتابها الناس وهى رباط اسمها فراوة ليس بها قرية ولا تتّصل بها عمارة ولهم عين ماء تجرى للشرب فى وسط القرية وليست لهم بساتين ولا زروع الّا مباقل على هذا الماء واهلها * دون الف[e] رجل ❀

وقُوهِسْتان من خراسان على مفازة فارس وليست بها مدينة بهذا الاسم وقصبتها قَايِن[f] ولها من المدن يُنَابِذ[g] والطَّبَسَيْن وتعرف بكُرِى[h] وخُور[i] وطَبَس

---

a) C. من الابار ut Abulfeda l. l. b) A. et B. ولهم; vid. quoque Abulfeda, p. ٤٥١. c) C. الترك. d) A. et B. وهى. e) C. نحو الفى. f) C. et E. addunt: واهلها شيعة. g) A. et B. h. l. ددسادد, infra دسادد (دمايد); C. مابد et دماددد; D. نبابذ s. بنابذ; Abulfeda, p. ٤٤٤ et ٤٥٣ نيـابـذ; E. بنابيد sed deinde كبادة; Edrîsî, I, p. 452 seq. بيانـد. *Djih.-Numa*, p. ٣٢٦, 15 ﺫ بيابـد. Vulgo scribitur جنابذ s. كنابـذ. In mappis hodiernis *Gunabad* exstat. h) In A. et B. بكروكرند, in D. بكرند legitur, dum C. et E. habent بكرى. Videtur mihi apud D. restituendum esse بكُرين, et apud A. et B. بكرى وكرين; cujus vero pars posterior tamquam nota marginalis e textu rejici debet. Nempe nomen oppiduli *Korín*, quod 3 Par. distat a Tabas, quoque *Korí* scribitur. Vid. supra p. ٣٣٢ et ann. *a*. Duo sunt oppida nomine طبس, unum طبس كيلكى s. طبس التمر, alterum طبس مسينان s. طبس العناب vocatum. Prius quoque طبس كرى (كرين) appellatur, propter vicinitatem. Utrumque oppidum vocatur الطبسين, teste Jacut, III, p. ٥١٤, 20 et 21, sed طبس كرى potius quam alterum. i) A.

ويعرف بطَبَس مَسِينَان[a] ، فامّا قاين فهى من الكبر نحو سَرَخس وبنآؤهم من طين ولها[b] قهندز وعليه خندق ومسجد الجامع ودار الامارة فى القهندز ومآؤهم من القنىّ[c] وبساتينهم قليلة وقراها متفرّقة وهى من الصرود ۞ وامّا الطَّبَسَيْن فانّها مدينة اصغر من قاين وهى من الجروم وبها نخيل وعليها حصن ولا قلعة[d] لها وبنآؤها طين ومآؤها من القنىّ ونخيلها اكثر من بساتين قاين ۞ وامّا خُور فانّها اصغر من الطبسين وهى بقرب خَوْسْت[e] وليس بخوست منبر وانّما المنبر بخُور[f] وبنآؤها من طين وليس لها حصن ولا قلعة[g] ولها بساتين قليلة ومآؤهم[h] من القنىّ وبها ضيق فى المآء واهلها اهل سوائم وهى على طرف المفازة *وليس لهم بساتين[i] ۞ وامّا يُنَابِذ فانّها اكبر من خور وبنآؤها من طين ولها قرى ورساتيق ومآؤها من قنىّ، والطَّبَس[k] اكبر من ينابذ ومآؤها من القنىّ وبنآؤها طين ولها حصن خراب وليس لها قلعة ۞ والنخيل[l] بقوهستان بالطبسين[m] وسائر ما ذكرناه من الصرود، وهذه المدن والقرى الّتى بقوهستان هى متباعدة[n] فى اعراضها مفاوز وليس العمارة بقوهستان مشتبكة اشتباكها بسائر نواحى خراسان وفى اضعاف هذه المدن مفاوز يسكنها الاكراد واصحاب السوائم من الابل والغنم، وفى حدّ قاين منها على يومين ممّا يلى نيسابور هذا الطين المحاحى[o] الّذى يحمل الى

---

et B. وحور ; infra A. حوز , B. جوز ; C. جور ; D. خوز et خور ; E. چون. Cf. Edrîsî, I, p. 435 et 453 et vid. supra p. ٢٣٤. In mappis hodiernis etiam memoratur. *a*) A. et B. مسار, C. مسان. *b*) B. وبها. *c*) A. et B. العين; cf. quoque Abulfeda, p. ٤٥٣ et Mokaddasí apud Jacut, IV, p. ٢٣, 4. *d*) C., D. et E. قهندز. *e*) A. et B. حوست, D. خوشب, sed Epit. Paris. خُواست. Vid. supra p. ٢٣٤ et ann. *b*, Jacut, IV, p. ٢٣, 6. *f*) A. نحوار, B. بجوار. *g*) C. et D. قهندز. *h*) C. et D. ومآؤها. *i*) Haec verba pugnant cum iis quae paullo ante dicta sunt ولها بساتين قليلة. Utra delenda sint, efficere nequeo. *k*) A. والطبسين. *l*) C. والحمل. *m*) B. بالطبس, C. والطبسين. *n*) B. مشاهدة. Cf. quoque Jacut, IV, p. ٢٠٩, 12. *o*) A. et D. المحاحى,

الآفاق للاكل، وليس بقوهستان فيما علمته نهر جارٍ الّا القنىّ والآبار ويرتفع منها شىء من الكرابيس يحمل الى الآفاق ومُسُوح ونِخاخ وليس بها امتعة مرتفعة ✿

واما بَلْخ فانّ الّذى يتّصل بها[a] طُخَيْرِسْتان والخُتّل وبَنْجَهير[b] وبَذَخْشان وعمل بَامِيَان وما يتّصل بها، فامّا مدن طخيرستان فانّها خُلْم وسِمِنْجَان[c] وبَغْلان وسَكْلَكَنْد[d] ووَرْوَالِيز[e] وآرْهَنْ[f] ورَاوَن والطايَقان[g] وسكيمشت[h]

B. المحاوى, Epit. Par. النجاحى, Edrîsî, I, p. 452 المحاجى. Appellatur ويبروزن طين الاكل a Thaälibí, *Latâïf*, p. ١١٤ seq., Mokaddasí MS. p. 158 طين الاكل. E. habet گل خوردنى, de quo nomine cf. Vullers sub گل. *a*) A. et B. به. Deinde A. et B. طخمرستان; C. طبرستان; E. تخاراستان. Vulgo طخارستان. *b*) A. et B. sine punctis. *c*) A. et B. وسمنحان; E. وسمنكان. *d*) Mokaddasi اسكلكند, quae forma quoque apud Jacut exstat et pronuntiatur اِسْكَلْكَنْد. Jacut, III, p. ٥١٨, 14 male وسكاكند. *e*) A. et B. ووروا كيير. Nomen hujus urbis scribitur quoque ولوالج (ut habent Jacut et Mokaddasí) et ولوالش, vid. Abulfeda, p. ٤٧٢ seq. et *Djih.-Numa*, p. ٢٥٣, 18, ubi praeterea forma ولج datur (cf. والج apud Jacut, II, p. ٤١٠, 2 et والجة, IV, p. ٨٩٤, 7). A Jacut quoque memoratur sub forma vitiosa وزوالين (et hinc sic editum est III, p. ٤٩١, 19 et ٥١٨, 15) et in *Djih.-Numa* l. l. paen. sub forma ذوالين. Cf. Jakubí, p. ٩٧٠. *f*) Vid. Jacut in v. et Jakubí l. l.; C. et E. ازهر, Edrîsî, I, p. 474 ازهر; Mokaddasí, p. 26 ادهن, p. 145 ازهر. E. infra in itinerario Transoxaniae آرِهن. *g*) C. والطايكان, de qua forma vid. Abulfeda, p. ٤٧٢. *Djih.-Numa*, p. ٢٥٣, 20 tres formas memorat طانقان, طايقان et طايكان, sed in itinerario p. ٢٥٤ urbem طالقان appellat. Mokaddasí semel habet الطادقان (*sic*), bis الطالقان et sic Jacut in v. et sub طخارستان, licet formam طايقان (in voce) adnotaverit. Anderson Talkan recepit et sic in mappis hodiernis exstat. Sed tamen suspicio manet Jacutum non de consulto طالقان scripsisse, sed male legisse. Apud Jakubí l. l. طاركان est, quod in طايكان mutare non audeo. Nempe haec forma juxta طانقان et طالقان non ita singularis esse videtur. *h*) Secundum C.; A. وسكمست,

وروب [a] وسَرَاى عاصم وخَشْت [b] أَنْدَرَاب واندراب ومَدْر وكاه [c] ۞ وامّا الخُتّل فانّ مدنها هَلَاوَرْد ولَاوَكَنْد وهما مدينتا الوَخْش وكاربنك [d] وتمليات [e] وهُلْبُك [f]

B. وسكمست, E. سمكيمست, Edrîsî سكيمست, Mokaddasí, p. 26 اسكيمست, p. 145 اسكيست, p. 147 اسكمشت. *Djih.-Numa*, p. ٢٥٣, 10 اشكمش, p. ٢٥٤, 15 اسكمش. Anderson, p. 176: "Eshkemesh is probably the Scassem of Polo, but Wood places an Eschkassem to the East of Fuezabad, which better answers the position required by the Venetian." *a*) A. et B. وورا, C. رورا, E. ورا, Ous. وروا, Edrîsî hoc nomine cum seq. conjuncto ورواسر. Conjectura recepi روب, quod inter urbes hujus regionis enumerant Mokaddasí et Jacut. *b*) A. et C. sine punctis; B. et Edrîsî حسب; E. حشب; Mokaddasí حسيب et خشب. Vid. Abulfeda, p. ٤٥٣ et ٤٦٣, qui enumerat formas خَوْست, خَشْت et خاست; Jakubí, p. ٩٨. Jacut habet خَوْست, خواشت et خاشْت; vid. in vv.; II, p. ٤١٠, 2 خَشْت. Recte A. et C. (quod B. et E. om.) addunt اندراب, in cujus vicinia jacet, ut distinguatur ab aliis ejusdem nominis locis. *c*) Infra in itinerario Codd. كه ut in mappa C. *d*) A. وكاويل ut Sprenger, p. 37, B. وىكاويل, C. وكاربك; Edrîsî, I, p. 479 كاربك, p. 480 كاونج. In capite seq. A. et B. فاربح et حانح, D. كاونبج et كابنج, C. كاردمح, E. كاوسح et كادمن. Mokaddasí, p. 26 كاوكند, p. 141 كاوند ut hinc Jacut, II, p. ٤٠٢, 18. Jakubí, p. ٩٩ فامنك. Cf. Sprenger, p. 45. *e*) A. et B. ودملاب ut quoque infra in itinerario Transoxaniae. C. ودملات, et infra ململات; E. hic يميليات, infra دملات; D. ملمك et دمليات; Edrîsî, p. 479 نهلتان, p. 480 تملتان; Mokaddasí دملات. Jacut, II, p. ٤٠٢, 18 e libro Mokaddasii dat تمليات ut recepi. Juynboll (Ann. ad *Merácid*, V, p. 134) proponit بملتان. *f*) A. et B. sine punctis; C. هملك, E. هايل.

وسكندرة [a] ومُنْك [b] واندبجاراغ [c] وفارغر [d] ورستاق بنك [e] وقد جعلت الخُتَّل فى ما وراء النهر ۞ وامّا عمل الباميان وما يتّصل بها فانّ مدنها الباميان وبسغورْفنْد [f] وسكاوند وكابُل ولجرا [g] وفَرْوان وغزْنة ' وبَنْجَهير * هى مدينة

---

a) Mokaddasí اسكندرة et sic Jacut, II, p. ٤٠٢, 18. b) A. et B. ومنبل, in capite seq. ومبل, دمل et مبل. c) A. et B. h. l. وبحاواغ, C. ومحاراغ. Vid. infra in capite seq., Edrîsî, I, p. 479, Jakubí, p. ٦٩ ubi اندبشاراغ. Jacut, II, p. ٤٠٢, 18 e Mokaddasí بنجاراغ. In hujus Codice legitur نجارع. d) A. وباقعر; in B. tantum superest وبا; C. وباسعه, E. om.; Mokaddasi فغر et بافعر. In capite seq. A. et B. يازعر et فارغر, C. فارىحى, D. فارغن, E. فارغى. Cf. ann. ad *Merâcid*, V, p. 133. Nomen fluvii cui hoc oppidum adjacet, in cap. seq. apud A. est دمارعد, B. مارعد, C. فارعى, E. فارغى, D. وارغن, Jacut, II, p. ١٧١, 17 فارعى, *Djih.-Numa*, p. ٣٦٠, 11 فارغى. Mappa C. et Mokaddasí فارعر, sed Mokaddasí MS. p. 13 فارعن. e) A. ورسمان دىك, B. ورستان دىك. In capite seq. A. et B. دمل. C., D., E. et Edrîsî habent منك (cf. ann. ad *Merâcid*, V, p. 133), sed hoc recipere non audeo, nihil enim cum nomine urbis مُنْك commune habere videtur. Originem nominis discimus e Jakubí, p. ٦٩ seq. ubi legimus ومدينة يقال لها روسادمل وهى مملكة الحرث بن اسد بن دمك صاحب الدواب الممكية. Nominis روسادمل prior pars legatur روستا pro رستاق. Annotatio editoris *h* quantocius deleatur. — Habet igitur locus nomen a viro principe. Mokaddasí, p. 26 habet نمك, p. 141 دمك; Jacut, II, p. ٤١٠, 3 رستاق بيل. f) A. et C. ودسعور دمد, B. ولسعور فند; Edrîsî, I, p. 477 سيغور قند; Mokaddasí, p. 145 دسعور قند (p. 26 لسم غور فند). In E. deëst, ut quoque apud Abulf., p. ٤٦٧, sed p. ٤٦٤ ult. hic habet بغشور ut *Djihán-Numa*, p. ٢٣٨, 16 بغچور. Jakubí, p. ٦٨ et p. ٧١ غورمد, semel غوروند ex var. lect., quod editor recepit, relegans ad Reinaud, *Mémoire sur l'Inde*, ubi fluvius Kabulistáni appellatur غُوْرْوَنْد. Haec quoque lectio in *Djih.-Numa* exstat, p. ٢٣٧ ult., ubi غوربند tamquam nomen provinciae memoratur. g) A. et B. ولجوا,

واحدة تسمّى بنجهير[a] ، وبَذَخْشان اقليم له رساتيق ومدينتها بذخشان وهى مملكة ابى الفتح ۞

فامّا بَلْخ فانّها مدينة فى مستو وبينها وبين اقرب الجبال اليها نحو اربعة فراسخ ويسمّى جبل كو[b] وعليها سور ولها[c] ربض ومسجد الجامع فى المدينة فى وسطها واسواقها حوالى مسجد الجامع ومسجدها[d] معمور بالناس على دوام الايّام كلّها وهى نحو من نصف فرسخ فى مثله وبناؤها الطين[e] ولها ابواب منها باب النُّوبَهار وباب رحبة[f] وباب الحديد وباب الهندوان وباب اليهود وباب شست بند[g] وباب يحيى[h] ولها نهر يسمّى دِهَآس[i] يجرى فى ربضها على باب النوبهار وهو نهر يدير عشرة ارحية ويسقى رساتيق الى سياه جرد[k] ويحفّ بابوابها كلّها البساتينُ والكروم وليس على سور المدينة خندق والسور من طين ۞

وامّا طُخَيْرِسْتان[l] فانّ اكبر مدينة بها الطايَقان وهى مدينة فى مستو

C. وبحما, E. دبحرا; infra A. et B. لبحرا, D. لجرياب, E. مرنبار. Edrisi l. l. بُخرا, p. 460 خرياب (لخرياب); Abulfeda, p. ٤٩٤ الجرا; Mokaddasí, p. 26 لبحراب, p. 145 لجوا; *Djih.-Numa*, p. ٢٣٨, 16 البحرا. *a*) Pro his E. (cf. Ous. p. 223) شاطير وما اين شهرهاى كوهستان در ما ورا النهر نهاديم وبعضى ازآن از اعمال باميان است et ante verba واما الختل in hac versione praecedit inscriptio شهرهاى كوهستان. Nempe hic et in seqq. الختل semper per كوه reddidit, legens الجبل. *b*) Sic A. et B.; C. كرم sed omissa copula sequenti. *c*) A. et B. وعليها. *d*) Ope C. correxi; A. et B. بينها وبين مسجدها. *e*) C. من طين. Deinde B. وبها. Ante ابواب E. addit ستة; C. habet ابواب كثيرة. *f*) Secundum E. et *Djih.-Numa*, p. ٣١٥ ult.; A. زحنه, B. رخنه. *g*) A. et B. نمسب (ست) هن. Secutus sum E. (ubi شصت). In *Djih.-Numa* بند deëst. *h*) A. دبحر, B. لبحر. *i*) A. دهادس, B. دهابيس. Gloss. in Epit. Par. ده آس يعنى عشرة ارحية. In *Djih.-Numa* l. l. 19 fluvius دروها appellatur. *k*) E. سياه كرد. Deinde D. et Abulfeda, p. ٤٩١ ويكتنف. *l*) B.

وبينها وبين الجبل غلوة ولها نهر كبير وبساتين وكروم ومقدار الطايقان نحو الثلث[a] من بلخ ثمّ يليها فى الكبر وَرْوَالِيز، ويلى وروالبز فى الكبر أَنْدَرَابَة وهى مدينة فى شعب[b] جبال وبها تجمع الفضّة الّتى تقع من جاربايه[c] وبنجهير بها نهران احدهما يسمّى نهر اندراب والآخر نهر كاسان ولها[d] كروم واشجار كثيرة، وجميع ما بقى من مدن طخيرستان متقارب فى الكبر وهى كلّها دون الطايقان ووروالبز واندرابة وهى ذات انهار واشجار وزروع كثيرة عامرة خصبة ❁

وامّا مدن الخُتّل[e] فانّها كلّها ذوات انهار واشجار وهى على غاية الخصب[f] وكلّها فى مستوى[g] الّا سكندرة[h] فانّها فى جبال على انّ الختّل كلّها جبال الّا الوَخْش، واكبر مدينة بالختّل مُنْك[i] يليها هُلْبُك والسلطان بهلبك، والخُتّل بين نهر وَخْشاب وبين نهر بَذَخْشان ويسمّى جرياب[k] وفى اضعافها[l] انهاركثيرة تجتمع كلّها قبل الترمذ بقرب القَوَاذِيَان فتصير كلّها جيحون ❁ ومُنْك تكون نحوًا من اندرابة وهُلْبُك اصغر منها، وابنية هذه المدن من طين وسور منك من جصّ وحجارة، يليها من دور الكفر وَخّان وكَرّان ❁ وبَذَخْشان مدينة اصغر من منك ولها رستاق كبير عامر جدًّا خصب وبها كروم وانهار وهى على نهر جرياب[m] من غربيّه ❁ ويكون بالختّل دوابّ كثيرة[n] تجلب الى الآفاق ويرتفع من بَذَخْشان[o] البجاذى واللازورد ولها معادن فى الجبال

طخرستان, C. et D. طخارستان, E. تخارستان. Quae sequuntur Jacut sub طالقان transcripsit. Et الطالقان in seqq. quoque habent C. et Edrîsî, I, p. 475.
a) E. نصف; D. الربع. *b*) C. addit بين. *c*) A., B. et E. plerumque جاريانه; C. uno loco خاربانه, D. semel خازدانه. Secutus sum Epit. Paris. ubi in marg. legitur يعنى ذوو اربعة ارجل, et Mokaddasí qui habet خاربايه et جاربادة; cf. Sprenger, p. 37. *d*) B. وبها. *e*) A. et B. saepius الخُتَل. *f*) A. et B. om. *g*) B. مستو. *h*) A. et B. سكرده, D. شكنده. Ceteri om. *i*) A. et B. ميل. Pro هلبك iidem saepius هلبيل. *k*) A. et B. h. l. حريك. *l*) C. تضاعيفها. *m*) A. et B. حوبان. *n*) B. لذيذة. *o*) Sic h. l. A. et B.; cf. Jacut, I. p. ٥٣٨, 17.

* تخرج منها[a] ويقع اليها مسك من طريق وخّان من تُبَّت ✿

واما بَنْجَهير فانّها مدينة على[b] جبل تشتمل على نحو عشرة آلاف رجل والغالب على اهلها العبث[c] والفساد ولهم نهر وبساتين وليست لهم مزارع، واما جاربايه فانّها اصغر من بنجهير وكلاهما معدن الفضّة ومقام[d] اهلها على تلك المعادن وليس بجاربايه بساتين ولا زروع ويشقّ وسط المدينة نهر بنجهير وهو نهر بنجهير وجاربايه جميعًا وينتهى الى فَرْوان[e] حتّى يقع فى ارض الهند ✿

واما عمل الباميان فانّ اكبر مدنها الباميان وتكون نحوًا من نصف بلخ[f] وتنسب تلك المملكة الى شير باميان وليس لها[g] سور وهى على جبل *ويجرى بين[h] مدنها نهر كبير يقع الى غرجستان وفواكههم تجتلب[i] اليهم وليس بها بساتين وليس بنواحى الباميان مدينة على جبل سوى الباميان وكلّها ذوات انهار واشجار وثمار الّا غَزْنَة[k] فانّها لا بساتين لها ولها نهر وليس فى هذه المدن الّتى فى نواحى بلخ اكثر مالًا وتجارة من غزنة فانّها فرضة الهند ✿ وكابُل لها *قلعة حصينة[l] واليها طريق واحد وفيها المسلمون ولها ربض به الكفّار من الهنود ويزعمون انّ الشاه لا يستحقّ[m] المُلْك الّا بان يعقد له الملك بكابل وان كان منها على بعد ولا يستحقّه حتّى يصل اليه فيعقد الشاهيّة له هناك وهى فرضة الهند ايضًا ✿ ويرتفع من بلخ النوق من البخاتى المقدّمة على سائر البخت بالنواحى وبها الاترجّ والنيلوفر وقصب السكّر وما لا يكون الّا بالبلدان الحارّة الّا انّه لا نخيل بها ويقع فيها وفى نواحيها الثلوج، ولجرا وسكاوند وكابل جروم حارّة غير انّه لا نخيل بها ✿

---

a) Ex C. et D.; A. et B. om. *b*) C. addit سفح. *c*) Abulfeda, p. ٤٦٩ العبث. *d*) A. et B. ومقابر. *e*) A. et B. فروار. *f*) C. addit فى الكبر. *g*) B. بها, *h*) A. et B. وبين. Deinde C. يديها pro مدنها. E. ودر پيش كوه رودى بزرگ ميرود. *i*) Abulfeda, p. ٤٦٧ تجلب. *k*) E. غزنى. Deinde A. et B. فانه. *l*) C. et D. قهندز موصوف بالتحصن. Deinde A. et B. واليه. *m*) A. et B. تستحق.

وامّا الغُور فانّها جبال يحيط بها من كلّ جانب دار الاسلام واهلها كفّار الّا نفرًا يسيرًا مسلمين[a] وهى جبال منيعة ولسانهم غير[b] لسان اهل خراسان وجبالهم خصبة كثيرة الزروع والمواشى والمراعى وادخلناها فى جملة خراسان لانّ ثلاثة من حدودها يحيط بها خراسان وحدّ لها يلى نواحى سجستان واكثر رقيق الغور يقع الى هراة وسجستان ونواحيها، ويمتدّ من ظهر الغور جبال فى[c] حدّ خراسان على حدود الباميان الى[d] البنجهير حتّى تدخل[e] بلاد وَخّان[f] وتفترق فى ما ورآء النهر الى داخل الترك على حدود * ايلاق والشاش[g] الى قرب خرخيز وفى هذا الجبل من اوّله الى آخره معادن الفضّة والذهب واغزرها ما قرب من بلاد خرخيز حتّى ينتهى الى ما ورآء النهر من فرغانة والشاش واغزر هذه المعادن فى دار الاسلام فى ناحية بنجهير وما والاها ✿

وامّا سواحل جيحون وخوارزم فانّا نذكرها[h] فى صفة ما ورآء النهر، وآمُل وزمّ هما مدينتان متقاربتان فى الكبر على شطّ جيحون ولهما مآء جارٍ وبساتين وزروع وآمل مجمع طرق[i] خراسان الى ما ورآء النهر وخوارزم على الساحل وزمّ دون آمل فى العمارة الّا انّ بها[k] معبر ما ورآء النهر الى خراسان ويحيط بهما جميعًا مفازة تتصل من حدود بلخ الى بحر خوارزم والغالب على هذه المفازة الرمال وليس بها عيون ولا[k] انهار الّا آبار ومراعٍ الى ان ينتهى الى طريق مرو الى آمل ثمّ يصير بينها وبين خوارزم وبلاد الغزيّة مفاوز تقلّ آبارها والسوائم بها ✿ واكثر السوائم بخراسان من الابل بناحية سرخس وبلخ فامّا الغنم فانّ اكثرها يجلب اليهم من بلاد الغُزيّة[l] ومن الغُور والخَلْج، وبخراسان من الدوابّ والرقيق والاطعمة والملبوس وسائر ما يحتاج الناس

---

a) A. et B. نفر يسير مسلمون. b) E. چون ut *Djih.-Numa*, p. ٢٥٤, 3. Anderson, p. 177: „Ghor is now occupied by Huzaruhs who speak very pure Persian." c) C. من. d) A. et B. على; vid. quoque Abulfeda, p. ٤٩٤. e) A. et B. يدخل et mox ويفترق. f) A. ىرحان. g) A. et B. لنان (لمار) والسوس; D. اليار والشاش. h) A. et B. نذكره. i) A. et B. طريق. k) A. et B. om. l) E. غزنى.

اليه ما يسعهم[a] فانفس الدوابّ ما يرتفع[b] من نواحى بلخ وانفس الرقيق ما يرتفع[b] من بلاد الترك اليهم وانفس ثياب القطن والابريسم ما يقع[c] من نيسابور ومرو[d] واطيب البزّ ما يرتفع من مرو وخير لحمان الغنم ما يجلب من بلاد الغُزّيّة[e] واعذب المياه واخفّها مآء جيحون وايسر اهل خراسان اهل نيسابور * وانجب اهل خراسان اهل بلخ ومرو فى الفقه والدين والنظر والكلام وازكى ارض خراسان السقى نيسابور والاعذآء ما بين هراة ومرو الروذ[f] وليس بخراسان جروم الّا ما كان بناحية قُوهستان فيما يلى فارس وكرمان[g] واشدّ خراسان بردًا وثلوجًا الباميان وخوارزم * الّا انّا[h] جعلنا خوارزم من ورآء النهر۞

المسافات بخراسان[i] * فمن نيسابور الى آخر حدّها ممّا يلى قُومِس الى قرية الاكراد بقرب أَسَدَاباذ ٧ ايّام ومن قرية الاكراد الى الدامَغان ٥ منازل ومن نيسابور الى سرخس ٦ مراحل ومن سرخس الى مرو ٥[k] مراحل ومن مرو الى آمل على شطّ نهر جيحون ٦ مراحل[l] فمن اوّل عمل نيسابور ممّا يلى قومس الى وادى جيحون على السمت ٢٣ مرحلة، ومن نيسابور الى اسفرائين وهو آخر عمل نيسابور[m] ٥ مراحل، ومن نيسابور الى بُوزْجَان ٤[n] مراحل ومن بوزجان الى بُوشَنْج ٤ مراحل ومن بوشنج الى هراة مرحلة[o] ومن هراة الى اسفزار ٣ مراحل ومن اسفزار الى دَرَه وهى آخر عمل هراة

---

a) Quae hinc usque ad المسافات sequuntur, in A. et B. desiderantur. *b*) D. يقع. *c*) D. يرتفع. *d*) E. بلخ. Verba واطيب — مرو ex solo C. (ubi البر). *e*) E. غزنى. *f*) E. ومردم مرو وبلخ فقيه واديب باشند وعاقل ودين دار وخوشترين زمينهاى خراسان آنچه اينت (l. آبستة) وآنچه بود هراة ومرو (videtur excidisse vocabulum ante بود), addens ومردم نشاپور نشاطى باشند. *g*) Haec inde a وانجب in C. desunt. *h*) C. لانا, E. زيراكه. *i*) E. praemittit fere ut D.: در مسافات خراسان استقصا نكرديم زيراكه معروفست اما نشاپوررا ياد كردن مهم بود. *k*) E. ٦. *l*) Ex C., D. et E. In A. et B. haec desunt. *m*) A. et B. addunt الى نيسابور. *n*) Sic A., B. et D.; C. خمس. *o*) C. اربع مراحل.

مرحلتان ومن دره الى سجستان ٧ ايّام فمن اسفرائين الى دره ١٩ مرحلة، ومن نيسابور الى طوس ٣ مراحل ومن نيسابور الى نَسَا ٦ مراحل ومن نسا الى فَرَاوة ۴ مراحل ومن نيسابور الى قاين قصبة قوهستان نحو ٩ مراحل ومن قاين الى هراة نحو ٨ مراحل ۞ ومن مرو الى مرو الروذ ٦ مراحل ومن مرو الى هراة ١٢ مرحلة ومن مرو الى ابيورد ٦ مراحل ومنها[a] الى نسا ۴ مراحل وقد ذكرنا ما بين مرو وآمل *وبين مرو[b] وسرخس، ومن هراة[c] الى مرو الروذ وهو طريق بلخ ٦ مراحل ومن هراة الى سرخس ٥ مراحل وقد ذكرنا الطريق من هراة الى نيسابور والى آخر حدّها ممّا يلى سجستان والى قصبة قوهستان ۞ والطريق من بلخ الى مرو الروذ ١٢ يومًا ومن بلخ الى شطّ الوادى طريق الترمذ يومان ومن بلخ الى اندرابة ٩ مراحل ومن بلخ الى الباميان ١٠ مراحل ومن الباميان الى غزنة نحو ٨ مراحل ومن بلخ الى بذخشان نحو ١٣ مرحلة ومن بلخ الى شطّ الوادى على طريق الخُتّل[d] بمكان يعرف بميلة ٣ مراحل ۞ وامّا عرض خراسان من بذخشان على شطّ وادى جيحون الى بحيرة خوارزم فمن بذخشان الى الترمذ على سمت النهر نحو ١٣ مرحلة *ومن الترمذ الى زمّ ٥ مراحل ومن زمّ الى آمُل ۴ مراحل ومن آمل الى مدينة خوارزم ١٢ مرحلة[e] ومن مدينة خوارزم الى بحيرة خوارزم ٦ مراحل ۞

قد ذكرنا المسافات الّتى بين المدن المشهورة بخراسان وسنذكر لكلّ مدينة مشهورة جوامع من المسافات بين المدن الّتى فى عملها فامّا نيسابور فانّ منها الى بُوزْجَان ۴ مراحل ومن بوزجان عن يسار الجائى من هراة الى نيسابور على رحلة مالِنْ ويعرف بمالن كواخرز[f] وليس بمالن هراة ومن مالن الى جايمند مرحلة ومن جايمند الى سَنْكان[g] يوم ومن سنكان الى يُنَابِذ[h] يومان ومن ينابذ الى قاين يومان وسلومك[i] اذا عدلتَ عن يسار سنكان

---

a) E. از مرو. *b*) A. et B. ومرو. *c*) C. مرو. *d*) B. et D. الجبل, A. sine punctis. *e*) Addidi ex C., D. et E.; A. et B. om. *f*) Vid. supra ad p. ٢٥٦ ann. *d*. Locum transcripsit Jacut, IV, p. ٣٩٨, 8 seqq. *g*) A. et B. سكان. *h*) A., B. et E. sine punctis. *i*) A. et B. وسكومل.

على يومين ومن سلومك الى زُوزَن يوم ومن زوزن الى قاين ٣ ايّام، ومن نيسابور الى تُرْشِيز[a] ۴ مراحل ومن ترشيز الى كُنْدُر[b] يوم ومن كندر الى ينابذ يومان ومن ينابذ الى قاين يومان، ومن نيسابور الى خُسْرُوجِرْد[c] ۴ مراحل وسابْزَوَار[d] قبل خسروجرد بنحو فرسخين ومن خسروجرد الى بهمناباذ مرحلة كبيرة ومن بهمناباذ الى مزنيان[e] على طريق قومس نحو فرسخ، ومن نيسابور الى خان روان[f] مرحلة ومن خان روان الى مِهْرَجَان يومان ومن مهرجان الى اسفرائين يومان، واذا خرجتَ من بهمناباذ الى مهرجان فالى أَزَاذْوَار يوم ومن ازاذوار الى دِيوَارَه يوم ومن ديواره الى مهرجان يومان ۞ وامّا مسافات مدن مَرْوَ فانّ من مرو الى كُشْمَيْهَن[g] منزل وهُرْمُزْفَرَّه بحذاء كشميهن على مقدار فرسخ عن يسارها وعليها طريق مفازة سيفانة[h] الّتى تؤدّى الى خوارزم وباشان قبل هرمزفرّه بفرسخ على طريقها وسِنْج[i] على مرحلة من المدينة بين طريق سرخس وطريق مرو وجِيرَنْج[k] على ٦ فراسخ من المدينة قبل زَرْق[l] بفرسخ على الوادى ومرورم[m] على هذا الطريق على ۴ فراسخ من مرو على الوادى والدَّنْدَانَقان[n] على مرحلة[o] من مرو على طريق

---

a) A., B. et E. برشير. Mokaddasí طرىست i. e. طريثيث. b) D. et E. كندرم et sic quoque Edrîsî, I, p. 454. c) E. خسروكرد. d) A. et B. وساروار. e) A. مرردان, B. موردان. f) Mokaddasí ريوند. E. خاوران. g) A. et B. كشيمهن. h) A. et B. omisso مفازة h. l. سكانه, sed infra in itiner. Khowarezmiae سمعانه. E. ibi سغانه (Ous. p. 275 مغانه); C. p. 123 سمداه; Edrisi, I, p. 467 et II, p. 187 seq. سنقا s. سنفاية. Ibn Batuta, III, p. 21 سيباية. D. utroque loco ut recepi. In mappa *Djih.-Numa* صغانه. i) A., B. et C. sine punctis. k) A. et B. sine punctis. l) A. et B. زُرق. D. habet درق et درق. Vid. supra ad p. ٢٦١ ann. *m*. m) *Sic* A.; B. fortasse ومرودم. In D. corrupte ومن دزق; ceteri om. n) A. et B. والددداوعان. o) C., D. et E. مرحلتين et sic quoque Mokaddasí qui habet (p. 165) وتاخذ من مرو الى حروجرد مرحلة ثم الى الدندانقان مرحلة. Statio intermedia fortasse est خرق, quae in mappa C. ad hanc viam inter مرو et الدندانقان ponitur.

سرخس والقَرِينَيْن[a] على ٤ مراحل من مرو على وادى مرو وخَرَق[b] على نحو ٣ فراسخ من المدينة بين طريق سرخس وابيورد وسَوْسَقَان[c] على نسق خرق الّا انّها[d] ابعد منها بنحو فرسخ ✿ وامّا مسافات مدن هَرَاة وما يتّصل بها من بُوشَنْج وباذَغيس وكَنْج رستاق فانّ من هراة الى أَسْفِزَار ٣ مراحل ومدن اسفزار هى اربع[e] قد سمّيناها وهى كلّها فى اقلّ من مرحلة وبين[f] هراة ومالن هراة نصف يوم وبين[g] هراة وكَرُوخ[h] ٣ ايّام وبين هراة وبوشنج يوم وبين بوشنج وكره[i] ٤ فراسخ عن يسار الذاهب الى نيسابور وبينها وبين الطريق نحو فرسخين[k] ومن بوشنج الى فركرده[l] يومان ومن فركرده الى خَرْكِرْده يومان[m] ومن خركرده الى زُوزَن يوم، ومن هراة الى باشان هراة مرحلة ومن باشان الى خَيْسَار[n] مرحلة خفيفة ومن خيسار الى استربيان[o] مرحلة ومن استربيان الى ماراباذ مرحلة خفيفة ومن ماراباذ الى أَوْفَة مرحلة خفيفة ومن اوفه الى خَشْت[p] يومان ويدخل من خشت فى حدّ الغُور، ومن هراة الى بَبْنَة مرحلتان ومن ببنة الى كيف مرحلة ومن كيف الى بَغْشُور[q] يوم ✿

---

a) A. والعرىمن, B. والعرس. Eodem modo corrupte apud ceteros. b) A. et B. وحرق, D. ودرق. c) A. et B. وسوسمان. d) A. et B. انه. e) A., B. et C. اربعة. Deinde A. et B. om. قد. f) A. et B. ومن. g) A. ومن. h) C. وكنج. Secundum Mokaddasí distantia inter Karukh et Herát est 1 diei, secundum Jacut (IV, p. ٢٧٠) 10 Paras. Si cum C. urbs regionis كنج رستاق legenda est, intelligitur كيف quam inter et Herát 3 dies sunt (Jacut, IV, p. ٣٠٨, 14). i) A. et B. كوه, D. كو, ut supra habent pro كوفا (vid. p. ٢٦٨ ann. d), sed de hac urbe h. l. cogitari non videtur. C. recte habet كره quae est urbs regionis بوشنج. Cf. Sprenger, p. 48. k) D. فرسخ. l) Supra p. ٢٦٧ ult. فركرد et خركرد. Pro خركرده. A. et B. h. l. جوكرده. m) C. يوم. n) A. et B. حسان, حسيان et حبيان, C. et D. حنمان, Mokaddasí حنبار. o) A. et B. sine punctis, C. سمربان, D. اسبارباذ. p) A. حسب, B. حسب. q) A. مغشور.

واما مسافات مدن بلخ فمن بلخ الى خُلْم يومان ومن خلم الى وَرْوالِيز[a]
يومان ومن وروالیز الى الطايَقان يومان ومن الطايقان الى بَذَخْشان[b] ٧ ايّام،
ومن خلم الى سِمِنْجان يومان ومن سمنجان الى أَنْدَرابَة ٥ ايّام ومن اندرابة
الى جاربايه[c] ٣ مراحل ومن جاربايه الى بَنْجَهِير يوم ومن عسكر بنجهير الى
فَرْوان مرحلتان، ومن بلخ الى بَغْلان ٦ مراحل منها الى سمنجان ٤ مراحل
والى بغلان مرحلتان ومن بلخ الى مَذْر ٦ مراحل ومن مذر الى كه[d] منزل
ومن كه الى البامیان ٣ مراحل، ومن بلخ الى اشْبُورقان[e] ٣ مراحل ومن
اشبورقان الى الفارياب ٣ مراحل ومن الفارياب الى الطايَقان[f] ٣ مراحل ومن
الطايقان[f] الى مَرْوَرُوذ ٣ مراحل ❁ والمسافة بين مدن قوهستان فمن قاين
الى زوزن ٣ مراحل ومن قاين الى طَبَس مَسِينان يومان ومن قاين الى
خُور يوم ومن خور الى خَوْسْت فرسخان ومن قاين الى الطَّبَسَيْن ٣ مراحل ❁
*فهذه جمل مسافات خراسان[g] ❁

ما وراء النهر

واما ما وراء النهر فيحيط به من شرقيّه فَامِر[h] وراشْت وما يتاخم الخُتَّل[i]
من ارض الهند على خطّ مستقيم وغربيّه بلاد الغُزِّيَّة والخَزْلَجيّة[k] من حدّ
طَرَاز[l] ممتدّا على التقويس حتّى ينتهى الى فَارَاب وبيسكند[m] وسُغْد سَمَرْقَنْد

---

a) A. et B. ورواكير, C. ورالس, E. والبس. *b*) Mokaddasí h. l. habet بلخشان.
c) A. sine punctis; B. حاريانه, C. خاربادة, D. حازبابه, E. جاريانه, Mokaddasí كاردانه. Vid. supra p. ٢٧٩ ann. *c*. *d*) Supra p. ٢٧٦ كاه. *e*) Supra (p. ٢٧٠) شبورقان. A. sine punctis, B. اسورقان, Mokaddasí استورقان. Cf. ad hunc locum Jacut, III, p. ٨٤٠, 21 seqq. *f*) H. l. quoque A. et B. الطالقان. D. vero habet ut recepi. *g*) Addidi ex D. et E. *h*) A. et B. sine punctis; C. قاس, D. ثغام. Vid. Vullers in v. Etiam nunc regio idem nomen habet (Pamir). Deinde A. et B. وراست. *i*) A. et B. fere semper الختل. *k*) A. et B. والخرلجية. *l*) C. حلوان, E. طوران. *m*) A. et B. وسكمد, E.

ونواحى بُخَارَا الى خُوَارِزْم حتّى ينتهى الى بحيرتها وشماليّه الترك، الخزلجيّة a من اقصى بلد فَرْغانة الى الطَّرَاز على خطّ b مستقيم وجنوبيّه نهر جيحون من لدن بَذَخْشان الى بحيرة خُوارزم على خطّ مستقيم وجعلنا خوارزم والخُتّل فيما ورآء النهر لانّ الختّل بين نهر جرياب c ووَخْشاب d وعمود جيحون جرياب e وما دونه من ورآء النهر وخوارزم مدينتها ورآء النهر وهى الى مدن ما ورآء النهر اقرب منها الى مدن خراسان f ۞

ما ورآء النهر من اخصب اقاليم الاسلام وانزهها واكثرها خيرًا واهلها يرجعون الى رغبة فى الخير g واستجابة لمن دعاهم اليه مع قلّة غائلة وسلامة ناحية وسماحة بما ملكت ايديهم مع شدّة شوكة ومنعة وبأس وعُدّة وآلة وكراع وسلاح، فامّا الخصب بها فانّه ليس من اقليم ذكرناه الّا يقحط اهله مرارًا قبل ان يقحط ما ورآء النهر ثمّ ان اصيبوا ببَرَد او جراد او آفة تاتى على زروعهم ففى فضل ما يسلم فى عرض بلادهم ما يقوم باودهم حتّى يستغنوا عن نقل شىء اليهم من غير بلادهم وليس بما ورآء النهر مكان يخلو *من مدن او قرى او مباخس او مراعٍ لسائمة h وليس شىء لا بدّ للناس منه الّا وعندهم منه ما يُقيم اودهم ويفضل عنهم لغيرهم، فامّا اطعمتهم فمن السعة

وسيركند (Ous. مركند), D. سكندر. Praeterea in seqq. hae lectiones in Codd. exhibentur: D. سكند, bis سنكند, اسماكند, سبناكند; E. bis سكند, سنكند, سمكند (Ous. سليكند); A. سمل كند, B. سمل كند; Edrîsî, II, p. 208 سيكند, 211 سناكند; Jacut, II, p. ٤٠٥, 4 بيكند, ut *Djih.-Numa*, p. ٤٩٠, 20 et ٤٩٣, 14. Dubium esse nequit eandem urbem significari quam Jacut, I, p. v٩٠ appellet بيسكند. Sed Sprenger, p. 32, secundum al-Bíruní eam appellat سنكند.

a) A. et B. male والخزلجية cum copula. *b*) A. حدّ. *c*) A. hic et mox حوناب, B. حرىاب. Vid. infra. *d*) A. ورحمساب. *e*) A. et B. male copulam addunt. *f*) Hic in A. et B. sequitur وهذه صورة ما وراء النهر. *g*) Jacut, IV, p. ٤٠٠, 16 addit والسخاء. *h*) Jacut: من العمارة من مدينة او قرى او مياه او زروع او مراع لسوائمهم.

والكثرة على ما ذكرناه، وامّا مياههم فانّها اعذب المياه واخفّها قد عمّت المياه العذبة جبالها وضواحيها ومدنها، وامّا الدوابّ ففيها[a] من النتاج ما فيه كفاية لهم مع كثرة ارتباطهم لها وكذلك البغال والحمر والابل، وامّا لحومهم فانّ بها من النتاج[b] ما يجلبونه من الغُزّيّة والخزلجيّة وما يتّصل بهم من حواليها ما يفضل عن كفايتهم، وامّا الملبوس ففيها[a] من * ثياب القطن ما يفضل عنهم حتّى ينقل عنهم الى الآفاق ولهم الفرآء[c] والصوف والاوبار[d]، وببلادهم من معادن الحديد ما يفضل عن حاجتهم فى الاسلحة والادوات وبها معدن[e] الفضّة والذهب والزيبق[f] الّذى لا يقاربه فى الغزارة والكثرة معدن فى سائر بلدان الاسلام الّا بَنْجَهِير فى الفضّة وامّا الزيبق والذهب[g] وسائر ما يكون فى المعادن فاغزرها ما يرتفع من ما[h] ورآء النهر وليس فى شىء من بلدان الاسلام النوشادر والكاغد الّا فيما ورآء النهر، وامّا فواكههم فانّك اذا تبطّنت السغد وأُشْرُوسَنَة[i] وفرغانة والشاش رايتَ من كثرتها ما يزيد على سائر الآفاق حتّى يرعاها[k] لكثرتها دوابّهم، وامّا الرقيق فانّه يقع اليهم من الاتراك المحيطة بهم ما يفضل عن كفايتهم وينقل[l] الى الآفاق من بلادهم وهو خير رقيق يحيط[m] بالمشرق كلّه، وبها من المسك الّذى يجلب اليهم من تُبّت وخرخيز ما ينقل الى سائر الامصار منها[n]، ويرتفع من الصَّغانيَان الى[o] واشَجِرْد من الزعفران ما ينتقل[p] الى الآفاق وكذلك الاوبار من السمّور والسِنجاب والثعالب وغيرها ممّا يحمل الى اقصى الغرب مع طرائف من

---

a) A. et B. فيها. b) Jacut, IV, p. ۴۰۱, 4 الغنم. Epit. Paris. نتاج الغنم. c) E. كرباس سمرقندى با كاغذ وقز et Jacut l. l. 5 القز pro الفراء, ut quoque D. d) Jacut addit والابريسم الخجندى ولا يفضّل عليه ابريسم البتّة. e) D. et Jac. معادن. f) E. addit وآهن. g) Jacut addit والنحاس. h) Addidi ما ex Jacut. i) A. et B. h. l. اسروشنه; cf. Jacut, I, p. ٢۴٥. k) A. يرعيها. l) E. وينتقل. m) C. et Jacut om., D. دىحط. n) B. فيها. Apud Jacut post الامصار additur الاسلامية. o) Jacut الى الصغانيان والى. Significant autem verba textus: „in tota regione quae inde a Saghánián ad Wáschdjird se extendit, colitur crocus." p) Jacut ينقل.

الحديد[a] والختو والبُزاة وغير ذلك ممّا يحتاج اليه الملوك ✿ وامّا سماحتهم فانّ الناس فى اكثر ما وراء النهر كانّهم فى دار واحدة ما ينزل احد باحد الّا كانّه رجل دخل دار نفسه[b] لا يجد المُضيِّف[c] من طارق فى نفسه كَرَاهة بل يستفرغ مجهوده[d] فى اقامة اوده من غير معرفة تقدّمت ولا توقُّعَ مكافاة بل[e] اعتقادًا للسماحة فى اموالهم وهمّة كلّ امرئ منهم على قدره فيما ملكت يده من القيام على نفسه ومن يطرقه وبحسبك[f] انّك لا تجد فيهم صاحب ضيعة[g] الّا كانت همّته ابتناء قصر فسيح ومنزل للاضياف فتراه عامّة دهره متانّقا فى إعداد ما يصلح لمن طرقه فاذا حلّ بينهم طارق تنافسوا فيه وتنازعوه فليس احد يتصرّف[h] بما وراء النهر فى مكان به ناس[i] يخاف الضِّباع فى ليل او نهار فهم فيما بينهم يتبارون فى مثل هذا حتّى يُجْحِف باموالهم ويقدح فى املاكهم[k] كما يتبارى سائر الناس فى الجمع ويتباهون بالملك والمكاثرة فى المال، ولقد[l] شهدتُ منزلًا بالسُّغْد قد ضُرِب الاوتاد[m] على باب داره فبلغنى[n] انّ بابها لم يُرَدّ[o] منذ مائة سنة واكثر لا يمنع من نزولها طارق وربّما نزل بالليل بغتةً من غير استعداد المائة والمائتان والاكثر بدوابّهم وحشمهم فيجدون من علف دوابّهم وطعامهم ودثارهم ما يعمّهم[p] من غير ان يتكلّف صاحب المنزل امرًا لذلك لدوام ذلك منهم *قد اقيم على كلّ عمل من يستقلّ به واعدّ[q] ما يحتاج اليه على دوام الاوقات بحيث لا

---

a) Sic A., B. et D. Videtur legendum الخدنك, vid. Thaálibí, *Latáif*, p. ١٢٨. Jacut l. l. 17 habet الحديد والحتر. b) Jacut صديقه. c) Sic A.; B. المُضيف. d) Jacut فى غاية. e) Ex D. et Jacut addidi بل. f) A. et B. sine punctis, D. ويحسبك. g) B. et D. صنعة; E. دهقانى. h) A. ينصرف. i) A. دناس, B. بناس. Deinde A. ىحاف الصباع. k) A. addit دم. l) In E. praecedit مصنف كتاب گويد كه در آن بديوار باز. m) B. et D. الاوتار. E. از سبب آن پوسيده بود بميخها دوخته. Jacut ضربت pro ضرب. n) E. گفتند. o) Jacut, IV, p. ٤٠٢, 1 يغلق ut D. Post منذ iidem زيادة على مائة omisso واكثر. p) C. يكفيهم, D. يغنيهم. q) B. اعدّ به استقل — اسم ند.

يحتاج معه الى تجديد امر عند طروقهم وصاحب المنزل من البشاشة[a] والاقبال والمساواة لاضيافه بحيث يعلم كلُّ من شاهده سروره بذلك وسماحته ولم ار مثل هذا ولم اسمع به فى شىء من بلدان الاسلام لرعيَّة ومع ذلك فانَّك لا تجد فى بلدان الاسلام اهل الثروة الَّا والغالب على اكثرهم صرف نفقاتهم الى خاصّ انفسهم فى الملاهى وما لا يرضاه الله والى المنافسات فيما بينهم فى الاشيآء المذمومة الَّا القليل وترى الغالب على اهل الاموال بما ورآء النهر *صرف نفقاتهم الى الرباطات وعمارة الطرق والوقوف على سبيل الجهاد ووجوه الخير الَّا القليل منهم[b] وليس من بلد ولا منهل ولا مفازة مطروقة ولا قرية آهلة الَّا بها[c] من الرباطات[d] ما يفضل عن نزول مَنْ طَرَقَه[e]، وبلغنى انَّ بما ورآء النهر زيادة على عشرة آلاف رباط فى كثير منها اذا نزل *النازل اقيم علف دابَّته وطعام نفسه ان احتاج الى ذلك[f]، وقلَّ ما رايتُ خانًا او طرف سكَّة او محلَّة او مجمع ناس فى الحائط بسمرقند يخلو من مآء جمد مسبَّل[g] ولقد اخبرنى من يُرْجَع الى خبره انَّ بسمرقند فى المدينة وحائطها فيما يشتمل عليه السور الخارج زيادة على الفى[h] مكان يسقى فيها مآء الجمد مُسَبَّلًا من بين سقاية[i] مبنيَّة وجباب منصوبة ✿ وامَّا بأسهم وشوكتهم فانَّه[k] ليس فى الاسلام ناحية اكبر حظًّا فى الجهاد منهم وذلك انَّ جميع حدود ما ورآء النهر *الى دار الحرب[l] امَّا خوارزم الى ناحية أَسْبِيجَاب فهم الترك الغُزِّيَّة ومن اسبيجاب الى اقصى فرغانة الترك الخَزْلجيَّة ثمّ يطوف بحدود ما ورآء النهر[m] من السفينة[n] وبلد الهند من

---

a) A. المشاسه. b) E. كه اكثر اموال خود صرف پُل ورباط كنند ومساجد بسيار سازند et in Epit. Par. (vid. locum laud. in ed. Abulfedae, p. ۴۸۷) post الخير additur: وعقد القناطر quod in Codd. Ibn Hauc. deëst. c) C., D. et Jacut l. l. 6 وبها. d) C. السرباطات *sic.* e) E. addit: واوقاف وارزاق بران بديد كرده اند. f) Jacut: الناس اقيم لهم علف دوابهم وطعام انفسهم الى ان يرحلون. g) B. مسبل et infra مُسْبَلًا. h) C. الف. i) A. شعابه. k) B. om. l) Jacut دار حرب. m) A. et B. om. n) A. et B. h. l. sine punctis; infra

ظهر الخُتّل الى حدّ الترك فى ظهر فرغانة فهم القاهرون لاهل هذه النواحى ومستفيض انّه ليس فى الاسلام دار حرب هم اشدّ شوكة من الترك فهم ثغر المسلمين فى وجه الترك يمنعونهم من دار الاسلام وجميع ما ورآء النهر ثغر يبلغهم نفير العدوّ، ولقد اخبرنى من كان مع نصر بن احمد رحمه فى غزاة شاوَغَر[a] انّهم كانوا يحزرون[b] ثلاث مائة الف وانّ اربعة آلاف رجل انقطعوا عن العسكر فضلّوا ايّامًا قبل ان يتهيّأ لهم الرجوع وما كان منهم[c] من غير ما ورآء النهر كثير عدد يعرفون باعيانهم، وبلغنى انّ المعتصم كتب الى عبد الله بن طاهر كتابًا *عرّض تهدّده[d] فيه وانفذ الكتاب الى نُوح بن أَسَد فكتب اليه انّ بما ورآء النهر[e] ثلاث مائة الف قرية[f] ليس من قرية الّا خرج[g] منها فارس وراجل *لا يبين على اهلها فقدهم[h]، وبلغنى انّ بالشاش[i] وفرغانة من الاستعداد ما لا يوصف مثله عن ثغر من الثغور حتّى انّ الرجل الواحد من الرعيّة عنده من بين مائة دابّة الى *خمس مائة[k] وليس بسلطان وهم على بعد دارهم اوّل سابق الى الحجّ[l] لا يدخل البادية قبلهم احد ولا يخرج منها بعدهم احد وهم مع ذلك احسن الناس طاعةً لكبرآئهم والطفهم خدمةً لعظمآئهم وفيما بينهم حتّى دعا ذلك الخلفآء الى ان استدعوا ممّا ورآء النهر رجالها وكانت الاتراك *جيوشهم لفضلهم[m] على سائر الاجناس فى البأس والجرأة والشجاعة والاقدام ودهاقين ما ورآء النهر قوّادهم وحاشيتهم

A. السغمه, B. et C. sine punctis; E. سقينه ut *Djih.-Numa*, p. ٣٩٠, 15. Edrîsî, I, p. 483 السقنيا. Jacut l. l. 12 الصغدية. Secutus sum D. *a*) Secundum D.; A. et B. سروءده, E. سروفنده, Jacut l. l. 16 اشروسنة. *b*) B. تحذرون, D. يحوزون. *c*) Jacut فيهم et كبير احد pro كثير عدد. *d*) B. ؛عرض بتهدده C. et Jacut يتهدده. *e*) A. et B. om. *f*) E. كلات. *g*) C. et Jacut ويخرج. Post منها Jacut addit كذا وكذا. *h*) A. et B. لا سى على اهله فتذمم. Cf. praeter C., D. et Jacut, Ous. p. 235. C. et Jacut يتبين. *i*) E. semper جاج. *k*) C. et Jacut مائتى دابة. Pro بسلطان, A. habet سلطان, C. بامير. *l*) A. et B. الجمع. *m*) Jacut, IV, p. ٤٠٣, 3 جيوشا تفضّلهم.

وخواصّ خدمهم للطفهم[a] فى الخدمة وحسن الطاعة والهيئة فى الملبس والزىّ السلطانيّة فصاروا حاشية الخلافة وثقاتهم ورؤساء عساكرهم مثل الفراغنة والاتراك الّذين هم شحنة دار الخلافة والاتراك الّذين كانوا لبأسهم ونجدتهم غلبوا على الخلافة مثل الأفْشين وآل ابى الساج من أُشْرُوسَنَة والاخْشيد من سمرقند والمرزبان بن تركسفى[b] وعُجَيْف بن عَنْبَسَة من السغد والبُخاراخذاه[c] وغيرهم من امراء الحضرة *وقوّادها وجيوشها[d]، والملوك على هذا الاقليم وعلى سائر خراسان آل سامان وهم من اولاد بَهْرام جُوبين الّذى سار ذكره فى العجم بالبأس والنجدة فلمثل هذه الاسباب ليس فى الاسلام ملك امنع جانبًا ولا اوفر عدّة ولا اكمل اسبابًا للملك منهم لانّه ليس فى الاسلام جيش الّا وهم شُذّاذ القبائل والبلدان والاطراف اذا تفرّقوا فى هزيمة وتمزّقوا فى حادثة لم يُلْتفق[e] منهم جمع بعده غير جيش هاؤلاء الملوك فانّ جيوشهم الاتراك المملوكون ومن الاحرار من يعرف داره ومكانه اذا[f] قُتل منهم قوم او ماتوا ففى وفور عددهم ما يعاد من بين ظهرانيهم مثلهم وان تفرّقوا فى حادثة تراجعوا كلّهم الى مكان واحد *فلا يقدح[g] فيهم ما يقدح فى سائر عساكر الاطراف ولا سبيل لهم الى التفرّق فى العساكر والتنقّل فى الممالك كما يكون عليه رسوم صعاليك العساكر وشحنة البلدان، ولقد خرج بارس[h] غلام لاسماعيل بن احمد رحه فى فتنة عبد الله بن المعتزّ هاربًا من احمد ابن اسماعيل رحه فخرج فى عدّة هالت الخلافة وظهر اثرها بقدومه من العدد والآلة والكراع والسلاح ولم يكن بحضرة الخلافة جيش مثله وانّما كان

---

a) A. et B. الطفهم. b) A. et B. نركسعى; D. كيسفى s. كسيفى. Infra A. et B. دركسفى, E. كسفى, Ous. (p. 258) كشفى, D. كسيدعى *sic;* Ibn Khaldun, III, f. ٥٢ v. (ed. Bul. III, p. ٣٦٨ ult. seq.) تـركش, Ibn Maskowaih, MS. p. 136 تركس. c) A. et B. والنحار حداده (ceteri om.). Intelligi autem videtur dux al-Motacimi de quo vid. Ibno 'l-Athír sub anno 222. Cf. V, p. ١٧٧ et Ibn Maskowaih sub anno 196. d) B. وقوادهم وجيوشهم. e) A. et B. يلمق. f) C. واذا, D. فان. g) B. بلا تقدح. h) E. فـارس, A. et B. دارس. Vid. Ibno 'l-Athír, VIII, p. ٥, ٦, ٤٢.

عبدًا للسَّمَانيَّة لم يتبيّن على اهل خراسان[a] فقده، وليس فى بلدان الاسلام ملوك قد اعرقوا فى الملك يتوارثونه من ايّام العجم مثلهم وقد بيّنّا ايّام آل سامان فى صفة فارس لانّهم من الفرس فبيّنّا مكانهم من فارس وسبب وقوعهم الى خراسان وجوامع من سيرتهم وايّامهم ممّا يغنى عن اعادته ✿ وامّا نزاهة ما وراء النهر فانّى لم ار ولا بلغنى فى الاسلام[b] بلد احسن خارجًا من بخارا لانّك اذا علوتَ قلعتها[c] لم يقع بصرك من جميع النواحى الّا على خضرة تتّصل خضرتها بلون السماء فكانّ السماء بها مُكبّة خضراء مكبوبة على بساط اخضر تلوح القصور فيما بينها كالنوائر[d] فيها واراضى ضياعهم مقوّمة[e] بالاستواء كانّها المرآة وليس بما وراء النهر وخراسان بلد[f] احسن قيامًا بالعمارة على ضياعهم من اهل بخارا ولا اكثر[g] عددًا على قدرها فى المساحة منهم وذلك مخصوص بهذه[h] البلدة لانّ الموصوف[i] من متنزّهات الارض سُغْد سمرقند ونهر الأُبُلّة وغُوطة دمشق على انّ سابُور[k] وجُور من[l] فارس لا تقصران[m] عن غوطة دمشق ونهر الابلّة ولاكنّ الذكر لهذه الاماكن، فامّا غوطة دمشق فانّك اذا كنتَ بدمشق *ترى بعينيك[n] على فرسخ واقلّ جبالًا قرعًا عن[o] النبات والشجر وامكنة خالية عن العمارة والخضرة واكمل النزهة ما ملأ البصر وسدّ الافق، وامّا نهر الابلّة فليس فيها ولا نواحيها مكان يستوقف[p] النظر الّا نحو فرسخ منها وليس بها مكان عالٍ[q] فيدركَ البصر اكثر[r]

---

a) E. در لشكر اسمعيل بن احمد male, nam hic jam mortuus erat. *b*) In C. sequitur حمر (i. e. خبر). Cf. Jacut, I, p. ٥١٧, 15 seqq. ubi بلدًا. *c*) C., D., E. et Jacut قهندزها. *d*) A. et B. كالفرادس. Fortasse legendum كالنرامس quod *Qámus* explicat per جُمان „globuli argenti." D. كالنراس اللمطية. Jacut كالنواوير. *e*) D. مقسومة, Jacut منعوتة. *f*) Jacut بلدة اهلها. *g*) A. اكبر. *h*) A. et B. به هذه. *i*) C. المخصوص. *k*) A. دسابور, B. et D. نيسابور. *l*) A. et B. دمن. *m*) A. تقصر, B. يقصر, C. دعصر, D. يقصران. *n*) A. ندى نعمك. *o*) B. et D. من. Cf. Jacut, III, p. ٣٩٤, ult. seqq. *p*) Jacut يستطرف. *q*) A. عادل. Deinde Jacut فلا يدرك. *r*) B. اكبر, A. sine punctis.

من فرسخ ولا يستوى المكان المستتر[a] ٱلّذى لا يرى منه ٱلّا[b] مقدار ما يرى *من مكان[c] ليس بالمستتر بالنزهة[d] ومكان يستعطف[e] البصر منه سعةً فى العيان وسفرًا فى المنظر، وامّا سغد سمرقند فلا اعرف بها ولا بسمرقند[f] مكانًا اذا علا الناظر *فيه على شرف الّا[g] وقع بصره على جبال خالية من الشجر او صحرآء غبرآء وان كان مزرعًا على انّ غبرة[h] المزارع فى اضعاف خضرة النبات من الزينة غير انّ الارض الغبرآء المنتشرة[i] عن تقويمها فى العمارة فى العيان تسلب بهجة الخضرة وتذهب بزينة الغبرة، ويحيط ببُخارا وقراها ومزارعها سور قطره عشرة[k] فراسخ فى مثلها كلّها عامرة[l]، وامّا سغد سمرقند فانّها انزه الاماكن الثلاثة ٱلّتى ذكرنا لانّها من حدّ بخارا على وادى السُّغْد يمينًا وشمالًا تتّصل[m] الى حدّ البُتّم لا تنقطع ومقدارها فى المسافة ثمانية ايّام مشتبكة الخضرة والبساتين *فهى ميادين وبساتين ورياض مشتبكة[n] قد حُفّت بالانهار الدائم جريها والحياض فى صدور رياضها وميادينها[o] فخضرة الاشجار والزروع ممتدّة على جانبَىْ[p] واديها ومن ورآء الخضرة من جانبيها[q] مزارع تحرسها ومن ورآء هذه المزارع مراعى سوائمها والقلعة[r] من كلّ مدينة وقرية بها تبصّ[s] فى اضعاف خضرتها كانّها ثوب ديباج اخضر قد سُيّرت[t]

---

a) B. المستر et mox بالمستر. b) D. om. الا, recte ut videtur. Sed cf. Jacut. c) Jacut ومكان. d) D. فى النزهة. e) D. يستوقف. f) B. سمرقنذ. g) D. recte omisso الا habet قهندزها; Jacut قهندزها ان يقع. h) A. غيره, B. غبير. i) B. المشرة. k) D. اثنا عشر. E. سيزده فرسنگ در دوازده فرسنگ. Vid. infra. l) A. عامر. Haec h. l. vix suo loco sunt. m) A. et B. يتصل, mox ينقطع ومقداره et مشتبك et sic Jacut (ubi تشتبك). Cf. quoque Abulfeda, p. ۴٨۴. n) D. والرياض والميادين, Jacut والرياض. o) A. وميازبها. p) Jacut حائطى. q) A. et D. جانبها. r) D. et Jacut وقصورها والقهندزات. s) Sic recte Epit. Paris.; A. et B. تبيض, D. تبصر; Jacut تلوح. t) A. et B. سمرت. Deinde supplevi ex D. et Jacut بمجارى مياهها. Pro سيّرت Jacut habet طرزت.

بمجارى مياهها وزُيّنت بتبصبيص[a] قصورها وهى ازكى بلاد الله واحسنها اشجارًا وثمارًا وفى عامّة مساكنهم البساتين والحياض والمياه الجارية قلّ ما تخلو سكّة او دار من نهر جارٍ، وبفرغانة والشاش واشروسنة وسائر ما ورآء النهر من الاشجار الملتفّة والثمار الكثيرة والرياض المتّصلة ما لا يوجد مثله فى سائر الامصار وبفرغانة فى الجبال الممتدّة بينها وبين بلاد الاتراك من الاعناب والجوز والتفّاح وسائر الفواكه مع الورد والبنفسج وانواع من الرياحين كلّ ذلك مباح لا مالك له ولا مانع دونه وكذلك فى جبالها وجبال ما ورآء النهر من الفستق المباح ما ليس فى بلد[b] وباشروسنة ورد يتّصل الى آخر الخريف ۞

ولما ورآء النهر كُوَرٌ اوّلها فيما يصاقب جَيْحُون على معبر[c] خراسان كورة بخارا ويتّصل بها سائر السغد المنسوب الى سمرقند وأُشْرُوسَنَة والشاش وفَرْغانة وكِشّ[d] ونَسَف والصَّغَانيَان واعمالها والخُتَّل وما يمتدّ على نهر جيحون من التِرْمِذ والقُوَاذِيَان[e] وأَخْسِيسَك[f] وخوارزم وامّا فاراب وأَسْبِيجَاب الى الطَّرَاز وايلَاق فمجموع الى الشاش وامّا خُجَنْدَة فمضمومة الى فرغانة وجمعنا ما بين[g] وَاشَجِرْد والصغانيان الى عمل الصغانيان وجعلنا الخُتّل فيما ورآء النهر لانّها ما بين وَخْشاب وجرياب[h] وجعلنا خوارزم ممّا ورآء النهر لانّ مدينتها ورآء النهر وهى الى كور ما ورآء النهر اقرب فامّا بخارا وكشّ ونسف فقد كان يجوز ان نجمعها كلّها الى[i] السغد ولكنّا فرّقناها[k] ليكون ايسر فى

---

a) A. et B. بمصمص, D. بترصيف, Jacut بتبييض. b) B. بلدة; Jacut, III, p. ٨٧٩, 11 ببلد غيره. c) A. معبّر. d) A. semper, B. plerumque كس. Deinde E. ونخشب l. وبشخب. e) A. والعرادداب, B. والعواردباب. E. وقباديان. f) A. h. l. واحشيسك, B. واحشمك, D. واخشنكث; C. sine punctis. Vid. Jacut in v. g) A. et B. دمن, D. دبين. h) A. et B. sine punctis. E. وحردات et mox حردان, C. حرباب. Juynboll, Ann. ad *Merácid*, V, p. 132 proponit legere خرياب, sed secutus sum D. et Jacut, II, p. ١٧١, 14; ۴٠٢, 20. Dimaschki, p. ٩۴ جرباب. i) A. et B. على. k) Pronomen addidi.

التفصيل a واخفّ وليس فى جمع الاطراف بعضها الى بعض ولا فى تفريقها كبير درك غير الابانة عمّا فى اعراضها من المدن والانهار وموضوعات المدن فى صفاتها فلا فرق بين الجمع فى ذلك والتفريق الّا لسهولة العبارة عنها فى التفصيل فنبدأُ ممّا وراء النهر بجيحون فنذكره ونذكر ما عليه من الكور

فاما جَيْحُونُ فانّ عموده نهر يعرف بجرياب يخرج من بلاد وَخّان b * فى حدود بذخشان c فيجتمع اليه انهار فى حدود الختّل والوَخْش فيصير منه d هذا النهر العظيم فمن هذه الانهار نهر يلى جرياب e يسمّى بأَخْشوا f وهو نهر هُلْبُك g ويليه نهر بربان h والثالث فارغر i والرابع نهر * انديجاراغ والخامس نهر وَخْشاب k وهو اعظم l هذه الانهار فتجتمع m هذه المياه قبل آرْهَنَ n ثم يجتمع مع o وَخْشاب قبل القُوَاذِيَان ثمّ يقع p اليه بعد ذلك انهار تخرج من البُتّم q فمنها انهار بالصَّغَانِيَان r وانهار بالقُوَاذِيَان * فيجتمع كلّه ويقع فى s جيحون بقرب القواذيان، * وامّا وَخْشاب فيخرج t من بلاد الترك حتّى يظهر فى ارض الوخش ويضيق u فى جبل هناك حتّى يعبر على v قنطرة ولا يعلم ماءً فى

---

a) A. et B. المعصل. *b*) Jacut, II, p. ١٧١, 15 وَخّاب et sic in v.; C. وجان. *c*) A. et B. om. Deindè B. فيجمع, C. فيخرج, Jacut وينصمّ. *d*) D. منها, Jacut من تلك الانهار. *e*) A. h. l. حوباب. *f*) A. et B. باخشوا; C. et E. باخس; Jacut بأَخْش. Cf. Juynboll l. l. *g*) A. et B. هليل. *h*) A., B. et C. دربان, D. بليان. Vid. Jacut et cf. Juynboll l. l. Mokaddasí, p. 13 سودسار. *i*) A. بمارعد, B. مارعد; vid. supra ad p. ٢٧٧, ann. *d*. *k*) Haec in A. et B. desunt; cf. supra p. ٢٧٧ et Juynboll, p. 133, ibi vero adde: E. اندنجاراع (mox اندىحراغ), Jacut انديخارع, Mokaddasí دنجاراع. D. habet ابذاجاراغ l. اندجاراغ. *l*) C. D. et Jacut اغزر. *m*) A. فيجتمع, B. فيجمع; D. addit اليه. *n*) A. et B. اوهن, D. ابهر; ceteri om. *o*) A. om. مع. *p*) C. et Jacut ترتفع. *q*) C. et Jacut addunt وغيره, D. وغيرها. *r*) C., D. et Jacut الصغانيان et القواذيان. *s*) C. et Jacut: فتجتمع كلها وتقع الى. *t*) C., D. et Jacut ومآء وخشاب يخرج. *u*) C., D. et Jacut ويصير. *v*) D. et Jacut om.

كثرته يضيق مثل ضيقه فى هذا الموضع وهذه القنطرة الحدّ بين الوخش[a] وبين وَاشْجِرْد، ثمّ يجرى هذا الوادى فى حدود بلخ الى التِرْمِذ ثمّ[b] على كالف ثمّ الى[c] زمّ ثمّ الى آمل حتّى ينتهى الى خوارزم ثمّ الى بحيرتها ولا ينتفع بمآء هذا الوادى بالختّل والترمذ الى ناحية زمّ احد فتعمر به زمّ وآمل وفرَبْر[d] ثمّ ينتهى الى خوارزم فيعمر خوارزم وعامّة نفعه لاهل خوارزم ✿

فاوّل كورة على جيحون ممّا ورآء النهر الخُتّل والوَخْش وهما كورتان غير انّهما مجموعتان فى عمل واحد وهما ما بين نهر * جَرْياب ووَخْشاب[e] فمن مدن الختّل هُلْبُك[f] ومُنْك * وتمليات وفارغر[g] وكاربنج[h] واندبجاراغ[i] ورستاق بنك[k] ومدن الوخش هَلَاوَرْد[l] ولَاوَكَنْد * ومقام السلطان بهُلْبُك[m] ومُنْك وهلاورد هما اكبر من هلبك غير انّ مقام السلطان بهلبك، والّذى يتاخم الوخش والختّل وَخّان والسغينة[n] وكَرّان وهى[o] دور كفر يقع منها المسك والرقيق وبوَخّان[p] معادن من الفضّة غزيرة وفى اودية الختّل ذهب يجمع فى السيول[q] من بلاد وخّان وبين وخّان وتُبّت[r] قريب، وارض الختّل ذات زروع كثيرة ومياه وثمار وهى على غاية الخصب والسعة وبها دوابّ ومواش كثيرة ✿

فاذا جزتَ الختّل والوخش الى نواحى واشجِرْد والقواذيان والترمذ

---

a) C., D. et Jacut الختل, E. ختلان. Deinde A. واسنجرد, E. بشكرد. (A., B. et D. om. وبين). b) Jacut addit بيهر. c) Jacut على. Deinde A. et B. الزم cum articulo et sic quoque C., sed ibi ex الى زم est ortum. d) A. et B. وفرس. e) A. حريلب ووخشان. f) A. et B. semper هليل. Deinde A. et B. ومنك. g) A. et B. habent وتملماب وبيازعر, vid. supra ad p. ٢٧٦ *e* et ٢٧٧ *d*. h) A. وحادح, B. ودادح. Vid. supra p. ٢٧٦ *d*. i) A. et B. sine punctis. k) B. h. l. ورساودمل, E. زوستاى مكه. Vid. supra p. ٢٧٧ *e*. l) A. et B. h. l. هلاورك. m) Haec delenda esse videntur. Deinde A. et B. ومنل. n) Vid. supra p. ٢٩٠ ann. *n*. o) A. et B. وهو. p) A. et B. ودورحان et habent من ante معادن. E. ودر وخشاب. C. deinde الفضة. C. semper habet وجان, Jacut وَخّاب. q) C. et D. addunt يجرى. r) Jacut, IV, p. ٩٠٩, 5 والبُسْت.

والصغانيان وما فى اضعافها فانّها كور مفردة بالاعمال، وامّا التِرْمِذُ فانّها مدينة على وادى جيحون لها قلعة[a] ومدينة وربض[b] ويحيط بالربض ايضًا سور ودار الامارة فى القلعة والحبس خارج القهندز فى المدينة فى السوق ومسجد الجامع فى المدينة والمصلّى داخل السور فى الربض واسواقها فى مدينتها وابنيتها طين ومعظم سككها واسواقها مفروشة بالآجُرّ وهى عامرة آهلة وفرضة تلك النواحى على جيحون واقرب الجبال اليها على نحو مرحلة وماؤهم[c] للشرب من جيحون ونهر يجرى من الصغانيان وليس لضياعهم من جيحون شرب وشرب ضياعهم من نهر الصغانيان، ولها من المدن صَرْمَنْجَن[d] وهاشم جرد[e] ۞ والقُواذِيان مدينة لها كورة وهى اصغر من الترمذ ولها من المدن نودز[f] ۞ والواشَجِرْد[g] نحو الترمذ فى الكبر وشُومَانُ[h] اصغر منها ويرتفع من واشجرد وشومان الى[i] قرب الصغانيان زعفران كثير يحمل الى الآفاق ويرتفع من القواذيان الفُوّة، والصَّغَانِيَانُ[k] مدينة اكبر من ترمذ الّا انّ الترمذ اكثر اهلًا ومالًا وللصغانيان قلعة ۞ وامّا أَخْسِيسَك[l] فهى بحذآء زَمّ وزمُّ فى ارض خراسان الّا انّهما مجموعتان[m] فى العمل والمنبر بالزمّ وهى مدينة خصبة صغيرة والغالب على اطرافها السوائم من الابل والغنم وعلى ظهر كلّ واحدة منهما مفاوز وآبار ومراعٍ ۞ وامّا فَرَبْر[n] فهى مدينة من بخارا قد وصفناها فى جملة بخارا ۞

---

a) C., D., E. et Jacut (I, p. ٨٤٣, 15) قهندز. *b*) A. et B. om. وربض, C, et D. om. ومدينة, sed E. recte شارستان وربض دارد. *c*) B. وماوها. *d*) A. et B. صرمىحن, C. صرمبحر, E. صرحىحر. Jacut in v. صرمنجان. D. vero habet صرمنجى. *e*) A. et B. sine punctis, C. et E. هاشم كرد. *f*) A. دودر, B. يودر, C. et D. دوذن, Edrîsî, I, .p. 480 سورن. Conjectura scripsi. Eadem urbs memoratur a Jakubío, p. ٩٩, ubi Codex دوز. *g*) E. ويشكرد. *h*) A. h. l. وسمرمان. *i*) A. et B. om. الى. Cf. quoque Jacut, IV, p. ٨٩٢, 1. *k*) E. جغانيان. *l*) E. اخسىكت, D. اخشنكث. *m*) A., B. et C. محموعان. *n*) Vocales in A.; B. فَرَبْر.

واما خُوَارِزمُ فانّه اسم الاقليم وهو اقليم منقطع عن خراسان وعمّا ورآء النهر وتحيط به المفاوز من كلّ جانب وحدُّها متّصل بحدّ الغُزّيّة فيما يلى الشمال والمغرب وجنوبيّه وشرقيّه خراسان وما ورآء النهر وهى فى آخر نهر جيحون وليس بعدها على النهر عمارة الى ان يقع فى بحيرة خوارزم وهى على جانبى جيحون ومدينتها فى الجانب الشمالىّ من جيحون ولها فى الجانب الجنوبىّ مدينة كبيرة تسمّى الجُرْجَانيّة [a] وهى اكبر مدينة بخوارزم بعد قصبتها وهى متجر الغُزّيّة ومنها تخرج القوافل الى جرجان والخزر والى خراسان، وقد كان فى التقدير ان نصوّر نصف خوارزم فى صورة خراسان ونصفها فى صورة ما ورآء النهر غير انّ الغرض فى هذا الكتاب معرفة صور هذه الاقاليم ومدنها فاخترتُ ان تكون خوارزم مجموعة فى الصورة وجعلتُها فى صورة ما ورآء النهر فابلغ بذلك غرضى من غير تكرار فى الصورتين، وبخوارزم من المدن سوى القَصَبَة دَرْغان [b] وهَزَارَاسب وخَيْوَه وخُشْمِيثَن [c] وأَرْدَخُشْمِيثَن [d] وسافَرْدَز [e] ونُوزوار [f] * وكردران خواش [g] وكُرْدَر [h] وقرية براتكين [i]

---

a) E. كركانج. *b*) A. et B. درعان. *c*) A. et B. وحمسں, ceteri om. Vid. Jacut in v. Mokaddasí habet خشمنن et خشمش. *d*) A. وارحس ممں, B. و.... دمنں, C. واردخمشيس, D. واردخسمش, E. ارد حسمس (Ous. اردجر). Infra in itin. A. et B. ارد حسمش, D. ازدخشمين, E. ارد حسر, *Djih.-Numa*, p. ٣٤٧, 11 a f. مار حسميں. Dimaschkí, p. ٢٢٣ أو حشمين. Jacut habet ارتخشميثن. Vid. I, p. ١٩١. *e*) A. et B. وساىرق, C. et E. وسافرون, D. وسافرزن. Vid. Jacut in v. *f*) A. ونعزروار, B. وىغدروار, C. ونوزان, D. et Edrîsî, II, p. 189 وبوران, E. نوران. Infra in itin. A. et B. نورواز, E. وردان (Ous. وركان), D. دوروام, Edrîsî, II, p. 192 بوروزم. Mokaddasí habet بوزابان, بوردان, دورادار et بوراوان. *Djih.-Numa* l. l. سوراوان. Sprenger pronunciat Burâwân (p. 34 et 35), sed edidi نوزوار, opinatus locum non differre a نوزكاث, sive aliud ejusdem loci sit nomen, sive ex hoc nomine corruptum. Certum est inter Djordjaniam et نوزوار candem distantiam (unius diei) esse, quae inter Djordjaniam et Nuzkáth. *g*) A. h. l. وكودار حواس; B. حوادس; C.

ومَدْمينية [a] ومرداجقان [b] والجُرْجانيّة ۞

فامّا قصبتها فانّها تسمّى بالخوارزميّة كاث [c] ولها قلعة ليست بعامرة وكانت

وكُردان حاس وحواس ;D. وكرّدان وخواش ;E. كردان خواس. Infra in itin. A. كردان حواس, B. كرزدان حواس et وكوران حراس; D. كردكان خاس, E. كردان خاس, C. كردان (كرداس) حاش. Mokaddasí كردرانحاس. Fluvius appellatur ab A. كودار حواس et كودران حواش, B. كردران خواش, C. كردان حواس, E. كرد خواس, D. كردران حواش (Edrîsî كردوان خواس), Mokaddasí كردران خاس. — Lectio C. sic explicanda est: dici aeque bene كردران خاش ac كردران خواش. *h*) A. et B. وكودر, C. وكردن, D. وكردى. Vid. Jacut in v. ubi كَرْدَر. *Merácid* praescribit كُرْدَر ut est in ed. Ibno 'l-Athír, V, p. ١١٥. *i*) A. et B. hic et infra درانكين (B. semel (براتكين); C. h. l. ابراتكين, infra براتكين et درانكتن; E. h. l. ادرادكمن, infra براتكين. Mokaddasí نرانكين, درانكين, برانكين, قرانكمن et دراذكمن. D. semper قراتكين.

a) A. et B. h. l. sine punctis, infra in itin. semel مدينه; C. ومذمينمه h. l., infra sine punctis; D. ومَدْمينية, infra semel مذمينيه; E. مدمنيه et مدىمه; Edrîsî, II, p. 192 مدنيثه; Sprenger, p. 32 et 36 confert Mokaddasii مدامثين, sed potius conferatur ejusdem مدكمينية s. مدكمينه. *b*) A. h. l. habet مرداجفان, B. مرداجغان ut Edrîsî, II, p. 189. Infra A. et B. مرداحعان et مرداحمان. C. مرداحقان; E. h. l. مرداحعان, infra مردايكان; Ous. p. 241 مرداجفان, p. 278 مردانكان. D. h. l. مردخوان (مردخوالن), infra مرداحفان, مرداخكان, مزداخكان, مزداخقان et مردى حعان et مرداجعان دزدادعان. Deinde C. addit وممهاد وخاس, E. (cf. Ous. p. 241) ممهاد حاحاسكر كانحك. Quae urbs primo loco intelligatur nescio. خاس inter urbes Khowarizmiae numeratur a Mokaddasîo. Ultima pars lectionis E. videtur legenda كركانجك i. e. كركانج صغرى (*Djih.-Numa*, p. ٣٤٥, 4 a f., Jacut, IV, p. ٢٦٠ seq.). *c*) A. et B. كار, E. كاب.

لها مدينة فخرّبها النهر وبنا الناس من ورآء المدينة وقد قارب النهر القلعة ويُخاف على تهدّمها [a] ومسجد الجامع على ظهر القلعة ودار خوارزم شاه عند مسجد الجامع والحبس عند القلعة وفى وسط المدينة . نهر جردور [b] يشقّ المدينة والسوق على جانبى هذا النهر وطولها نحو ثُلُث فرسخ [c] فى نحوه وامّا ابوابها فقد تهدّم بعض المدينة وذهب ابواب ما تهدّم منها والباقى قد بنى خلف [d] ما تهدّم على الوادى ✿ واوّل حدّ خوارزم يسمّى الطّاهريّة ممّا يلى آمُل فيمتدّ هذه العمارة فى جنوبىّ [e] جَيْحُونَ وليس فى شماليّه عمارة الى ان ينتهى الى قرية تسمّى غارابخشنة [f] ثمّ يكون من غارابخشنة الى مدينة خوارزم عامرًا *من جانبى جيحون جميعًا [g] وقبل غارابخشنة بستّة فراسخ نهر ياخذ من جيحون فيه عمارة الرستاق الى المدينة ويسمّى هذا النهر كَاوخُوَارِه [h] وتفسيره اكل البقر وهو نهر عرضه نحو خمسة اذرع [i] وعمقه نحو قامتين فيحمل السفن ويتشعّب من كاوخواره بعد ان يجرى [k] خمسة فراسخ نهر يسمّى كريه [l] يعمر به بعض الرساتيق ✿ وليس للعمارة على شطّ جيحون من الطاهريّة الى هَزَارَاسْب كبير عرض ثمّ يعرض بهزاراسب

---

a) Quod tempore Ibn Haucalis jam acciderat. Mokaddasí quoque dicit ولهم قهندز قد خرّبه النهر. A. et B. تهدّمه. C. القهندز ويخاف على هدمه. *b*) ? A., B. et C. حردور, D. جركور, Ous. خردور. *c*) E. سه فرسخ, D. ثلاثة فراسخ. *d*) Secundum C.; A. et B. خلفه. *e*) A. et B. inverso ordine, sed male, h. l. شمالى, deinde جنوبيه. *f*) A. عارابحسمه, et غارابحسمه; B. غارابحسه et ut C. غارابحنسه; D. عارامحسنه, Epit. Paris. غارامخشه; E. غار البحبية. Conjectura scripsi. Fortasse autem legendum est غار أَخَشَّه, coll. Jacut, II, p. ١٣٣, 14. *g*) A. et B. om. et habent عامر. *h*) D. غاوخواره ut supra C. habuit عالف l. غالف pro كالف. *i*) C. ابوع, D. ابواع, E. كز. (Pro خمسة A. et B. خمس). *k*) C. يصير, D. يجوز. Deinde C. et E. فرسخين (دو فرسنك). *l*) A. كوده, B. كرده, D. كونه. Secutus sum C. et E.; Ous. et *Djih.-Numa*, p. ٣٤٩, 4 a f. كربه. Mokaddasi MS. p. 142, con-

فيصير عرضه نحوًا من مرحلة الى مقابل[a] المدينة ثمّ لا يزال يضيق حتّى يصير بالجُرْجانيّة نحو فرسخين ثمّ ينتهى الى قرية تسمّى كيت[b] * على خمسة فراسخ من كوجاغ[c] وهى قرية بقرب جبل وليس[d] فى العرض عمارة غيرها وورآء[e] هذا الجبل المفازة ✿ ومن هزاراسب الى ساثر ما على غربىّ جيحون انهار منها نهر هزاراسب ياخذ من جيحون ممّا يلى آمُل وهو نحو نصف كاوخواره ويحتمل[f] السفن ؛ ثمّ على نحو فرسخين من هزاراسب نهر يعرف بكردران خواش[g] وهو اكبر من نهر هزارسب ؛ وبعده نهر خَيْوَه وهو نهر اكبر[h] من كردران خواش وتجرى فيه السفن الى خيوه ؛ وبعده نهر مدرا[i] وهو نهر اكبر[k] من كاوخُوَاره مرّتين[l] تجرى فيه السفن الى مدرا وبين نهر مدرا ونهر خيوه نحو ميل ؛ ومن نهر مدرا الى نهر وداك[m] تجرى فيه السفن الى الجرجانيّة وبين نهر وداك ونهر مدرا نحو ميل ومن نهر وداك الى مدينة خوارزم نحو فرسخين ؛ واسفل المدينة من ناحية الجرجانيّة نهر يسمّى بوه[n] فيجتمع مآء بوه ومآء وداك فى حدّ قرية تعرف باندراستان[o]

---

fundens كريه et كاوخواره habet كردر. *a*) A. دعابل. *b*) A. et B. bis حمب, semel خبث; D. et Edrîsî (I, p. 189) h. l. خبت, infra كيث. E. حمب. Mokaddasí ter خنب, semel حمب. Vámbéry, *Reise in Mittelasien*, p. 273 inter pagos urbis Kungrat memorat Kiet. Recepi igitur lectionem D., licet apud nostrum et Mokaddasí praeferendum videatur جيت. *c*) Haec addidi ex D. et E. (ubi كرادح, Ous. كراغ). *d*) A. om. *e*) A. et B. وذرا. *f*) A. sine punctis, B. وتحمل. *g*) A. بكودار حوىمن, B. بكردران حوس. Vid. supra ad p. ٢٩٩ ann. *g*. *h*) A. et B. om. *i*) D. et Mokaddasí مدرى. Hic semel وردى. *k*) A. om. *l*) A. et B. مدرمن. *m*) A. et B. (omissis verbis الى نهر) h. l. وذلك, infra ter ودان, semel وذاك. E. دال, *Djih.-Numa*, p. ٣٤٩, 3 a f. رودال. Edrîsî ودال, Mokaddasí ورال. Secutus sum D. *n*) A. et B. دره (semel دوه); D. سويه; Edrîsî بورة. Secutus sum C., E., Mokaddasí (ubi بوّه) et *Djih.-Numa*. *o*) B. باندراسان, D. باندرستان, E. اندرباز; Mokaddasí اندرسمان.

اسفل منها الى ما يلى الجرجانيّة وذاك اكبر من بوه وتجرى فيهما السفن الى الجرجانيّة على غلوة ثمّ يكون هناك سكر يمنع السفن ومن مجتمع[a] هذين المآءين الى الجرجانيّة نحو مرحلة، وبين نهر كاوخواره والمدينة اثنا عشر فرسخًا وعرض نهر[b] خوارزم عند المدينة نحو فرسخين، ولكُرْدَر[c] نهر ياخذ من اسفل مدينة خوارزم على اربعة فراسخ من اربعة مواضع متقاربة فيصير نهرًا واحدًا مثل * بوه ووداك[d] اذا اجتمعا، ويقال انّ جيحون كان مجراه فى هذا الموضع واذا قلّ مآء جيحون يقلّ[e] الماء فى هذا النهر ✿ وبحذآء كيت[f] فى المفازة بفرسخ من الجانب الشمالىّ المدينة الّتى تسمّى مَذْمِينية[g] وهى من جيحون على اربعة فراسخ الّا انّها من الجرجانيّة وانّما صار هكذا لانّ النهر يحول[h] من كُرْدَر يقطع ما بين كيت ومذمينية وليس على الشطّ بعد مذمينية عمارة، وبين جيحون وكردر رستاق مرداجقان[h] وبين مرداجقان وجيحون فرسخان وهى بحذآء الجرجانيّة، ولكلّ[i] قرية بين كردر[k] والمدينة نهر يقع[l] من جيحون وجميع هذه الانهار كلّها من جيحون ✿ ثمّ ينتهى جيحون الى بحيرة خوارزم بموضع[m] فيه صيّادون ليس به قرية ولا بنآء ويسمّى هذا الموضع خليجان[n] وعلى شطّ هذا البحر * ممّا يلى[o] خليجان ارض الغُزيّة فاذا كان الصلح جآءوا من هذا الجانب الى قرية براتكين[p] ومن الجانب الآخر الى الجرجانيّة وكلتاهما[q] ثغر، وفى جيحون

---

a) B. مجمع. *b*) A. et B. om. Supplevi ex D.; C. وعرض الكاوخواره. *c*) A. et B. ولكودر, C. ولكوص, D. ولكردى. Recte Mokaddasí. *d*) A. et B. دره وذاك, C. بوه ودلك. *e*) E. زياد شود. *f*) Vid. supra p. ٣٠٢ ann. *b*. *g*) A. et B. sine punctis, E. h. l. مدمينه, D. مدينيه, mox ومدينيّة. Vid supra ad p. ٣٠٠ ann. *a*. *h*) A. et B. sine punctis. *i*) A. et B. وذلك. *k*) A. كودر, C. كردن, D. كردى. *l*) C. et D. يرتفع. *m*) A. et B. موضع. *n*) A. حلحار, B. حلحاب, C. حملجان, D. ut recepi; E. خلجان ut Jacut, I, p. ٥١۴, 14. *o*) C. et Jacut من مقابل, D. مما يقابل. *p*) A. sine punctis; D. فراتكين. E. h. l. ده آب كير. *q*) Secundum C.; A. et B. وكلاهما; D. وهى.

قبل ان يبلغ كاوخواره بنحو *ثلاثة فراسخ[a] جبل يقطع جيحون وسطه قطعًا[b] فيضيق النهر حتّى يعود عرض الماء الى نحو من الثُلْث ويسمّى هذا الموضع ابوقشة[c] وهو موضع يخاف على السفن منه من شدّة جريه والهور الّذى عند مخرجه، وبين الموضع الّذى يقع فيه نهر جيحون الى الموضع الّذى يقع فيه نهر الشاش من هذه البحيرة نحو اربعة[d] ايّام، ووادى جيحون ربّما جمد فى الشتاء حتّى يعبر عليه بالاثقال ويبتدئ جموده من ناحية خوارزم حتّى يعلو الى حيث انتهى الجمد[e] وابرد ما على جيحون من البقاع خوارزم، وعلى شطّ بحيرة خوارزم جبل جغراغز[f] يجمد عنده الماء حتّى يبقى الى الصيف وهو اجمة قصباء ودور هذه البحيرة فيما بلغنى نحو من مائة فرسخ ومآؤها مالح وليس لها مغيض ظاهر ويقع فيها نهر جيحون ونهر الشاش وانهار غيرهما[g] فلا يعذب مآؤها ولا يزيد[h] على صغرها ويشبه والله اعلم ان يكون بينها وبين بحر الخزر خروق[i] يتّصل بها مآءهما وبين البحرين نحو من *عشرين مرحلة[k] على السمت[l] ۞ وخوارزم مدينة خصبة كثيرة الطعام والفواكه الّا انّها لا جوز بها ويرتفع منها من ثياب القطن والصوف امتعة كثيرة تنقل الى الآفاق وفى خواصّ اهلها يسار وقيام على انفسهم بالمروّة وهم[m] اكثر اهل خراسان انتشارًا وسفرًا فليس بخراسان مدينة كبيرة الّا وبها من اهل خوارزم جمع كبير ولسانهم لسان مفرد وليس بخراسان

---

a) C. et D. مرحلة. *b*) A. om. *c*) A. ردرمنه, B. ددرمسه, C. ودوىسه, E. بوقسه. Secutus sum D. *d*) C., D. et Ous. عشرة. Jacut, I, p. ٥١۴, 17 et 18 سوى عدة ايام. *e*) A. الحمل, B. انتهاء الجبل. Secutus sum C., D. et E. *f*) A. حفواغر, B. حعواغر, C. حعراعر, D. حفراعن, (كه غايت سردسير باشد). E. جغاغر, Ous. چغاغر. *g*) Ex C.; A., B. et D. غيرها. *h*) C. et Jacut addunt فيها. *i*) Jacut addit وذ-زوز et habet deinde ويستمد ماءها. Pro بها C. منها. *k*) Jacut عشر مراحل. *l*) A. et B. addunt: فاما ما بين خوارزم والبحيرة, quae fortasse e Jacut l. l. compleri debent verbis: فهو رمال وسبع لا يمنع من النفر. *m*) A. et B. وهى.

بلد على لسانهم وزيّهم القراطق والقلانس وخلقهم لا يخفى فيما بين اهل خراسان ولهم بأس على الغُزيّة ومنعة وليس ببلدهم معادن ذهب ولا فضّة ولا شىء من جواهر الارض وعامّة يسارهم من متاجرة الترك واقتناء المواشى [a] ويقع اليهم اكثر رقيق الصقالبة والخزر وما والاها مع رقيق الاتراك والاوبار من الفنك والسمّور *والثعالب والخزّ [b] وغير ذلك من اصناف الوبر ٭

فهذا ما على جيحون من الكور فنبدأ ممّا وراء النهر فى كورة بخارا لانّها اوّل الكور وبها دار امارة خراسان وهى مستقيمة على ترصيف كور ما وراء النهر ثمّ يتبع [c] ما يليها على الاتّصال ان شاء الله ٭ امّا بُخارا واسمها بُومِجْكَث [d] فهى مدينة فى مستوى [e] وبناؤها خشب مشتبك [f] ويحيط ببنائها [g] قصور وبساتين وسكك وقرى تكون اثنى عشر فرسخا فى مثلها ويحيط *بجميع ذلك [h] سور يجمع هذه القصور والابنية والقرى والقصبة فلا يرى [i] فى اضعاف [k] ذلك كلّه مفازة [l] ولا خراب ومن دون هذا السور على [m] قصبة المدينة وما يتّصل بها [n] من القصور والمساكن والمحالّ والبساتين الّتى تعدُّ مع [o] القصبة ويسكنها من يكون *فى جملة [p] القصبة شتاء وصيفا سور آخر *قطْرُه فرسخ [q] ولها مدينة داخل هذا السور يحيط بها سور حصين ولها قلعة [r] خارج المدينة تتّصل [s] بها مقدار مدينة صغيرة *وفى داخلها قلعة

---

a) A. للمواشى. *b*) E. وسنجاب. *c*) B. نتبع, D. يتبعها. *d*) Mokaddasí نموجكث. *Djih.-Numa*, p. ٣٦٢ ult. بو محكث. *e*) C. et D. مستواة; Jacut, I, p. ٥١٧ ult. على ارض مستوية. *f*) B., D. et Jacut, I, p. ٥١٨ مشبّك. *g*) B. بهذا البناء المشبك, C. بها البنا المشتبك, Jacut بهذا البناء. Deinde habent من القصور والبساتين والمحال والسكك المفترشة والقرى المتصلة ما يكون. Textus apud Jacut corruptus est. *h*) C. et D. بها كلها. Deinde C. et Abulfeda, p. ٤٨٩ habent سور واحد. *i*) A. et Jacut ترى. *k*) Jacut خلال. *l*) D. et Jacut قفار. *m*) D. ét Jacut add. خاص. *n*) A. et B. به. *o*) D. et Jacut ممن. *p*) D. من اهل; cf. Jacut. B. om. جملة. *q*) C., D. et Jacut نحو فرسخ فى مثله. *r*) C., D. et Jacut قهندز. *s*) A. et B. يتصل; C.,

اخرى[a] ومسكن ولاة خراسان من آل سامان فى هذه القلعة ولها ربض ومسجد الجامع على باب القلعة[b] فى المدينة وحبسها فى القلعة واسواقها فى ربضها، وليس بخراسان وما ورآء النهر مدينة اشدّ اشتباكًا من[c] بخارا ولا اكثر اهلًا على قدرها[d] ولهم فى الربض نهر السُّغْد يشقّ[e] الربض واسواقها وهو آخر نهر السغد فيُفْضى[f] الى طواحين وضياع ومزارع ويسقط فاضله[g] فى مجمع ماء يجاور[h] بِيكَنْد الى قرب فَرَبْر يعرف بسام خواش[i]، وامّا المدينة فلها سبعة ابواب حديد منها باب يعرف بباب المدينة[k] وباب يعرف *بباب نور[l] وباب يعرف بباب حُفْرَة[m] وباب يعرف بباب الحديد وباب يعرف بباب القهندز وباب يعرف بباب بنى اسد وهو باب مهر وباب يعرف بباب بنى سعد، ولقلعتها بابان احدهما باب الرِيكِسْتان[n] والآخر باب الجامع يشرع الى مسجد الجامع، وعلى الربض دروب[o] فمنها درب يخرج منه الى خراسان يعرف بباب[p] الميدان ويليه[q] ممّا يلى المشرق درب يعرف بدرب ابراهيم ويلى هذا الدرب درب يعرف بدرب الرِّيْو[r] ويليه درب يعرف بالمردقشة[s] ويليه درب يعرف بدرب

D. et Jacut متصل. Deinde C. et Jacut ومقداره, D. وهو فى مقدار. *a*) C., D. et Jacut وفيه قلعة. E. وحصنى دران هست. *b*) D. et Jacut القهندز, et sic apud Mokaddasí, MS. p. 137. *c*) C. addit ابنية. *d*) C. et D. addunt منها. *e*) A. يشتق. Mox C. واسواقه. *f*) B. ويفضى. *g*) D. et Jacut الفاضل منه. *h*) A. دحاور, B. يجاوز; D. ببخارا وبيكند; Jacut بحذاء; Ous. p. 245: „to the borders of;" Edrisí, II, p. 194: „dans le canton de." Deinde A. et B. بكمد. *i*) A. et B. حواش, D. et Jacut خاس, E. كاس. *k*) E. در شارستان. *l*) A. et B. دبون. E. در درد. Secutus sum D. et partim Mokaddasí, qui habet دور. Epit. Paris. habet بون s. بوان. *m*) A. et B. sine punctis. E. كنده. Epit. Paris. جهره et sic inde *Masálik al-absár* in *Not. et Extr.* XIII, p. 248. D. et Mokaddasí ut recepi. *n*) A. الدكستان. Mokaddasí السهلة. In E. hic multa desunt, et adscribitur مختصر شد. *o*) A. درب. *p*) D. et Mokaddasí بدرب. *q*) A. ولاده, B. وثلاثة, D. وباب. *r*) A. et B. الديو, D. الربو, Ous. p. 246 خديق. Cf. Jacut in v.; Mokaddasí om. *s*) D.

كلاباذ وهذا الباب وباب المردقشة يخرج منهما الى نَسَف وبلخ ويلى درب كلاباذ درب يعرف بالنُّوبهار ويليه درب يسمّى سمرقند يُفْضى الى سمرقند وسائر ما ورآء النهر ويليه درب فغاسكون[a] ثمّ درب الرّامِيثَنيّة[b] ثمّ يليه درب حدشرون[c] وهو طريق خوارزم ثمّ باب غُشَج[d] وفى وسط الربض على اسواقها دروب فمنها باب الحديد ويليه باب قنطرة حسّان ويليه بابان عند مسجد ماج[e] ويليهما باب يعرف بباب رخنة[f] ويليه باب عند قصر ابى هشام[g] الكنانىّ ويليه باب عند قنطرة السُّوَيقة ويليه باب فارْجَك[h] ويليه باب دروازجه[i] ويليه باب سكّة مُغَان[k] ويليه درب سمرقند الداخل، وليس فى مدينتها ولا قهندزها مآء جارٍ لارتفاعها ومياههم من النهر الاعظم وينشعب[l] من هذا النهر فى المدينة انهار منها نهر يعرف بنهر فَشِيذيزَه[m] ياخذ من نهر بخارا فى مكان يعرف بالورغ[n] يجرى فى درب المردقشة على جُوبَار ابى ابراهيم حتّى ينتهى الى باب الشيخ الجليل ابى الفضل[o] ويقع فى نهر نُوكَنْدَه[p] وعلى

المردكشان, Mokaddasi مردكسان. Ous. بروقسه et infra سروقسه. Fortasse ex hoc nomine depravatum est nomen hodiernum partis urbis, Malkuschan; vid. Vámbéry, *Reise*, p. 289. *a*) D. نعاشكور, Mokaddasí معاسكون. *b*) A. الرامتينيه, B. الرامتينيه, D. الراميثيه, Epit. Paris. الرامثينه, Ous. امينه, Mokaddasí الرامثنه. Derivatum videtur nomen a راميثنة pago unde quoque نهر الراميثنة nomen duxit. *c*) Sic A. et B.; D. حدشرون, Ous. حدسرور, Mokaddasí حدرسون. Epit. Paris. خدمتك. *d*) Sic cum vocal. A. et B.; D. عشج, Epit. Paris. عشنج, Ous. غنج, Mokaddasí عج. *e*) Sic D. et Ous.; A. et B. ماخ. *f*) A. et B. رحنه, D. زخنه, Ous. رجيه. *g*) D. et Ous. هاشم. *h*) A. et B. فارجل, D. فارحل (مارحل). Vid. Jacut in v. *i*) Sic A. et B.; D. حدفارجه. *k*) A., B. et D. معان. *l*) A. et C. sine punctis, D. وتنشعب. *m*) A. وسردده, B. وسرددده, E. غسرده, D. ودمردده, Epit. Paris. فيرديزه (s. فتيرديزه). Conjectura scripsi, coll. Jacut in v. *n*) A., B. et Ous. hic et infra sine punctis, D. بالوزغ. *o*) Vid. supra ad p. ٢٩٦, ann. *a*. *p*) A. et B. fere semper نركنده,

هذا النهر نحو الفى بستان وقصر سوى الارضين ومن فى[a] هذا النهر الى
مغيضه نحو من نصف فرسخ، ونهر يعرف بجُوَيْبَار بكار[b] ياخذ من هذا
النهر[c] فى وسط المدينة بموضع يعرف بمسجد احيد[d] ويغيض بنُوكَنْده وعلى
هذا النهر شرب بعض الربض ونحو من الف بستان وقصر سوى الارضين،
ونهر يعرف بجُوَيْبَار القَوَارِيرِيّين[e] ياخذ من النهر فى المدينة بموضع يعرف
بمسجد العارض فيسقى بعض الربض وهى اغزر واعمّ[f] للاراضى والبساتين من
نهر بُكار، ونهر يعرف بجوغشج[g] ياخذ من النهر عند مسجد العارض
* فيسقى بعض الربض حتّى يخرج الى نوكنده وهو يعرف بجويبار العارض[h]،
ونهر يعرف بنهر بَيْكَنْد[i] ياخذ من نهر المدينة عند راس سكّة ختع[k] فيسقى
بعض الربض ويغيض بنوكنده، ونهر نُوكَنْدَه ياخذ من النهر عند دار حمدونة[l]
وهو مغيض للمياه عليه شرب بعض الربض ويفضى الى المفازة وليس عليه
شرب ضياع، ويليه نهر الطاحونة ياخذ من النهر فى المدينة بموضع يعرف
بالنُّوبَهَار وعليه شرب[m] بعض الربض ويدير ارحية وينتهى الى بَيْكَنْد ومنه شرب

E. بركنده. Ous. et Epit. Paris. ut recepi. In D. lectio sibi non constat. *Djihân-Numa*, p. ٣٥١, 7 a f. et 4 a f. نوكند. *a*) D. فم. *b*) A. دىحودمار ىكار, B. دمادمار دكان, D. بخويبار يكان, Epit. Paris. بجويبار نكار, E. رود با سكان. Infra A., B. et D. بكار, Epit. Paris. نكار; *Masâlik al-absâr* جوباريكان s. بيكار. Jacut, II, ١٣٨, 21 praescribit جُويَبَار, sed vid. in voce. *c*) E. از رود بزرگ; cf. *Djih.-Numa* l. l. *d*) B. الحيد deinde in احيد correctum; D. احمد, Epit. Paris. et hinc *Masâlik al-absâr* احمد. *e*) A. et B. دىحويبار العواردىن. Ous. قزازان. *f*) D. واعمر. *g*) A. بغروحمج, B. بغدرحمج, D. بجوعسج. Ultimam lectionem recepi, coll. nomine درب غشج (vid. supra). *h*) Haec ex D. coll. Ous. inserui. Cf. *Djih.-Numa* l. l. ubi جويبار عارض, quod Norberg (I, p. 474, 4) vertit per »ille amplus." *i*) A. بيبكند, B. دىكند, Epit. Paris. بنوكند, *k*) B. حتع, D. جمع, Epit. Paris. خَتَع. *l*) A. et B. h. l. حمدويه. *m*) A. et B. سوق شرب et in A. duo puncta sub ش. D. et Epit. Paris. (cf.

اهل بـبـيـكـنـد، ونهر يعرف بنهر كُشنة[a] ياخذ من النهر فى المدينة عند النوبهار وعليه شرب النوبهار مـن الربض فيفضى الى قصور وضياع كـثـيـرة وبساتين حتّى يجاوز[b] كشنة الى مايَمْرْغ[c]، ونهر يعرف بنهر رَبّاح[d] ياخذ من النهر بقرب الريكستان[e] فـيـسـقـى بعض الربض وينتهى الى قصر رباح فيسقى نحو الف من البساتين والقصور سوى الارضين، ونهر الريكستان ياخذ من النهر بقرب الريكستان ومنه شرب الريكستان والقلعة ودار الامارة حتّى ينتهى الى قصر *جَلَال دِيزَه[f]، ونهر ياخذ من النهر فى المدينة بقرب قنطرة حمدونة تحت الارض الى حياض بباب بنى اسد وتقع فضلته فى فارقين القهندز، ونهر يعرف بنهر زغاركنده[g] ياخذ من النهر بمكان يعرف بورغ[h] فيجرى على باب دروازجه وعليه سوق[i] دروازجه الى باب سمرقند حتّى ينتهى الى سبيد[k] ماشه ويجاوزه نحوًا من فرسخ وعليه قصور وبـسـاتـيـن واراضى كثيرة ۞ وامّا رساتيق بخارا فـمـنـهـا الذر[l] وفرغيدد[m] وسخر[n] ورستاق الطَّوَاوِيس وبورق[o] وخَرْغَانَة[p] السفلى وبومة[q] ونجار جفر[r] ورستاق كاخُشْتُوَان[s] وانديار كندمان[t]

Masâlik al-absâr, p. 250) ببيوت. *a*) A. h. l. كسمه, B. et E. كسه, D. كشبه, Epit. Paris. كُشنة; Ous. كيثه. Infra B. كشمه, A. uti recepi. *b*) A. ىاحاور, B. تجاوز. *c*) A. et B. داردمرع, D. فاريمرغ. *d*) A. sine punctis, B. رداح et رياح, D. باج et رباح, Epit. Paris. دباح et ريـاح (*Masâlik al-absâr* ريـاح). E. et Ous. ut recepi. *e*) A. et B. semper الددكستان. *f*) A. et B. sine punctis. *g*) Sic A. et B.; D. رىحاركند. *h*) A., B. et Ous. sine punctis, D. بوزغ. *i*) D. ببيوت. *k*) A. et B. سيبد. *l*) Sic A. et B.; D. الدن. Infra A., B. et D. نهر الدر. E. duobus vicis seqq. omissis دروفن. *m*) D. وبرعندى. *n*) A. وسحر, D. وسجن. *o*) A. et B. ودورق, D. وبروق, *Djih.-Numa*, p. ٣٥١, 12 a f. تورق. Secutus sum E. et Ous. *p*) A. et B. hic et mox deinde جوغـانه; D. جرعابه; E. فرغانه; Ous. فرغانه et جرغانه; *Djih.-Numa* قرغانه. Infra occurrit forma خرغانكث quam habet Jacut. *q*) A. et B. sine punctis. *Djih.-Numa* بوم. Secutus sum E. In D. nomen corruptum est. *r*) A. h. l. ونحار صفر,

وسامجن[a] ما دون وسامجن ما ورآء وفراور[b] السفلى واروان[c] وفراور العليا فهذه الرساتيق داخل الحائط، وخارج الحائط جَزَّة[d] وشابخش[e] ويسبير[f] رستاق كَرْمينيَة وخرغانة[g] العليا ورامند[h] وبِيْكَنْد وفَرَبْر؟ ويتشعّب[i] من عمود نهر السغد فى حدّ بخارا خارجًا عن القصبة من الحائط الخارج بناحية الطَّواوِيس الى ان ينتهى الى باب المدينة انهار كثيرة تتفرّق فى القرى والمزارع فى الحائط وعليها عمارة قرى بخارا فمنها نهر يعرف بسافرى[k] كام ياخذ من النهر فيسقى القرى حتّى ينتهى الى وَرْدانة وعليه شربهم، ونهر يعرف بخَرْغان[l] رود ياخذ من النهر فيسقى القرى حتّى ينتهى الى راوس[m] وعليه شربهم، ونهر يعرف بنجار جفر ياخذ من النهر فيسقى القرى حتّى ينتهى الى خَرْمَيْثَن[n] وعليه شربهم، ونهر يعرف بنهر جُرْغ[o] ياخذ من النهر

B. وبخار صفر, D. (خمص) وبحان حمف, E. نجاحمر, Ous. بخاجمر. Infra A. نحار حمص, B. نجار حمص, D. بحان حعر. s) A. et B. كاشختوان; *Djih.-Numa* روستا كاوحستوان. Vid. Jacut in v. t) A. et B. اندبار كيذبان, D. واندما كندمان, E. ابددان كمدمان, Ous. انديدان كندمان.

a) E. ut Ous. سامجر et post ما وراء addit النهر. b) E. فرادن, Ous. فرادن, *Djih.-Numa* فراوان. c) A. et B. h. l. اروات, E. اردان, D. اروار. *Djih.-Numa* recte اروان. Vid. infra. d) A. حره, B. حده, D. حدّه, Ous. چند. Conjectura scripsi. e) A. et B. وسادخس, D. وشابحش, Ous. مباخس, *Djih.-Numa* مناحس. f) A. وبسير, B. وبسر, D. ونشر, Ous. ويس. Conjectura scripsi, coll. يسبركث apud Jacut. g) A. h. l. وحوعانه. h) Sic A. et B.; D. (عركند) ورستاق عركيد, Ous. عرقند, *Djih.-Numa* عرقيد. i) A., B. et C. sine punctis. k) A. et B. بساوى, Ous. كافرى كام, *Djih.-Numa* l. l. 5 a f. كافر كام. l) A. دحرغان, B. بحزغان. m) D. رواش. Fortasse conf. زَوش apud Jacut. n) A. حرمس, B. حومنس, D. حومس, Ous. جرمش. o) A. h. l. حودح, B. حردح; mox B. الجزع. Vocalis secundum A.; D. sine articulo.

حتّى ينتهى الى الجُرْغ وعليه شربهم فيعود الفاضل فى النهر، ونهر يعرف بنُوكَنْدَه[a] ياخذ من النهر فيسقى القرى حتّى ينتهى الى فرانة[b] وعليه شربهم، ونهر يعرف بنهر فَرْخْشَة[c] ياخذ من النهر فيسقى القرى حتّى ينتهى الى فَرْخْشَة ومنه شربهم، ونهر يعرف بنهر كشنة ياخذ من النهر فيسقى القرى حتّى ينتهى الى كشنة وعليه شربهم، ونهر يعرف بنهر الرّامِيثَنَة[d] ياخذ من النهر فيسقى القرى حتّى ينتهى الى الراميثنة وعليه شربهم، ونهر فراور[e] السفلى ياخذ من النهر فيسقى القرى حتّى ينتهى الى فاراب[f] وعليه شربهم، ومنها نهر يعرف بأروان ياخذ من النهر فيسقى القرى حتّى ينتهى الى بَانَب[g] وعليه شربهم، ونهر يعرف بفراور العليا ياخذ من النهر فيسقى القرى حتّى ينتهى الى اوبوقار[h] وعليه شربهم، ونهر يعرف بنهر خامة[i] ياخذ من النهر فيسقى القرى حتّى ينتهى الى خامة وعليه شربهم، ونهر يعرف بتنكان[k] ياخذ من النهر فيسقى القرى حتّى ينتهى الى وَرْكَة وعليه شربهم، ونهر يعرف بنهر نُوكَنْدَه[l] ياخذ من النهر فيسقى القرى حتّى ينتهى الى نُوباغ[m] الامير وعليه شربهم، وما فضل من ماء نهر السغد فانّه يجرى فى نهر يعرف بالذر[n] وهو النهر الّذى يشقّ ربض بخارا ومنه انهار المدينة الّتى ذكرناها، واكثر هذه الانهار تحمل السفن كبرًا[o] وغزارة وكلّها تاخذ من النهر داخل

---

a) Secundum D. et Ous.; A. et B. بدكنده. *b*) A. et D. sine punctis; B. فرانه; vid. Ous. *c*) Ous. برجد, *Djih.-Numa* برجه. Pro كشنة habent hi بستة. *d*) A. et B. sine punctis; D. الرامينه. Jacut auctoritate Imránii زاميثنة, sed cf. sub راميثن. Hinc درب الراميثنية nomen habere videtur. Vid. supra p. ٣٠٧, ann *b*. *e*) A. h. l. فرادر, mox فراور; B. فراور et قراور; D. h. l. فروان ut *Djih.-Numa;* Ous. فراذن. Vid. ٣١٠ ann. *b*. *f*) B. فاراب, D. فارادة; cf. Jacut sub فَرَاب. *g*) A., B. et D. sine punctis. Vid. Jacut in v. *h*) A. اوبوفار, B. sine punctis; D. نوقان. *i*) A. et B. حامه, D. خاتمه (خاحمه). *k*) A. et B. بمكان, Ous. تلنكان; D. om. *l*) A. دوكنده, B. دوكيده, D. بوكنده. *m*) A. et B. دوباغ, Ous. برقاع. *n*) A., B. et D. الدر. *o*) C. كثره.

حائط بخارا من حدّ الطَّوَاوِيس الى ان ينتهى الى المدينة ۞ وابنية قرى بخارا كلّها على اشتباك البناء والتقدير فى المساكن وارتفاع اراضى الابنية وهى محصّنة بالقلاع بالابنية[a] المجموعة وليس فى داخل الحائط جبل ولا مفازة[b] واقرب الجبال اليها جبل وَرْكة[c] ومنه حجارة بلدهم للفرش والابنية ومنه طين الاوانى والنورة والجصّ ولهم خارج الحائط ملّاحات ومحتطبهم من بساتينهم وما يحمل اليهم من المفاوز من الغضا[d] والطرفاء واراضى بخارا كلّها قريبة الى الماء لانّها مغيض ماء السغد ولذلك لا تنبت الاشجار العالية فيها مثل الجوز والدُّلْب والخَوْر وما اشبهه فاذا كان منه شجر فهو قصير غير نامٍ، وفواكه بخارا اصحّ فواكه ما وراء النهر والذّها طعمًا ومن عمارة بخارا انّ الرجل ربّما قام على الجريب الواحد من الارض فيكون منه معاشه[e] ومن كثرة عددهم انّ ما يرتفع من بلادهم يقصر عن كفايتهم لوفور عددهم وتضاعفهم[f] على ما يخرج من اراضيهم فيحمل اليهم الميرَ من الطعام وسائر ما يحتاجون اليه من سائر ما وراء النهر، والجبل الّذى يتّصل ذيله[g] بقرية وَرْكة[h] جبل يمتدّ الى سمرقند فيما بين كِشّ وسمرقند حتّى يتّصل بجبال البُتّم عاطفًا على أُشْرُوسَنَة[i] فى عرض فَرْغانة حتّى يخرج على ناحية شَلْجِى[k] والطَّرَاز ثمّ يمتدّ فيما اخبرنى به من سلك تلك[l] السُّبُل الى حدّ الصين وهذه المعادن الّتى باشروسنة وفرغانة وايلاق وشَلْجِى ولبان[m] الى ارض خرخيز

---

a) D. والابنية. Apud E. h. l. leguntur ودر حايط بخارا دوازده دروازه است.
b) E. addit : همه قهندز وكوشك وباغ وزرعت. c) D. h. l. وزكه. d) A. et B. العصى, D. القصبا et sic E. نى وكز. *Masálik al-absár* ut Epit. Paris. et Abulfeda, p. ۴۸۳, uti recepi. e) E. هرچه احتياج او ومتعلقان او است از آن حاصل شود. f) A. وتصاعقهم. g) A. ذدمله, B. ذبله, ceteri om. h) C. et D. addunt هو. i) E. ستروشه. k) A., B. et C. سلحى, D. سلحن et سلجن. Cf. Jacut sub شلجيكث et شلج. C. mox deinde ut recepi. l) A. et B. ذلك. m) B. habet ولمار, C. ولسان, D. والباميان quod aperte vitiosum est. Apud E., Ous. et *Djih.-Numa* desideratur. Est urbs in provincia Asbídjáb, quam Mokaddasi appellat لبان (لسان).

كلّها فى عمود هذا الجبل وما يتّصل به من الجبال والنوشادر الّذى فى عمل البُتّم والزاج والحديد والزيبق والنحاس والآنك [a] والذهب * والچرَاغ سَنْك [b] والنفط والقير والزفت والفيروزج والنوشادر الّذى بفرغانة والجبل الّذى ذكرتُه بفرغانة انّه يحترق حجارته مثل الفحم والثمار المباحة [c] الّتى وصفتُها بفرغانة كلّ ذلك فى هذا الجبل [d] فى سنامه او سفحه او ما يتّصل به وفى هذا الجبل بناحية البتّم وجبال الشّاوذَار [e] بسمرقند * مياه حرٌّ وبردٌ [f] غير انّ فيها عيونًا تجمد فى الصيف اذا اشتدّ الحرُّ حتّى تصير كالاعمدة وتنقطع ويكون مآؤها فى الشتآء حارًّا وتاوى اليها السوائم لدفء [g] موضعها فى الشتآء ۞ ولبخارا مدن داخل حائطها وخارجًا عنه [h] فامّا داخل حائطها فالطَّوَاوِيسُ وهى اكبر منبر بعد القصبة وبَمْجْكَث [i] وزَنْدَنَة [k] ومَغْكان وخُجَادة [l] وخارج الحائط بِيْكَنْد وفَرَبْر وكَرْمِينِيَة وخُدِيمَنْكَن [m] وخَرْغَانْكَث [n] ومَذْيَامْجْكَث ۞ فامّا الطَّوَاوِيسُ فانّها مدينة لها سوق ومجمع عظيم ينتابه الناس من اقطار ما ورآء النهر فى وقت معلوم من السنة ويرتفع منها من الثياب القطن ما ينقل الى سائر المواضع وهى مدينة كثيرة [o] البساتين والمآء الجارى خصبة

---

a) E. سُرب. *b*) D. والفضة; ceteri om. (B. الجاع سنك). *c*) Secundum C. et E. (cf. supra p. ٢٩٥, vs. 7). A. et B. المثلوجة *sic*. *d*) A. et B. الجبال. *e*) A. et B. النوشادر, C. الساق, D. الزانج. Vid. infra. Lectio A. et B. h. l. videtur suadere شاوذار. *f*) Haec ex D. addidi. *g*) A. et B. لدفاء. *h*) A. et B. عنها. *i*) A. et B. ونمعلب, E. محكث, D. بمحكث; Sprenger, p. 20 secundum Mokaddasí نمجكث. Edrîsî, II, p. 194 منجكث. Cf. *Masálik al-absár* l. l. p. 251 et Jacut in v. Locus apud Jacut, I, p. ٥١٨, 13 corruptus est. *k*) A. et B. وردقة. *l*) A. et B. خجاره, D. et E. sine punctis. Vulgo scribitur جخادة, vid. Jacut, II, p. ٣٦ et ٤٠٣, sed Mokaddasí habet خجادى et vid. porro Edrisi, II, p. 194 et *Masálik al-absár* l.l. *m*) A. et B. حدىمكر, E. حدىمكر, D. حدمكر. Vid. Jacut in v. *n*) A. وغرغانكث, B. وجدغانكث, Mokaddasí جرغانكث. *o*) A. كبيرة.

ولها قلعة[a] ومدينة ومسجد جامعها فى المدينة، وامّا المدن الّتى داخل الحائط فهى متقاربة فى الكبر والعمارة *ولكلّ منها حصن[b] ٭ وامّا كَرْمِينِيَة فهى اكبر من الطواويس واعمر واكثر عددًا واخصب وخُدِيمَنْكَن[c] من كرمينية وبحذآئها خَرْغَانْكَث[d] ومَذْيَامَجْكَث وهى متقاربة فى الكبر والعمارة، ولكرمينية قرى كثيرة وكذلك لكلّ منبر قرى ومزارع الّا بيكند فانّها وحدها غير انّ بها من الرباطات ما لا اعلم فى بلدان ما ورآء النهر اكثر عددًا منها وبلغنى انّ عددها نحو من الف رباط ولها سور حصين ومسجد جامع وتُأُنِّق[e] فى بنائه وزُخْرِف محرابه فليس بما ورآء النهر محراب احسن زخرفًا[f] منه ٭ وفَرَبْر مدينة قريبة من جيحون ولها قرى وهى عامرة خصبة ٭ وامّا لسان بخارا فانّها لسان السغد الّا انّه يحرّف[g] بعضه ولهم لسان الدَرِيّة[h] واهلها يرجعون من الادب الى[i] ما يفضلون به *ما ورآء النهر[k]، ونقودهم الدرهم ولا يتعاملون بالدينار فيما بينهم وهى كالعَرْض الّا انّ لهم درهمًا يسمّونه الغِطْرِيفىّ وهى دراهم من حديد وصفر وآنك وغير ذلك من جواهر مختلفة قد رُكّبت فلا يجوز هذا الدرهم الّا فى عمل بخارا وَحْده وسكّته تصوير وهو من ضرب الاسلام وكذلك المُسَيّبِيّة[l] والمحمّديّة من ضرب الاسلام، وامّا زيّهم فالغالب عليهم الاقبية والقلانس على زىّ اهل ما ورآء النهر، ولهم داخل الحائط وخارجه اسواق متّصلة معلومة فى اوقات من الشهر دارّة يجرى فيها من الشرآء والبيع للثياب والرقيق والمواشى وغير ذلك ممّا يتّسع به اهلها، ويرتفع من بخارا ونواحيها من ثياب القطن ما ينقل الى الآفاق وكذلك البسط

---

a) C. et D. قهندز et sic Jacut, III, ٥٥٥ ult. *b*) C. ولكلها قهندز وحصن, D. ولكلها قهندزات. *c*) A. وحدّدمكر, B. وحددمكر. *d*) A. خرعانكث, B. حرغادكث. *e*) D. habet synon. قد تنوّق, quod pro تنوف rescribatur apud Jacut, I, p. ٧٩٧, 9 et Abulfeda, p. ٤٨٩. *f*) D. et Jacut زخرفة. *g*) A., B. et D. بحرف, C. بحرّف. *h*) Sic A.; B. الدُرِيّة, C. الدريّة, D. sine punctis; cf. locum Hamzae apud Jacut, III, p. ٩٣٥ sub فهلو. *i*) Ex C. addidi. *k*) C. اقاليمهم. *l*) A., B. et D. sine punctis. Vid. Jacut, I, p. ٥١٩, 6.

والمصلّيات وثياب من الصوف تستحسن، ويتحدّث اهل بخارا انّ من بَرَكَة القلعة a انّه لم تخرج منها جنازة والٍ b قطّ وما عُقدت فيه راية خرجت فهُزمت وهذا من الاتّفاق العجيب ان صحّ، ويقال انّ اصل c اهل بخارا فى قديم الايّام ناقلة اصطخر ✿ وسكن ولاة خراسان السّامانيّة بخارا لانّها اقرب مدن ما ورآء النهر الى خراسان فمن كان بها فخراسان امامه وما ورآء النهر ورآءه ولهم من حسن الطاعة وقلّة الخلاف على الولاة ما يؤدّى الى اختيار المقام بينهم على سائر ما ورآء النهر، واوّل ولاة خراسان من آل سامان اسماعيل ابن احمد جآءته ولاية خراسان وهو ببخارا فاستدام المقام بها فبقيت الولاية بها فى اولاده وقد كان ولاة ما ورآء النهر يقيمون قبل ذلك امّا بسمرقند وامّا بالشاش وفرغانة *فى وجوه الترك d وكان عمل e وُلاة بخارا يُحزر مفردًا من خراسان الى ان زالت ايّام آل طاهر ✿ وامّا خُجَادة f فهى على يمين الذاهب من بخارا الى بيكند على ثلاثة فراسخ وبينها وبين الطريق نحو فرسخ ✿ وامّا مَغْكان فانّها من المدينة على خمسة فراسخ عن يمين طريق بيكند وبينها وبين الطريق نحو ثلاثة فراسخ ✿ وامّا زَنْدَنَة g فانّها من المدينة على اربعة فراسخ شمالىّ المدينة ✿ وامّا بومجْكَث h فانّها على يسار الذاهب الى الطَّواوِيس على اربعة فراسخ i وبينها وبين الطريق نحو نصف فرسخ ✿

---

a) C. القلعة بالقهندز. *b*) C. ميت. *c*) A. et B. om. Addidi ex C., D. et E. *d*) A. et B. ووجوه الترك, ceteri om. *e*) Conjectura addidi, quia lectio ولاة a D. confirmatur. E. ولايت بخارا از خراسان جدا بودى. Pro يحزر A. habet يحرز, B. نحرد, D. يردون. *f*) A. ححاره, B. ححاده, E. نجاره, D. حجاره. *g*) Codd. sine punctis. *h*) A. et B. بومحكث, E. بومحكث. Patet hic et infra in itinerario, ubi eadem repetuntur, non intelligi بومجكث قصبة بخارا, sed verosimillime locum supra بمجكث appellatum. Idem jam observavit Jacut, I, p. ٧٣٧. Fortasse legendum est تُومُجْكَث, coll. Beládsorí, p. ٤٢٠ et ann. *f* et تُمُشْكَث apud Jacut in v.; Ibno 'l-Athír, IV, p. ٤٢٢ تُومشكث.
i) Jacut addit من بخارا.

ومن كَرْمِينِيَة الى خُدِيمَنْكَن [a] فرسخ فيما يلى السغد وبين خديمنكن وطريق سمرقند غلوة عن يسار الذاهب الى سمرقند ✿ ومَدْيامَجْكَث ورآء وادى السغد اعلى من خديمنكن بمقدار فرسخ ✿ وخَرْغانْكَث بحذآء كرمينية على فرسخ من ورآء الوادى ✿

ويتّصل ببخارا من شرقيّها السُّغْد واوّلها اذا جزتَ كرمينية [b] الدَّبُوسِيَة [c] ثمّ رَبِنْجَن [d] والكَشانِيَة [e] واشْتِيخَن وسَمَرْقَنْد وكلّ هذا قلب السغد على انّ من الناس من يزعم انّ بخارا وكِشّ ونَسَف من السغد ولكنّا افردناها ✿ وقصبة السُّغْد سَمَرْقَنْد وهى مدينة على جنوبىّ وادى السغد مرتفعة عَلِيّة [f] ولها قلعة [g] ومدينة وربض فامّا القلعة ففيها الحبس ودار الامارة عامران وامّا المدينة *فلها سور [h] واربعة ابواب باب الصِّين فى جهة المشرق وباب نُوبَهار فى جهة المغرب وباب بخارا فى جهة الشمال وباب كِشّ فى جهة الجنوب، ولها اسواق ومساكن ومآء جارٍ يدخل اليها فى [i] نهر من رصاص [k] وهو نهر قد *بُنِيت له [l] مُسَنّاة عالية من حجارة يجرى عليها المآء من الصفّارين *حتّى يدخل [m] من باب كِشّ ووجه هذا النهر رصاص كلّه وذلك انّ حوالى المدينة خندقًا [n] قد تسفّل لانّه استعمل طينه فى سور المدينة فبقى حواليها خندق عظيم فاحتيج الى مسنّاة فى هذا الخندق يجرى المآء عليها الى المدينة

---

a) A. et B. sine punctis. b) C. كرمانيه ut quoque Jacut, III, p. ١٣٥, 17, cf. IV, p. ٢٦٨, 9. c) A. et B. الددسونه. E. دبوسى. d) A. زنجر, B. رىنحر. Jacut, II. p. ٧٥٣ male ربيخن; vide sub اربنجن (I, p. ١٨٩ seq.). e) A. والكيسانيّة, E. كسانى. f) A. et B. عليه. Secutus sum Epit. Paris. g) C., D. et E. قهندز. h) C. et D. فيشتمل عليها حصن. i) A. من. k) A. et B. hic et infra رضراض pro رصاص. Cf. praeter C., D. et E. (ارزيز), Jacut, III, p. ١٣٤, 20 et 21, Abulfeda, p. ٤٩٣ et Mokaddasí MS. p. 136 والماء يدخل اليها فى قناة من رصاص فوق الخندق. l) D. et Jacut بنى عليه. Pro مسناة Abulfeda habet حمالات. m) D. et Jacut الى ان يدخل المدينة. n) A. et B. خندق.

وهو نهر جاهلىّ فى وسط السوق بموضع يعرف براس[a] الطاق وهو اعمر موضع بسمرقند وعلى جنبات[b] هذا النهر غلال[c] موقوفة على مرمّات هذا النهر وعليه حَفَظة من المجوس عليهم حفظُهُ شتآء وصيفًا[d]، والمسجد الجامع فى المدينة بينه وبين القلعة عرض الطريق وفى المدينة مياه من هذا النهر وبساتين[e] وفيها دار الامارة لآل سامان غير دار الامارة بالقلعة *والمدينة من الربض على جانبه[f] قريب من وادى السغد *الّذى هو بين الربض والمدينة وذلك انّ سور الربض ممتدّ من ورآء وادى السغد من مكان يعرف بافشينة[g] على باب كُوهَك حتّى يطوف بوَرْسَنِين ثمّ يطوف على باب فَنَك[h] وعلى باب رِيوَدَد[i] ثمّ الى باب فرخشيذ[k] ثمّ الى باب غَدَاوَد ثمّ يمتدّ الى الوادى والوادى للربض كالخندق ممّا يلى الشمال[l] *ويكون قطر السور المحيط بربض سمرقند فرسخين[m] غير انّ *الربض شربه[n] ومجمع اسواقه راس الطاق ثمّ تتّصل به الاسواق والسكك والمحالّ وفى تضاعيف[o] ذلك قصور وبساتين فليس من سكّة ولا دار الّا وفيها مآء جارٍ الّا القليل وقلّ[p] دار تخلو من بستان حتّى انّك اذا صعدتَ اعلى القلعة لم تبدُ[q] المدينة للمنظر لاستتارها بالبساتين والاشجار واكثر الاسواق والتجارات فى الربض الّا شيئًا يسيرًا[r] فى

---

a) Jacut بباب et sic C. b) Jacut حافّات, D. حاشيتى. c) A. et B. علال; D. et Jacut غلات. d) Jacut addit: فى شرط عليهم, D. مستغرض ذلك عليهم. e) Jacut عليها بساتين. f) E. وشارستان بر يك گوشه ربض است. بذلك. g) E. بادسه, mox افشينه; D. با قسينه, mox ادسه; infra A. ادسمه, B. السمته; Mokaddasí l. l. افشيه. Cf. Ibno 'l-Athír, V, p. ١١٣. Deinde E. از حد كوهك. h) E. فتك omisso باب, D. قند. i) D. h. l. الدبودر, infra ربورد. E. hic et infra ديودد; A. et B. infra ربودد. Mokaddasí ut recepi. Vid. Jacut in v. k) D. قصر اسد, E. فنحنيد. Infra D. فرخشيذ, E. قرحند (Ous. فوجند), A. et B. فرخشد. Mokaddasi فرخشيد. l) Haec omnia in A. et B. ut in C. desunt. m) E. ومقدار اين ديوار دو فرسنك در دو فرسنك بود. n) E. چشمهٔ شهر. o) A. تضاعف. p) A. om. q) A. et B. تبدا. r) A. et B. شىء يسير.

المدينة وهى فرضة ما ورآء النهر ومجمع التجار ومعظم جهاز ما ورآء النهر يقع بسمرقند ثمّ يتفرّق الى سائر الكور وكانت دار امارة ما ورآء النهر بها الى ايّام اسماعيل بن احمد فنقلها الى بخارا، ولسور ربضها ابواب منها باب غَدَاوَد[a] وباب اِسْبَسْك[b] وباب سُوخَشِين[c] وباب افشينة وباب وَرْسَنِين[d] وباب كُوهَك وباب رِيوَدَد وباب فرخشيذ، ويزعم الناس[e] انّ تُبَّعًا بنا مدينتها وانّ ذا القَرْنَيْن اتمّ بعض بنآئها ورايتُ على باب كَشّ صفيحة من حديد قد كُتب عليها كتابة[f] زعم اهلها انّها بالحميريّة وانّهم يتوارثون علم ذلك بانّه بنآء تُبّع وكتب عليه انّ من صنعآء الى سمرقند الف فرسخ[g] وانّ كتابته من ايّام تُبّع فوقعت فتنة بسمرقند فى ايّام مقامى بها واحرق الباب وذهبت الكتابة واعاد ذلك الباب ابو المظفّر محمّد بن لُقْمان[h] بن نصر بن احمد بن اسد كما كان من حديد من غير تلك الكتابة، وتربة سمرقند من اصحّ تربة وايبسها[i] ولولا كثرة البُخارات من المياه الجارية فى سككهم ودورهم وكثرة اشجار الخَلّاف[k] بينهم لاضرّ بهم *فرط يبسها[l] وبنآؤها طين وخشب، واهلها يرجعون الى جمال بارع ورزانة وهم من الافراط فى اظهار المروّة وتكلّف القيام على انفسهم ما يزيدون على سائر بلاد خراسان حتّى يجحف ذلك باموالهم، وبسمرقند مجمع رقيق ما ورآء النهر وخير الرقيق بما ورآء النهر تربية[m] سمرقند، وبينها وبين اقرب الجبال نحو مرحلة خفيفة الّا انّه يتّصل بها جبل صغير يعرف بكُوهَك يمتدّ طرفه الى سور سمرقند وهو مقدار نصف ميل فى الطول ومنه احجار بلدهم والطين المستعمل فى الاوانى والنورة

---

a) A. et B. عداود, Mokaddasî عداوذ. *b*) A. et B. اسبسك, D. استشك, Mokaddasí اسيسك. Jacut اسبسكث. *c*) A. سو حشبن, B. سو حشحمن, D. سو حسمن, Mokaddasí سو خشمن. Videtur compositum ex سو et خشين (vid. Vullers in v.v.). *d*) A. et B. ورستن. *e*) C. بعضهم, D. بعض الناس. *f*) Abulfeda, p. ٤٩٣ كتيبة. *g*) Plura apud Jacut, III, p. ١٣٦, 11 seqq. *h*) E. احمد. *i*) A. et D. وادسهها, B. وابنيتها. *k*) *Teschdíd* in A. et Epit. Paris. *l*) C. تقارب ابنيتهم واشتباكها. *m*) A. دردة, B. برية.

والزجاج وغير ذلك وبلغنى انّ به[a] ذهبًا وفضّة غير انّه لا يتسوّغ[b] العمل فيه، والبلد كلّه طرقه ومحالّه وسككه الّا قليلًا مفترش[c] بالحجارة، ومياههم من وادى السغد وهذا الوادى مبدأه من جبال البُتّم على ظهر الصَّغانيّان وله مجمع ماء يعرف بجن[d] مثل بحيرة[e] حوالبها قرى وتعرف الناحية ببُرْغر[f] فينصبّ منها بين جبال حتّى ينتهى الى بُنْجيكَث[g] ثمّ ينتهى الى مكان يعرف بوَرَغْسَر وتفسيره راس السكر[h] ومنه تنشعب[i] انهار سمرقند ورساتيق تتّصل بها من غربىّ الوادى من جانب سمرقند فامّا انهار الجانب الشرقىّ على الوادى فانّها تاخذ بحذاء وَرَغْسَر بمكان يعرف بغوبار[k]، وذلك انّ بهذا المكان تنفسح الجبال وتظهر الاراضى الّتى يمكن فيها الزرع وجرى الانهار فتاخذ من وَرَغْسَر[l] انهار منها نهر برش[m] ونهر بارمش[n] ونهر بشمين[o]، فامّا

---

a) A. بها. *b*) B. et D. يسوغ. *c*) Abulfeda, p. ۴۹۳ مفروشة, C. et D. مفروش. *d*) A. et B. omisso يعرف habent بماجى. D. بمحن, sed Epit. Paris. ut *Masálik al-absár*, p. 255 بماجى. Infra D. يعرف بجن, C. يسمى حى. H. l. C. بوى et Jacut, III, p. ٣٩۴, 18 يقال له وى. *Djih.-Numa*, p. ٣٦١, 2 totum fluvium نهر بوى appellat. In mappa Mokaddasí بحيرة دى legitur. Cum nomine جن fortasse componendum est nomen ورغجن apud Jacut, qui minus recte habet ورعجن et وزغجن. *e*) C., D. et Jacut البحيرة. *f*) A. بمرغو, B. دمرعر, D. بورغى. Secutus sum E. h. l. (infra برغز habet) et Jacut l. l. et I, p. ۴٩٠, 15. — Deinde B. فينصب. *g*) A. et B. نحكث, E. سحكث. *h*) Pers. ورغ = برغ. *i*) A. et Mokaddasí MS. p. 160 ىنشعب, B. ينشعب, D. تشعب, Jacut تتشعب. *k*) A. et B. دعرىان, D. دعودان, Ous. p. 255, ىمىمه. Infra A. عور, B. et D. عوىار et عوبار, Ous. غربان. — Deinde A. ولذلك. *l*) D. ورغى. A. hic et deinde semper ورغسر, B. interdum ورغشر. *m*) A. et B. h. l. درىىن, deinde برس; D. ترش; E. بوس; Edrîsî, II, p. 198 برسن. *n*) A. دارمسن, دارمسن, دارمس ut B., et بارمس; D. تارمش, E. بارمش (Ous. بارمس); Edrîsî بارمس. *o*) A. دسمىن et دسيمن, B. سيمن et دسمن, D. بسمنى, Edrîsî بشمى.

نهر برش فانّه نهر يمتدّ على ظهر سمرقند فمنه انهار المدينة والحائط والقرى الّتى تتّصل بها من مبتدأه الى منتهاه، وامّا نهر[a] بارمش فانّه يلى هذا النهر من ناحية الجنوب وعليه القرى من اوّله الى آخره نحو مرحلة، وامّا نهر[b] بشمين فانّه من بارمش ممّا يلى الجنوب ويسقى من اوّله الى آخره قرى كثيرة غير انّ انقطاعه دون انقطاع هذين النهرين واكبر هذه الانهار برش ثمّ بارمش وهما يحتملان[c] السفن، وينشعب[d] من هذه الانهار انهار يكثر احصاؤها حتّى يعمر بها من القرى والمزارع من وَرَغْسَر الى آخره رستاق يعرف بالدَّرْغَم على عشرة فراسخ فى الطول وعرضه نحو اربعة فراسخ الى نحو فرسخ وهذه الرساتيق كلّها تعرف بوَرَغْسَر ومَايَمْرغ وسنجرفغن[e] والدرغم، وامّا الانهار الّتى تاخذ من غوبار فانّها نهر اشْتيخَن[f] والسناواب[g] ونهر بوزماجز[h]، وينشعب[i] من وادى السغد انهار كثيرة[k] على امتداده بحذآء كلّ بلدة وكلّ رستاق فمنها انهار رِبِنْجَن[l] وانهار الدَّبُوسِيَة وانهار كرمينية حتّى ينتهى الى بخارا ويكثر عدد الانهار برستاق سمرقند لكثرة عدد قراها وتعدّدها وربّما كان للقرية الواحدة منها نهران وثلاثة ويكثر فى المدينة انشعاب الانهار الصغار[m] بحسب عدد الدور والبرك والبساتين والقصور ومن اطلّ من شَرَف على وادى السغد لم ير الّا خضرة ممتدّة لا يتخلّلها الّا قصر او قلعة، وبوَرَغْسَر كروم وضياع

---

a) A. om. b) A. et B. om. c) B. et D. يحملان. d) A. وىمشعب, D. وينتشعب. e) A. وسحردعن, B. وسحرىعد, Ous. بيبخرو. Infra D. سنجرفغن, E. دمحر; Mokaddasí شحرفغر et سحرقعر; Edrisi, II, p. 202 سيحن. Nomina loci in فغن multa sunt in hac regione e. g. خشوفغن. f) A. et B. sine punctis. g) A. والمساواب, B. والمماواب, D. والسماواب, Ous. ساوات. Conjectura puncta adscripsi. h) A. et B. بور ما حر, D. دون ما جنى, E. مور ما جن (Ous. مور ما جز) Edrîsî بوزماخن. Infra A. et B. دورماحر, D. بوهاجر, Ous. بورماخر. Mokaddasí بورماجر et بورماحر. *Djih.-Numa*, p. ٣٥٠, 13 باجر. Conjectura scripsi. i) A. sine punctis. k) Addidi ex D. et E. l) A. ىىحن, B. ردىحن. m) A. et B. الاصغار.

وبساتين قد ازيل عنها الخراج وجُعل على اهلها عوض الخراج اصلاح سكور ذلك المآء واحكام بثوقه وامتداد الوادى فى الصيف يكون من ثلوج جبال البتّم[a] واشروسنة وسمرقند[b] ۞ وامّا رساتيق سمرقند فانّ اوّلها بَنْجِيكَث[c] ومدينتها بنجيكث ثم تليها[d] وَرَغْسَر ومدينتها ورغسر ويلى بنجيكث جبال السّاوَدَار[e] وليس بها منبر *وبين الساودار وورغسر فيما يلى سمرقند رستاق مايَمْرْغ وسنجرفغن وليس بهما منبر[f] غير انّ بمايَمْرْغ مكانًا يعرف بالرِيوَدَد[g] كان بها مقام الاخْشِيذ ملك سمرقند وهى قرية فيها قصور الاخشيذيّة *وسنجرفغن وورغسر كانا من مايمرغ فافردا عنها[h]، ويتّصل برستاق مايمرغ رستاق الدَّرْغَم وليس به منبر، ويتّصل برستاق الدرغم رستاق أَبْغَرْ[i] وليس به منبر[k]، والسّاوَدَار هو الجبل الّذى عن جنوبىّ سمرقند وليس بنواحى سمرقند رستاق اصحّ هوآء ولا زرعًا وفواكهَ منه واهلها اصحُّ الناس *الوانًا وابدانًا[l] وطوله زيادة على عشرة فراسخ وبالساودار عمْرٌ للنصارى[m] يعرف بوزكرد، ورستاق الدَّرْغَم ازكى هذه الرساتيق فى الزروع ويفضل من اعنابها[n] ما يحمل الى غيرها من

---

a) E. غرجستان. b) Pro his omnibus inde a وينشعب, E. (ut Ous.) redactionem aliam, multo locupletiorem habet (cf. Abulfeda, p. ٤٨٤), feré consentientem cum D., ad cujus textum collatio dabitur. c) A. بحمكث, B. بسحىكث, D. بيحنكث. d) A. et B. يليه. e) Jacut nomen bis sive ter memorat, nempe sub شاوذار, sub شاذوان (ساذوان) et, ut videtur, sub ساروان, quae est lectio C., E. et *Djih.-Numa*, p. ٣٩١, 15 a f. (B. semel habet لساوران). Cf. supra p. ٣١٣ ann. *e*. f) Ex D. restitui. g) A. sine punctis, B. بالريردد. h) Haec addidi ex D. et E. i) A., B. et D. sine punctis. k) Hic D. plura addit, e quibus in C. leguntur (cf. Ous.): واشدّ رساتيقها اشتباكًا فى الابنية رستاق مايمرغ (Cod. عانمرع). l) C. ابدانا واحسنهم الوانا. m) E. addit كه ترسايان آنجا جمع شوند وعبادت كنند واوقاف بسيار بران است, quae cum aliis quoque D. habet. Nomen ecclesiae in D. est وركوده, in E. زركرد, in B. وركرد. Fortasse legendum est زركرد. n) A. اغنايها, B. اعيانها.

الرساتيق، واما أَبْغَرْ[a] فانّها مباخس غير انّ قراها اكثر[b] عددًا من رساتيق[c] سمرقند *واراضيها منجبة[d] بلغنى انّ القفيز *البَذْر يُريع بها[e] مائة قفيز[f] وبها مراع كثيرة، فهذه رساتيق سمرقند عن جنوبىّ الوادى ۞ فامّا شماليّه فانّ اعلاها يَارْكَث[g] وهى متاخمة لأُشْرُوسَنَة وليس بها منبر ومآؤها ليس من مآء السغد وانّما هى عيون والمباخس بها كثيرة ومراعيها واسعة خصبة، ورستاق بُورْنَمَذ[h] ممّا يلى اشروسنة وليس به منبر وقراها يسيرة، ويتّصل بيارکث رستاق بوزماجز[i] ممّا يلى سمرقند[k] ومدينتها بَارْكَث[l]، ويتّصل بها رستاق كَبُوذَنْجَكَث وهو رستاق مشتبك القرى والاشجار ومدينتها كَبُوذَنْجَكَث، وعلى ظهر هذا الرستاق رستاق وَذَار ومدينتها وذار وهو رستاق خصب كثير الزروع له سهل وجبل وسقى ومزارع ومراعٍ ووذار وكثير من قرى هذه الرساتيق

---

a) Pro ابغر ut A., B. et D. habent, E., Jacut, I, p. ٥٩٤ et *Djih.-Numa*, p. ٣٥٠, 11 بَرْم dant. b) D., E. et Jacut واكثر اعمر. c) Jacut رستاق, D. سائر رساتيق الواحد ربما. d) D., E. et Jacut واموالهم المواشى. e) Jacut ربما (B. habet يربع). f) Jacut addit واهلها اصحّ الناس اجساما اخرج زيادة على deinde D., E. et Jacut: (ربما) و(من) مرحلتين نحو (رستاق البرم) وطول ابغر يكون (كان) للقرية الواحدة من الحوزة (الحدود) نحو فرسخين (الفرسخين) واكثر (او اكثر). His denique addit E. وچراكاه اين ناحيت به از چراكاه ما ورا النهر است. g) A. et B. بارکنت, *Djih.-Numa*, p. ٣٥٠, 12 بارکنت. h) A. فورنمد, B. دوريمدّ, Ous. قورغد, E. فورغد, Mokaddasí وريمد, *Djih.-Numa* l.l. 13 قوزغ. Vid. Jacut in v. Fortasse autem in textu nostro legendum فورنمذ. i) A. et B. sine punctis. k) A. et B. om. l) Mokaddasí ابارکث. E. addit: پر از دو جايكاهى تنك بود از حد عردان (غوبار) تا سمرقند يكمرحله. D. habet وهو اعرض رستاق فى شمالىّ وادى السغد واكثره قرى يمتدّ من غوبار الى قرب سمرقند مرحلة فى مثلها. Quae olim in nostro textu fuisse scripta, docemur quoque collatione Mokaddasii qui habet: وهو اعرض رساتيق هذا الوجه واكثرها قرى يكون مرحلة فى مثلها.

لقوم من بكر بن وائل يعرفون بالسباعيّة كانت لهم بسمرقند ولايات وكانت لهم بها دور ضيافات واخلاق حسنة، ويتّصل به رستاق المرزبان وهو المرزبان بن تركسفى [a] الّذى كان استدعى الى العراق فى جملة دهاقين السغد [b] ✿ ونقود سمرقند الدراهم الاسماعيليّة والمكسّرة والدنانير ولهم دراهم تعرف بالمحمّديّة تركّب من جواهر شتّى من حديد ونحاس وفضّة وغير ذلك [c]، واشْتيخَن مدينة مفردة *فى العمل [d] عن سمرقند ذات [e] رساتيق وقرى *كثيرة البساتين [f] والمتنزّهات ولها مدينة وقلعة [g] وربض وانهار مطّردة [h] ومن بعض قراها عُجَيْف بن عَنْبَسَة *واسواق اشتيخن هى الّتى [i] استصفاها المعتصم ثمّ اقطعها المعتمد محمّد بن طاهر بن عبد الله بن طاهر ✿ والكُشَانيّة اعمر مدن السغد [k] مقاربة لاشتيخن فى الكبر ولها قرى ورستاق دون اشتيخن فى المقدار والدَّبُوسِيَة وأَرْبِنْجَن من جنوبىّ الوادى على جادّة خراسان ورِبِنْجَن اكبر رستاقًا من الدبوسيَة وقلب مدن السغد الكُشَانيّة [k] ✿

---

a) Vid. supra ad p. ٢٩٢ ann. *b*. *b*) E. addit: دوازده روستاى معمور دارد سمرقند اكثر انرا ذكر كرديم شش روستا بر جانب راست وادى سغد وشش وروستاى: بر چپ, quae itidem habet D. et quoque *Djihán-Numa*. Deinde sequitur: .دعمان (لعيان .Ous) در قديم از سمرقند بوده اكنون از ستروشنه (اشروسنة) است *c*) E. addit: .جز در اعمال سمرقند جائى ديگر نرود *d*) A. et B. om.; addidi ex E., Jacut in v. et Abulfeda, p. ٤٩١ qui habet .منفردة *e*) Jacut et Abulfeda .ولها *f*) Abulfeda وهى فى غاية النزهة والخصب والاشجار والثمار وكثرة وروستاها وبساتين البساتين et sic fere D. et Jacut (q. v.). E. والقرى والرياض .ومزارع بسيار دارد *g*) Sic A. et B.; ceteri .قهندز *h*) D. et Jacut addunt .وضياع *i*) Jacut عجيف بن عنبسة دريين ناحيت .E .وبها قراه الى ان .مقام داشتى تا انكه معتصم اورا بگرفت *k*) In seqq. A. et B. textum in compendium redactum offerunt. Collatio versionis Persicae, Abulfedae et Jacuti dabitur ad textum Ibn Haucalis. *l*) A. الكشانيّة ut quoque .الدبوسيّة C. et E. h. l. .الكيسانية

* وكِشّ مدينة ما ورآء النهر[a] وهى مقدار ثُلْث فرسخ فى مثله[b] بنآؤها من طين وخشب *وفواكهها كثيرة تدرك قبل غيرها لانّها من الجروم[c] ولها اربعة[d] ابواب باب الحديد باب عبيد الله باب القصّابين باب المدينة الداخلة[e] وهى مدينتان داخلة وخارجة ولها نهران كبيران نهر القصّارين[f] ونهر اسرود[g] وهما يجريان على باب المدينة[h] وبها يسقط التَرَنْجُبين* الّذى يحمل الى

a) A. et B. om. Addidi ex Abulfeda, p. ٤٩١, qui deinde habet وقدرها. E. كش با قهندز وشهرستان وجامع. Ous. (p. 259) ut Jacut (IV, p. ٢٧٤) cum D. faciunt. Jacut praefert pronunciare كِسّ. b) A. et B. مثلها. E., Jacut, Kazwînî, II, p. ٣٧٢ et *Djih.-Numa*, p. ٣٥٣ ثلاثة فراسخ habent. Sed Abulfeda et D. cum A. et B. consentiunt. c) Abulfeda: وهى خصبة وفواكهها تدرك قبل فواكه غيرها من بلاد ما وراء النهر وهى مدينة وبيّة غورية. Jacut وهى مدينة حصينة (l. خصيبة) تُدْرِك فيها الفواكهُ اسرع ما تدرك بسائر ما وراء النهر غير انها وبئة على ما يكون عليه بلاد الغور. E. كرمسير است وميوه پيشتر از همه ما ورا النهر دارد. D. habet (l. خصيبة) وهى مدينة حصينه جدًّا جروميّة تدرك فيها الفواكه اسرع ممّا تدرك فى سائر ما ورآء النهر وتاتى بواكرها الى بخارا وهى وبيّة. Denique Mokaddasí, p. 138 وهى خصبة ومنها تحمل البواكير — وهى مدينة شريّة لولم تكن وبيّة. d) E. شش, quodammodo recte, nam Ous., D. et Mokaddasí quatuor portas urbis interioris, duas exterioris memorant. e) Sic quoque Mokaddasí. Ous. et *Djih.-Numa* شارستان. D. المدينة الخارجة. f) A. العصادين, B. الفصادين, Ous. قصابان. Secutus sum Abulfedam, D., Mokaddasî, *Djih.-Numa* et Edrîsî, II, p. 200. g) Abulfeda اشور, *Djih.-Numa* اسورد. Num huc pertinet اسروت quod memorat Abulfeda, p. ٤٩٩? h) Sequitur hic apud D. et Ous. descriptio canalium quae pagos irrigant. Deinde D., Ous., Jacut, Kazwini et partim E. addunt: وفى المدينة والربض فى عامّة دورها مياه جارية وبساتين حسنة et porro quoque Abulfeda: وطول عملها (Jacut اعمارها) نحو اربعة ايام فى مثلها. E. his adjungit وفى

الآفاق ☆ [a] وامّا نَسَف [b] فمدينة لها * ربض وسور واربعة [c] ابواب باب النجارية [d] وباب سمرقند وباب كِش وباب غُوبَذِين [e] ولنسف قرى كثيرة ونواح ولها منبران سوى منبر المدينة والغالب على قراها المباخس والخصب والسعة ونهرها ينقطع فى بعض السنة فيسقون بساتينهم ومباقلهم ومباطخهم بالآبار حتّى يعود المآء فى النهر ☆

وامّا أُشْرُوسَنَة [f] فاسم الاقليم كما انّ السغد اسم الاقليم وليس ثَمّ [g] مدينة بهذا الاسم والغالب عليها الجبال حدود [h] اشروسنة غربيّها حدود سمرقند شماليّها الشاش [i] وبعض فرغانة جنوبيّها بعض حدود كشّ والصغانيان وشُومان ووَاشْجِرْد [k] وراشْت شرقيّها بعض فرغانة [l] ومدنها ارسيانيكث [m] وكركث [n] وغزق [o]

جبالها العقاقير الكثيرة, quae quoque habet D., dum Kazwînî et D. supplent: ومنها يرتفع الملح المستحجر. *i*) A. sine punctis, B. البرّنجمن. Kazwînî بها شوك الترنجبين.

a) E. addit: وروستاى آنرا از بسيارى نشمرديم زيراكه بى حد وحصر است. Apud Ous. et D. regiones enumerantur. *b*) E. نخشب. *c*) D. قهندز خراب قهندز خراب دارد وربض را چهار .E ,وربض له اربعة et sic quoque fere Mokaddasí in suo Istakhrii Cod. legit. Jacut, IV, p. ٧٨١, 16 قهندز وربض ولها ابواب اربعة. *d*) A. et B. sine punctis; Ous. بخارى ut quoque *Djih.-Numa*, p. ٣٥٣, 2. Secutus sum D. *e*) A. et B. عوبدىن, D. عوديين et عوبدىن, Ous. غزنين. Vid. Jacut in v. *f*) Abulfeda, p. ٤٩٧ nostro tribuit lectionem اسروشنة, quae interdum in Codd. occurrit. *g*) D. et Jacut (I, p. ٢٧٨) بها. Pro الاقليم Abulfeda bis للاقليم. *h*) Abulfeda ويحيط باسروشنة من الشرق; Jacut ut D. والذى يطوف بها من اقاليم ما وراء النهر من شرقيها, et sic E. orditur a confinio orientali. *i*) A. المسادمى, B. (a prima manu) et D. البساتين, quae ridicula lectio بساتين infra quoque in E. recurrit. E. h. l. جاج, B. ab altera manu السأس. *k*) A. واسحرد; Jacut وولاشجرد; E. وويشكرد. *l*) Hic D., Jacut (sed imperfecte) et E. inserunt: ومدينتها الكبرى تسمّى بلسان الاشروسنة

ووغكث [a] وسَابَاط وزَامِين وديزَك [b] ونُوجَكَث [c] وخَرَقَانَة [d] ومدينتها الّتى يسكنها الولاة هى بُونْجِكَث وبناؤها طين وخشب وهى مدينة داخلها مدينة اخرى على كلّ منهما سور [e] وللمدينة الداخلة بابان ويجرى فى المدينة الداخلة نهر كبير عليه فيها رحى ويشتمل حائطها [f] على دور وبساتين وقصور وكروم [g]

(D.) دومحكث. Nomen hujus urbis in A. deinde scribitur بنوبحكث, in B. دودمحكث, in C. بمودىحكث. E. habet بومحكث, Ous. بومحكث, Abulfeda, p. ۴۹۹ نوبنجكث, Jacut l.l. et in v. بنجيكث. Mokaddasí dat tres lectiones ينحكث, بيحكث et نويحكث, *Djih.-Numa*, p. ٣٥٥, 20 يومحكث. Cf. porro Juynboll Ann. ad *Merácid*, IV, p. 275 seq. *m*) A. ارساينكث, B. ارستانيكث, D. ارسمادمكث, E. اران بيامكث (Ous. اران سامكث, *Djih.-Numa* اران بنامكث). Mokaddasí اسبانمكث (male) et ارسبانيكث. Infra A. ارسادمكث, B. ارسادمكث, D. ارسبانكث, E. ازنيكث, C. ارسادمكث; Edrîsî, II, p. 203 ارسانيكث, p. 206 ارسبانكث; Epit. Paris. ارسامكث. *n*) A. et B. كدكد; D. hic et infra ut recepi (semel كوكب); E. et *Djih.-Numa* كوكب; Edrisi كوكث et كركث; Mokaddasí كردكث. Infra A. et B. كوكب, C. كوكب, Ous. p. 263 كوكث. *o*) A., B., C. et D. semper عرق, E. غرق (semel عرك).

a) A. et B. ودكب, infra فعكب; D. فنكث, infra فغكث; E. وعكث (Ous. وعكث); C. معكمب et موكث. Mokaddasi فعكث, et in mappa فعكث. Fortasse conferendum est فنكد apud Jacut. *b*) A. دول, B. درك, Jacut, I, p. ٢٧٩, 1 دارك. *c*) A. دودىحكب, B. بودىحكب, mox ambo ut recepi; D. نوحكث, E. دومحكث. Infra A. et B. دىكاحب, D. دوحكث ut C., E. دوكحث. Mokaddasí بوجكث, برجكث et نوجكث. Edrîsî بوخيكث. *Djih.-Numa* infra pro باب نوجكث habet باب ابن حكمث (p. ٣٥٥, 6 a f.). Vid. Jacut in v. *d*) A. et B. حردانه, infra ut Ous. حرقانه; D. حرفانه, infra خرفانه; E. خرفانه, infra حرفانه; C. حروانه; Edrisi خرقانه. Mokaddasí خرقانه et خرفانه. Jacut, I, p. ٢٧٩, 1 خَرْقانة, sed cf. II, p. ۴٢٥, 3. *e*) C. ولها ربض وعليه سور. *f*) Jacut خندقها; cf. Ous. *g*) Jacut et D. addunt وزروع.

وقطرها نحو فرسخ وابوابها اربعة باب زامين باب مرسمندة[a] باب نوجكت باب كهلباذ[b] ولها ستّة انهار كلّها من منبع واحد هو من المدينة على اقلّ من نصف فرسخ ۞ وتليها فى الكبر زامين وهى على طريق فرغانة الى السغد *وتسمّى المدينة[c] سوسندة ۞ وديزك مدينة فى السهل بها رباطات وخانات ومآء ينبع من عين وهى كثيرة النزه والبساتين والمياه وليس بجميع اشروسنة نهر تجرى فيه سفينة ولا بها بحيرة ۞ *والبتّم جبال شاهقة منيعة[d] واكثرها تغلّب عليها البرد[e] وبالبتّم حصون منيعة جدًّا وفيه معدن[f] الذهب والفضّة والزاج والنوشادر[g] وهو جبل فيه مثل الغار يبنى[h] عليه بيت ويستوثق من ابوابه وكواه *فيرتفع من الغار[i] بخار يشبه بالنهار الدخان وبالليل النار فاذا تلبّد هذا *البخار قلع منه وهو[k] النوشادر *ولا يتهيّأ لاحد ان يدخله من

---

a) مرسمندة est una ex urbibus hujus provinciae, ut patet ex Mokaddasí, D. et Jakubí, p. ٧۴ paen. ubi ارسمندة. D. h. l. سمرقند, deinde مرسندة et سمندة; Ous. h. l. ابن سمندر (*Djih.-Numa* ابن شمند), deinde ut E. سمندة. Mokaddasí h. l. سمندة, sed ter ut h. l. ex A. et B. recepi. b) B. دهلباد, D. et *Djih.-Numa* كهليان, Ous. كهلبان, Mokaddasí كهاباد. c) D. et Jacut: ولها اسم آخر وهو, quod nomen quoque in D. scribitur سوسندة (B. سوسنة), a Jacut سبذة. E. سليسدة بليس خوانند. وبزبان ايشان اين شهررا سلفسدة بليس. Ous. سليسدة بليس. Fortasse igitur in textu scribendum est سرسندة, sive cum B. سرسنة (cf. Vullers, II, p. 277 *b*). d) Desunt haec non tantum apud A. et B., sed quoque apud E. (et Ous.), ubi verba seqq. de frigore Oschrusanam spectare videntur. Supplevi ex Abulfeda, p. ۴٨۴, et D. e) Abulfeda والغالب عليها شدة البرد. D. والغالب عليها النزهة, quam lectionem confirmat Epit. Paris. Addunt وبها قرى آهلة. Deinde C. et Jacut, I, p. ۴٩. والبتم حصن منيع جدًّا. f) D. et E. معادن. g) C., D. et Jacut addunt الذى يحمل الى الافاق. h) Iidem قد بنى. Deinde C. et Jacut يستوثق من بابه. i) C. et Jacut يرتفع من هذا الموضع, D. فيجتمع فى ذلك البيت من الغار, Abulfeda وفيه عين يرتفع منه. k) C., D. et Jacut om. وهو. C. الدخان فيه صار, Jacut البخار كان منه مثل.

شدّة حرّه[a] الّا ان يلبس لبودًا[b] ويدخل بها كالمختلس[c] وهذا البخار ينتقّل[d] من مكان الى مكان فيحفر عليه حتّى يظهر فاذا *انقطع من مكان[e] حُفر عليه من[f] مكان آخر *فظهر منه[g]، والبُتّم جبال تسمّى البتّم *الاوّل والاوسط[h] والداخل وماء سمرقند والسغد وبخارا من البتّم الوسطى[i] ✿ ومينك[k] الموضع الّذى قاتل فيه قُتَيْبَة بن مسلم *وحَصَر الأَفْشين هناك[l] ✿

واما الشاش[m] وايلاق فانّ مقدار عرضهما مسيرة يومين فى ثلاثة[n] وهى كثيرة القرى والعمارات والمنابر وهى فى ارض سهلة كثيرة المراعى والرياض وبالشاش وبايلاق مدن كثيرة ذوات ابواب واسوار وارباض وقلاع واسواق وانهار تخترق بعض المدن، مدن الشاش بِنْكَث[o] ودنفغانكث[p] وجينَانْجَكَث[q]

---

a) Jacut ferme ut A. et B. فلا — يدخل هذا البيت لشدة, cf. Abulfeda et Kazwînî, II, p. ٣٤٢; C. et D. وداخل هذا البيت من شدّة الحرّ بحيث لا يتهيّأ لاحد ان يدخله الّا احترق (D. ما). *b*) C., D., Jacut et Abulfeda addunt يرطّبها بالماء. Deinde C. ويدخله, Jacut ثم يدخله, D. ويدخل omisso بها. Pro كالمختلس, B. كالمتخلس, Abulfeda بسرعة. *c*) Addunt C., D., Jacut et Abulfeda فياخذ ما يقدر عليه من ذلك, et solus Jacut (quem sequitur Kazwînî) ويسرع الخروج. *d*) Secundum A.; D. et Jacut ينتقل, B. et C. sine punctis. *e*) C. خفى فى هذا المكان; cf. D. *f*) C. فى. *g*) C. et D. حتى يظهر. *h*) C. الاولى والوسطى. Deinde D. والبتم الخارج. *i*) D. et Jacut الاوسط. *k*) A. منبل, B. منبل, D. مثيل et ممل. Vid. ad textum Ibn Haucalis. *l*) D. المسوّدة وهناك حصن يعرف بالافشين الاكبر صاحب المعتصم الخ. *m*) E. بساتين (Ous. نشامين). Vid. supra p. ٣٢٥ ann. *i*. *n*) C., D. et E. addunt ايام. *o*) A. et B. دمكث. Nunc Taschkend appellatur. *p*) A. et B. ددمعادمكث, D. دنفعاكث, E. ددوماكث, Ous. ديوماكث, Edrîsî, II, p. 206 دهناكث. In itinerario A. دمعانكث, B. ددمعادكث, D. يفغانكث, C. زنفقالكث, E. دهقان نكث, *Djih.-Numa*, p. ٣٦٣, 9 دهقانكث. Mokaddasí بكث et نمانكث. Cf. Sprenger, p. 21 ult. In mappa C. ددممانكث, in mappa Mokad-

وذاجاكث وبنّاكث [a] وخَرشْكَث [b] واشبينغو [c] واردلانكث [d] وخُذينكث [e] وكنكراك [f] وكلشاجك [g] وغرجند [h] وغنّاج [i] وجبوزن [k] وورذوك [l] وكبرنه [m] وغدرانك [n] ونُوجَكث [o]

---

dasíi زبعانكر. q) A. حسادححكث, B. حساككث, D. حسابححكث. Vid. Jacut in v., Edrîsî l.l. et Mokaddasí. Deinde A. et B. ذبحاكث, D. حاكث.

a) A. et B. ببلانكث *sic*. D. ثماكث, E. ساكب, Ous. ساكث, Edrîsî نياكث, Mokaddasi نباكث et بياكث. b) A. حوسنكث, B. حوسنكث, E. حوسكث. Vide Jacut in v. c) ? A. اسمعو, B. اسممعو, D. سممعوا, Edrîsî شينقو, E. سعو. In itinerario A. اسنمغوا, B. اسسمعوا, D. اسمعوا, C. اسبعوا, E. اسمعوا, Edrîsî, II, p. 213 اسفينان. d) A. et B. اردلادكث, D. ادررلاكث s. ادرولاكث, E. coll. Ous. ادرلانكث; infra A., B. et E. اردلانكث et اودلادكث, D. bis اردلانكث, semel اردولاكث, C. اردلادكب; Edrîsî, II, p. 211 اردولانكث, sed p. 213 اردلانكث. e) A. حددنكث, B. et E. sine punctis, D. حدبيكث, Ous. حدينكث, Edrîsî خذنيكث. Infra A. حدنتكث, B. حدنكث, C. حددمكب, D. حدينكث, E. سكث. f) A. كنكدال, B. كنلرال, D. كنكراك, E. كمكوان (Ous. كنكوان), Edrîsî كنكرات. Infra A. et B. كمرال, C. كمراك, D. كيماك, E. كيراك. In mappa C. دىحممال, in mappa Mokaddasíi كراك. g) A. كلشححك, B. فلشحل, D. كالشاجك s. كلساجك, E. indist. كليخمث, Ous. كلنجد, Edrisi l.l. اكلساجك, sed p. 213 كاخشك. Infra A. كسحك, B. كلساحك, D. كالسحاك. C. كلحمنك, E. كلحان. In mappa Mokaddasíi طسحك. In textu hujus hoc et quatuor nomina quae praecedunt desiderantur. h) A. et B. عرجند, D. عركنده, E. et Ous. عرصل, Edrisi غركنده. Infra A. et B. عرجند, D. عوجند, E. غرقند. Mokaddasí in textu et in mappa عرجند. Fortasse legendum est غزجند s. غزكند. i) Hoc et quinque nomina seqq. in A. et B. desunt. D. عماج, E. عاج, infra A., B. et D. عناج. Vid. praeter Edrîsî et Mokaddasí, Jacut in v. k) D. حيوزن, E. ut Ous. جنودر, Edrîsî حبورن. Infra A. et B. حمورن, D. حمودن, E. ضورذن. Mokaddasi جيمورن et حمورن, in

وغَزَك a وأَنُوذكث b وبغنكث c وبركوش d وخاتونكث e وجيغوكث f وفرنكث g وكداك h

---

mappa حموری. *Djih.-Numa*, p. ٣٩٣, 11 اجبوربين. *l*) D. وردوک ut infra A. et B.; infra D. وورودک. E. وورودل (Ous. وردل), infra دورووک. C. infra دروزک. Edrisî وردوک; Mokaddasí وردک (et دردک). *m*) D. کبیرپه, Edrisî کبیرثه, E. کمرنه (Ous. کرنه). Infra A. کمرده, B. et D. کرده; C. in textu کعرده, in mappa کمرده; E. کفرده; Mokaddasí in textu کبرنه (et کبیرپه), in mappa کمرپه. *Djih.-Numa*, p. ٣٩٣, 11 کفرنه. *n*) D. نمدزانک, E. نمدد دالک (Ous. عدد دالک), Edrisî نمورانك. Infra A. et B. عدوانکر, D. غدوانك, E. عروال, C. عدرانك; *Djih.-Numa* l.l. عدوال, Mokaddasí نمددوانك (et پمدوانک) et in mappa بادردوانک. In mappa C. نمرداو. Hic ut saepius in hisce nominibus prorsus conjectura edidi. *o*) D. دوحکث, E. نوحکب, Ous. بوحکت, Edrisî بوجکث, Mokaddasí برجکث et نوجکت. Vid. Jacut in v.

a) A. et B. عرک s. ut infra عرل; D. عرک, infra عرک وتدعی عرق; E. ut Ous. غزل, infra غزل; Edrîsî غزک; Mokaddasí in textu et mappa عرک, C. in mappa عرق. Jacut in v. habet غزق. *b*) Sic infra in A. et B.; h. l. habent انرذکث s. انذذکث; D. ابردکث et infra ایردکث; E. ادودکب (Ous. الودکت) et infra ادردکب; Edrîsî ایسورکث; Mokaddasí ابزدکث (et ایردکث), in mappa ادرکث; C. in mappa ادورکث. *Djih.-Numa*, p. ٣٩٣, 11 ابروکث. *c*) A. et B. هنکث, D. نعمکث, E. نعمک (Ous. لعبک), Edrisî بغنکث, Mokaddasí بشکث. Infra A. et B. نعونکب, D. يغونکث. *d*) A. دوکش, B. دوىش, D. et E. نرکوس; Ous. برکوس, Mokaddasí برکوس (et درکوس), Edrisî برکوش. Infra A., B. et C. درکوس, D. ترکوش, E. برکوس. In mappa C. رکوس, in mappa Mokaddasii رکوص. *e*) A. et B. حانوبکث, D. حادونکث, E. sine punctis (Ous. حانرکث); Edrisî خاتونکث. Infra A. et B. حددونکث et حادونکف, C. حادونکب et حادورکب, E. حادورکب et حادورکب. D. infra et Mokaddasí p. 26 ut recepi. Mokadd. p. 131 جانونکث et in mappa حادونکث. *Djih.-Numa*, p. ٣٩٣, 13 حابورکث.

ونكالك *a* ٭ مدن ايلاق قصبتها *b* تعرف بتُونْكَت *c* ولها من المنابر سِكاكت *d* وبانجخاش *e* ونوكث *f* وبالايان *g* واربيلخ *h* ونمودلغ *i* وتُكَث *k* وخُمْرَك *l* وبِسْكَت *m*

---

f) A. صعوكث, B. et D. حمعوكث, E. حغركب, Ous. جغركت, Edrîsî جيغوكث, Mokaddasí جيغوكث (in mappa جغركت). Infra A. et B. حمعوكت, D. حيغوكث, E. حعركب, C. ...حمعو. *Djih.-Numa*, p. ٣٦٣, 10 جغركت. *g*) A. فرىلث, B. فرىلث, infra A. et B. فريكت; E. مرىكث (Ous. مرنكث) et infra مرىكب; Edrîsî فرنكث; Mokaddasí فرنكن et فريكر. D. hic et infra ut recepi; *Djih.-Numa* l. l. فرتكث. Vocales adscripsi coll. Jacut qui memorat فرنكد (quae lectio Mokaddasii esse videtur) tamquam locum propinquum Samarcando (vid. quoque *Lobbo 'l-lobáb* in v.), et hujus loci nomen a nostro diversum non videtur. *h*) A., B., E. et Ous. كدال; D. et Edrîsî ut recepi; infra C. كواك, E. كمال, A., B. et D. كذلك *sic*. Mokaddasí كرال et كزال. *Djih.-Numa* l.l. 11 كتال. Fortasse conferendum est كدك apud Jacut.

a) A. et D. sine punctis; B. يكالك, E. et Ous. كالك; Mokaddasí نكالك et بكالك. Secutus sum Sprenger, p. 19. Male sequitur in A. اربيلخ, in B. اربيلخ; vid. infra ann. *h*. *b*) A. et B. om. *c*) A. et B. سوكت et eodem modo Edrisi et Mokaddasí (توكث) ايلاق قصبتها نوكث. Confusa nempe videntur nomina تونكث et نوكث (cf. infra ann. *f*). Nomen autem تُونْكَث s. تُونْكَت quoque scribitur تُنْكَث s. تُنْكَت, vid. Abulfeda, p. ٤٩٩ coll. ٤٩٥, *Djih.-Numa*, p. ٣٥٤, 5 a f., Dimaschkí, p. ٢٢١. Jacut, I, p. ٨٨٠, 9 تَنْكُث, sed hoc minus recte videtur, si conferatur تونكث (I, p. ٤٢١, 16 et ٩٠٠, 7). *d*) A. سكاكف, B. سكايف, D. اسكاكث, E. et Ous. سكاكت, Edrîsî سكاكند. Infra A. سكاكب, B. سكاكت, D. سكالت, C. سكاحكث. Mokaddasí شاوكث (vid. Jacut in v.), sed in mappa شكاكت. *e*) A., B. et E. دالحخاس, D. الجخاش, Edrisi الخجاش. Infra A. داحخاف, B. داحخاب, C. دادخاس, D. بانجخاش. Mokaddasí in textu دادخاس et تادخاس, in mappa دالخاس et alio loco

وكُهسيم [a] ودَخْكَت [b] وخـاش [c] وخرجانكث [d] ☆ والشاش وايلاق [e] متّصلة لا فصل

الحالس. *f*) A. دونكث, B. دودىت, D. دوكث, E. دكب, Mokaddasí تونكث et دونكث. Jacut نُكْث sed potius videtur legendum نُكث ut habet *Merácid*. Jacut (IV, p. ٨١١) dicit de hac urbe كانت قصبة ايلاق. *g*) A. ايلاق, B. ادلاق; D. دالادان ut infra A. et B.; E. بالادان; infra D. بالامان, C. مالان. Mokaddasí بالاباز et بالابان. Edrîsî ut recepi. *h*) A. et B. اربياج, اربيلج et اردلج; D. et E. اردلج. Infra A. ادىلج, B. اربلخ, D. اربطخ, C. ردىلخ. (Nempe C. ibi conjungit nomen cum praecedenti انها, ut fiat انهار بلخ, et sic in E. ورودهاى بلخ). Edrîsî ايلخ, Mokaddasí ابلخ et ابلج. In mappa C. اداج, in mappa Mokadd. املج. *i*) A. دمودكع, B. نمودىع, D. بمـوداج, E. دمودلع; infra A. دموذلغ, B. et D. نمودلغ, C. دمودلغ. Edrîsî يصوداخ, Mokaddasí ثولغ et بمولوغ. In mappa C. دوربع, in mappa Mokadd. دمورلع. *k*) A. et E. دكب, B. دكث, E. نكث ut Mokaddasí, Edrîsî بكث. Infra A., B. et C. دكث, D. نيكث, E. نكث. Legi تكث coll. Jacut in v., sed haesitans, nam omnino fieri potest تُكَث ortum esse e contractione nominis تُنْكَث = تونكث. Jacut male تكث componit cum نكث. *l*) A. et B. h. l. حمرل ut E. et Ous., infra حمول; D. الحمرك, infra حموك ut Mokaddasí; E. infra حمرك, C. حمرك. Vid. Jacut in v. *m*) A. et B. دسلف, D. دوحكث, E. دسكب, Ous. بسكث, Edrîsî بوجكث, sed p. 213 بسكث. Infra A. دسكث, B. سكث, D. بسكث, E. ساب (cf. Dimaschkí, p. ٢٢١ سلث). Mokaddasí h. l. in catalogo habet سيكث (cf. Jacut in v.) sed mox deinde بسكث s. دسكث. Vid. Jacut in v.

a) A., B. et Mokaddasí ut recepi; D. هسمم, E. كهثم, Ous. كهشم, Edrîsi كهشيم. Infra A. كره سيمر, B. كره سر, D. كوة سيم, C. كوه سهم, E. recte كوه سيم. Nomen significat „Mons argenti." *b*) A. et B. جكث, D. دحكث, E. دحكب, Ous. وحكث, Edrisi زجكث, Mokaddasí ادجكث et ادحكث, et in mappa حركب. Infra A., B. et E. محكب, D. فحكث; C. in mappa حكث.

بينهما وبايلاق معدن ذهب وفضّة واكبر مدن ايلاق نوكث وتونكث وليس بما ورآء النهر دار ضرب الّا بسمرقند وتُونكث [a] ۞ وامّا أَسْبِيجَاب فمدينة نحو الثُّلث من تونكث [b] وفى ربضها بساتين ومياه وابنيتها طين ولها اسواق مشحونة وهى خصبة كثيرة الغلّات والمنافع وليس [c] بما ورآء النهر مدينة [d] لا خراج عليها الّا اسبيجاب وحولها مدن وقرى كثيرة ۞ وامّا خُجَنْدَة فمتاخمة لفرغانة وهى على غربىّ نهر الشاش ليس فى عملها مدينة غير كَنْد ولها نهر عظيم يسافر فيه بالمتاجر والمِيَر ۞ وفَرْغَانَة اسم الاقليم وقصبتها أَخْسِيكَث وهى مدينة على شطّ نهر الشاش يحيط بها سور وخارجه ربض يحيط به سور آخرُ ۞ وتليها فى الكبر مدينة قُبَا وهى مدينة من انزه تلك المدن لها قلعة وربض وجامع واسواق ۞ ثم مدينة أُوش وهى عامرة مسوّرة بها قلعة ودار امارة وهى ملاصقة للجبل الّذى عليه مرقب الاحراس على الترك ۞ وأُوزْكَنْد آخر [e] مدن فرغانة ممّا يلى دار الحرب ولها سور وربض وقلعة [f] ومياه جارية وبساتين وليس بما ورآء النهر اكبر من قرى فرغانة ربّما بلغت [g] القرية مرحلة لكثرة اهلها

Vid. Jacut in v. et sub ذخكث. *c*) A. et D. حاس, B. حانس, E. خاص (حاص). Infra A. حانس, B. حابس, E. جانبين, C. حاس, D. ut recepi. Mokaddasî خاس et حاس. *d*) A. et B. حرخابكث, D. حركاكث, E. حرجاكث, Ous. حركات, Infra A. خوكاكب, B. حدكاكب, C. خرحادكث, E. حومانكب (*Djih.-Numa*, p. ٣٦٣, 13 حومانكث), D. . . . حد. Edrîsî خرخكث, Mokaddasî حاكاكث et خجاكث. *e*) D. et Abulfeda, p. ۴۹۹ addunt جميعا.

a) Sic habent A. et B.; E. om., Ous. دينكث, C. ودمكث. *b*) D. بيكث, C. دمكث i. e. بنكث; Abulfeda, p. ۴۹٥ تنكث i. e. تونكث. In E. haec desiderantur. *c*) D. et E. addunt بخراسان ولا ut quoque Jacut, I, p. ٢۴٩, 17. *d*) D. et Jacut بلد. *e*) Ex hoc vocabulo in احرٌ corrupto, ortum est كرمسيراست versionis Persicae (Ous. p. 271). *f*) D., Ous. et Jacut, I, p. ۴٠۴ habent قهندز pro قلعة. *g*) A. addit سب. Num legendum سعة? D. et Jacut, III, p. ٨٧٩, 13 وربما بلغ حد القرية. E. habet دراز. Pro مرحلة E. ut Ous. فرسخ.

وانتشار مواشيهم[a] ومزارعهم، ولفرغانة كور لكلّ كورة منها عدّة مدن لكلّ مدينة منها رستاق فيه عدّة قرى منها كورة كَاسَان وكورة جِدغِل ومِيَان رُوذَان ومدينتها خَيْلَام وبها مولد الامير ابى الحسن[b] نصر بن احمد فى دار خير بن ابى الخير، ويرتفع من فرغانة اكثر ما فى ايدى الناس من الذهب والفضّة والزيبق ويخرج من جبالها الجراغ سنك والفيروزج والحديد والصفر والذهب والآنك، وبأُسْبَرَة جبل حجارة سود *تحترق كما يحترق[c] الفحم تباع منه *ثلاثة اوقار[d] بدرهم *ورماده يبيّض الثياب[e] ٭

المسافات بما وراء النهر[f] الطريق من وادى جيحون بفَرَبْر الى فرغانة من فربر الى بِيكَنْد مرحلة كبيرة[g] من بيكند الى بُخَارَا مرحلة من بخارا الى الطَّوَاوِيس مرحلة[h] من الطواويس الى كَرْمِينِيَة مرحلة من كرمينية الى الدَّبُوسِيَة مرحلة خفيفة من الدبوسية الى رَبِنْجَن مرحلة خفيفة من ربنجن الى زَرْمان[i] مرحلة من زرمان الى سمرقند مرحلة من سمرقند الى أَبَارْكَث[k]

---

a) A. حواشيهم. Deinde D. مراعيهم, Jacut زروعهم. b) A. الحسن, D. الجيش; vid. Ibno 'l-Athír, VIII, p. ٥٨. c) C., D., Abulfeda et Jacut, I, p. ٢٣٨ تحترق كما يحترق. d) In C. praecedit كل. Jacut حمل; Abulfeda, p. ٤٨٦ منها pro منه. e) C., Jacut et partim Abulfeda ورماده اذا احترق اشتدّ, اشتد بياض, Jacut ويستعمل فى تبييض الثياب. (Pro اشتد, C. et Abulf. استد, Ous. p. 272 „moistened with water"). f) In C. praecedit: من وادى جيحون الى فرغانة وهو اقصى ما وراء النهر ٢٣ مرحلة واما طريق الشاش (Cod. الستام) فان من وادى جيحون الى الطراز ٢٢ مرحلة ومن بخارا الى بلخ ١٣ مرحلة ومن اراد ان يسير من بخارا الى خوارزم فى العمارة فانه يسير ١٢ مرحلة. g) Mokaddasí habet 2 dies, interposita statione مياركال, de qua vid. p. ٣٣٧ ann. f. h) Idem 2 dies, interposita statione دِيمَس (cf. Jacut in v.). Kodáma habet ومن مدينة بخارا الى ينبوع ٤ فراسخ ومن الينبوع الى الطواويس ٣ فراسخ. i) A. et C. sine punctis, D. om., E. زردان, Mokaddasí رربان. Vid. Jacut in v. Idem quoque habet formas زرماز et رزمان. Kodáma درمان. Inter hanc stationem

مرحلة من اباركث الى رباط سَعْد a مرحلة وفى هذه المرحلة اذا *جزت
برباط b ابى احمد مفرق c طريق فرغانة والشاش ومن رباط سعد الى بُورْنَمَذ d
مرحلة من بورنمذ الى زَامِين مرحلة من زامين الى سَابَاط مرحلة من ساباط
الى اركَنْد e مرحلة من اركند الى شَاوَكَث f مرحلة من شاوكث الى خُجَنْدَة
مرحلة من خجندة الى كَنْد g مرحلة من كند الى سُوج h مرحلة من سوج
الى رِشْتَان i مرحلة من رشتان الى زَنْدَرَامش k مرحلة من زندرامش الى قُبَا
مرحلة من قبا الى أُوش مرحلة كبيرة من اوش الى أُوزْكَنْد مرحلة كبيرة
هذا هو الطريق القصد من فربر الى أُوزكند وهى آخر ما وراء النهر، ومن
اراد من خجندة الى أَخْسِيكَث قصبة فرغانة خرج من كَنْد الى سُوج مرحلة
ومن سوج الى خُوَاكَنْد مرحلة كبيرة ومن خواكند الى أَخْسِيكَث مرحلة l
وهناك طريقان احدهما فى المفازة والرمال v فراسخ الى باب اخسيكث ثمّ

---

et Samarcand hic interponit قصر علقمة, ut Ibn Khordâdbeh, p. 47 (ed. Barbier de Meynard). *k*) A. ادارکت, B., C. et E. sine punctis, D. ادارکث, Ous. اباركت. Supra p. ٣٢٢ ut vulgo باركث. Cf. ibi ann. *l*. In mappa C. ادارکند, in mappa Mokaddasii امارك. *Djihân-Numa*, p. ٣٦٢, 19 اثاركيت et ٣٦٣, 2 اثاربكث.

a) D. et E. سغد. Vocales in A. Post مرحلة in E. sequitur جمله ده مرحله بود.
b) C. صرت من, D. صرت الى. *c*) C. تتفرق الطرق. *d*) A. et B. فوردمد, C. دوردمد, D. فرونمد, E. مزرعه, *Djih.-Numa*, p. ٣٦٢, 20 مرزغد; Kodáma فورهه. Vid. supra p. ٣٢٢ ann. *h*. *e*) Secundum A. et B.; E. ازكند; C. et Ous. اوركند; *Djih.-Numa* l.l. male اوزكند; D. et Mokaddasí om.; Kodáma habet ركيد. *f*) A., B., E., Mokaddasí et *Djih.-Numa* ساوكت (ساوكث); C. شاركث. Vid. Jacut in v. *g*) D., E. et *Djih.-Numa* h. l. كنده. Paullo inferius A. habet كُند. *h*) A. نسوخ, B. et C. نسوج, D. مسوج, E. et *Djih.-Numa* سوخ. *i*) A. et B. sine punctis, C. رنشاف, D. رستناق, et sic E. et *Djih.-Numa*. In mappa Mokaddasii رشتار. *k*) A. et B. ندرامس et رندرامس, C. زبيروامش et رروامش. In mappa Mokaddasii رندااراس. *l*) D. et E. addunt كبيرة.

تعبر نهر الشاش الى اخسيكث والآخر تعبر النهر الى باب[a] ٥ فراسخ ومن باب الى اخسيكث ۴ فراسخ فجميع المسافة من فربر الى أوزكند ٢٣ مرحلة ✿ وامّا طريق الشاش الى اقصى بلد الاسلام فانّك تخرج من اباركث[b] الى قَطْوان ديزه[c] مرحلة وطريق الشاش وفرغانة واحد الى رباط ابى احمد ثمّ تعدل عن يسارك الى الشاش اذا خرجتَ من رباط ابى احمد فتنزل قطوان ديزه وان شئتَ نزلتَ خَرَقانَة[d] ومنها الى ديزك ومنها الى بئر[e] الحسين ثمّ بئر[f] حميد ثم وينكرد[g] ثمّ أُسْتُورْكث[h] ثمّ تُونْكَث[i] ثمّ الى رباط بالقلاص

a) C. دادان et داب. Secundum Mokáddasí inter Akhsíketh et Báb sunt بريدان quod eodem redit ac 4 Paras. Kodáma dicit de باب: وهى مدينة عظيمة من مدائن فرغانة كثيرة الخير على نهر شاش. b) A. et E. اداركث, B. sine punctis, C. ادادكث, D. اياركث, Ous. ut recepi. c) A. et B. وطران ددوه et قطران دبره; C. دطمان et دره ut Edrísí, II, p. 212; D. قطران et وطوان دبره ut E. et *Djih.-Numa.* Mokaddasí inter urbes provinciae Oschrusanae enumerat قطران دزه, inter urbes Çoghdí قطوانة. Cum hac potissimum componendus videtur locus قطوان quem memorat Jacut. Attamen lectionem cum و recepi. Kodáma habet itinerarium paullo diversum: ومن تباركث (l. باركث) الى جسر دعى (بغى Ibn Khord. p. 48, 5 *novus?* an l. ينغى) فى مفازة رطوان (l. قطوان) ۴ فراسخ ومن جسر يغى (ينغى?) الى فورهه (بورنمد s. فورنمد l.) وهى جبال ٥ فراسخ ومن فورهه الى زامين فى مفازة ۴ فراسخ وزامين هذه مفرّق الطريق الى شاس وطريق الى فرغانة. d) A. حرقانه, B. sine punctis, D. et E. حرفانه. Recte scribitur in C. e) C. et E. شق et sic *Djih.-Numa.* f) A. ودمكرد, B. ودمكدد, C. ردمكرد, D. ددمكردد, E. فنكرد, Ous. فيك, Edrîsî ديبكرد. Infra A. وننكبرد, B. وسكبرد, C. et E. ويشكرد, D. وينكود. Conjectura edidi. g) A. et B. اسموركث, mox استوركث ut h. l. C. et D. Scribitur quoque ستوركث. Variis modis nomen corruptum legitur in Codd. e. g. سبوركث, اسوىث, استونكث, سبوركث, اشوركث cet. Sprenger, p. 21 pronunciat شُيُوركث.

يسمّى انفرن[a] ثمّ الى غُزكرد[b] قرية ثمّ الى اسْبِيجاب ثمّ الى بُذَخْكَث[c] ومن بذخكت الى الطّراز يومان لا رباط بينهما ولا عمارة، ومن اراد طريق بَنَاكِث فانّه ينزل من أَبَارْكَث رباط سعد ومنه الى زامين ومن زامين الى خاوَس[d] ومن خاوس الى بناكث ثمّ الى استوركث فذلك من وادى جيحون الى الطراز ٢٣ مرحلة ۞

الطريق من بخارا الى بلخ من بخارا الى قراجون[e] مرحلة ثمّ الى ميانكال[f] مرحلة ثمّ الى مايْمُرغ مرحلة كبيرة ثمّ الى نَسَف مرحلة ثمّ الى سُونَج[g] مرحلة ومن سونج الى الدادكى وهو ديدجى[h] الى كَنْدَك[i] مرحلة ومن كندك الى باب الحديد[k] مرحلة ومن باب الحديد الى رباط رازيك[l]

---

i) A. et B. تونْكث; C. دمكث, D. بنكث, Ous. بيكث, Edrîsi نيكث i. e. بِنْكَث, *Djih.-Numa* سكث. Mokaddasi دىكث. Cf. Sprenger, p. 22.

a) A. et B. ادعرن, C. ادعردن, D. ان قون, Ous. ايقرن, *Djih.-Numa* ايقر (Pro رباط بالقلاص hic habet رباط ابو العباس). Edrîsî البرق. Rursus conjectura scripsi. *b*) Secundum A. et B.; C. et D. sine punctis; Mokaddasí غركرد et عركرد. *Djih.-Numa* عركز. *c*) A. et B. ىدحكث. *Djih.-Numa* باحكث. *d*) A. et B. جاوس, Mokaddasí خاوص (semel كاوس). Vide Jacut in v. *e*) A., C. et E. sine punctis; D. قراحون ut Edrisi, I, p. 485; Ous. فراجون et sic Sprenger, p. 29. *f*) Secundum A. et B.; C. مبارکاك, D. et E. مايكال, Ous. مسالكال, Edrîsî منابكاك, Mokaddasi ميارکال. *g*) A. et B. سوىح, C. سوح, D. سوح, E. دسوخ, Ous. نسوخ; Edrîsî, I, p. 484 سورنج, sed II, p. 200 recte سونج. Vid. Jacut in v. D. supra سوىحر. *h*) A. et B. الدادكى (B. الداركى); C. ردىكن, D. الودىكن, E. ددكن, Ous. ديركن, — ردحى Edrisi رقدكند. *i*) A. et B. كندل et كىدل; C. كىدك, D. كيدك et كندك, Ous. كندل, E. كىدل. Mokaddasi habet ut recepi. Cf. Sprenger, p. 28 et 29. *k*) C. الحمردن. *l*) A. et B. راريك omisso رباط, C. اردك, D. دارنك, Ous. ازيك, Edrisi l.l. داراك, sed secundum Sprenger, p. 29 quoque دارنك. Fortasse

مرحلة ثمّ الى هاشم جرد مرحلة ثمّ الى الترمذ مرحلة ومن الترمذ تعبر الى سياه جرد مرحلة ومنها الى بلخ مرحلة وذلك من بخارا الى بلخ ١٣ مرحلة ٭ الطريق من سمرقند الى بلخ تخرج من سمرقند الى كشّ يومين ثمّ الى كندك ٣ مراحل ويتّصل طريق بخارا وسمرقند الى بلخ ٭

الطريق من بخارا الى خوارزم الطريق فى المفازة تخرج من بخارا مرحلة الى فَرْخَشَة عامرة [a] ثمّ تسير ٨ مراحل كلّها فى مفازة لا منزل بها ولا رباط ولا ساكن انّما هو سير على المرعى فلذلك لم يكتب له منازل فامّا من اراد ان يعبر جيحون الى آمُل ويسير الى خوارزم فانّ من بخارا الى فربر مرحلتين ومن فربر تعبر الوادى الى آمل فتسير من آمل فى حدّ [b] آمل الى وِيزَة [c] مرحلة ومن ويزة الى مردوس [d] مرحلة ومن مردوس الى اسباس [e] مرحلة ومن اسباس الى سيفانة [f] مرحلة ثمّ الى الطاهريّة مرحلة ثمّ الى جكربند [g] مرحلة ثمّ الى دَرْغان مرحلة ثمّ الى سدور [h] مرحلة ثمّ الى هَزَارَسْب مرحلة ثمّ الى مدينة خوارزم مرحلة فذلك من بخارا الى خوارزم على العمارة ١٢ مرحلة، هذه المسافات بين مشاهير المدن بما وراء النهر والطريق الى أُشْرُوسَنَة قد دخل فى طريق فرغانة لانّك اذا دخلتَ الى خَرَقَانَة وزامين فهى من مدن اشروسنة ٭

---

conferendum est زاذك apud Jacut, II, p. ٩٠٦. Loco hujus stationis et sequentis Mokaddasí habet قرنة.

a) A. et B. عامر. *b*) A. om. *c*) A. et B. ونزه, C. sine punctis, D. وبره, E. ويره. Cf. Jacut in v. et *Djih.-Numa*, p. ٣٦٢, 6 a f. *d*) C. مردوش, E. مردومين, *Djih.-Numa* مردوپين. *e*) A. et C. sine punctis, E. ut recepi; B. et Ous. اساس, D. استياس, Edrîsí, II, p. 88 اسفاس, *Djih.-Numa* اسياس. *f*) Secundum D.; A. et B. sine punctis; C. سمداه, E. سغانه, Ous. مغانه, *Djih.-Numa* مغان. Vid. supra ad p. ٢٨٤ ann. *h*. *g*) A., B., C. et E. sine punctis. Ous. جربند; in D. et Edrîsí lacuna est. *Djih.-Numa* جكربند. Vid. Sprenger, p. 32 et 33. C., E. et *Djih.-Numa* hanc stationem post درغان ponunt. *h*) Secundum A.; B. سدود, C. سدرس, E. et Ous. سدون, *Djih.-Numa* سدو.

وسنذكر المسافات بين اقاليم ما وراء النهر فنبدأُ من الخُتّل الى خوارزم ثمّ باقاليم ما وراء النهر   مسافات الختّل والصّغانيان وما بينهما من معبر بَذَخْشان على نهر جرياب[a] الى مُنْك[b] ٦ مراحل ومن منك الى قنطرة الحجر على وَخْشاب مرحلتان[c] فاذا نزلتَ على نهر وَخْشاب فالى ليوَكَنْد[d] مرحلتان وتنزل على المآء ايضًا الى هَلَاوَرْد مرحلة وهلاورد وليوكند على شطّ وخشاب وهما مدينتا الوخش ومن معبر آرْهَن الى هلاورد مرحلتان ومن المعبر الى هُلْبُك[e] يومان ومن هلبك الى منك يومان وكاربنج[f] فوق معبر آرهن على نهر جرياب بنحو من فرسخ وتمليات[g] من قنطرة الحجر على ٤ فراسخ فى طريق منك ومن معبر بَذَخْشان الى رستاق بنك[h] مرحلتان ومن رستاق بنك تعبر نهر انديجاراغ[i] ثمّ تدخلها وبين رستاق بنك وانديجاراغ مرحلة ومن انديجاراغ تعبر نهر فارغر ثمّ تدخل فارغر بينهما يوم تعبر نهر برسان[k] الى هلبك فهذه مسافة ما بين الوخش والخُتّل ٭ والطريق[l] من التِّرْمذ الى الصّغانيان من الترمذ الى جَرْمَنْكان[m] مرحلة ثمّ الى دارزَنْجى[n] مرحلة ثمّ

---

a) A. et C. حرداب. *b*) A. et B. ميل. *c*) E. از منك تا پولى كه پيش كوبرى. *Djih.-Numa* ازين گفتيم شش مرحله. *d*) A. et B. لىوكند, C. ادركند, D. الموكنده et دودكند, E. ابركند, Ous. ابركند ut *Djih.-Numa*. Supra p. ٢٧٩ et ٢٩٧ scribitur ut vulgo لاوكند. *e*) A. et B. هليل, E. ملنك. *f*) A. فاريح, B. ساريح. Vid. supra p. ٢٧٩ ubi كاربنك receptum est, et cf. ann. *d*, et p. ٢٩٧. *g*) A. sine punctis, B. دىملماب; vid. supra p. ٢٧٩ ann. *e* et p. ٢٩٧. *h*) A. et B. دمل; vid. supra p. ٢٧٧ ann. *c* et p. ٢٧٩. *i*) A. اددىخاراع, B. sine punctis; vid. supra l.l. *k*) A. et A. درسان, C. درفان, D. شومان, E. بوقان, Edrîsî, I, p. 480 بلسان. *l*) A. om. *m*) A. et B. جرممكان, C. حرمفان, E. جرمنكان (Ous. حرميكان), *Djih.-Numa* l.l. 3 a f. حرمنكان. D. pro una duas stationes habet ut Edrîsî, nempe (Edrisi خرميقار) الى جرميقان مرحلة ومن جرميقان الى صرمنجى (Edr. صرمنخى) مرحلة, dum in catalogo regionum quem D. in initio capitis habet, enumerantur صرمان وصرمنجى. Quod

الى الصغانيان مرحلتان ☆ والطريق [a] من الصغانيان الى واشْجِرْد [b] من
الصغانيان الى شُومان مرحلتان ثمّ الى انديان [c] يوم ثمّ الى واشجرد يوم
ومن [d] الواشجرد الى ايلاق [e] يوم ومن ايلاق الى دربند [f] يوم ومن دربند الى
جاوكان [g] يوم ومن جاوكان الى القلعة يومان والقلعة من راشْت ¹ ومن
الصغانيان الى باسَنْد مرحلتان ومن الصغانيان الى زينور [h] مرحلة ومن
الصغانيان الى بوراب [i] مرحلة ومن الصغانيان الى ريكدشت [k] ٩ فراسخ

---

locum priorem attinet, Jacut, III, p. ٣٨٣ scribit صَرْمَنْجان s. صرمنكان, Abulfeda, p. ٤٥٩ ex العزيزى habet الشرماجان distantia 6 Paras. a Tirmidh, 12 Paras. a Saghánián. P. ٤٨٥ seq. memorat eundem locum sub formis صَرْمَنْجان et جرمنكان. Jakubí, p. ٦٩ habet شرمكان (Cod. سرمكان s. سرمكان). Ibn Khordádbeh, p. 52, 5 صريم خان, Kodáma موسجان (l. صرمنجان). Utrum hic an alter locus intelligatur supra p. ٣٩٨ ubi A. et B. صرممحس, D. vero صرمنجى, efficere nequeo. *n*) A. et B. sine punctis. E. دار زنكى ut *Djih.-Numa*, Kodáma (ubi corrupte داركى) et Jakubí l.l. (ubi دار زنكا). Jacut habet (II, p. ٥٢.) دارَزَنْج. Ibn Khordádbeh خان زنجى. Mokaddasí ut recepi.

a) A. et B. om. *b*) A. et B. واسنجرد et واشنجرد (B. semel الواسنجرد). E. ويشكرد et sic *Djih.-Numa* l.l. paen. ubi pro ويشكو وويل legatur ويشكرد ووُيل. *c*) A. افدىان, B. ادىان, C. امريان, D. امديار, E. فريان (Ous. الوبان), Edrisi, I, p. 482 انديان, *Djih.-Numa* l.l. paen. ايونان. Mokaddasí inter urbes provinciae Saghánián memorat ارديان (vid. Sprenger, p. 21), in mappa habet ودىار. *d*) A. et B. male الطريق من. *e*) D. ايلان; C. utramque habet lectionem. *f*) A. et B. دربىد et دربند. Edrisi درنيك. *g*) A. حادكان, B. حاركان, C. جاركان, D. خاركان; E. et Ous. جاوكان, *Djih.-Numa* كاوكاند; Edrisi كاركان. *h*) A. et B. ردمور, C. رىموں, D. زىنور, E. ريتون (Ous. زيتون), Edrisi, I, p. 484 زينون, Mokaddasí دينور ,زىنور et رىسوں et in mappa دىنور. *i*) A. et B. دوران et دوراب; C. حوراب et دوراب; D. بوارث, E. كوراب (Ous. كوراست);

والطريق من بوراب يجاوزها بفرسخين ثمّ يجاوز ريكدشت بثلاثة فراسخ على سمت الطريق الى بادماب[a] ومن الترمذ الى القُوَاذِيَان مرحلتان ومن القواذيان الى الصغانيان ٣ مراحل ومن واشجرد الى قنطرة الحجارة يوم فهذه مسافات ما بين الصغانيان الى اقصى الخُتَّل[b] ❁

مسافات خوارزم من مدينة خوارزم الى خَيْوَه مرحلة ومن خيوه الى هَزَارَسْب مرحلة ومن المدينة الى الجُرْجانيّة[c] ٣ مراحل منها الى أرْدَخُشْمِيثَن[d] مرحلة ومن اردخشميثن الى نُوزوار[e] مرحلة ومنها الى الجرجانيّة مرحلة وبين هزارسب وكردران خواش ٣ فراسخ ومن كردران خواش الى خيوه[f] ٥ فراسخ ومن خيوه الى سافَرْدَز[g] ٥ فراسخ ومن سافردز الى المدينة ٣ فراسخ ومن المدينة الى درجاش[h] مرحلتان ومن درجاش الى كُرْدَر[i] مرحلة ومن كردر الى قرية براتكين[k] يومان ومَذْمِينية[l] وقرية براتكين متقاربتان غير انّ الاقرب

Edrisí توراب; Mokaddasí بوراب et سوارب et in mappa فوران. *k*) A. ربكد et ربكند; B. ربكو et ربكند; C. ربكردشت; D. ريكا et كند; E. رمكدست, Ous. ريك دشت "the sand-desert." Si ست bis legitur A. habet ربكدست. Mokaddasí دنكرست. Pro فراسخ, C. et E. مراحل.

a) A. et B. sine punctis, C. داراب, Mokaddasí بادماب et بادمات. Apud ceteros deëst. *b*) E. كوهستان (legit الجبل). *c*) E. ut solet كركانج. *d*) A. et B. اردحسمش. *e*) A. et B. نوروار. Vid. supra p. ٢٩٩ ann. *f*. *f*) A. et B. h. l. حمره, C. كيوه, E. حمره, Ous. حيره. *g*) A., B. et D. ساغرزن, C. سامرون, E. سافرون. Vid. supra l.l. ann. *e*. *h*) A. et B. h. l. كردران, quae lectio explicanda est e C.: ومن المدينة الى كردن تمضى الى درجاس مرحلتين, coll. Edrisi, II, p. 192. Secutus sum E. Pro درجاش ut B. habet, A. ررجاش, C. درجاس, E. درخاس, Edrisi دخارس. Ibn Haucal in descriptione Khowarezmiae dicit nomen antiquae metropolis fuisse ورحاش. Idem, ut videtur, locus a Mokaddasí appellatur درسان. *i*) C. كردن, D. كردب, E. كردان. Mox deinde A. ودرد, B. قرذر. *k*) A. براكبين, B. براكيين. *l*) A., B. et C. sine punctis.

الى جيحون مذمينية ومن مذمينية الى وادى جيحون ۴ فراسخ وبين مرداجقان[a] ونهر جيحون فرسخان وهى بحذآء الجرجانيّة وبين الجرجانيّة وجيحون فرسخ ❁

وامّا مسافات المدن ببخارا فانّ من بُومجْكَث وهى قصبة بخارا الى بِيكَنْد مرحلة ومن بومجكث الى خُجَادَة[b] ٣ فراسخ على يمين الذاهب من بخارا الى بيكند وبينها وبين الطريق نحو فرسخ، وامّا مَغْكان فانّها من المدينة على ٥ فراسخ عن يمين طريق بيكند وبينها وبين الطريق نحو ٣ فراسخ، وامّا زَنْدَنَة[c] فانّها من المدينة على ۴ فراسخ فى شمالىّ المدينة، وامّا بومجكث[d] فانّها على يسار الذاهب الى الطّواويس على ۴ فراسخ وبينها وبين الطريق نحو نصف فرسخ، وبين كَرْمينِيَة وخُديمَنْكَن[e] فرسخ ممّا يلى الصُّغد وبين خديمنكن وطريق سمرقند غلوة عن يسار الذاهب الى سمرقند، ومَذْيَامَجْكَث ورآء وادى السغد اعلى من خديمنكن مقدار فرسخ، وخَرْغانْكَث[f] بحذاء كرمينية على فرسخ من ورآء الوادى وخرغانكث عند مذيامجكث ❁

وامّا مسافات مدن سمرقند والسغد فانّ من سمرقند الى أَبَارْكَث[g] ۴ فراسخ ومن سمرقند الى وَرْغَسَر[h] ۴ فراسخ ومن ورغسر الى بَنْجِيكَث[i] ٥ فراسخ فمن سمرقند الى بنجيكث ٩ فراسخ، ومن سمرقند الى وَذَار[k] فرسخان ومن سمرقند الى كَبُوذَنْجَكَث فرسخان ومن سمرقند الى اِشْتِيخَن ٧ فراسخ على شمال سمرقند ومن اشتيخن الى الكُشَانِيَة غربىّ اشتيخن ٥ فراسخ ومن

---

a) A. et B. مرداحمان. b) A. et C. حجاره, B. حجاده. c) A. et B. رردده, C. et D. ردىده, E. زيديّه, *Djih.-Numa*, p. ٣٦٣, 1 زبده. d) A. et B. دومحكث, ceteri om. (in C. tantum superest ..وحكب). Supra p. ٣١٥ ann. *h.* observavi legendum videri تومجكث. e) A. حدىمكىن et حدنمكين, B., D. et E. sine punctis. f) A. et B. حرعابكث, C. دعاكب, E. دعادكث. g) A. et B. ادارکث, D. ابارکث. h) A. et B. ورغسر. i) A. et B. ىىحىكث. Deinde A. et B. ىىىى. k) A. et B. ودار, C. et E. ودان. D. habet ويذار quae forma quoque a Jacut memoratur.

اشتيخن الى زرّمان a فرسخ واحد ومن الكشانية الى رَبِنْجَن b فرسخان ☆
والمسافات c بكِشّ ونَسَف فمن كِشّ الى نسف d ٣ مراحل ممّا يلى المغرب
ومن كشّ الى الصغانيان ٦ مراحل ومن كشّ الى نَوْقَد قُرَيْش e ٥ فراسخ f
على طريق نسف ومن كشّ الى سُونَج فرسخان g يعدل اليها من نوقد
قريش واسكيفغن h على فرسخ من سونج وسونج اقرب الى نسف من اسكيفغن،
ومن نسف الى كَسْبَة i ٤ فراسخ على طريق لبخارا اسفل من الطريق الّتى
ذكرنا وبين نسف وبين بَزْدَة ٦ فراسخ فهذه مسافات مدن نسف وكشّ ☆
وامّا مسافات مدن أُشْرُوسَنَة فانّ من خَرَقَانَة الى دِيزَك k ٥ فراسخ ومن خرقانة
الى زَامين ٩ فراسخ ومن زامين الى ساباط ٣ فراسخ ومن زامين على طريق
خاوَس الى كركث l ١٣ فرسخا عن يسار الذاهب الى فرغانة، وبين مدينة
اشروسنة وساباط ٣ فراسخ فيما بين الجنوب والمشرق، وبين نُوجَكَث m وخرقانة
فرسخان فيما بين المشرق والجنوب من خرقانة، وارسيانيكث n على حدّ
فرغانة من شرقىّ مدينة اشروسنة على ٩ o فراسخ، وفغكث p على ٣ فراسخ

---

a) A. et B. رومان, C. رزقان, E. زرقان ut *Djih.-Numa* l.l. 4, Ous. ارقان. *b*) A. et B. sine punctis, C. et E. اربنجن. *c*) A. et B. والمفازات (A. والمعاراب). Deinde A. دكس, B. بكفين et sic porro كفس pro كش. *d*) E. نخشب. *e*) A. et B. نوقد قرشين, mox A. نوقد نوسمن, B. نوقد قريس; C. نوقد قرس, D. نوقد, E. بوكث (Ous. بوكث ut *Djih.-Numa* l.l. 5). D. alio loco نوقد قرس, Edrisi, II, p. 200 بوبر فريس (Sprenger, p. 30). Male Sprenger p. 21 duo nomina locorum ex uno fecit; vid. Jacut in v. *f*) C. et E. مراحل. *g*) Correxi ex D.; A., B., C. et E. مرحلتان. *h*) A. et B. اسكيفغن, C. اسكغعر, D. اسكنفغن et اسكتفغن; Edrîsî اسكيفغن. Jacut merum nomen habet. Mokaddasí ut recepi. *i*) A., B. et D. كسنة, C. كسمة, E. كشمة. Deinde A., B., C. et D. نردة, E. نبردة. *k*) A. et B. ددرل s. ددرك; C. et E. درك, D. حرك, Edrisi, II, p. 203 et 205 جبزك. *l*) A. et B. كوكب, C. كوكث. *m*) A. et B. نوكجكث. Vid. supra p. ٣٣٦ ann. *c*. *n*) A. sine punctis, B. ارسانكث. *o*) C. et E. سبعة. *p*) A. et B. وعكث et فغكث. Supra وغكث, vid. p. ٣٣٦ *a*.

من المدينة فى طريق خُجَنْدَة ومن فغكث الى غَزَق[a] فرسخان ومن غزق الى خجندة ٩ فراسخ ❊

المسافات بين مدن الشاش وايلاق[b] واسْبيجاب وما يتّصل بها بَنَاكث على نهر الشاش ومنها الى خَرَشْكَث[c] فرسخ ومن خرشكث الى خدينكث[d] فرسخ ومنها الى استوركث[e] ٣ فراسخ ومنها الى دنفغانكث[f] فرسخان ومنها الى زالثيكث[g] فرسخ ومنها الى بِنْكَث[h] فرسخان فهذه المدن على طريق بناكث الى بِنْكَث، وامّا المدن الّتى على طريق تُونْكَث[i] وتونكث قصبة ايلاق فانّ من تونكث الى نوغكث[k] فرسخا ومنها الى بالايان[l] فرسخان ومنها الى بانجخاش[m] فرسخان ومنها الى سكاكث[n] فرسخ ومنها الى نوكث[o] فرسخ، فامّا ما بين نهر تُرك[p] ونهر ايلاق ممّا يلى المشرق من طريق ايلاق فانّ

---

a) A., B., C. et D. عرق. *b*) A. وبايلاق. *c*) A. sine punctis, B. حرسكث, C. et E. حرسكث. *Djih.-Numa* l.l. 9 حرسيكث. *d*) A. et B. حرىمكث, C. حدىمكث, E. دىنكث. *e*) A. et B. اسوركث, *Djih.-Numa* سوركث. *f*) A. et B. دىععانكث, *g*) Lectio incertissima est. A. habet النتكث, B. الشكث, C. indistincte رىمدكث, E. زالنيكث, *Djih.-Numa* بارانكث (cf. Sprenger, p. 22, 1). Non videtur differre ab urbe البيكث quam Mokaddasí, p. 131, inter urbes hujus provinciae memorat. Apud D. non exstat. *h*) A. et B. ىىكث et ىمكث. *i*) A. et B. دوىكث, D. نيكث وثويكث et mox deinde habet فان من نيكث الى نويكث فرسخا. C. على طريق ىكمث ودوىكث قصبة الخ. *k*) Lectio incerta est. A. دوغلث, B. دوعلث, C. دوحكث, D. نويكث. Cf. نوبخكث apud Dimaschkí, p. ٢٢١. — Deinde A., B. et C. فرسخ. *l*) A. et B. sine punctis. D. hanc stationem praecedenti anteponit. *m*) Vid. supra ad p. ٣٣١ *e*. *n*) Vid. p. ٣٣١ *d*. Deinde A. addit فرسخ ومنها الى دلمكاكث quae verba mera repetitio esse videntur. *o*) A. نوكث, B. توكث, D. بونكث. *p*) A. et B. دىل (بىرك), semel بُىرك et semel نزل; D. تيبرك; C. النترك et النىرك; Edrîsî بىرك; E. ترك et in descriptione Scháschi (cf. p. ٣٢٣ *k*) نهر تركستان. Vid. quoque Ibn Khordádbeh, p. 48.

على ترك من بِنْكث[a] على فرسخين جيغوكث وتليها على فرسخين فرنكث
وتليها على فرسخ بغونكث وتليها على فرسخين أُنوذكث وكداك[b] وغدرانك
وكبرنه وغَزَك ووردوك وجبوزن كلّها متقاربة فى مقدار يوم او نحوه، وما بين
نهر تُرك ونهر ايلاق من غربىّ طريق ايلاق فانّها اشبينغو[c] وكلشجك
واردلانكث وبِسْكَث وسامسيرك[d] وخُمْرَك وغَنّاج كلّها فى مقدار مرحلة * فى
نحوها[e]، وامّا ما بين بَنَاكِث ونوكث[f] ونهر الشاش ونهر ايلاق فانّها غرجند
وخاش ودَخْكَث ونُكَّث وكوه سيم فى مقدار يومين فى اقلّ من يوم، وامّا
ما بين نهر ايلاق ونهر الشاش عن غربىّ نوكث فانّها اربيلخ ونمودلغ فى
مقدار ٥ فراسخ وجِينَانْجَكَث[g] على طريق وينكرد الى بنكث[h] وبينها وبين
نهر الشاش فرسخان ونجاكث[i] على وادى الشاش ويجتمع عندها بنهر[k] ترك
وبينها وبين بناكث[l] ٣ فراسخ وكنكراك على نهر ترك بقرب خدينكث على
فرسخ، فامّا ما بين نهر ترك وحائط الشاش الّذى من وراء القلاص فخاتونكث[m]
على فرسخين من المدينة وبركوش على ٣ فراسخ من خاتونكث على سمتها
ومنها الى خرِكانكث[n] ٤ فراسخ على سمت المشرق ومن بنكث الى اسبيجاب

---

a) A. et B. دىكث. *b*) A., B. et D. وكذلك, C. وكواك, E. وكمال. Vid. supra ad p. ٣٣٣, ann. *h*. *c*) Vid. supra ad p. ٣٣٩ ann. *c*. Praecedit h. l. in A. et B. copula quam delevi. *d*) Secundum A. et B.; C. ودساممكث, D. وسرك; E. وساميك, Edrisî, II, p. 213 سيسوك. Mokaddasí inter urbes provinciae Ilák memorat سام سرك s. سام دمرك. *e*) C. et D. ونحوها. *f*) A. دوكب, B. توكث, C. بودكب, E. بونكث, D. ثويكث. Mox A. نوكث, B. توكث, C. نوكب, E. نوكث, D. تونكث. *g*) A. حمادحكث, B. حماككث, D. حمادحكث, E. حيادحكث, *Djih.-Numa*, p. ٣٦٣, 12 حيابحكث. Vid. supra p. ٣٢٨ et ann. *q*. *h*) A. et B. نيكف, C. نيكث, D. ينكث, A. دىكث. *i*) A. et B. دىحاىب. *k*) Secundum C.; A. et B. ذنهر. *l*) E. دىكث. Secundum Jacut, IV; p. ٧٤٣ 15 distantia est tantum 2 Paras. *m*) Conjunctionem ف ex C. addidi. *n*) Supra p. ٣٣٣ خرجانكث; vid. ann. *d*.

٤ مراحل ومن اسبيجاب الى أُسْبَانيكَث[a] مرحلتان ومن اسبانيكث الى كدر[b] قصبة باراب مرحلتان خفيفتان ومن كدر الى شَاوَغَر[c] مرحلة ومن شاوغر الى صَبْرَان[d] مرحلة خفيفة ووَسِيج[e] على غربىّ النهر على الشطّ اسفل من كدر بفرسخين وباراب عن شرقىّ الوادى وبين كدر والنهر نصف فرسخ ۞ الطريق من أَخْسيكَث الى شِكِث[f] ٩ فراسخ وهى اوّل مِيَان رُوذَان ومن اخسيكث الى سلات[g] آخر ميان روذان نحو ٥ مراحل ومن اخسيكث الى كاسان[h] فى شماليّها ٥ فراسخ ومن كاسان الى اردلانكث[i] منزلتان ومن كاسان الى نجم[k] فى سمت الشمال يوم ومن اخسيكث الى حدّ كَرْوان[l] نحو ٧ فراسخ * والى وانكث من[m] اخسيكث نحو ٧ فراسخ وحدُّها يتّصل بايلاق وهى بين المغرب

---

a) A. et B. sine punctis (B. اسمادمكث). *Djih.-Numa*, p. ٣٩٣, 14 بيكث (apud eundem بنكث corruptum est in بيكند). Mokaddasí semel recte اسبانيكث, sed bis male ارسبانيكث (ارسمادمكث). *b*) A. كدر et كدر, B. كندر, C. كزر, D. كدر ut Abulfeda, p. ٤٩٣, E. كزر. Mokaddasí MS. p. 134 كدر. Pro باراب, C., D. et E. فاراب. *c*) A., B. et D. ساوعر, C. شاوع., E. مياوغز, *Djih.-Numa*, p. ٣٩٣, 15 ساوغر. Vid. Jacut in v. et Mokaddasí. *d*) A. صمراف, B. صنمراو, C. صمرار, D. صبران ut *Djih.-Numa* l.l., E. صران. Vid. Jacut in v.; Mokaddasí habet سوران. *e*) A. et D. وسخ, B. et E. وسح. Vid. Abulfeda, p. ٤٩٣ et Jacut in v. *f*) A. et B. تونكث, D. سلب, E. دمكب, Edrîsî, II, p. 212 شكث; vid. Jacut in v. Mokaddasí MS. p. 131 et 133 بكث et بكب. *g*) A. سلاب semel سلات, B. سلاب et سلات, D. شلاث et شيلاث, E. سلار; *Djih.-Numa*, p. ٣٩٣, 15 سيلاب. *h*) Vid. Jacut in v. A. et B. واساب i. e. قاسان. Ibno 'l-Athír, IV, p. ٤٦. كاشان. *i*) A. et B. اودلادكث. Vid. supra ad p. ٣٢٩ ann. *d*. Est urbs provinciae Schásch. *k*) A. et B. sine punctis. D. محم et مجم. Vid. Edrîsî, II, p. 211. *l*) D. h. l. et Edrîsî l.l. كوران, sed E., D. infra, Edrisi, p. 212 et Dimaschkí, p. ٢٢١ ut A. et B. *m*) A. والى اوسدس, B. والى اعمد من, D. والزاشت من اخسيكث على. Quod h. l. conjectura

والشمال من اخسيكث وكروان بينها وبين كاسان *[a] ۴ فراسخ* ومن اخسيكث الى كروان نحو ۹ فراسخ، *وباراب واخسيكث[b] على شطّ نهر الشاش وكَنْد *بينها وبين[c] نهر الشاش زيادة على فرسخ وكذلك بين وانكث والوادى زيادة على فرسخ وبين خُوَاكَنْد والوادى ۵ فراسخ ومن قُبَا الى رِشْتَان[d] بينه وبين نهر الشاش كلّه نحو مرحلة.ومن قُبَا الى اشتيقان[e] ۳ فراسخ ومن اشتيقان الى الوادى ۷ فراسخ وهى على طريق قبا الى اخسيكث، ومن سُوج[f] الى بامكاخش[g] ۵ فراسخ ومن بامكاخش الى طماخس نحو من ميل ومن بامكاخش الى سوج ۵ فراسخ ومن سوج الى اوال على طريق اوجنه[h] نحو ۱۰ فراسخ، ومن قبا الى نَقَاد[i] نحو المشرق نحو ۷ فراسخ وحدودهما متّصلة، ومن أُوش[k] الى مدوا فرسخان، ومن وانكث[l] الى خَيْلَام ۳ فراسخ ومن خيلام

---

scripsi وانكث (ab Ibn Haucal et Mokaddasí (ادكث) inter urbes regionis Nasá in Farghána memorata), infra in A. et B. scribitur وادكث et ونكث, in D. نكث.

a) D. et E. om. Deinde A. et B. om. اخسيكث et B. ex الى كروان facit الكروان. *b*) A. وداروحسكث, B. وداروحشكث (interdum nempe in Codd. اخشيكث scribitur). *c*) A. et B. بينه وبينها على شط. *d*) A., B., D. et Edrîsî, II, p. 213 رستاق. In D. deëst الى et deinde om. كله. Corrupta haec videntur, nisi fortasse verba significent iter a Kobá ad Rischtán semper distantia unius diei a flumine pergere. *e*) A. et B. اسبعاب, D. اسبعسان. In catalogo urbium provinciae Farghánae D. اسنعان, Edrîsî, II, p. 213 اسفينان, E. اشنيقان, Ous. اسبيقان. Mokaddasi nomen ter memorat اشنقان, استيقان (اشتيقان) et اشتيقان. *Djih.-Numa*, p. ۳۹۳, 16 اسيقان. Edrîsî hanc urbem confundere videtur cum ea cujus nomen supra اشبينغو edidimus; minus recte. *f*) A. et B. دسرح et سرح, D. دسوج. *g*) A. دامكاخش. Mokaddasî in mappa باركاحس, C. in mappa مكاحس ut *Djih.-Numa* l.l. 17. *h*) Sic A.; B. sine punctis, D. اوخبه. Fortasse legendum est اوزجند = اوزكند. *i*) A. et B. دعاد. *k*) A. اودمن. *l*) A. et B. ونكث.

الى سلات ٧ فراسخ وسلات وبيسكند a ليس بهما b منبر ولكنّهما ثغران فلذلك ذكرناهما ۞

اخر كتاب الاشكال c

a) Vid. de hoc nomine p. ٢٨٦ ann. *m.* b) A. et B. بها. c) In utroque Codice librarius hic adnotavit: نَقَلْتُهُ مِنْ نُسْخَةٍ فى غَايَةِ الصِّحَّةِ وَٱلسَّدَادِ.